Defects in Semiconductors

Defects in Semiconductors

Proceedings of the Materials Research Society Annual Meeting,
November 1980, Copley Plaza Hotel, Boston, Massachusetts, U.S.A.

EDITORS:

J. Narayan
Solid State Division, Oak Ridge National Laboratory, Oak Ridge,
Tennessee, U.S.A.

and

T.Y. Tan
Thomas J. Watson Research Center, Yorktown Heights, New York,
U.S.A.

NORTH HOLLAND
NEW YORK · OXFORD

PHYS
.0831/3787

Published by:

Elsevier North Holland, Inc.
52 Vanderbilt Avenue, New York, New York, 10017

Sole distributors outside USA and Canada:

North-Holland Publishing Company
P.O. Box 103
Amsterdam, The Netherlands

Library of Congress Cataloging in Publication Data

Main entry under title:

Defects in semiconductors.

    (Materials Research Society symposia proceedings;
    v. 2 ISSN 0272-0-9172)

    Bibliography: p.
    Includes index.
    1. Semiconductors—Defects—Congresses. I. Narayan, Jaydish.
    II. Tan, T.Y. III. Materials Research Society. IV. Series: Materials
    Research Society. Materials Research Society symposia proceedings ; v. 2.
QC611.6.D4D43   537.6'22   80-28826
ISBN 0-444-00596-X

Manufactured in the United States of America

Contents

QC 611
.6
D4
D43
1981
copy 2
PHYS

v

*Invited

CHAPTER III. PROPERTIES OF DISLOCATIONS AND INTERFACES

*Invited

*Invited

Preface

This book contains selected papers from the symposium on "Defects in Semiconductors" held in Boston, Massachusetts, November 16-20, 1980, sponsored by the Materials Research Society. The symposium represented an international forum in which scientists from over ten foreign countries participated. Although this conference had many similarities with previous international conferences on lattice defects in semiconductors, notably Freiburg (1974), Dubrovnik (1976), Nice (1978), Oiso (1980), one of the distinguishing features of this conference was the emphasis on the electronic properties of dislocations and the point defect-dislocation relationships. The scope of the present conference was to address the new advances in defect characterization techniques, and then focus attention on point defects, dislocations, and point defect-dislocation interrelationships in elemental as well as compound semiconductors. Properties of these defects were discussed in light of various physical processes and phenomena including nucleation and formation of macroscopic defects, thermal and laser annealing of ion implantation and neutron damage, thermal oxidation and oxygen precipitation, and crystal growth. It was heartening to see a significant progress being made in the understanding of the physical properties of point defects, in the correlations of atomic structures with the electronic properties of dislocations, in identifying the role of silicon self-interstitials in impurity diffusion, and in bridging the gap between the structures of point and extended defects. Substantial progress in the field of defects in compound semiconductors was noteworthy.

Manuscripts were submitted by symposium participants in camera-ready form. Each manuscript was reviewed by at least one referee and modified accordingly. We would like to thank the referees for their conscientious efforts and speedy return of their comments. Finally we wish to express our sincere gratitude to many of our colleagues for providing their advice, guidance and encouragement in organizing the meeting.

J. Narayan
T. Y. Tan

x

Acknowledgments

A large part of the credit for the success of the symposium is due to conference participants, particularly the invited speakers who provided excellent summaries of specific areas and set the tone of the meeting. They are:

B. R. Appleton	B. C. Larson
J. Chikawa	S. Mahajan
W. K. Chu	D. M. Maher
J. W. Corbett	J. M. Meese
A. G. Cullis	J. R. Patel
A.J.R. DeKock	P. M. Petroff
W. Frank	S. T. Picraux
U. Gösele	A. Seeger
P. B. Hirsch	W. K. Tice
S. M. Hu	J. Washburn
L. C. Kimerling	G. D. Watkins

We are also grateful to session chairmen for the smooth running of the symposium.

R. W. Balluffi	R. Sinclair
W. L. Brown	M. L. Swanson
J. H. Crawford, Jr.	G. Thomas
G. W. Cullen	K. N. Tu
A. G. Cullis	G. D. Watkins
P. B. Hirsch	C. W. White
P. K. Iyengar	F. W. Young, Jr.
K. A. Jackson	

It is our pleasure to acknowledge with gratitude the financial assistance provided by ONR (Dr. L. R. Cooper), DARPA (Dr. R. A. Reynolds), and DOE (Dr. M. C. Wittels). We offer our sincere thanks to Mrs. Sylvia G. Evans, Miss Susan E. Himes and Mrs. Julia T. Luck for their secretarial services before and after the conference.

PREVIOUSLY PUBLISHED MRS SYMPOSIA

Phase Transitions 1973
H.K. Henisch, R. Roy, L.E. Cross, editors
Pergamon, New York 1973

Scientific Basis for Nuclear Waste Management-I
G.J. McCarthy, editor
Plenum Press, New York 1979

Conference on In-Situ Composites-III
J.L. Walter, M.F. Gigliotti, B.F. Oliver, H. Bibring, editors
Ginn & Co., Lexington, Massachusetts 1979

Laser Solid Interactions and Laser Processing 1978
S.D. Ferris, H.J. Leamy, J.M. Poate, editors
American Institute of Physics, New York 1979

Scientific Basis for Nuclear Waste Management-II
C.J. Northrup, editor
Plenum Press, New York 1980

Ion Implantation Metallurgy
C.H. Preece and J.K. Hirvonen, editors
The Metallurgical Society--A.I.M.E., New York 1980

Laser and Electron Beam Processing of Materials
C.W. White and P.S. Peercy, editors
Academic Press, New York 1980

Scientific Basis for Nuclear Waste Management-III
J.G. Moore, editor
Plenum Press, New York 1981

Defects in Semiconductors

Published 1981 by North-Holland, Inc.
Narayan, and Tan, eds.
Defects in Semiconductors.

ELECTRON PARAMAGNETIC RESONANCE IN SEMICONDUCTORS

JAMES W. CORBETT, RICHARD L. KLEINHENZ AND NEAL D. WILSEY[*]
Institute for the Study of Defects in Solids, Physics Department, State University of New York at Albany, Albany, New York 12222
[* Permanent address: Naval Research Laboratory, Washington, DC 20375]

ABSTRACT

The use of electron paramagnetic resonance (EPR) in the study of defects in semiconductors is briefly reviewed, including group IV (C-diamond, Si, Ge, SiC), III-V (AℓSb, GaAs, GaSb, GaP, InAs, InP, InSb), II-VI (BaO, BaS, BeO, CaO, CaS, CaSe, CdO, CdS, CdSe, CdTe, MgO, SrO, SrS, ZnO, ZnS, ZnSe, ZnTe) and miscellaneous systems. The identification of defects via EPR is described as is the exploitation of that identification as a tool in future studies. Particular attention is paid to Si, where is emerging an integrated panorama of identified defects ranging from point defects to aggregates through intermediate defect configurations (as discussed by Tan) to dislocations and stacking faults; EPR results in Si as a testing ground for the theory of shallow donors, in the understanding of diffusion at high temperatures and in the study of heat-treatment defects are discussed as examples of the use of EPR as a tool in defect studies.

INTRODUCTION

We have been asked to try to give a perspective view of the use of electron paramagnetic resonance (EPR) in semiconductors. Our space is too limited to give details or much depth, but we feel we can give a proper view of the past and future contributions of this field of research.

Spectroscopy in general makes contributions to defect studies by the identification of defects and by the use of that identification as a tool in further studies. Characteristically the sharper the spectroscopic lines the more incisive the technique can be. *Central to defect identification is the determination of the defect symmetry*, which may be manifest by a spectral variation with respect to crystal orientation and which may be probed by the use of polarized light in absorption and in luminescence, by the response to an external field, etc.

EPR is simply a form of spectroscopy, specifically Zeeman spectroscopy, except that the usual Zeeman effect involves the splitting of electronic transitions due to a magnetic field; EPR observes the direct transition between the Zeeman-split electronic levels. To observe EPR there must be an unpaired spin, i.e., a non-zero paramagnetism, such as may be due to electrons in the conduction band [and we will see that conduction electron spin resonance (CESR) has been done in many semiconductors] or a suitably charged point-, line-, or surface-defect state. For a free electron in space the energy difference between the two Zeeman-split levels of the spin 1/2 system is given by $g\beta H$ with H- the magnetic field, β- the Bohr magneton and g- a constant determined by the Dirac theory of the electron. If that electron is bound to a free atom in space, the Zeeman levels of the electron may be split and shifted due to various effects such as the hyperfine-interaction and the quadrupole interaction with the nucleus of that atom. If the electron is localized at a defect in a solid the possible interactions are even richer since the electron may not reside on just one atom, but on several or even many. Further the quantities describing those interactions are no longer scalars but are tensors reflecting the crystal en-

vironment. These additional interactions can be a severe disadvantage since
the resultant lines may be so broadened as to be effectively unobservable. The
overall situation may be seen in Table I where we list the isotopes making up
the several semiconductors that we consider, the isotopic abundances, their
spins, magnetic moments and quadrupole moments.

TABLE I
Relevant Properties of Elements [1]

Isotope	Natural Abundance (%)	Spin	Magnetic Moment (μN)	Electric Quadrupole Moment ($10^{-24} cm^2$)
Al^{27}	100	5/2	+3.6414	+0.15
As^{75}	100	3/2	+1.439	+0.29
B^{10}	19.78	3	+1.8007	±0.08
B^{11}	80.22	3/2	+2.6885	+0.04
Ba^{132}	0.097	—	—	—
Ba^{134}	2.42	0	0	0
Ba^{135}	6.59	3/2	+0.8365	+0.18
Ba^{136}	7.81	—	—	—
Ba^{137}	11.32	3/2	+0.9357	+0.28
Ba^{138}	71.66	0	0	0
Be^{9}	100	3/2	−1.1776	+0.05
C^{12}	98.89	0	0	0
C^{13}	1.11	1/2	+0.7024	—
Ca^{40}	96.97	0	0	0
Ca^{42}	0.64	—	—	—
Ca^{43}	0.145	7/2	−1.317	—
Ca^{44}	2.06	—	—	—
Ca^{46}	0.0033	—	—	—
Ca^{48}	0.18	—	—	—
Cd^{106}	1.22	—	—	—
Cd^{108}	0.88	0	0	0
Cd^{110}	12.39	0	0	0
Cd^{111}	12.75	1/2	−0.5943	—
Cd^{112}	24.07	0	0	0
Cd^{113}	12.26	1/2	−0.6217	—
Cd^{114}	28.86	0	0	0
Cd^{116}	7.58	0	0	0
Ga^{69}	60.4	3/2	+2.016	+0.19
Ga^{71}	39.6	3/2	+2.562	+0.12
Ge^{70}	20.53	0	0	0
Ge^{72}	27.43	0	0	0
Ge^{73}	7.76	9/2	−0.8792	−0.22
Ge^{74}	36.54	0	0	0
Ge^{76}	7.76	0	0	0
In^{113}	4.28	9/2	+5.523	+0.82
In^{115}	95.72	9/2	+5.534	+0.83
Mg^{24}	78.70	0	0	0
Mg^{25}	10.13	5/2	−0.8553	+0.22
Mg^{26}	11.17	0	0	0
O^{16}	99.759	0	0	0
O^{17}	0.037	5/2	−1.8937	−0.026
O^{18}	0.204	0	0	0
P^{31}	100	1/2	+1.1317	—
S^{32}	95.0	0	0	0

Isotope	Natural Abundance (%) (continued)	Spin	Magnetic Moment (μ_N)	Electric Quadrupole Moment (10^{-24}cm^2)
S^{33}	0.76	3/2	+0.6433	−0.055
S^{34}	4.22	0	0	0
S^{36}	0.014	0	0	0
Sb^{121}	57.25	5/2	+3.359	−0.29
Sb^{123}	42.75	7/2	+2.547	−0.37
Se^{74}	0.87	0	0	0
Se^{76}	9.02	0	0	0
Se^{77}	7.58	1/2	+0.534	—
Se^{78}	23.52	0	0	0
Se^{80}	49.82	0	0	0
Sr^{84}	0.56	—	—	—
Sr^{86}	9.86	0	0	0
Sr^{87}	7.02	9/2	−1.093	+0.36
Sr^{88}	82.56	0	0	0
Te^{120}	0.089	—	—	—
Te^{122}	2.46	—	—	—
Te^{123}	0.87	1/2	−0.7359	—
Te^{124}	4.61	—	—	—
Te^{125}	6.99	1/2	−0.8871	—
Te^{126}	18.71	0	0	0
Te^{128}	31.79	0	0	0
Te^{130}	34.48	0	0	0
Zn^{64}	48.89	0	0	0
Zn^{66}	27.81	0	0	0
Zn^{67}	4.11	5/2	+0.8755	+0.17
Zn^{68}	18.57	0	0	0
Zn^{70}	0.62	—	—	—

For silicon we see that the predominant isotope (Si^{28} = 95.3% abundant) has spin zero and so contributes nothing in the way of spectral splittings; the remaining isotope (Si^{29}) has spin 1/2 so that it creates the simplest of splittings and its abundance (4.7%) is just about ideal to provide a dilute label on lattice sites. In the III-V compounds on the other hand we find the lattice constituents often have several magnetically active isotopes with large spins which split the lines into many sub-lines with strong (and over-lapping) anisotropies; little wonder that the lines (when observed) in III-V compounds tend to be very broad.

When the lines are resolved, all these interactions can prove very incisive. For example in ordinary spectroscopy it is not unusual to determine the overall defect symmetry and perhaps, via an isotope effect, the identity of the dominant elements involved. In EPR the overall symmetry is encompassed in the anisotropy of the g-tensor; for the silicon defects it is not uncommon to find this g-tensor anisotropy varying and thereby revealing changes in the defect environment, e.g., the hopping of electrons from one bonding configuration to another, or the changing influence of lattice strains at the defect. In addition in silicon it is not uncommon for the defect spectra to exhibit resolved hyperfine lines, which reveal the extent to which the wave-function extends over the neighboring atoms of the lattice, and may reveal the presence of impurity atoms, which ordinary spectroscopy would miss. Further there are double resonance techniques which have already had considerable impact and have promise of much, much more, e.g., electron-nuclear double resonance (ENDOR), electron-electron double resonance (ELDOR), optical detection of magnetic resonance (ODMR). For example, the extensive ENDOR studies [2-8] of the

shallow donors in silicon, which have mapped the electron wave-function over
thirty shells of atoms in the lattice surrounding the donor atom, has proven to
be a challenging testing ground for the theory of shallow donors [9-18],a chal-
lenge which theory has largely, but not completely met. Although EPR of the
shallow acceptors has been observed [19-22], ENDOR measurements to test the
more difficult theory [12,23-26] have not been carried out. Other ENDOR measurements have been carried out on deep-level systems and stand as a formidable
testing-ground for the more difficult theory of deep-levels. Specifically
Ludwig [27] has mapped out the wave-function on eight shells of neighboring
atoms to the sulfur deep-donor, and Ammerlaan and co-workers have mapped out
the wave-function for nineteen shells of atoms for the positive divacancy spectrum [28] and thirty-three shells of atoms for the negative divacancy spectrum
[29]. Although substantial work on the theory of the divacancy has been done
[30-32], there remain substantial discrepancies to be reconciled.

SURVEY OF DEFECTS IN SEMICONDUCTORS

In this section we present lists of defects for which an at least tentative identification exists in the EPR literature. Certainly there are defects
which have been well identified without the help of EPR, many of which are in-
accessible to EPR because they do not involve an unpaired spin. For example,
the oxygen interstitial in silicon has a well-established configuration, ar-
rived at by studies of the vibrational absorbtion lines [33-36]. As a further
example, one should realize that the defects which we list represent a specific
charge state of the defect; many of the defects have several charge states in
the forbidden gap and typically a charge state change converts a paramagnetic
state to a non-paramagnetic state (or vice-versa) and the latter is inacces-
sible to EPR. Like other forms of spectroscopy, EPR provides a limited "window" with which to view defects; if the defects are not visible through that
window the spectroscopy is no help, but if it is visible, EPR may provide extensive information about the atomic and electronic configuration of the defect.
In Tables II, III and IV we list the defects for the group II-V, III-V and
IV semiconductors [37].

TABLE II

Defects with Identified EPR Spectra in II-VI Compounds in the Literature [37].
In each system there may be spectra due to unidentified defects. V_x denotes a
vacancy at an X site; X_{INT} – the element X in an interstitial site; X – the
element (X) in a substitutional site; (X·Y) – an association of X and Y; CESR –
conduction electron spin resonance; a-X,Surface – a resonance attributed to the
amorphous X and to the surface, resp.; etc.

BaO

V_O
Gd, Mn

BaS
V_S

TABLE II (CONTINUED)

BeO

V_{Be}, V_O

Aℓ, B, Cu, F, Li

B_{INT}

CaO

V_{Ca}, V_O; $(V_O \cdot V_O)$; $(V_O \cdot V_{Ca} \cdot V_O)$

O_{INT}

Ag, Ce, Co, Cr, Cu, Eu, F, Fe, Gd, Ir, K, Li, Mg, Mn, Mo

Na, Nd, Ni, Pb, Ti, Tm, V, Yb

$(Li \cdot H)$, $(Li \cdot Li)$, (OH), $(V_O \cdot Mo)$, $(Mn \cdot Mn)$

Surface

CaS

Eu, Mn, Sn, Yb

Surface

CaSe

Pb, Sn

CdO

Cu, Ni

$(V_{Ca} \cdot V_O + OH)$

CESR

CdS

V_{Cd}, V_S

Ag, Br, Cℓ, Co, Cr, Cu, Dy, Er, Eu, Fe, Ga, Gd, Ge, I

Mn, Ni, Nd, Pb, Sn, Ti, Tm, V, Yb

$(Fe \cdot Ag)$, $(Fe \cdot Ag)$, $(Fe \cdot Cu)$, $(Fe \cdot Li)$, $(Gd \cdot ?)$, $(N \cdot N)$

CESR, Surface

CdSe

V_{Cd}

Co, Eu, Fe, Gd, Mn, Nd, Sn, Ti, V

CESR, Surface

TABLE II (CONTINUED)

$\underline{CdTe}$

V_{Cd}

Ag, Cr, Co, Cu, Er, Eu, Fe, Gd, Li, Mn, Nd,

        Ti, Tl, Tm, Yb

$(Fe \cdot F_6)$ $(Yb \cdot Ag_3)$ $(Yb \cdot Ag_4)$ $(Yb \cdot Au_3)$ $(Yb \cdot Au_4)$ $(Yb \cdot Cu_4)$ $(Yb \cdot Li)$

CESR

$\underline{MgO}$

V_{Mg}, V_O; $(V_{Mg} \cdot V_O)$; $(V_O \cdot V_{Mg} \cdot V_O)$

O_{INT}

Ag, Al, Co, Cr, Cu, Dy, Er, Eu, F, Fe, Gd, H, Ir,

    Li, Mn, Na, Ni, Pd, Rh, Ru, Sn, V, Yb

$(V_{Mg} \cdot Al)$, $(V_{Mg} \cdot Cr)$, $(V_{Mg} \cdot F)$, $(V_{Mg} \cdot Fe)$, $(V_{Mg} \cdot Li)$, $(V_{Mg} \cdot Mn)$, $(V_{Mg} \cdot Na)$

      $(V_{Mg} \cdot OH)$, (OH), $(Cr \cdot Cr)$, $(Mn \cdot Mn)$, $(V \cdot V)$

Surface

$\underline{SrO}$

V_{Sr}, V_O

Ag, Ce, Co, Cr, Cu, Er, Eu, F, Fe, Gd, K, Li, Mn, Na, Ni, V

$(V_{Sr} \cdot OH)$, (OH)

Surface

$\underline{SrS}$

Mn

$\underline{ZnO}$

V_{Zn}, V_O; $(V_{Zn} \cdot V_{Zn})$

Zn_{INT}, O_{INT}

Al, Cl, Co, Cr, Cu, Fe, Ga, Gd, In, Li, Mn, Nb

      Ni, Pb, Sn, V, Yb

$(Cu \cdot H)$, $(Cu \cdot OH)$, $(Fe \cdot Ag)$, $(Fe \cdot Cu)$, $(Fe \cdot Li)$, (OH)

CESR, Surface

$\underline{ZnS}$

V_{Zn}, V_S

Al, Br, Co, Cr, Cu, Dy, Er, Eu, F, Fe, Ga, Gd, Ge,

    I, In, Mn, Nd, Ni, Pb, Sc, Si, Sn, Tl, Ti, V, Yb

$(Y_{Zn} \cdot Al)$, $(V_{Zn} \cdot Br)$, $(V_{Zn} \cdot Cl)$, $(V_{Zn} \cdot Cu)$, $(V_{Zn} \cdot Fe)$

      $(V_{Zn} \cdot Ga)$, $(V_{Zn} \cdot In)$, $(Ag \cdot Fe)$, $(Ag \cdot Ga)$, $(Cu \cdot Fe)$, $(Cu \cdot Ga)$, $(Fe \cdot Li)$

CESR, Surface

TABLE II (CONTINUED)

ZnSe

V_{Zn} ; $(V_{Zn} \cdot Zn_{INT})$

As, Co, Cr, Dy, Er, Eu, Fe, Ge, Mn, Ni, P,
    Pb, Si, Sn, V, Yb
$(V_{Zn} \cdot Te)$; (Ag·Fe), (As·Fe), (Cu·Fe), (Fe·Li), (P_4)

Surface

ZnTe

Co, Cr, Er, Fe, Ge, Mn, Pb, V, Yb

(Ag·Fe), (Cu·Fe), (Fe·Li)

TABLE III
Defects with Identified EPR Spectra in III–V Compounds in the Literature [37].
[See TABLE II for explanation of symbols.]

AℓSb

Surface

GaAs

Cd, Co, Cd, Cr, Cu, Fe, Mn, Ni, S, Si, Te, Ti, V, Zn

(Mn·O), As_{Ga}

CESR, Surface, a–GaAs

GaSb

CESR, Surface

GaP

V_{Ga}

P_{Ga}

C, Cd, Co, Cr, Fe, Ge, Mn, N, Ni, O, S, Se, Si, Sn, Te, Zn

(Cd·O), (Li_3) (Li·Mn), (N_{INT}) (Mn·S)

Surface

InAs

Fe
CESR, Surface

InP

Co, Cr, Fe, Ni
$(Fe \cdot In_{INT})$
CESR

InSb
CESR, Surface

TABLE IV

Defects with Identified EPR Spectra in   IV Systems   in the Literature [37].
[See TABLE II for explanation of symbols.]

C (Diamond)

  V

  C_{INT}

  B, Co, N, Ni

  $(V \cdot N)$, (N_2) (N_3), $(N \cdot A\ell)$, $(N \cdot Si)$

  Surface, a-C

Si

  V, V_2, V_3, 2 V_4, V_5

  Si_{INT}, $(Si_{INT} \cdot Si_{INT})$

  $A\ell$, As, Au, B, Cr, Fe, Ga, Li, Mg, Mn, P, Pd, Pt,
               S, Sb, V, Zn

  $A\ell_{INT}$, B_{INT}, C_{INT}, P_{INT}

  $(V \cdot A\ell)$, $(V \cdot As)$, $(V \cdot B)$, $(V \cdot Fe)$, $(V \cdot Ge)$, $(V \cdot Li)$, $(V \cdot O)$, $(V \cdot P)$, $(V \cdot Sb)$, $(V \cdot Sn)$

  $(V \cdot C \cdot O)$, $(V \cdot P_2)$

  $(V_2 \cdot O)$, $(V_2 \cdot O_2)$, $(V_3 \cdot O)$, $(V_3 \cdot O_2)$, $(V_3 \cdot O_3)$

  $(A\ell \cdot A\ell)$, $(A\ell \cdot Cr)$, $(A\ell \cdot Fe)$, $(A\ell \cdot Mn)$, $(As \cdot As)$, Au $\cdot$ Cr), $(Au \cdot Fe)$, $(Au \cdot Mn)$,
         $(B \cdot B)$, $(B \cdot Cr)$, $(B \cdot Fe)$, $(B \cdot Mn)$, $(Cr \cdot Ga)$, $(Cr \cdot In)$,
         $(Fe \cdot Ga)$ $(Fe \cdot Fe)$, $(Fe \cdot In)$, $(Fe \cdot S)$, $(Ga \cdot Ga)$, $(Ga \cdot Mn)$, $(In \cdot Mn)$,
         $(Li \cdot O)$, $(Li \cdot O \cdot Si_{INT})$, (Mn_4), $(S \cdot S)$, $(P \cdot P)$, $(Pt \cdot O)$

  CESR, Surface, a-Si, Disl.

Ge

  As, Bi, Mn, Ni, P, Sb

  P_{INT}

  $(V \cdot O)$

  CESR, Surface, a-Ge

SiC

  V_C , $(V_C \cdot V_C)$

  $A\ell$, B, N

  $(V_{Si} \cdot B)$

  Surface

In addition there is some information in the literature on EPR studies in other semiconductor systems: As, As_2S_3, As_2Se_3, B, BN, $Cd_xHg_{1-x}Se$, $Cd_xHg_{1-x}Te$, $CdIn_2S_4$, $Cd_xMg_{1-x}Te$, $CdS_{1-x}Se_x$, $CdSiP_2$, GaSe, a (for amorphous) $Ge-A\ell$, aGe-Cu, aGe-S-Ag, aGe-S, aGe-Te-Si, HgSe, HgTe, In_2Te_3, PbTe, $Pb_xSn_{1-x}Te$, Se, SiTe, SnTe, Te, aTe-As-Ge-S, $Zn_xHg_{1-x}Se$, $Zn_xHg_{1-x}Te$, $ZnSiAs_2$, $ZnSiP_2$.

What is included as a "semiconductor" is somewhat arbitrary. We have included work on diamond, with its $\sim$ 5.5 eV forbidden gap, because it can under some circumstances be a semiconductor; we include $Pb_x Sn_{1-x} Te$, which may have a zero-energy band gap, because it is termed a "small-band gap semiconductor." We include purely covalent systems, highly ionic systems, and systems with a wide variety of lattice structures (and in the amorphous cases, no "lattice" at all!). We include them because the technique unifies the systems, and we expect that the physics of defects in the different systems may have common features (as do semiconductor defects)with defects in purely ionic systems and even metals [38].

It will be seen in Tables II, III and IV that there is a wide variation in the amount of information available, Si the most, followed by MgO, etc. The disparities reflect a combination of factors (e.g., Is the system widely studied? Are high quality, suitably-doped, single crystals available?) besides the questions related to the ease of EPR studies discussed above. It will be seen that information on some impurity centers are available, less on intrinsic defects such as vacancies and interstitials and varying amounts on defect complexes. The tables give a perspective view of the past effort and probably of balance of effort and impact of EPR studies in the future. We are optimistic, for example, about the opportunities for EPR work in III-V compounds, and fully persuaded about the importance of such work, but Table III clearly reflects the difficulty of EPR in these compounds, particularly in contrast to Tables II and IV.

We should include a warning about the immutability of the identifications listed in these tables, for the benefit of the non-expert (or perhaps even more as a reminder for the expert, for we all may be snared by feelings of priority, paternity and pride!). Any identification is like a "house of cards"; it is made up of a lot of bits and pieces of information which assemble into a (more or less) consistent whole, but which may collapse under the impact of "one-too-many" experiments. Some identifications are so buttressed that they might as well be cast in concrete, but not all parts of that edifice are as well established. Any assessment of the validity of a model would be subjective. Instead we simply say *caveat canem*.

It should also be noted that there are many EPR spectra in the literature for which even a tentative defect identification has not been advanced. No attempt was made to include such information.

The full perspective concerning the utility of EPR studies in semiconductors cannot be seen from the listings in these tables. We will try to amplify that perspective by considering the defects in silicon in the next section.

SILICON

At this point we will assume that the identifications shown in Tables II, III and IV are established and try to give a perspective of the uses of these identified defects (via EPR measurements) as tools in related studies. [In practice the separation of identification and use is rarely clean, one feeding the other.] There are many examples which could be given, however, space limits our treatment. We will cover aspects of topics not covered in the conference.

The first topic exploits the special nature of EPR spectra. It is the study [2,39-68] of the EPR of shallow donors (say, phosphorus) in silicon as a function of phosphorus concentration which ranges from ca. $10^{15}/cm^3$ in which case the donors are completely isolated up to ca. $10^{19}/cm^3$ in which regime the donors are strongly interacting in a metallic system. In the dilute regime the electron resides on the phosphorus atom which is 100% P^{31} (see Table I) with a nuclear spin 1/2 and consequently the electron resonance is split into two lines by this hyperfine interaction. As the phosphorus concentration is increased a new line [39] occurs mid-way between these lines which is due to the anti-ferromagnetic   [42] exchange coupling between pairs of lines and in more concentrated samples additional lines due to transitions in clusters of three and four nearby donors [40]. In more concentrated samples these lines broaden and shift in ways which can be correlated [59,62,65] with the formation of an impurity band by the hopping of electrons from phosphorus to phosphorus, which process can be followed with increasing phosphorus concentration through the metal-insulator (or semiconductor) transition [69,70] (which occurs [59] at a donor concentration of $\sim 4 \times 10^{18}/cm^3$) on into the metallic regime. In fact the spectral behavior is a complicated function of donor concentration, sample temperature, microwave measuring power, etc., and although the general features are established Si:P remains a simple model system in which carrier the percolation of charge carriers among disordered centers may still fruitfully be studied [59].

Let us turn now to questions of diffusion in silicon. As shown in Table IV, EPR studies have established some properties of the vacancy (V) [71,72], the divacancy (V·V) [73,74], and the vacancy-phosphorus (V·P) [75-77], the vacancy-arsenic (V·As) [78], the vacancy-antimony (V·Sb) [78], and the vacancy-aluminum (V·Aℓ) [79] pairs; less is known about the intrinsic interstitial [80], although a di-interstitial has been described [83]. The normal expectation would be that this defect information obtained at low temperatures could be integrated with defect studies at high temperatures where self-diffusion and impurity studies are carried out. But there has been a substantial controversy on what is the dominant native defect in equilibrium at high temperatures. The controversy has been reviewed in detail recently [84-88] and we expect will be treated in other papers at this meeting, so we will not dwell upon it. In most elemental systems, the vacancy is the dominant intrinsic defect in equilibrium at high temperatures. It has been suggested that in silicon the corresponding defect is the divacancy [89-91] or the split-vacancy [92], but the main contender versus the vacancy has been the interstitial, a point of view vigorously championed by Seeger and his co-workers [88,93-96]. The problem is that the direct observation of the high-temperature defect has not been feasible and quenching studies may be troubled by impurity effects [97]. There are a wide range of data which must be accommodated: self diffusion under intrinsic and extrinsic Fermi level conditions (i.e., in the presence of high concentrations of boron, phosphorus or arsenic); impurity diffusion as a function of the impurity concentration, Fermi level, temperature, stress, and electric field, including the diffusion parameters and profiles; the "emitter-dip" effect; the "kink" in the phosphorus profile; etc. Remarkably Fair [84] has shown that employing all the charge states of the vacancy [$V^=$, V^-, V^x (i.e., x = neutral), V^+, V^{++}], including the V^{++} charge state which is now thought to occur [98,99], all of the above data can be reconciled with the vacancy diffusion mechism; for example, in self-diffusion, in intrinsic silicon, V^- and V^+ (or V^{++}) dominate at high temperatures (T > 1150°C) and V^- and V^x dominate at low temperatures in a way that is accord with the low temperature work on the vacancy [77].

Let us turn now to what we consider to be another exciting aspect of defect studies in silicon. As we have discussed earlier, EPR studies have established the identity of the single vacancy [71,72], the divacancy [73,74], the trivacancy [100], a planar- [100-103] and a non-planar tetravacancy [100] and a non-planar pentavacancy [100,103,104], and have determined some of the properties of these defects. Less is known about the intrinsic interstitial although theory [105-107] agrues that it should be a split-<100> interstitial, a conclusion supported by carbon interstitials which have the equivalent configuration [108]. It has been proposed [81] that these split-interstitials might aggregate in a <110> direction and constitute the <110> rod-like defects found in electron microscope studies [109-120]. Tan [82,121] amplified and extended this proposal. The chain of aggregated interstitials he refers to as an intermediate defect configuration (IDC), which may form as a step in the aggregation of point defects, but he recognized that this IDC still can lower its energy further by a lattice shear to yield a pair of dislocation dipoles in which there are no broken bonds. He realized also that the planar vacancy defects can be extended along a <110> axis, so that they too constitute an IDC which can also rearrange by a shear to form a (different) pair of dislocation dipoles in which again there are no broken bonds. We will see much more of these results, and their extensions, in papers of this conference. We note here simply that these same kinds of configurations can occur in precipitation processes which may be found in ion implantation and/or in heat-treatment. And here again there is a role for EPR studies. For example, in studies of the oxygen-related heat-treatment defects, it has been established [122] that precipitation of oxygen gives rise to shallow donors; these donors are thought to involve four oxygen atoms but the defect configuration has not been established. As discussed in other papers at this meeting, by infrared, electrical and electron-microscope studies, it has been established that oxygen precipitation proceeds through a sequence much like that envisioned by Tan as discussed above. Ammerlaan and co-workers [123,124] have carried out EPR studies on heat-treatment centers in oxygen-rich silicon. They observed ten different EPR spectra, one for which a model has been previously proposed [125], namely a substitutional carbon next to a vacancy-oxygen pair, the (V·C·O) in Table IV, or equivalently a carbon-oxygen pair in a divacancy. No model has been established yet for the remaining nine spectra, but it is interesting to note that while two of the spectra are isotropic the remaining ones, including the dominant ones have 2mm (or C_{2v}) symmetry with the <110> axis a principal axis, as would be characteristic of the intermediate-defect configurations discussed above. Moreover, the values of the g tensor parameters vary in novel ways suggesting interactions between defects (e.g., via a changing state of strain in the lattice) during precipitation. We are confident that as defect models are established for these spectra, this understanding will contribute greatly to the understanding of oxygen-related defects in silicon.

SUMMARY

EPR studies have been widely applied in semiconductors, in some cases with substantial success already achieved, in other cases with progress coming with difficulty. Looking at the identifications listed in Tables II, III and IV may lead the reader to believe that essentially everything has been done; we will now present some of that data in a different light which we hope will give a more balanced perspective. In Fig. 1 we show a periodic table in which the elements are marked for which some EPR information is available in silicon; in Fig. 2 we show comparable data for germanium; in Fig. 3 we show the comparable data for elements in MgO. Recalling that the most is known in Si and in MgO, it can be seen that much remains to be done.

ACKNOWLEDGMENTS

The work of the first two authors was partially supported by ONR (Grant N00014-75-C-0919), which support is gratefully acknowledged. The last author gratefully acknowledges the sabbatical leave from NRL which initiated this work.

ELEMENTS WITH ASSOCIATED EPR SPECTRA IN
SILICON

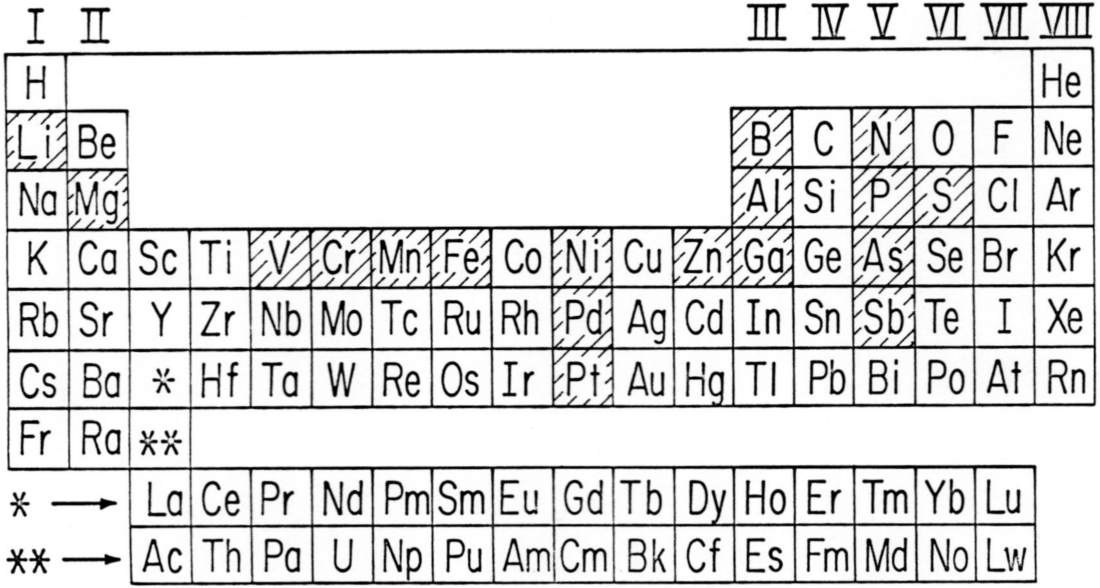

FIGURE 1

ELEMENTS WITH ASSOCIATED EPR SPECTRA IN GERMANIUM

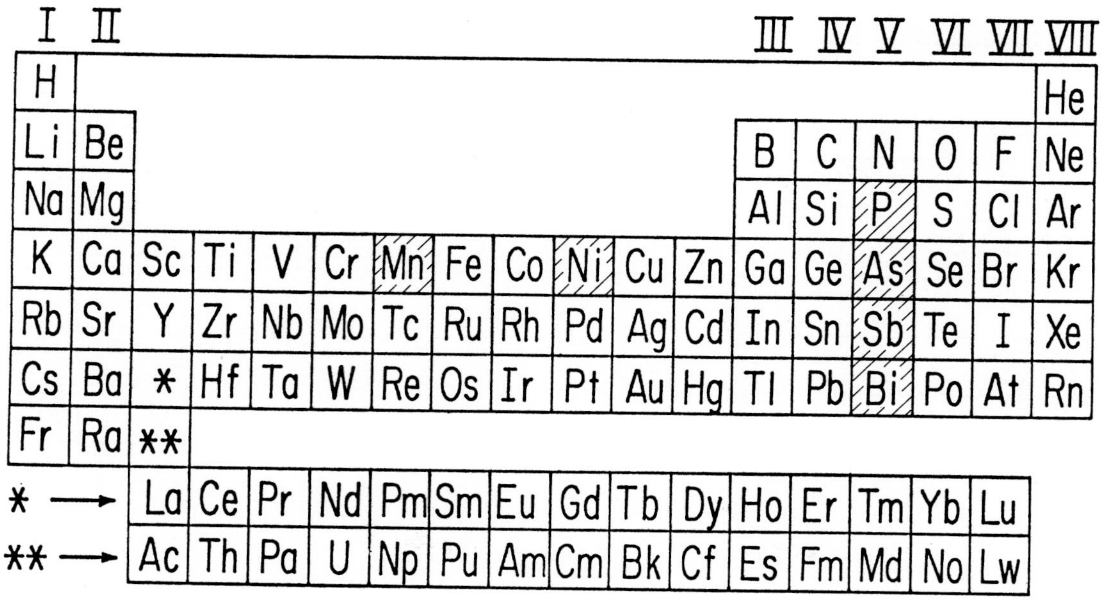

FIGURE 2

14

ELEMENTS WITH ASSOCIATED EPR SPECTRA IN
MgO

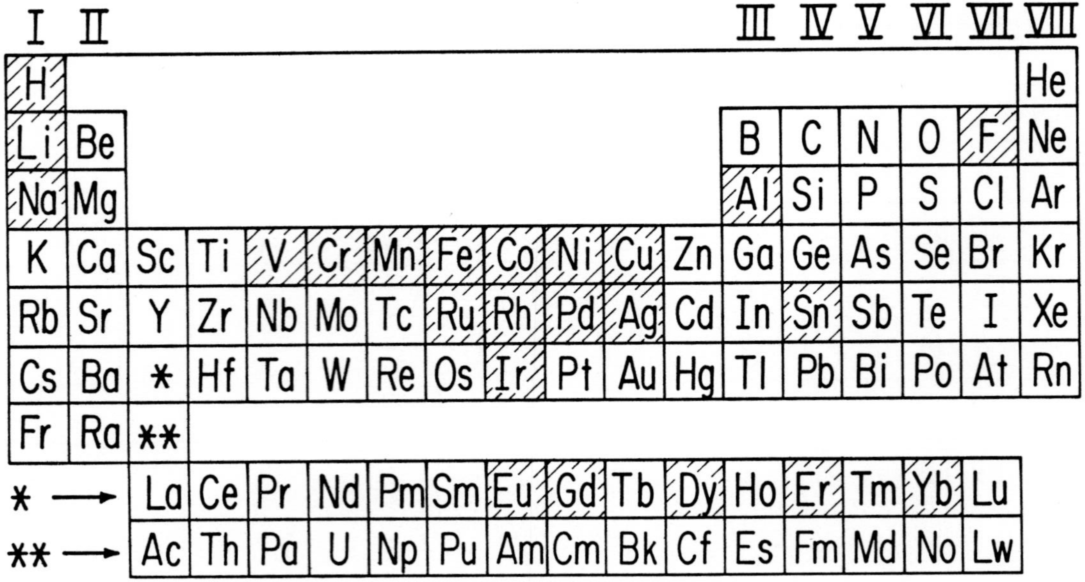

FIGURE 3

REFERENCES

1. Handbook of Chemistry and Physics (Chem. Rubber Co., Cleveland, 1971) 52nd edition, p. B245 et seq.

2. G. Feher, Phys. Rev. 114, 1219, 1245 (1959).

3. E. B. Hale and R. L. Mieher, Phys. Letters A29, 30 (1969).

4. E. B. Hale and R. L. Mieher, Phys. Rev. 184, 739 (1969).

5. E. B. Hale and R. L. Mieher, Phys. Rev. 184, 751 (1969).

6. E. B. Hale and T. G. Castner, Phys. Rev. B 1, 4763 (1970).

7. J. L. Ivey and R. L. Mieher, Phys. Rev. B 11, 822 (1975).

8. J. L. Ivey and R. L. Mieher, Phys. Rev. B 11, 849 (1975).

9. C. Kittel and A. H. Mitchell, Phys. Rev. 96, 1488 (1954).

10. W. Kohn and J. M. Luttinger, Phys. Rev. 97, 883 (1955).

11. W. Kohn and J. M. Lottinger, Phys. Rev. 98, 915 (1955).

12. W. Kohn, Solid State Physics 5, 257 (1957).

13. R. A. Faulkner, Phys. Rev. 184, 713 (1969).

14. E. B. Hale, J. Phys. Chem. Solids 34, 621 (1973).

15. S. T. Pantelides and C. T. Sah, Phys. Rev. B 10, 621,638 (1974).

16. S. T. Pantelides, Festkörperprobleme 15, 149 (1975).

17. J. R. Onffroy, Phys. Rev. B 17, 2062 (1978).

18. R. L. Mieher, Phys. Rev. B 17, 2069 (1978).

19. G. Feher, J. C. Hensel and E. A. Gere, Phys. Rev. Letters 5, 309 (1960).

20. T. Shimizu and N. Tanaka, Phys. Letters A 45, 5 (1973).

21. J. M. Cherlow, R. L. Aggarwal and B. Lax, Phys. Rev. B 7, 4547 (1973).

22. H. Neubrand, Phys. Stat. Sol. (b) 86, 269 (1978).

23. W. Kohn and J. M. Luttinger, Phys. Rev. 97, 869 (1955).

24. W. Kohn and D. Schechter, Phys. Rev. 99, 1903 (1955).

25. D. Schechter, J. Phys. Chem. Solids 23, 237 (1962).

26. A. Baldereschi and N. O. Lipari, Phys. Rev. B 9, 1525 (1974).

27. G. W. Ludwig, Phys. Rev. 137, A1520 (1965).

28. J. G. de Wit, E. G. Sieverts and C. A. J. Ammerlaan, Phys. Rev. B 14, 3494 (1976).

29. E. G. Sieverts, Ph.D. Thesis, University of Amsterdam, 1978.

30. T. F. Lee and T. C. McGill, J. Phys. C: Solid State Phys. 6, 3438 (1973).

31. C. A. J. Ammerlaan and C. Weigel in: Lattice Defects in Semiconductors — 1976, N. B. Urli and J. W. Corbett, eds. (Inst. Phys., Bristol-London, 1977) pp. 448-457.

32. C. A. J. Ammerlaan and J. C. Wolfrat, Phys. Stat. Sol. (b) 89, 85 (1978).

33. W. Kaiser, P. H. Keck and C. F. Lange, Phys. Rev. 101, 1264 (1956).

34. H. J. Hrostowski and B. J. Alder, J. Chem. Phys. 33, 980 (1960).

35. J. W. Corbett, R. S. McDonald and G. W. Watkins, J. Phys. Chem. Solids 25, 873 (1964).

36. D. R. Bosomworth, W. Hayes, A. R. L. Spray and G. D. Watkins, Proc. Roy. Soc. Lond. A 317, 133 (1970).

37. There are over one thousand references which should be cited for the information in Tables II, III and IV. Space does not permit our including these references. Questions about the origin of the information listed may be addressed directly to the authors.

38. J. W. Corbett, Surface Sci. 90, 205 (1979).

39. C. P. Slichter, Phys. Rev. 99, 479 (1955).

40. G. Feher, R. C. Fletcher and E. A. Gere, Phys. Rev. 100, 1784 (1955).

41. G. Feher in: Paramagnetic Resonance, W. Low, ed. (Academic Press, NY, 1963) Vol. 2, pp. 715-740.

42. D. Jerome and J. M. Winter, Phys. Rev. 154, A1001 (1964).

43. C. Robert, C. R. Acad. Sc. Paris 260, 6337 (1965).

44. S. Maekawa and N. Kinoshita, J. Phys. Soc. Japan 20, 1447 (1965).

45. Y. Sugiura, J. Phys. Soc. Japan 20, 2098 (1965).

46. S. Maekawa, J. Phys. Soc. Japan 21, 1221 (1966).'

47. S. Maekawa, J. Phys. Soc. Japan, 21 Suppl., 574 (1966).

48. H. Kodera, J. Phys. Soc. Japan, 21 Suppl., 578 (1966).

49. N. Kumar and K. P. Sinha, Zeit. F. Phys. 197, 26 (1966).

50. P. R. Cullis and J. R. Marko, Phys. Rev. B 1, 632 (1970).

51. F. T. Hedgcock and T. W. Randorf, Canad. J. Phys. 48, 2930 (1970).

52. D. J. Lépine, Phys. Rev. B $\underline{2}$, 2429 (1970).

53. J. D. Quirt and J. R. Marko, Phys. Rev. Letters $\underline{26}$, 318 (1971).

54. H. Ue and S. Maekawa, Phys. Rev. $\underline{3}$, 4232 (1971).

55. J. D. Quirt and J. R. Marko, Phys. Rev. B $\underline{5}$, 1716 (1972).

56. Y. Toyoda and Y. Hayashi, J. Phys. Soc. Japan $\underline{33}$, 718 (1972).

57. H. Nagashima, Y. Ochiai and E. Matsuura, Sol. State Comm. $\underline{11}$, 733 (1972).

58. J. D. Quirt and J. R. Marko, Phys. Rev. B $\underline{7}$, 3842 (1973).

59. M. Rosso, J. Phys. Soc. Japan $\underline{38}$, 780 (1975).

60. P. R. Cullis and J. R. Marko, Phys. Rev. B $\underline{11}$, 4184 (1975).

61. K. Morigaki, N. Kishimoto and D. J. Lépine, Sol. State Comm. $\underline{17}$, 1017 (1975).

62. K. Morigaki and M. Rosso, J. de Phys. $\underline{36}$, 1131 (1975).

63. Y. Ochiai and E. Matsuura, Phys. Stat. Sol. (a) $\underline{38}$, 243 (1976).

64. H. Kamimura and N. F. Mott, J. Phys. Soc. Japan $\underline{40}$, 1351 (1976).

65. N. Kishimoto and K. Morigaki, J. Phys. Soc. Japan $\underline{42}$, 137 (1977).

66. Y. Toyoda and K. Morigaki, J. Phys. Soc. Japan $\underline{43}$, 118 (1977).

67. Y. Toyoda, N. Kishimoto, K. Murakami and K. Morigaki, J. Phys. Soc. Japan $\underline{43}$, 114 (1977).

68. Y. Ochiai and E. Matsuura, Phys. Stat. Sol. (a) $\underline{45}$, K101 (1978).

69. N. F. Mott, Advances in Phys. $\underline{21}$, 785 (1972).

70. N. F. Mott, Metal-Insulator Transitions (Taylor & Francis, London, 1974)

71. G. D. Watkins, J. Phys. Soc. Japan $\underline{18}$, Suppl. II, 22 (1963).

72. G. D. Watkins in: Radiation Damage and Defects in Semiconductors, J. E. Whitehouse, ed. (Inst. of Phys. Bristol-London, 1973) pp. 228-237.

73. G. D. Watkins and J. W. Corbett, Phys. Rev. $\underline{138}$, A543 (1965).

74. J. W. Corbett and G. D. Watkins, Phys. Rev. $\underline{138}$, A555 (1965).

75. G. D. Watkins, J. W. Corbett and R. M. Walker, J. Appl. Phys. $\underline{30}$, 1198 (1959).

76. G. D. Watkins and J. W. Corbett, Disc. Faraday Soc. $\underline{31}$, 86 (1961).

77. G. D. Watkins and J. W. Corbett, Phys. Rev. $\underline{134}$, A1359 (1964).

78. E. L. Elkin and G. D. Watkins, Phys. Rev. $\underline{174}$, 881 (1968).

18

79. G. D. Watkins, Phys. Rev. 155, 802 (1967).

80. See, for example, the discussion in refs. 81 and 82.

81. J. W. Corbett and J. C. Bourgoin in: Point Defects in Solids, J. H. Crawford, Jr. and L. M. Slifkin, eds. (Plenum, NY, 1975) Vol. 2, pp. 1–161.

82. J. W. Corbett, J. P. Karins and T. Y. Tan, Proc. of the Int'l. Conf. on Ion Beam Modification of Materials – 80, in press.

83. Y. H. Lee, N. N. Gerasimenko and J. W. Corbett, Phys. Rev. B 14 4506 (1976).

84. R. B. Fair in: Processing Technologies, D. Kahng, ed. (Academic Press, NY, 1981) in press.

85. J. A. Van Vechten, Phys. Rev. B 17, 3197 (1978).

86. J. A. Van Vechten and C. D. Thurmond, Phys. Rev. B 14, 3551 (1976).

87. J. C. Bourgoin and M. Lannoo, Rad. Effects 46, 157 (1980).

88. A. Seeger, W. Frank and U. Gösele in: Defects and Radiation Effects in Semiconductors, 1978, J. H. Albany, ed. (Inst. of Phys., Bristol-London, 1979) pp. 148–170.

89. R. F. Peart, Phys. Stat. Sol. 15, K119 (1966).

90. R. N. Ghoshtagore, Phys. Rev. Letters 16, 890 (1966).

91. B. L. Kendall and D. B. DeVries in: Semiconductor Silicon, R. R. Haberecht, ed. (Electrochem. Soc., NY, 1969) pp. 358–421.

92. B. J. Masters, Sol. State Comm. 9, 283 (1971).

93. A. Seeger and M. L. Swanson in: Lattice Defects in Semiconductors, R. R. Hasiguti, ed. (Univ. of Tokyo Press, Tokyo, 1966) pp. 93–130.

94. A. Seeger and K. P. Chik, Phys. Stat. Solidi 29, 455 (1968).

95. A. Seeger in: Radiation Effects in Semiconductors, J. W. Corbett and G. D. Watkins (Gordon and Breach, NY, 1971) pp. 29–38.

96. W. Frank in: Lattice Defects in Semiconductors, 1974, F. A. Huntley (Inst. of Phys., Bristol-London, 1975) pp. 23–48.

97. Y. H. Lee, R. L. Kleinhenz and J. W. Corbett in: Defects and Radiation Effects in Semiconductors, 1978, J. H. Albany, ed. (Inst. of Phys., Bristol-London, 1979) pp. 521–531.

98. G. A. Baraff, E. O. Kane and M. Schlüter, Phy. Rev. Letters 43, 956 (1979).

99. G. D. Watkins and J. R. Troxell, Phys. Rev. Letters 44, 593 (1980).

100. Y. H. Lee and J. W. Corbett, Phys. Rev. B 9, 4351 (1974).

101. W. Jung and G. S. Newell, Phys. Rev. 132, 648 (1963).

102. K. L. Brower in: Radiation Effects in Semiconductors, J. W. Corbett and G. D. Watkins, eds. (Gordon and Breach, NY, 1971) pp. 189-195.

103. M. Nisenoff and H. Y. Fan, Phys. Rev. 128, 1605 (1962).

104. Y. H. Lee and J. W. Corbett, Phys. Rev. B8, 2810 (1973).

105. G. D. Watkins, R. P. Messmer, C. Weigel, D. Peak and J. W. Corbett, Phys. Rev. Letters 27, 1573 (1971).

106. C. Weigel, D. Peak, J. W. Corbett, G. D. Watkins and R. P. Messmer, Phys. Rev. B 8, 2906 (1973).

107. C. Weigel and J. W. Corbett, Z. Physik B 23, 233 (1976).

108. G. D. Watkins and K. L. Brower, Phys. Rev. Letters 36, 1329 (1976).

109. D. J. Mazey, R. S. Barnes and R. S. Nelson in: Proc. 6th Int'l Conf. on Electron Microscopy, R. Uyeta, ed. (Maruzen Co. Ltd., Tokyo, 1966) pp. 363-368.

110. S. M. Davidson and G. R. Booker, Rad. Effects 6, 33 (1970).

111. R. W. Bicknell, Proc. Roy. Soc. A 31, 75 (1969).

112. L. T. Chadderton and F. H. Eisen, Rad. Effects 7, 129 (1971).

113. P. K. Madden and S. M. Davidson, Rad. Effects 14, 271 (1972).

114. M. D. Matthews and S. J. Ashby, Phil. Mag. 27, 1313 (1973).

115. W. K. Wu and J. Washburn, J. Appl. Phys. 48, 3742 (1977).

116. K. Seshan and J. Washburn, Rad. Effects 14, 267 (1972).

117. E. Nes and J. Washburn, J. Appl. Phys. 42, 3559 (1972).

118. A. L. Aseev, V. V. Bolotov, L. S. Smirnov and S. I. Stenin, Sov. Phys. Semicond. 13, 764 (1979).

119. I. G. Salisbury and M. H. Loretto, Phil. Mag. A 39, 317 (1979).

120. C. A. Ferreira-Lima and A. Howie, Phil. Mag. 34, 1057 (1976).

121. Teh-Yu Tan, Phil. Mag., in press.

122. W. Kaiser, H. L. Frisch and H. Reiss, Phys. Rev. 112, 1546 (1958).

123. S. H. Muller, M. Sprenger, E. G. Sieverts and C. A. J. Ammerlaan, Sol. State Comm. 25, 987 (1978).

124. S. H. Muller, E. G. Sieverts and C. A. J. Ammerlaan, to be published.

125. Y. H. Lee, J. W. Corbett and K. L. Brower, Phys. Stat. Sol. A 41, 637 (1977).

Published 1981 by North-Holland, Inc.
Narayan, and Tan, eds.
Defects in Semiconductors

NEGATIVE-U PROPERTIES FOR POINT DEFECTS IN SILICON*

G. D. WATKINS
Department of Physics and Sherman Fairchild Laboratory, Lehigh University,
Bethlehem, Pennsylvania 18015

ABSTRACT

A defect has negative-U properties if it can trap two
electrons (or holes) with the second bound more strongly
than the first. It is as if there were a net attraction
between the two carriers (negative Hubbard correlation
energy U) at the defect, and the defect energy levels in
the gap are therefore inverted from their normal order.
Experimental evidence is presented that interstitial
boron and the lattice vacancy, both common simple point
defects produced by electron irradiation of silicon,
have this unusual property. These defects represent the
first and only concrete examples of negative-U centers
in any material and serve as models for an understanding
of the phenomenon.

WHAT IS U?

An impurity, or other simple point defect, often introduces more than one
electrically active level in the forbidden gap of a semiconductor. For such
a defect, it is normally found that if one electron is trapped, the level
position for the next trapped electron is raised in energy, closer to the
conduction band. This is illustrated in Fig. 1(a) for a typical defect with
both a single donor (0/+) and acceptor (-/0) level. (The notation i/j means

Fig. 1(a). Normal positive-U level
ordering for a defect with its acceptor level (A) above the donor
level (B).
  (b) Inverted negative-U ordering.

that the defect has charge i if the level is filled with an electron, j if
empty.) The acceptor level is expected to be above the donor level because
the second electron, although possibly trapped into a similar orbital as the
first, and experiencing therefore the same attractive interaction to the core
of the defect, is <u>repelled</u> by the Coulomb interaction with the first electron.
This added Coulomb repulsion energy is defined as U and in Fig. 1(a) is the

---

*
Research supported by the U.S. Navy, ONR Electronics and Solid State Sciences
Program, Contract No. N00014-76-C-1097.

separation between the donor and acceptor levels. (U is often called the Hubbard "correlation energy," being first introduced by Hubbard in his treatment of conductivity in narrow band semiconductors [1]).

POSITIVE-U

All defects which introduce more than one level in a crystalline semi-conductor are either known to have positive-U behavior or have been assumed to have this property. For the defect example of Fig. 1a, for instance, the normal positive-U ordering of levels means that each of the three charge states D^+, D^O, D^- can be the thermodynamically stable one depending upon the position of the Fermi level, E_F. The donor and acceptor levels have unam-biguous meanings and can be measured either by a Hall measurement (Fermi level position determination under thermodynamic equilibrium conditions for the electronic system) or by direct ionization excitation from the levels as in DLTS and photoconductivity or photocapacitance studies.

NEGATIVE-U

Negative-U properties for a defect mean that the levels are inverted from their usual order, as shown in Fig. 1(b), with the acceptor level below the donor. Let us defer for the moment how this might come about and consider the unusual properties that such a defect would have, if it could exist. In the first place, D^O is no longer a stable charge state. There is no position of the Fermi level that makes it thermodynamically stable. Two isolated neutral defects, in thermal contact with each other (via the con-duction band), can lower their energy by ionizing, gaining the energy $|U|$

$$2D^O \rightarrow D^+ + D^- + |U| \qquad (1)$$

In particular, if $2E_F > E_A + E_D$ (the energy for two electrons at the Fermi energy exceeds the energy for the two electrons trapped at the defect), then in equilibrium all defects are in the D^- state. Conversely, if $2E_F < E_A + E_D$, all defects are in the D^+ state. With $2E_F = E_A + E_D$ we have a mixture of D^+ and D^-, but no D^O. A Hall measurement would therefore detect neither level. Instead it would indicate a single level halfway between the two levels, $(E_D+E_A)/2$, which is where the Fermi level becomes locked over the range of temperature or carrier densities that D^+ and D^- are simultaneously present. The levels of Fig. 1(b) can still be considered real, however, in the sense that direct transitions between the band edges and the levels can still be studied. In a properly designed experiment, DLTS and optical excitations should be able to measure each of the inverted level positions.

The remarkable feature of a defect with negative-U properties is that there is an effective attractive interaction between electrons at the defect. In a sense, this is an "extrinsic" Cooper pair [2], the electrons binding by pairs to the defect.

MECHANISM FOR NEGATIVE-U BEHAVIOR

The concept of negative-U properties in solids was first introduced by Anderson [3] in 1975 in order to explain the peculiar properties of chalco-genide glasses. The concept was subsequently extended to defects in these materials by Street and Mott [4]. Their idea was that the energy gained by electron pairing at dangling bonds near a defect, coupled with large lattice relaxation as the bonds pull together, might overcome the Coulombic repulsion of the two electrons, supplying a net attractive interaction between them at the site. An alternative microscopic model has been suggested by Kastner, Adler and Fritzsche [5], in which a change could occur in the bonding coordination of a defect with its neighbors as its charge state changes, thus supplying the driving force. They named such defects "valence alternation pairs." In either model, strong electron-lattice coupling and large lattice relaxations or rearrangements are essential ingredients for possible negative-U behavior. The analogy to Cooper pair formation in superconductors is therefore an intriguing one. There too it is the electron-phonon coupling that supplies the attractive force for pair formation. In the case of a defect with a soft polarizable "ligand" molecular framework, a readjustment of its geometry may allow an increase in the attractive electron-atom core interactions (the origin of molecular binding), while at the same time allowing the electrons to avoid each other, minimizing their repulsion.

The negative-U model appears to account for the dominant properties of the chalcogenide glasses: The thermodynamic instability of the single electron occupied state explains the curious absence of paramagnetism in doped materials. The failure to achieve low resistivity n- or p-type material vs. chemical doping is explained by assuming that the inverted donor and acceptor levels straddle the gap center, locking the Fermi level near midgap, etc. Indeed, it is generally accepted now that the negative-U model for defects is probably the correct one to describe the chalcogenide glasses. However, as yet no direct microscopic identification of any defect with these properties has been made in these materials. In fact, until the work that we will describe in this paper, no defect with negative-U properties had been identified to exist in <u>any</u> material.

NEGATIVE-U DEFECTS IN SILICON

Kastner, Adler and Fritzche [5] pointed out that there is nothing unique about glasses. They argued that negative-U centers such as their valence-alternation-pairs should also exist in crystalline materials. It is interesting to note, however, that they specifically excluded "tetrahedral materials" such as silicon. Recently, two examples of simple point defects in silicon have been reported to have negative-U properties [6-8]. In other words, so far, the <u>only</u> identified defects with these properties are actually in crystalline silicon.

The two defects are the isolated lattice vacancy and interstitial boron, both common defects produced in silicon by electron irradiation. The level positions proposed for these defects [6-8] are illustrated in Fig. 2. For the vacancy, Fig. 2(a), the first donor level (0/+) at $\sim E_v + 0.05$ eV lies below the second donor level (+/++) at $\sim E_v + 0.13$ eV. This

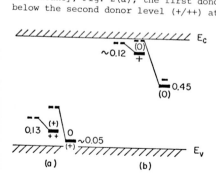

Fig. 2. Level positions suggested for (a) the vacancy and (b) interstitial boron in silicon. The dashed lines indicate schematically the level positions before relaxation.

inverted order for the levels is consistent with the suggestion by Baraff, Kane, and Schluter [9] who first pointed out from theoretical grounds that the greater Jahn-Teller relaxation for V^0 than for V^+ could supply the driving force for negative-U behavior. This relaxation can be viewed as a form of bond pairing between the four dangling bonds of the vacancy and, as such, can be considered an example of the Anderson-Street-Mott model [3,4].

For interstitial boron, Fig. 2(b), the acceptor level (-/0) at E_C - 0.45 eV lies, inverted, below the donor state (0/+) at E_C - 0.12 eV. As will be shown, there is evidence that the negative-U mechanism for interstitial boron may involve a configuration change in the lattice similar to the model of Kastner, et al. [5].

Interstitial boron

Interstitial boron was first observed by electron paramagnetic resonance (EPR) in the neutral B_i^0 state [10]. Consistent with the negative-U model, this charge state was found to be unstable in n-or p-type material and could only be generated by shining near band gap light on the sample at temperatures below carrier freezeout. In high resistivity material, the decay of the photogenerated signal was found to be strongly temperature dependent, with a time constant of 15 min at ∿50 K. This decay can be interpreted as electron emission from the neutral state

$$B_i^0 \rightarrow B_i^+ + e^- \quad , \qquad (2)$$

the temperature dependence being consistent with the donor level position indicated in Fig. 2(b) at ∿E_C - 0.12 eV.

DLTS studies initially detected only a single level at E_C - 0.45 eV that could be identified with interstitial boron [7]. This identification came from detailed correlation of production and annealing properties in DLTS and EPR studies and is considered a firm identification. In these studies, no evidence of the shallower level at E_C - 0.12 eV was detected.

The failure to detect the shallow level in DLTS is easily understood for the inverted negative-U ordering. The experiments were performed in n-type material (3 x 10^{16} P/cm^3; 1 x 10^{16} B/cm^3), the majority carrier trap filling pulses converting the interstitial boron atoms to B_i^- at the beginning of each reverse bias monitoring period. Subsequent electron emission decay then proceeds by

$$B_i^- \xrightarrow{0.45} B_i^0 + e^- \xrightarrow{0.12} B_i^+ + 2e^- \qquad (3)$$

Here we have indicated the thermal activation barrier associated with the level position (in eV) for each emission process. Note that the limiting process is the _first_ electron emission (the deeper level). At a temperature high enough for this to occur, the second more weakly bound electron emission should follow immediately. The result, therefore, is that only the deeper acceptor level is detected by DLTS. It appears therefore simply as a single peak at E_v + 0.45 eV. However, because two carriers are released, the amplitude of the peak will actually be twice its normal amplitude. This therefore serves as a unique signature of a negative-U defect: the release of carriers by _pairs_. If one could independently determine the concentration of the defects, then a measurement of the DLTS amplitude would provide a key test for negative-U behavior. This was tested in the following way:

In Fig. 3 we show the DLTS spectrum in n-type silicon partially counterdoped

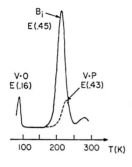

Fig. 3. DLTS of partially counterdoped n-type silicon (1 x 10^{16} B,
3 x 10^{16} P/cm^3) after 4.2 K electron
irradiation and 100 K anneal. The
dashed curve is after annealing at
240 K to remove the interstitial
boron.

with boron, which had been irradiated at 4.2 K by 1.5 MeV electrons and subsequently annealed to 100 K. The initial primary damage products should be
equal concentrations of simple lattice vacancies and interstitial silicon
atoms. From EPR studies [10,11] it has been concluded that the interstitial
silicon atoms are unstable, migrating at ∿4.2 K to be trapped by substitutional
boron atoms to produce interstitial boron. After the 100 K anneal, where
vacancies are known to migrate [11], the vacancies are trapped by oxygen to
form V·O pairs (at E_c - 0.16 eV [12]) and V·P pairs (at E_c - 0.43 eV [12]).
Since boron is expected to be the dominant trap for interstitials, and oxygen
and phosphorus those for the vacancy, the concentration of interstitial boron
should be approximately equal to the sum of the V·O and V·P pairs. Note in
Fig. 3 that all three can be monitored in the DLTS experiment, the presence
of the V·P pairs being revealed after annealing of the interstitial boron.
As seen in the figure, the intensity of the B_i peak is very close to <u>twice</u>
the sum of those for the two vacancy-associated levels. This can be taken
as strong evidence that B_i^- indeed does emit <u>two</u> electrons [6,7].

The final critical test for negative-U behavior is to detect the E_c - 0.12
eV level directly. The level is deep enough so that if it could be repe-
titively prepared in the metastable B_i^0 state it should be observable directly
by DLTS. Unfortunately, this won't work in the normal DLTS experiment because at the low temperatures required to see the shallow level, repetitive
trap filling pulses, no matter how short, will eventually convert all of the
interstitial boron to B_i^-, removing them from the experiment. In a recent
experiment [8], this difficulty has been avoided by illuminating the n-type
diodes with penetrating light (through a room temperature silicon filter) to
photoionize the B_i^-

$$B_i^- \xrightarrow{h\nu} B_i^0 + e^- , \qquad (4)$$

preventing accumulation of B_i^- during the DLTS experiment. The result is shown
in Fig. 4. In the presence of the light, a new peak is observed. A study of
the temperature dependence of its emission rate locates its level position at
E_c - 0.12 eV, as indicated in Fig. 2. This peak disappears upon thermal
annealing at ∿240 K, or injection annealing at ∿150 K in ∿1:1 correspondence
with the disappearance of the E_c - 0.45 eV interstitial boron peak. This
confirms its association with the interstitial boron.

The DLTS capacitance transient sequence for this experiment is illustrated
in Fig. 5. After a short trap filling pulse, some B_i^+ ions have captured only
one electron to become B_i^0, others have captured two to become B_i^-. The
B_i^0 decays rapidly by 0.12 eV thermally

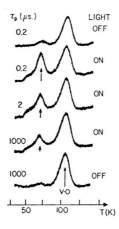

Fig. 4. The metastable B_i^0 donor level at $\sim E_c - 0.12$ eV observed in the presence of light.

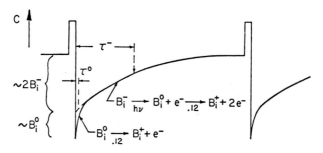

Fig. 5. DLTS capacitance transient sequence in the presence of light, showing both the thermally activated decay of B_i^0 and the slower photo-excited decay of B_i^-.

activated emission as indicated and is detected as the DLTS peak in Fig. 4 by tuning the DLTS spectrometer (the double boxcar window) to its time constant (τ^0 in Fig. 5). The B_i^- decays more slowly (τ^-) being limited by the photoionization to B_i^0 followed by rapid thermal emission to B_i^+. The pulse repetition period is adjusted to be long (~75 ms) with respect to the photo-ionization decay ($\tau^- \sim 25$ ms) so that all interstitial boron has been returned to B_i^+ for the start of the next pulse.

The amplitude of the 0.12 eV peak is therefore a direct measure of the concentration of B_i^0 at the end of a single trap filling pulse. We note in Fig. 4 that the height of the 0.12 eV peak decreases with increasing pulse width, an anomolous result for a normal level. On the other hand, this is a unique and necessary signature of a negative-U defect, the longer pulse producing in this case more of the deeper two-electron B_i^- state and less of the shallow one-electron B_i^0 state.

In this experiment, the concentration of B_i^- present after each pulse can also be measured directly by monitoring the amplitude of the photoinduced decay of the E_C - 0.45 eV level. (In Fig. 5, the DLTS spectrometer boxcar is tuned to the slower photo-decay time constant $\tau \sim 25$ ms, and the amplitude monitored. After correction for a small known background contribution due to photoionization of other defects, the concentration of B_i^- is taken as proportional to one half the photodecay amplitude because again for it two carriers are released per decay.) Figure 6 shows the concentration of B_i^- estimated in this way along with the concentration of B_i^0 estimated from the

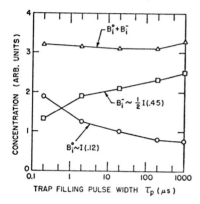

Fig. 6. The concentrations of B_i^0 and B_i^- vs. trap filling pulse width τ_p, estimated from the intensities I of the E_C - 0.12 eV peak and the E_C - 0.45 eV photo-decay peaks in the presence of light.

height of the E_C - 0.12 eV peak vs. pulse width. The complementary behavior and the constancy of the sum demonstrates dramatically the fact that they are inverted negative-U levels associated with the same defect [8].

Vacancy

The experimental observations for the vacancy are similar to those described in detail above for interstitial boron, and will be only briefly summarized here. Referring to the level structure of Fig. 2, they are:

1. EPR is observed for a photo-excited metastable V^+ state [11].

2. Decay of the photoexcited V^+ EPR signal reveals a thermal activation energy of ~ 0.05 eV [11], consistent with a location for the single donor state at $\sim E_V + 0.05$ eV and the reaction

$$V^+ \xrightarrow{0.05} V^0 + h^+. \tag{5}$$

3. DLTS reveals only a deeper hole trap at $E_V + 0.13$ eV [13,14]. Careful measurements of the relative amplitude of the DLTS peak indicates a two hole excitation [6], consistent with

$$V^{++} \xrightarrow{0.13} V^+ + h^+ \xrightarrow{0.05} V^0 + 2h^+. \tag{6}$$

4. EPR studies reveal that photogeneration of V^+ releases holes

$$V^{++} \xrightarrow{h\nu} V^+ + h^+, \tag{7}$$

confirming the existence of the second donor state and the stability of V^{++} in p-type material [8].

The single donor state is too shallow to detect directly in a simple DLTS experiment because its hole emission will occur in the same temperature region as majority carrier freezeout. Direct electrical confirmation of its existence and position therefore remains for a future, more ingenious experiment.

DISCUSSION

The experimental results for interstitial boron demonstrate conclusively that it has negative-U properties. As such, it is apparently the first identified defect known to have these properties in any material. The donor level is located at $E_c - 0.12$ eV and the acceptor level at $E_c - 0.45$ eV. The negative-U is therefore ~ 0.3 eV, a very large value.

The detailed microscopic mechanism is not known with certainty at this stage. However, EPR studies have allowed models to be suggested [7], one of which is illustrated in Fig. 7. The distorted configuration for B_i^0 is

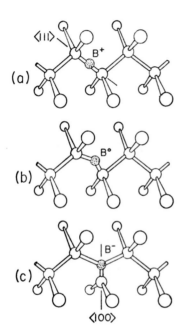

Fig. 7. Possible model for interstitial boron lattice rearrangement, associated with its negative-U behavior.

consistent with the hyperfine and g-tensor analysis of the EPR spectrum. The configurations for B_i^+ and B_i^- have been inferred indirectly [7,9]. In this model, interstitial boron occupies interstitialcy, bonded, configurations, but flips from a bond centered configuration (two-fold coordinated) for B_i^+ to a split-<100> configuration (three-fold coordinated) for B_i^-. Metastable B_i^0 is intermediate between the two positions. If correct, this is similar in concept to the "valence alternation pair" model of Kastner et al. [5]. Other models are also possible [7], and work is continuing to determine the exact mechanism.

The evidence for negative-U behavior for the vacancy is also strong. Remaining only for 100% certainty is a direct <u>electrical</u> confirmation of the existence of the shallow single donor state. If correct, the mechanism for the behavior has already been experimentally established from EPR studies [11], and confirmed from theoretical estimates [9], to be associated with the Jahn-Teller distortions of the vacancy. The symmetry of the distortion is tetragonal, which is a bond-like formation by pairs between the four dangling bonds of the neighbors. This would therefore be an example of the model conceived by Anderson [3] and Street and Mott [4].

This first observation of well identified simple point defects with negative-U behavior could be very important. I have already mentioned the possible importance in chalcogenide glasses. What we are learning in microscopic detail in the present studies in silicon can serve as a model for understanding negative-U defects in these more complex systems. We are also learning <u>how</u> to study and recognize them: photo-DLTS, two carrier release, different <u>Hall</u> and DLTS properties, etc.

How prevalent are negative-U defects in semiconductors? Have some such defects been here all along and simply undetected because we didn't know what to look for, or weren't suspicious? How about grain boundaries, dislocations, surfaces? Does this hold the key to some of their properties? Already surface reconstruction is being modeled by negative-U concepts [15]. Finally, what new exotic electronic properties might be available for a solid with such defects? Metastability implies memory and switching effects. How about "extrinsic Cooper pair" defects in a superconductor? This has also recently been considered [16,17].

This is an exciting new phenomenon and there are undoubtedly interesting surprises ahead.

ACKNOWLEDGEMENTS

The contributions of J. R. Troxell, R. D. Harris, and A. P. Chatterjee, who performed most of the experiments described in this paper, are gratefully acknowledged.

REFERENCES

1. J. Hubbard, Proc. Roy. Soc. A<u>276</u>, 238 (1963).

2. L. N. Cooper, Phys. Rev. <u>104</u>, 1189 (1956).

3. P. W. Anderson, Phys. Rev. Lett. <u>34</u>, 953 (1975).

4. R. A. Street and N. F. Mott, Phys. Rev. Lett. <u>35</u>, 1293 (1975).

5. Marc Kastner, David Adler, and H. Fritzsche, Phys. Rev. Lett. <u>43</u>, 1504 (1975).

6. G. D. Watkins and J. R. Troxell, Phys. Rev. Lett. <u>44</u>, 593 (1980).

7. J. R. Troxell and G. D. Watkins, Phys. Rev. B<u>22</u>, 921 (1980).

8. G. D. Watkins, A. P. Chatterjee, and R. D. Harris in: <u>Proceedings of the 11th International Conference on Defects and Radiation Effects in Semiconductors, Oiso, Japan 1980</u> (Inst. of Phys., London, in press).

9. G. A. Baraff, E. O. Kane, and M. Schlüter, Phys. Rev. Lett. $\underline{37}$, 1504 (1979).

10. G. D. Watkins, Phys. Rev. B$\underline{12}$, 5824 (1975).

11. G. D. Watkins in: Lattice Defects in Semiconductors 1974, F. A. Huntley, ed. (Inst. of Phys. Conference Ser. 23, London 1975) p. 1.

12. L. C. Kimerling in: Radiation Effects in Semiconductors, N. B. Urli, ed. (Inst. of Phys. Conference Ser. 31, London 1977) p. 221.

13. G. D. Watkins, J. R. Troxell, and A. P. Chatterjee in: Defects and Radiation Effects in Semiconductors 1978, J. H. Albany, ed. (Inst. of Phys. Conf. Ser. 46, London 1979) p. 16.

14. J. R. Troxell, Ph.D. Thesis, Lehigh University 1979 (unpublished).

15. C. T. White and K. L. Ngai, Phys. Rev. Lett. $\underline{41}$, 885 (1978).

16. E. Simanek, Solid State Comm. $\underline{32}$, 731 (1979).

17. C. S. Ting, D. N. Talwar, and K. L. Ngai, Phys. Rev. Lett. $\underline{45}$, 1213 (1980).

Published 1981 by North-Holland, Inc.
Narayan, and Tan, eds.
Defects in Semiconductors

FROM THE MYSTERY TO THE UNDERSTANDING OF THE SELF-INTERSTITIALS IN SILICON

W. FRANK, A. SEEGER, AND U. GÖSELE

Max-Planck-Institut für Metallforschung, Institut für Physik, Heisenbergstr. 1, and Universität Stuttgart, Institut für Theoretische und Angewandte Physik, Pfaffenwaldring 57, D-7000 Stuttgart 80, Germany

ABSTRACT

Our present knowledge on self-interstitials in silicon and the rôle these defects play under widely different experimental conditions are surveyed. In particular, the following phenomena involving self-interstitials either in supersaturations or under high-temperature thermal-equilibrium conditions are considered: mobility-enhanced diffusion of self-interstitials below liquid-helium temperature, thermally activated diffusion of self-interstitials at intermediate temperatures (140 K to 600 K), concentration-enhanced diffusion of Group-III or Group-V elements in silicon at higher temperatures, and — as examples for high-temperature equilibrium phenomena — self-diffusion and diffusion of gold in silicon. This leads to the picture that the self-interstitials in silicon may occur in different electrical charge states and possess dumbbell configurations or are extended over several atomic volumes at intermediate or high temperatures, respectively.

1. INTRODUCTION

Regarding the outstanding rôle played by semiconductors in modern technology our present knowledge on intrinsic point defects in these materials is disappointingly modest. In fact, today silicon is the only semiconductor for which this knowledge is approaching a level that may be called 'fairly satisfactory'.

The rates of progress in the understanding of the properties of vacancy-type and interstitial-type defects in silicon differed vastly during the last two decades: Whereas in 1970 essential information on monovacancies, divacancies, and vacancy—impurity complexes such as the A- or E- centre already existed, the seventies clearly were the 'decade of the self-interstitials'. Hence, after a brief summary of the early developments concerning self-interstitials in silicon (Sect. 2) this state-of-the-art report concentrates on the rapid progress in this field during the past ten years.

2. EARLY DEVELOPMENTS

2.1. Self-interstitials at extremely low temperatures

In 1964 Watkins [1, 2] discovered by means of EPR studies on Al-doped p-type silicon that during electron irradiation at 4 K equal concentrations of Al^{2+} interstitial atoms (Al_i) and vacancies (V)

are produced. He suggested that this observation may be understood
if the self-interstitials (I) created by electron irradiation can
migrate over long distances already at 4 K: If this is the case
the self-interstitials recombine with vacancies,

$$I + V \rightarrow \text{annihilation,} \qquad (1)$$

or, alternatively, replace substitutional Al atoms (Al_s)
according to

$$I + Al_s \rightarrow Al_i. \qquad (2)$$

Since electron irradiation introduces vacancies and self-interstitials in pairs, equal concentrations of Al interstitials and
vacancies are left behind after the elimination of the self-interstitials via the reactions (1) and (2).

Watkins' observation has been confirmed later by several authors,
who demonstrated self-interstitial migration to take place in
both n- and p- type silicon during electron irradiation down to
at least 0.5 K [3 - 5]. For germanium silimar evidences for the
migration of self-interstitials at extremely low temperatures
exist [6 - 7].

2.2. Self-interstitials under high-temperature equilibrium conditions

In 1968 Seeger and Chik [8] pointed out that, in contrast to
metals or germanium, the self-diffusion data on silicon above
about 1200 K cannot be understood in terms of a vacancy mechanism.
Hence, these authors [8] made the unconventional suggestion that
at high temperatures self-diffusion in silicon is governed by
self-interstitials. For further details the reader is referred
to the original literature [8, 9] and to review articles [10 - 13]
on this subject. As will be shown below, the dominance of self-
interstitials under high-temperature thermal-equilibrium conditions and the high mobility of self-interstitials at very low
temperatures do not exclude each other but rather belong to the
characteristic features of silicon.

3. ELECTRICAL CHARGE STATES AND GEOMETRICAL CONFIGURATIONS OF THE SELF-INTERSTITIALS

At the beginning of the past decade the knowledge on self-interstitials in silicon was scarce and unrelated, and the views put
forward by Watkins and collaborators on the one hand (Sect. 2.1)
and by Seeger and coworkers on the other hand (Sect. 2.2) seemed
to be contradictory. Thus, at that time one talked about the
'mystery of the interstitials in silicon'.

A period of revealing the 'mystery' set in after the 1972 Reading Conference. There Seeger and Frank [10] had focused attention
on the fact that in non-metals self-interstitials may occur not
only in different geometrical configurations but also in different
electrical charge states and that the properties of the self-interstitials in silicon may thus be both configuration- and charge-
dependent.

3.1. Electrical charge states

The first effort to obtain information on the electrical charge
states of self-interstitials in semiconductors was undertaken by
James and Lark-Horovitz [14], who considered a self-interstitial
as an essentially free atom embedded in the semiconducting matrix.
We shall not consider the predictions of this model, since they
have not been confirmed, neither by more rigorous theoretical
investigations nor by experiments.

Using the tight-binding approximation Blount [15] concluded
that self-interstitials in silicon or germanium should be
amphoteric with an acceptor level in the upper half and a donor
level in the lower half of the forbidden gap (Fig. 1). Calculations based on an atomic-orbital—molecular-orbital cluster technique by Watkins et al. [16] have supported Blount's results.

On the experimental side Blount's model is confirmed by the dependence of the self-diffusion in silicon on chemical doping.
Among the original arguments presented by Seeger and Chik [8, 9]
in favour of an interstitialcy self-diffusion mechanism in silicon
above about 1200 K are indications in early self-diffusion data
that the self-diffusion coefficient is increased by both n and p
doping — in agreement with Blount's model, but at variance with
what one would expect if self-diffusion were governed by acceptor-
type vacancies. The difficulties inherent in these measurements
involving the short-lived radioactive isotope ^{31}Si have been
avoided in recent experiments by Hettich et al. [17]. Following
McVay and DuCharme [18] these authors [17] simulated the self-
diffusion in silicon by the diffusion of the long-lived tracer
^{71}Ge in silicon. Indeed, in these experiments, which extend down
to 1130 K, the doping dependence was found to follow the predictions of Blount's model above approximately 1200 K (Fig. 2).
We take this as evidence that at high temperatures self-diffusion
in silicon is indeed governed by self-interstitials present in
thermal equilibrium and that the electrical charge state of the
self-interstitials is negative, neutral, or positive in n-doped,
intrinsic, or p-doped silicon, respectively.

Various authors [19 - 21] have attempted to account for the
doping dependence of the silicon self-diffusion coefficient above
about 1200 K in terms of a monovacancy mechanism. These attempts
have been shown to be unable to give quantitative agreement [13].
However, we point out that in principle a formal agreement of the
vacancy diffusion models with the doping dependence found by experiment can be enforced if, disregarding the position of the
donor level established by Watkins et al. [2, 22], the monovacancy is endowed with essentially the same electronic levels
that we attribute to the self-interstitial. Hence, experiments
giving direct information on the nature of the intrinsic defects
present at high temperatures, e. g., those on swirl defects
(Sect. 5.1) or the diffusion of gold in silicon (Sect. 7) are of
great importance.

Fig. 2 shows a change-over of the doping dependence at about
1200 K in the sense that below this temperature diffusion is enhanced or slowed down by doping with arsenic or boron, respectively. This doping dependence is compatible with the operation of a vacancy diffusion mechanism below about 1200 K, as
suggested originally by Seeger and Chik [8, 9]. We believe that
it cannot be decided at present whether under thermal-equilibrium
conditions — in spite of the dominant rôle played by self-inter-

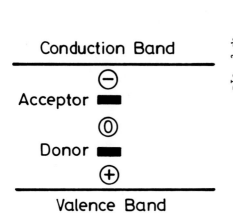

Fig. 1. Energy levels of the self-interstitial in silicon (according to Blount [15]).

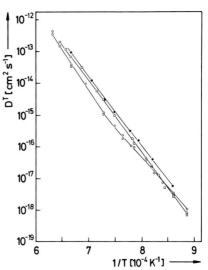

Fig. 2. Temperatue dependence of the diffusion coefficient of ^{71}Ge tracers in intrinsic (O), boron-doped (□), or arsenic-doped (●) silicon, respectively (according to Hettich et al. [17]).

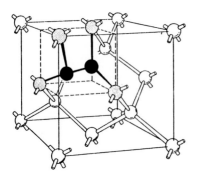

Fig. 3. Singly positively charged self-interstitial in silicon (according to Tan et al. [26]).

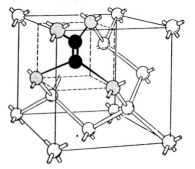

Fig. 4. Electrically neutral self-interstitial in silicon (according to Tan et al. [26]).

stitials — the influence of vacancies must <u>not</u> be neglected. This view has been adopted in the combined vacancy—interstitial models of Hu [23, 24] and Sirtl [25].

3.2. Geometrical configurations

At the 1974 Freiburg conference Frank [11] reviewed internal friction measurements on boron-implanted silicon by Tan et al. [26]. Mainly from these data it was concluded that the positive (I^+) or neutral (I^o) silicon self-interstitial possesses a <110> (Fig. 3) or <100> (Fig. 4) dumbbell configuration and anneals out with a migration enthalpy of H_I^M+ = 0.85 eV or H_I^Mo = 1.5 eV between 370 K and 420 K or 540 K and 600 K, respectively. At the 1976 Dubrovnik conference Seeger et al. [12] presented evidence that the negative (I^-) self-interstitial is a <100> dumbbell, gives rise to an acceptor level at E_c-0.39 eV and to the Si-G25 EPR signal, and possesses a migration enthalpy in the range 0.3 eV $\leq H_I^M$- $\leq$ 0.4eV, corresponding to an annealing temperature between 140 K and 170 K.

4. OVERVIEW

Here we present a scenario of the picture of the self-interstitials in silicon. Against this background details will be discussed in Sects. 5 and 6.

(i) The statement in Sect. 3.2 that the self-interstitials in silicon occur in various dumbbell configurations refers to temperatures T below their annealing temperatures, e.g., to T $\leq$ 570 K for I^o. <u>At higher temperatures</u>, at which supersaturations of self-interstitials can no longer be maintained, this <u>point-defect character</u> of the self-interstitials <u>is gradually lost</u> (Sect. 5.1).

(ii) The migration enthalpies presented in Sect. 3.2 characterize the <u>normal* thermally activated diffusion</u> of the self-interstitials in the corresponding charge states and dumbbell configurations. With the loss of the dumbbell nature <u>at high temperatures</u> the <u>migration enthalpies</u> of the self-interstitials <u>increase</u> (Sect. 5.1).

(iii) The high <u>mobility</u> of the self-interstitials <u>at extremely low temperatures</u> (Sect. 2.1) is <u>radiation-induced and essentially athermal</u> (Sect. 6.1.4), in contrast to the normal thermally activated diffusion considered in item (ii).

(iv) For convenience in the further discussion it will be distinguished between the following temperature regimes: (a) the <u>low-temperature regime</u> in which supersaturations of dumbbell-type interstitials may be eliminated by mobility enhancement as induced by irradiation; (b) the <u>regime of intermediate temperatures</u> in which the interstitials still possess dumbbell configurations and in which supersaturations of interstitials anneal out via normal thermally activated diffusion; (c) the <u>high-temperature regime</u> in which the interstitials have assumed extended configurations and are present in thermal equilibrium.

---

*Purely thermally activated diffusion is called 'normal diffusion'. For a more precise definition we refer to [27].

5. CHANGE OF THE SELF-INTERSTITIAL CONFIGURATIONS BETWEEN INTERMEDIATE AND HIGH TEMPERATURES

5.1. Experimental evidence

Precise measurements of the ^{31}Si self-diffusion over the temperature range 1320 K to 1658 K were recently performed by Mayer et al. [28]. These authors described their data (full circles in Fig. 5) by a tracer self-diffusion coefficient

$$D^T = 1460 \exp(-5.02 \text{ eV}/kT) \text{ cm}^2\text{s}^{-1}, \tag{3}$$

where k is Boltzmann's constant and T the absolute temperature. In agreement with previous self-diffusion data on silicon (for a review see [13]) the preexponential factor in (3) is extraordinarily large compared to a value of $0.1 \text{ cm}^2\text{s}^{-1}$ typical for metals. This outstanding feature has already been realized by Seeger and Chik [8, 9]. They ascribed it to a large value of the self-diffusion entropy and gave the following interpretation: The self-interstitials governing self-diffusion are extended over several, say, about ten atomic volumes. From Blount's model [15] and the fact that at high temperatures the Fermi level is close to the middle of the forbidden gap [29] it follows that these extended interstitials should be electrically neutral [12, 13]. It is clear that this conclusion can be reconciled with the statement that at intermediate temperatures the neutral self-interstitials in silicon are dumbbells (Sect. 3.2) only if these defects change their configuration between medium and high temperatures.

Seeger et al. [12] have shown that, allowing for a linear temperature dependence of the formation ($H_{I^o}^F$) and the migration ($H_{I^o}^M$) enthalpy of I^o and taking into account the thermodynamic relations

$$(\partial S_{I^o}^{F,M}/\partial T)_p = (1/T)(\partial H_{I^o}^{F,M}/\partial T)_p \tag{4}$$

valid for experiments performed at constant pressure p, the self-diffusion data of Mayer et al. [28] may be fitted quite well (curve in Fig. 5) by assuming $S_{I^o}^F = S_{I^o}^M = k$ and $H_{I^o}^M = 1.5$ eV at 570 K ($S_{I^o}^F$ [$S_{I^o}^M$] = formation [migration] entropy of I^o). The other parameters given in Table 1 follow from the fitting. These results support the picture that at intermediate temperatures I^o is a dumbbell (compare $S_{I^o}^F = S_{I^o}^M = k$ at 570 K, which is characteristic of an unextended defect) with a migration enthalpy of 1.5 eV (Sect. 3.2) and that at high temperatures I^o becomes more and more extended (compare the considerably larger values of $S_{I^o}^F$ and $S_{I^o}^M$ in the temperature range of the self-diffusion measurements [Table 1]).

Fairly direct evidence for the interstitial nature of the defects present in silicon under high-temperature equilibrium conditions as well as indirect support of the change of the interstitial configuration as a function of temperature come from the investigation of the so-called A-type swirl defects or (briefly) A-swirls found in dislocation-free silicon single crystals grown by float-zone techniques at suitable rates [30 - 33]. Using transmission electron microscopy Föll and Kolbesen [32] and Petroff and de Kock [33] have shown independently that A-swirls are interstitial-type dislocation loops. Föll et al. [32, 34] have suggested that A-swirls are formed by clustering of the high-temperature-equilibrium point defects that can neither reach the crystal surface nor disappear at grown-in dislocations during the cooling

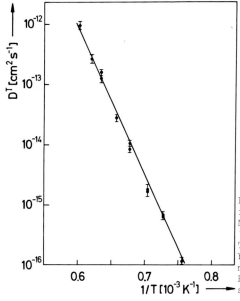

Fig. 5. [31]Si tracer self-diffusion coefficient measured by Mayer et al. [28] between 1320 K and 1658 K (full circles). The adjusted curve was obtained by assuming linear or logarithmic temperature dependencies of $H_I^F o$ and $H_I^M o$ or $S_I^F o$ and $S_I^M o$, respectively [12].

TABLE 1
Values of $S_I^F o$, $S_I^M o$, $H_I^F o$, and $H_I^M o$ obtained from fitting the self-diffusion data of Mayer et al. [28] (compare Fig. 5). $S_I^F o + S_I^M o$ or $H_I^F o + H_I^M o$ is the self-diffusion entropy or enthalpy, respectively.

T[k]	$S_I^F o[k]$	$S_I^M o[k]$	$H_I^F o[eV]$	$H_I^M o[eV]$
1658	6.11	6.96	3.04	2.02
1320	5.02	5.69	2.90	1.86
570	~ 1	~ 1	~ 2.6	~ 1.5

of the crystals. It then follows that the thermal-equilibrium defects dominating in Si at high temperatures must be self-interstitials. This conclusion is strengthened by the fact that the concentration of self-interstitials 'precipitated' as A-swirls is in good agreement with the self-interstitial concentration at high temperatures estimated with the aid of the values of $H_I^F o$ and $S_I^F o$ given in Table 1 [12, 13]. Moreover, this fact demonstrates the self-consistency of the picture that the configuration of I^0 is temperature-dependent, since the values in Table 1 have been deduced from self-diffusion data under this assumption.

5.2. Atomistic model

At low and intermediate temperatures the electrons forming the bonds between the two atoms of an I^o dumbbell are well localized in the centre of gravity of the defect (Fig. 4). When temperature increases, these electrons occupy excited states and thus become more and more delocalized. As a consequence of this spreading-out of electrons the dumbbell relaxes and spreads out, too. Kimerling [35] proposed that the interstitial may first be extended over a ring of five atoms. Since, independent of such details, the spread-out interstitial may be realized by more microstates and may perform diffusional jumps via more paths than a dumbbell-type interstitial, the spreading-out results in an increase of both $S_I^F o$ and $S_I^M o$ with increasing temperature. Analogous considerations apply to I^- and I^+, though, according to Blount's model [15], these play a minor rôle at high-temperature equilibrium conditions at which the Fermi level is near the middle of the forbidden gap [29].

Thermodynamics tells us that the increase of the entropies of formation (S_I^F) and migration (S_I^M) of I^o, I^+, and I^- with increasing temperature is accompanied by an increase of the corresponding enthalpies of formation (H_I^F) and migration (H_I^M)* [see, e.g., the relations (4) in the case of I^o]. An atomistic interpretation of the increase of these enthalpies is that for extended interstitials the formation or migration requires the breaking of a larger number of bonds than for dumbbells. In the case of the migration of the dumbbells Frank and coworkers [11, 12] have argued that H_I^M is determined by the strength of the intra-atomic bonds of the dumbbells. This view is supported by the decrease of H_I^M in the order I^o, I^+, I^- provided that the extra electron of I^- is in a fairly extended orbit, antibonding with respect to the dumbbell atoms.

6. ENHANCED DIFFUSION

Enhanced diffusion in semiconductors has been reviewed recently [27, 36]. Therefore, the present discussion of this subject is limited to an extent needed for a further understanding of the self-interstitials in silicon.

The description of enhanced diffusion is facilitated by distinguishing between mobility-enhanced diffusion and concentration-enhanced diffusion [27, 37]. In the first case the normal thermally activated diffusivity, say, of silicon self-interstitials,

$$D_I = g_I a^2 \nu_I \exp(S_I^M/k)\exp(-H_I^M/kT), \qquad (5)$$

is enhanced by external influences such as particle irradiation, illumination, or carrier injection (g_I = geometrical factor, a = lattice parameter, ν_I = attempt frequency). In the second case the transport of matter at high temperatures, say, the self-diffusion (diffusion of substitutional foreign atoms) in silicon

---

*The omission of +, o, or − from the quantities S_I^F, S_I^M, H_I^F, and H_I^M means that the corresponding statements are not restricted to a specific charge state of the self-interstitials.

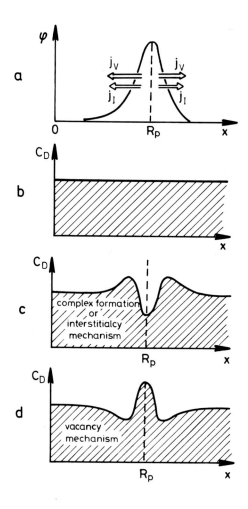

Fig. 8. Redistribution of dopants as a consequence of spatially inhomogeneous concentrations of intrinsic defects: Production rate φ of intrinsic defects as a function of depth x below the irradiated surface at x = 0 (a). Uniform dopant concentration C_D before irradiation (b). $C_D(x)$ after irradiation in the case of complex formation or of an interstitialcy mechanism (c). $C_D(x)$ after irradiation in the case of a vacancy mechanism (d).

The formation of intrinsic-defect—dopant complexes [52, 53] is
expected to be important if a strong attraction between the dopants and the intrinsic defects exists. In this case mobile vacancy—dopant and/or self-interstitial—dopant complexes may be
formed at a high rate in the vicinity of R_p, where the concentrations of intrinsic defects are high. By preferentially migrating
in the direction of their negative concentration gradients, these
complexes leave the region around R_p, thus producing there a dip
(between two smaller humps) in the dopant concentration (Fig. 8c).
The inverse Kirkendall effect [52, 53] can give rise to a redistribution of dopants without any attraction between the dopants
and the intrinsic defects. This effect is a consequence of the
conservation of matter during diffusion, which requires that, in
the case of a vacancy mechanism, a vacancy flux (j_V) must be balanced by fluxes of matrix atoms (j_M) and dopants (j_D) in the
opposite direction [54], i.e.,

$$j_V = - (j_M + j_D), \qquad (9)$$

whereas, in the case of an interstitialcy mechanism, a flux of
self-interstitials (j_I) produces a combined flux of matrix atoms
and dopants in the same direction and of the same magnitude,
i.e.,

$$j_I = + (j_M + j_D). \qquad (10)$$

As a consequence, the out-flux of intrinsic defects from the region around R_p, in which C_V and C_I are high, generates at R_p a
peak (between two smaller dips) or a dip (between two smaller
humps) in the dopant concentration, depending on whether the
dopants diffuse via a vacancy or an interstitialcy mechanism, respectively (Fig. 8d or c).
  Proton irradiation of boron- or phosphorus-doped silicon produces a dip in the dopant concentrations [49 - 51]. This means
either that the complex-formation mechanism operates or, if the
inverse Kirkendall effect is realized, that at high temperatures
boron and phosphorus in silicon diffuse preferentially via an
interstitialcy mechanism. The latter is more likely, since under
proton irradiation an attraction between dopants and intrinsic
defects, which is a prerequisite for the operation of the complex-formation mechanism, does not appear to exist. This is concluded [27, 37] from the observation that the concentration-enhanced diffusion of dopants in silicon induced by irradiation
becomes temperature-independent in the so-called sink-controlled
regime [47, 55, 56].
6.2.2. Oxidation-enhanced diffusion and anomalous diffusion
phenomena
  Oxidation-enhanced diffusion [57 - 63] and the so-called anomalous diffusion phenomena (e.g., the emitter push effect, the
movement of marked layers, and the kink-and-tail structure of
phosphorus diffusion profiles) [13, 53, 64 - 66] are thoroughly
discussed in a companion paper by Gösele and Frank [67]. Here we
elaborate on these effects only so far as to bring out the relationship to self-interstitials in silicon.
  A common feature of the oxidation-enhanced diffusion and of the
anomalous diffusion phenomena is the enhancement of the diffusion
of substitutional dopants (e.g., phosphorus or boron) due to a

supersaturation of intrinsic point defects injected into the
silicon from the surface as a result of surface oxidation or of a
high surface concentration of phosphorus (or, in some cases, of
boron), respectively. It is obvious that from these experiments
alone one can not decide whether the enhancement of the diffusion
of dopants is due to a supersaturation of vacancies or self-interstitials. However, by studying the climb of dislocations in a
transmission electron microscope under experimental conditions
giving rise to oxidation-enhanced or anomalous diffusion this
decision becomes possible. In all experiments of this type in
which the supersaturation of intrinsic point defects was sufficiently high (for details see [67]), dislocations were found to climb
in directions which are compatible with the emission of vacancies
or the absorption of self-interstitials.*Since in the presence of
a supersaturation of intrinsic point defects the thermodynamic
force driving dislocation climb is such that the point defects
are absorbed, one is led to the conclusion that the defects in-
ducing oxidation-enhanced or anomalous diffusion are self-interstitials. This implies that the Group-III and Group-V elements —
at least boron and phosphorus — diffuse in silicon at high temperatures predominantly via an interstitialcy mechanism. However,
it should be kept in mind that this conclusion is based on experiments in which self-interstitials were present in supersaturations.

7. THE DIFFUSION OF GOLD IN SILICON

The diffusion of gold in silicon shows extraordinary features,
e.g., unusual diffusion profiles in both thick specimens [70,71]
and wafers (Fig. 9), and an increase of the gold concentration in
the middle of gold-diffused wafers proportional to the square root
of the time of diffusion (Fig. 10). These findings are presumably
related to the following features [70,75 - 78]: (a) Gold atoms
occupy in silicon both substitutional (Au_s) and interstitial (Au_i)
sites, and therefore an interchange

$$Au_i \rightleftarrows Au_s \qquad\qquad (11)$$

takes place during the diffusion. (b) The solubility of Au_s is
essentially higher than that of Au_i. (c) The diffusivity of Au_i is
much larger than that of Au_s and exceeds the diffusivity of Group-
III or Group-V elements in silicon by several orders of magnitude.
For a long time it has been believed that the interchange (11)
is controlled by the so-called Frank—Turnbull mechanism [79],

$$V + Au_i \rightleftarrows Au_s. \qquad\qquad (12)$$

Recently, the present authors [80, 81] have shown that the predictions based on this assumption are at variance with the experimental data, whereas all observations may be accounted for in terms

---

*This holds for the change of the sizes of oxidation-induced inter-
stitial-type stacking faults during surface oxidation [23, 68] or
in-diffusion of phosphorus [69] as well as for the formation of
helical dislocations from straight screw dislocations in bipolar
transistors during the final emitter diffusion [53].

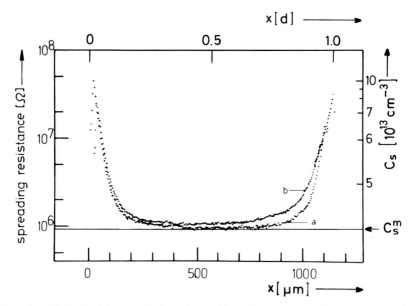

Fig. 9. Distributions of Au_s in a Si wafer measured by means of
the spreading-resistance technique after 5-hour diffusion-anneal-
ing treatments at about 1100 K [72]. In the case a, which is of
main interest in the present context, the Au was deposited on both
surfaces, in the case b on the left-hand surface only.

of the 'kick-out' mechanism,

$$Au_i \leftrightarrows Au_s + I. \tag{13}$$

We shall briefly summarize these considerations, since they strongly support the view that the intrinsic point defects dominating
in silicon under high-temperature equilibrium conditions are self-
interstitials.

7.1. Basic equations

Diffusion of gold in silicon via the kick-out mechanism is described by the following equations:

$$\partial C_I/\partial t = D_I \, \partial^2 C_I/\partial x^2 + \partial C_s/\partial t - A_I[\, C_I/C_I^{eq} - 1], \tag{14}$$

$$\partial C_i/\partial t = D_i \, \partial^2 C_i/\partial x^2 - \partial C_s/\partial t, \tag{15}$$

$$C_i/C_I C_s = C_i^{eq}/C_I^{eq} C_s^{eq}. \tag{16}$$

Here x is the space coordinate in the direction of diffusion, t
the diffusion time, C_n the concentration and D_n the diffusivity of
the defects of type n, where n = I, i, or s for self-interstitials,

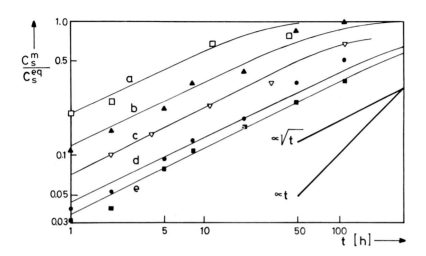

Fig. 10. C_s^m/C_s^{eq} measured as a function of time t on crystals with
different dislocation densities ϱ or/and at different diffusion
temperatures T (curve a [□]: ϱ unknown, T $\sim$ 1375 K [73]; curve b
[▲]: $\varrho \sim 10^4 cm^{-2}$, T $\sim$ 1110 K [74]; curve c [▽]: ϱ unknown, T $\sim$ 1275 K
[73]; curve d [●]: $\varrho \sim 10^2 cm^{-2}$, T$\sim$1110 K [74]; curve e [■]: $\varrho \sim 0$,
T $\sim$ 1110 K [74]. Except for very long times, all series of
measurements follow $\sqrt{t}$ laws.

Au_i, or Au_s, respectively. The equilibrium value of C_n is denoted
by C_n^{eq}. The quantity A_I means the efficiency of dislocations as
sinks (sources) for I and is proportional to the dislocation density ϱ. By the law of mass action (16), it is assumed that <u>local</u>
equilibrium between Au_i, Au_s, and I is established everywhere at any
time via the kick-out mechanism (13).

7.2. Survey of the predictions of the kick-out mechanism

7.2.1. High dislocation density
In this case, which is characterized by an instantaneous establishing of the self-interstitial equilibrium concentration
$(C_I = C_I^{eq})$, (15) and (16) yield for C_s a normal diffusion equation
with the effective diffusion coefficient

$$D_{eff}^{(1)} = C_i^{eq} D_i / (C_i^{eq} + C_s^{eq}). \qquad (17)$$

Since the same result is obtained in the frame-work of the Frank—
Turnbull mechanism, experiments involving high dislocation densities are not suitable to discriminate between the Frank—Turnbull
and the kick-out mechanism.

7.2.2. Dislocation density not high
By definition, the dislocation density is considered <u>not</u> to be

high if $C_I = C_I^{eq}$ is not fulfilled. Then (13) and the establishing
of the Au_i equilibrium concentration are the fastest reactions,
so that, in addition to the simplified description of (13) by
the mass-action law (16), as a further simplification (15) may be
replaced by $C_i = C_i^{eq}$. With this assumption (14) and (16) lead to

$$\partial C_s / \partial t = (\partial / \partial x)(D_{eff}^{(2)} \; \partial C_s / \partial x) + A_I \left[C_s^{eq}/C_s - 1 \right] \quad (18)$$

with the effective, C_s-dependent diffusion coefficient

$$D_{eff}^{(2)} = C_I^{eq} C_s^{eq} D_I / C_s^{2} \quad (19)$$

provided that $C_s > (C_I^{eq} C_s^{eq})^{1/2}$, which is fulfilled in all cases
of interest [80]. Under equivalent assumptions the Frank—Turnbull
mechanism yields a completely different equation for C_s: The
effective diffusion coefficient is C_s-independent, and the term
describing the influence of dislocations as sources (sinks) for
vacancies differs from the second term in eq.(18) [80, 82]. Hence,
it is not surprising that in all cases in which the dislocation
density is not high the Frank—Turnbull mechanism and the kick-out
mechanism make different predictions and may thus be distinguished
by comparison with experiments.

Wafer with medium dislocation density. In this case the second
term on the right-hand side of (18) governs the time dependence
of the Au_s concentration in the centre of the wafer, C_s^m, which is
implicitly given by

$$\ln[(1 - C_s^m/C_s^{eq})/(1 - C_s^o/C_s^{eq})] + (C_s^m - C_s^o)/C_s^{eq} +$$

$$+ A_I t/C_s^{eq} = 0 \quad (20)$$

and which reduces to

$$C_s^m = (2A_I C_s^{eq} t)^{1/2} \quad (21)$$

for $C_s^o << C_s^m < C_s^{eq} [C_s^o = C_s(x,t=0) = C_s^m(t=0)]$. Under these conditions
the Frank—Turnbull mechanism predicts $C_s^m \propto t$.

Wafer with low dislocation density. Seeger and Frank [83] have
shown that, if both terms on the right hand-side of (18) are important, for $C_s^o = 0$ and $C_s^m < C_s^{eq}$

$$C_s^m = (2A_I C_s^{eq} t)^{1/2} F, \quad (22)$$

where F is a function of $A_I d^2/C_s^{eq} D_I$ and d denotes the thickness
of the wafer.

Dislocation-free wafer. This case, in which $A_I = 0$ [eq.(18)],
has been treated for both $C_s^o = 0$ [80] and $C_s^o \neq 0$ [84] in approximations
valid for C_s^m values not too close to C_s^{eq}. In particular, for $C_s^o = 0$ the Au_s concentration in the middle of the wafer is given by

$$C_s^m = 2/d \left(\pi C_s^{eq} C_I^{eq} D_I t \right)^{1/2}. \quad (23)$$

Dislocation-free semi-infinite solid. The diffusion of gold from
the surface of a thick, dislocation-free silicon specimen into
the bulk corresponds mathematically to the boundary-value problem
defined by the differential equation (18) with the diffusion coefficient (19) and $A_I = 0$, the boundary conditions

$$C_s(x=0,t) = C_s(x=d,t) = C_s^{eq}, \qquad (24)$$

and the initial condition

$$C_s(0 < x < d, \ t = 0) = C_s^o. \qquad (25)$$

An exact analytical solution of this problem has recently been presented by Seeger [81].

7.3. Comparison with experiments

In a study of the penetration of gold into thick dislocation-free silicon specimens Wilcox and LaChapelle [70, 71] observed diffusion profiles that differ considerably from the error-function-type shape expected according to the Frank—Turnbull mechanism [79, 80, 82]. Seeger [81] demonstrated that the Wilcox—LaChapelle profiles are in good agreement with the prediction made by the kick-out mechanism for a dislocation-free semi-infinite solid (Sect. 7.2.2).

Except for $C_s^m/C_s^{eq} \gtrsim 0.5$ the concentration of gold in the middle of silicon wafers (Fig. 9) increases proportional to $\sqrt{t}$, independent of the dislocation density (Fig. 10). This time dependence is indeed predicted by the kick-out mechanism, irrespective of whether the dislocation density is medium, low, or zero (Sect. 7.2.2). Within the framework of the Frank—Turnbull mechanism (which, e.g., for wafers with medium dislocation density, predicts a $C_s^m \propto t$ law) this observation cannot be understood.

In order to exploit fully the possibilities to obtain information on self-interstitials in silicon, investigations of the diffusion of gold in highly perfect silicon crystals should be undertaken. From such measurements $C_I^{eq}D_I$ and C_s^{eq} may be deduced separately, as may be realized from an inspection of the effective diffusion coefficient $D_{eff}^{(2)}$ [eq.(19)] and of the boundary condition (24). In the following we shall estimate $C_I^{eq}D_I$ at two temperatures in two different ways from the data on the diffusion of gold in silicon available at present.

A fit of Seeger's theory [81] to the diffusion profiles measured on dislocation-free, thick silicon crystals by Wilcox and La-Chapelle [70] shows that their parameter K [71] is given by

$$K = 2.5 \times 10^{-9} C_I^{eq} D_I / C_s^{eq}. \qquad (26)$$

At 1273 K (K = 1.96 cm^{2}s^{-1} [71], $C_s^{eq} = 10^{16}$ cm^{-3}[85]) eq.(26) yields

$$C_I^{eq}D_I = 1.5 \times 10^{-16} cm^2 s^{-1}. \qquad (27)$$

From the Au$_s$ concentration $C_s^m = 4.15 \times 10^{13}$cm^{-3} in the middle of a dislocation-free, 1125 μm-thick silicon wafer gold-diffused for 5 hours at 1100 K (Fig. 9), with the aid of eq.(23) and with C_s^{eq}(T = 1100 K) = 10^{15}cm^{-3}[85] one finds

$$C_I^{eq}D_I = 1.9 \times 10^{-18} cm^2 s^{-1}. \qquad (28)$$

A comparison of (27) and (28) with values of the tracer self-diffusion coefficient D^T in silicon estimated from the diffusion of ^{71}Ge in silicon [17] yields a correlation factor

$$f_I = D^T / c_I^{eq} D_I \tag{29}$$

of approximately 0.5. This value is very reasonable for a self-diffusion mechanism via extended self-interstitials (Sect. 5.1). Hence, the data on the diffusion of gold in silicon confirm quantitatively the outstanding rôle of self-interstitials in silicon under high-temperature equilibrium conditions.

REFERENCES

1. G. D. Watkins, in: Radiation Damage in Semiconductors, P. Baruch ed. (Dunod, Paris 1964) p. 97.

2. G. D. Watkins, in: Lattice Defects in Semiconductors 1974, F. A. Huntley ed. (Institute of Physics, London and Bristol 1975), Inst. Phys. Conf. Ser. No. 23, p. 1.

3. K. L. Brower, Phys. Rev. B 1, 1908 (1970).

4. R. E. McKeighen and J. S. Koehler, Phys. Rev. B 4, 462 (1971).

5. P. S. Gwozdz and J. S. Koehler, Bull. Am. Phys. Soc. 17, 307 (1972).

6. J. Arimura and J. W. MacKay, in: Radiation Effects in Semiconductors, F. L. Vook ed. (Plenum Press, New York 1968) p. 204.

7. W.D. Hyatt and J. S. Koehler, Phys. Rev. B 4, 1903 (1971).

8. A. Seeger and K.P. Chik, phys. stat. sol. 29, 455 (1968).

9. A. Seeger, Rad. Effects 9, 15 (1971).

10. A. Seeger and W. Frank, in: Radiation Damage and Defects in Semiconductors, J. E. Whitehouse ed. (Institute of Physics, London and Bristol 1972) p. 262.

11. W. Frank, in: Lattice Defects in Semiconductors 1974, F. A. Huntley ed. (Institute of Physics, London and Bristol 1975), Inst. Phys. Conf. Ser. No. 23, p. 23.

12. A. Seeger, H. Föll, and W. Frank, in: Radiation Effects in Semiconductors 1976, N. B. Urli and J. W. Corbett eds. (Institute of Physics, Bristol and London 1977), Inst. Phys. Conf. Ser. No. 31, p. 12.

13. A. Seeger, W. Frank, and U. Gösele, in: Defects and Rad. Effects in Semiconductors 1978, J. H. Albany ed. (Institute of Physics, Bristol and London 1979), Inst. Phys. Conf. Ser. No. 46, p. 148.

14. H. M. James and K. Lark-Horovitz, Z. Phys. Chem. (Leipzig) 198, 107 (1951).

15. E. I. Blount, J. Appl. Phys. 30, 1218 (1959).

16. G. D. Watkins, R. P. Messmer, C. Weigel, D. Peak, and J. W. Corbett, Phys. Rev. Lett. 27, 1573 (1971).

17. G. Hettich, H. Mehrer, and K. Maier, in: Defects and Radiation Effects in Semiconductors 1978, J. H. Albany ed. (Institute of Physics, Bristol and London 1979), Inst. Phys. Conf. Ser. No. 46, p. 500.

18. G. L. McVay and A. R. DuCharme, in: Lattice Defects in Semiconductors 1974, F. A. Huntley ed. (Institute of Physics, Bristol and London 1975), Inst. Phys. Conf. Ser. No. 23, p. 91.

19. D. Shaw, phys. stat. sol. (b) 72, 11 (1975).

20. J. A. van Vechten and C. D. Thurmond, Phys. Rev. B 14, 3551 (1976).

21. R. B. Fair, in: Semiconductor Silicon 1977, H. R. Huff and E. Sirtl eds. (Electrochemical Society, Princeton 1977) p. 968.

22. G. D. Watkins, J. R. Troxell, and A. P. Chatterjee, in: Lattice Defects in Semiconductors 1978, J. H. Albany ed. (Institute of Physics, Bristol and London 1979), Inst. Phys. Conf. Ser. No. 46, p. 16.

23. S. M. Hu, J. Appl. Phys. 45, 1567 (1974).

24. S. M. Hu, J. Vac. Sci. Technol. 14, 17 (1977).

25. E. Sirtl, in: Semiconductor Silicon 1977, H. R. Huff and E. Sirtl eds. (Electrochemical Society, Princeton, 1977) p. 4.

26. S. I. Tan, B. S. Berry, and W. Frank, in: Ion Implantation in Semiconductors and Other Materials, B. L. Crowder ed. (Plenum Press, New York and London 1973) p. 19.

27. W. Frank, U. Gösele, and A. Seeger, in: Proceedings of the International Conference on Radiation Physics in Semiconductors and Related Materials (Tbilisi/USSR, September 1979), in the press.

28. H. J. Mayer, H. Mehrer, and K. Maier, in: Radiation Effects in Semiconductors 1976, N. B. Urli and J. W. Corbett eds. (Institute of Physics, Bristol and London 1977), Inst. Phys. Conf. Ser. No. 31, p. 186.

29. A. K. Jonscher, Principles of Semiconductor Device Operation (Wiley, New York 1960).

30. L. J. Bernewitz and K. R. Mayer, phys. stat. sol. (a) 16, 579 (1973).

31. A. J. R. de Kock, Phillips Res. Rep., Suppl. 4, 1 (1973).

32. H. Föll and B. O. Kolbesen, Appl. Phys. 8, 319 (1975).

33. P. M. Petroff and A. J. R. de Kock, J. Cryst. Growth 30, 117 (1975).

34. H. Föll, B. O. Kolbesen, and W. Frank, phys. stat. sol. (a) 29, K 83 (1975).

35. L. C. Kimerling, in: Defects and Radiation Effects in Semi-conductors 1978, J. H. Albany ed. (Institute of Physics, Bristol and London 1979), Inst. Phys. Conf. Ser. No. 46, p.56.

36. J. C. Bourgoin and J. W. Corbett, Rad. Effects 36, 157 (1978).

37. U. Gösele, W. Frank, and A. Seeger, in: Defects and Radiation Effects in Semiconductors 1978, J. H. Albany ed. (Institute of Physics, Bristol and London 1979), Inst. Phys. Conf. Ser. No. 46, p. 538.

38. A. Seeger and H. Mehrer, in: Vacancies and Interstitials in Metals, A. Seeger, D. Schumacher, W. Schilling, and J. Diehl eds. (North-Holland, Amsterdam 1970) p. 1.

39. H. Mehrer, J. Nuclear Materials 69/70, 38 (1978).

40. J. R. Troxell, A. P. Chatterjee, G. D. Watkins, and L. C. Kimerling, Phys. Rev. B 19, 5 336 (1979).

41. L. C. Kimerling and D. V. Lang, in: Lattice Defects in Semi-conductors 1974, F. A. Huntley ed. (Institute of Physics, London and Bristol 1975), Inst. Phys. Conf. Ser. No.23, page 589.

42. J. D. Weeks, J. C. Tully, and L. C. Kimerling, Phys. Rev. B.12, 3286 (1975).

43. J. C. Bourgoin and J. W. Corbett, Phys. Lett. 38A, 135 (1972).

44. J. R. Troxell and G. D. Watkins, Phys. Rev. B 22, 921 (1980).

45. B. J. Masters and E. F. Gorey, J. Appl. Phys. 49, 2717 (1978).

46. W. M. Lomer, Harwell Report 1540 (AERE, Harwell 1954).

47. J.G. Dienes and A. C. Damask, J. Appl. Phys. 29, 1713 (1958).

48. R. Sizmann, J. Nuclear Material 69/70, 386 (1978).

49. P. Baruch, J. Monnier, B. Blanchard, and C. Castaing, Appl. Phys. Lett. 26, 77 (1975).

50. P. Baruch, in: Radiation Effects in Semiconductors 1976, N. B. Urli and J. W. Corbett eds. (Institue of Physics, Bristol and London 1977), Inst. Phys. Conf. Ser. No. 31, p. 126 (1977).

51. W. Akutagawa, H. L. Dunlop. R. Hart, and O. J. Marsh, J. App. Phys. 50, 777 (1979).

52. P. R. Okamoto and L. E. Rehn, J. Nuclear Materials 83, 2 (1979).

53. U. Gösele and H. Strunk, Appl. Phys. 20, 265 (1979).

54. F. Seitz, Acta Cryst. $\underline{3}$, 355 (1950).

55. D. G. Nelson, J. F. Gibbson, and W. S. Johnson, Appl. Phys. Lett. $\underline{15}$, 246 (1969).

56. K. Gamo, K. Masuda, S. Namba, S. Ishihara, and I. Kimura, Appl. Phys. Lett. $\underline{17}$, 391 (1970).

57. W. G. Allen, Sol. State Electr. $\underline{16}$, 709 (1973).

58. D. A. Antoniadis, A. G. Gonzalez, and R. W. Dutton, J. Electrochem. Soc. $\underline{127}$, 2243 (1980).

59. K. H. Nickolas, Sol. State Electr. $\underline{9}$, 35 (1966).

60. G. Masetti, S. Solmi, and G. Soncini, Sol. State Electr. $\underline{16}$, 1419 (1973).

61. G. Masetti, S. Solmi, and G. Soncini, Phil. Mag. $\underline{33}$, 613 (1976).

62. R. Francis and P. S. Dobson, J. Appl Phys. $\underline{50}$, 280 (1979).

63. D. A. Antonidis, A. M. Lin, and R. W. Dutton, Appl. Phys. Lett. $\underline{33}$, 1030 (1978).

64. S. M. Hu, in: Diffusion in Semiconductors, D. Shaw ed. (Plenum Press, London 1973) p. 217.

65. A. F. W. Willoughby, Rep. Prog. Phys. $\underline{41}$, 1665 (1978).

66. A. F. W. Willoughby, J. Phys. D $\underline{10}$, 455 (1977).

67. U. Gösele and W. Frank, in these proceedings.

68. B. Leroy, J. Appl. Phys. $\underline{50}$, 7996 (1979).

69. C. L. Claeys, G. J. Declerck, and R. J. van Overstraeten, in: Semiconductor Characterization Techniques, P. A. Barnes and G. A. Rozgonyi eds. (Electrochemical Society, Princeton 1978) p. 366.

70. W. R. Wilcox and T. J. LaChapelle, J. Appl. Phys. $\underline{35}$, 240 (1964).

71. W. R. Wilcox, T. J. LaChapelle, and D. H. Forbes, J. Electrochem. Soc. $\underline{111}$, 1377 (1964).

72. M. J. Hill, M. Lietz, R. Sittig, W. Frank, U. Gösele, and A. Seeger, in: Proceedings of the International Conference on Radiation Physics in Semiconductors and Related Materials (Tbilisi/USSR, September 1979), in the press.

73. G. J. Sprokel and J. M. Fairfield, J. Electrochem. Soc. $\underline{112}$, 200 (1965).

74. J. L. Lambert, Wiss. Ber. AEG-Telefunken $\underline{45}$, 153 (1972).

75. W. C. Dash, J. App. Phys. <u>31</u>, 2275 (1960).

76. F. A. Huntley and A. F. W. Willoughby, Solid-State Electronics <u>13</u>, 1231 (1970).

77. F. A. Huntley and A. F. W. Willoughby, J. Electrochem. Soc. <u>120</u>, 414 (1973).

78. F. A. Huntley and A. F. W. Willoughby, Phil. Mag. <u>28</u>, 1319 (1973).

79. F. C. Frank and D. Turnbull, Phys. Rev. <u>104</u>, 617 (1956).

80. U. Gösele, W. Frank, and A. Seeger, Appl. Phys. <u>23</u>, 361 (1980).

81. A. Seeger, phys. stat. sol. (a) <u>61</u>, 521 (1980).

82. M. D. Sturge, Proc. Phys. Soc. (Lond.) <u>73</u>, 297 (1959).

83. A. Seeger and W. Frank, to be published.

84. W. Meyberg, unpublished.

85. W. M. Bullis, Solid-State Electronics <u>9</u>, 143 (1966).

POINT DEFECTS, DIFFUSION MECHANISMS, AND THE SHRINKAGE AND GROWTH OF
EXTENDED DEFECTS IN SILICON

Ulrich Gösele*
IBM Thomas J. Watson Research Center, Yorktown Heights, N.Y., USA

Werner Frank
MPI für Metallforschung, Institut für Physik, and Universität Stuttgart, Institut für Theoretische und
Angewandte Physik, Stuttgart, Fed. Rep. Germany

ABSTRACT

The paper introduces the basic point-defect models proposed for silicon,
which involve either vacancies or self-interstitials only, or both types of point
defects simultaneously under thermal-equilibrium conditions. The growth and
shrinkage kinetics of oxidation-induced stacking faults as well as oxidation-
enhanced or -retarded diffusion phenomena are discussed within the frame
work of these models. Whereas no unambiguous conclusions on the dominant
diffusion mechanism can be drawn from the available oxidation-related
experiments, recent investigations on so-called anomalous diffusion phenomena
(e.g., the 'emitter-push effect') and on the diffusion of gold in silicon demonstrate Si self-interstitials to be the point defects governing self- and impurity
diffusion. The possibility of a coexistence of vacancies and self-interstitials in
thermal equilibrium is discussed in this context. The paper concludes with
speculations on how carbon in conjunction with self-interstitials may influence
the nucleation process of oxygen precipitates in silicon.

INTRODUCTION

Point defects in silicon govern the diffusion of substitutional dopants and also the growth or
shrinkage of extended defects such as oxidation-induced stacking faults (OISF), which in turn may
adversely affect the yield of highly integrated silicon devices. Although most of the technically
relevant effects induced by and related to point defects have been investigated thoroughly, no
consistent and generally accepted picture has emerged. Even the simple and basic question which type
of point defects (vacancies or self-interstitials) dominates self- and impurity diffusion in silicon under
thermal-equilibrium conditions, is still a highly controversial matter. For this reason we will first
present, in a simplified manner, the main point-defect models proposed for silicon and then discuss the
shrinkage and growth kinetics of oxidation-induced stacking faults as well as oxidation-enhanced
diffusion phenomena in terms of these models. Hereby we concentrate on areas in which either recent
progress has been made or open questions can be defined properly. Subsequently we review investigations on the so-called anomalous diffusion effects (e.g., the 'emitter-push effect') which give a
clear-cut answer on the diffusion mechanism of, at least, phosphorus and boron in silicon. The

*On leave of absence from the Max-Planck-Institut für Metallforschung, Stuttgart, Fed. Rep.
Germany.

56

possibility of a coexistence of vacancies and self-interstitials in silicon under thermal-equilibrium conditions is one of the topics of the following section which deals with the diffusion of Au and Ni in silicon. Finally we shall address some questions related to the growth of oxygen precipitates in silicon and speculate on the role of carbon in the corresponding nucleation process.

POINT-DEFECT MODELS

General Remarks. Under thermal equilibrium conditions elemental crystals contain a certain number of point defects such as vacancies (V) and/or self-interstitials (I). Their equilibrium concentration C_V^{eq} or C_I^{eq} is determined by their free enthalpy of formation G_V or G_I, respectively, and (in atomic fractions) given by

$$C_X^{eq} = \exp(-G_X/kT), \tag{1}$$

where k is Boltmann's constant and T the absolute temperature. The index X stands for V or I. In semiconductors point defects may occur in various charge states. In this case (1) applies to the neutral state. The equilibrium concentrations of the charged point defects may be determined from the position of the Fermi level, provided the positions of the corresponding energy levels in the bandgap are known; see, e.g., [1-4].

It is by no means obvious whether the ratio C_I^{eq}/C_V^{eq} is such that one of the species may be completely neglected for all practical purposes, and if so, which one, vacancies or self-interstitials, occurs predominantly. Because of the low absolute values of the point-defect concentration in silicon, even near the melting point, a direct Simmons-Balluffi-type experiment[5], which would settle these questions, has not yet been performed successfully. In principle, both vacancies and self-interstitials may occur simultaneously. In this case the question arises whether a reaction of the type

$$I + V \rightleftharpoons 0 \tag{2}$$

where 0 denotes the ideal lattice, may occur or not.

As yet we have only considered the *concentrations* of point-defects. The concentration values determine which point defect species will most likely survive in a quenching experiment and may be observed in the form of point-defect agglomerates, e.g., dislocation loops of vacancy or interstitial type. However, in most experiments we are going to concentrate on in this paper it is the self-diffusion coefficient $D_V C_V^{eq}$ for vacancies or $D_I C_I^{eq}$ for self-interstitials, which determines the experimental result, e.g., the shrinkage kinetics of OISF. D_V and D_I are the diffusivities of vacancies and self-interstitials, respectively. Although diffusivities of V and I have been measured in low-temperature experiments after creation of V and I by appropriate electron irradiations, the observed high diffusivities, at least in the case of self-interstitials, are probably partly due to a diffusion enhancement by non-equilibrium electronic excitations induced by the electron beam [6-8]. The thus determined diffusivities cannot reliably be applied to high-temperature equilibrium situations. In a high temperature tracer self-diffusion experiment the measured quantity is

$$D^{SD} = f_V D_V C_V^{eq} + f_I D_I C_I^{eq} \tag{3}$$

where f_V and f_I are correlation factors, both of them lying between 0.5 and 1[1,9]. The question which contribution, $D_V C_V^{eq}$ or $D_I C_I^{eq}$, is larger will be discussed throughout this paper.

Pure Vacancy Models. Most point-defect models suggested for silicon assume - in analogy to metals - that the dominant point-defects are vacancies with $C_V^{eq}>>C_I^{eq}$ and $D_V C_V^{eq}>>D_I C_I^{eq}$ see, e.g., [3,10-14]. Under these circumstances self-interstitials can be completely neglected in thermal equilibrium situations. Self-diffusion and the diffusion of substitutional impurities (e.g., As, B, P) proceed consequently via a vacancy mechanism. The various models differ widely with respect to the diffusivities of vacancies in different charge states and the corresponding energy levels in the band gap. Self-interstitials may occur only in non-equilibrium situations, e.g., as the consequence of Frenkel pair creation during irradiation with particles of sufficiently high energies. Within these vacancy models the interstitial-type dislocation loops ('A-swirls') observed after the growth of Si single crystals[11,15] are consequently attributed to non-equilibrium processes during crystal growth.

<u>Pure</u> <u>Interstitial</u> <u>Model.</u> A pure interstitial model with $C_I^{eq} >> C_V^{eq}$ and $D_I C_I^{eq} >> D_V C_V^{eq}$ for all temperatures accessible to experiments, is the extreme opposite case to the pure vacancy models. Within such a model [16], self-diffusion and diffusion of substitutional impurities (e.g., group-III and -V elements: B, P) occur via an interstitalcy mechanism involving Si self-interstitials, as schematically illustrated for one diffusional jump in Fig. 1. Without restriction of generality in Fig. 1 the self-interstitial is depicted in a dumbbell configuration. Analogous to the vacancy models various charge states are attributed to the self-interstitials, and highly mobile complexes consisting of self-interstitials and substitutional dopants are assumed to exist.

<u>Combined</u> <u>Vacancy</u> <u>and</u> <u>Interstitial</u> <u>Models.</u> The increasing number of experimental results indicating the importance of self-interstitials in silicon has initiated a variety of point-defect models, the basic and common ingredients of all of which is that both vacancies and self-interstitials are taken into account[1,17-21]. The first model in this series was put forward by Seeger and Chik [1] in 1968. Based on the analysis of data on Ni precipitation in Si [22] they proposed that self-diffusion is carried by vacancies below 900°C. Self-diffusion above, this temperature is attributed, rather intuitively, to self-interstitials. In order to account for the unusually high values of the entropies of formation and migration indicated by the self-diffusion data available at that time[23,24], these authors suggested an interstitial configuration spread-out over about 10-15 lattice sites. For an updated version of the Seeger-Chik model see, e.g., [8,25].

Based on the experimental observations that oxidation of Si enhances boron and phosphorus diffusion and, under the same circumstances, leads to the growth of interstitial-type stacking faults, Hu[18] proposed in 1974 that self-diffusion and the diffusion of substitutional impurities occur by a 'dual mechanism', which means partly by a vacancy and partly by an interstitialcy mechanism. Hu assumes that the interstitialcy component for boron is highest, less for phosphorus and negligible for arsenic. In a later paper Hu [19] speculates that there could well be a reaction barrier preventing mutual recombination of V and I. In such a way vacancies and self-interstitials could be coexistent without interference[21]. The present authors consider negligible recombination as unlikely in view of the results of irradiation experiments in which vacancy-interstitial recombination processes readily occur[26].

In 1977 Sirtl[20] suggested that vacancies and self-interstitials maintain a local equilibrium, which includes vacancy-interstitial recombination and creation of vacancy-interstial pairs in the bulk. A deviation from the equilibrium values C_V^{eq} and C_I^{eq} induced, e.g., by external influences will again result in local point-defect equilibrium

$$C_V C_I = C_V^{eq} C_I^{eq} \qquad (4)$$

established via the reaction scheme (2). Sirtl does not investigate the energetic feasibility of the bulk creation process which probably requires an activation energy of 7-8eV.

In the following we will discuss the various experimental results mainly in terms of a pure vacancy or a pure interstitial model, but occasionally we will also comment on the combined models.

SHRINKAGE AND GROWTH KINETICS OF OXIDATION-INDUCED STACKING FAULTS AND OXIDATION-ENHANCED DIFFUSION

<u>General</u> <u>Remarks.</u> Thermal oxidation of Si induces nucleation and growth of interstitial-type stacking faults on (111) planes (for ref. see [18,27]) and leads to enhanced diffusion of substitutional dopants, such as boron [28-30], phosphorus [29-33] or arsenic[34]. As first pointed out by Hu[18] these effects are closely related to each other as, e.g., seen by the result that under the same oxidation conditions both the diffusion of boron and the growth rate of oxidation-induced stacking faults (OISF) increase with the surface orientation in the order (111), (110), (100). We summarize here some of the main results on which we will comment later on.

i) The length (or radius) of OISF depends on oxidation time with an exponent between 0.5 and 1, depending on the oxidation conditions; for references see, e.g., [18, 27, 35-37]. ii) During oxidation at temperatures above ~1200°C OISF first grow and then start to shrink away. This phenomenon has first been observed by Lawrence[38] and later on independently by Hu [39], who termed it 'retrogrowth'. A small percentage of chlorine in the oxidizing atmosphere leads to the same

58

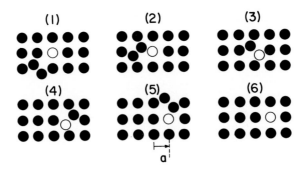

Fig. 1. Schematic of one diffusional jump of a substitutional atom indicated by an open circle via an interstitialcy mechanism. Time sequence (1)-(6); jump distance: a.

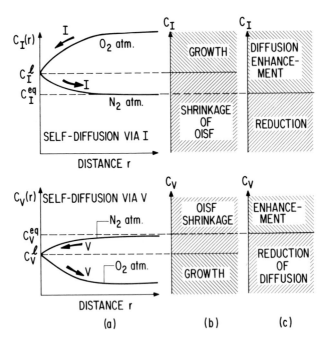

Fig. 2. Influence of point defect concentration C_I or C_V on the growth/shrinkage behavior of *interstitial-type* oxidation-induced stacking faults (b), and on the diffusion of substitutional dopants (c) assuming that self-interstitials I or vacancies V dominate self- and impurity diffusion. (a) indicates the corresponding point-defect fluxes in the two models. $C_I^\ell = C_I^{eq} \exp(\Delta G/kT)$ and $C_V^\ell = C_V^{eq} \exp(-\Delta G/kT)$ are the concentrations of I and V at the partial dislocation of the OISF, respectively.

behavior already at much lower temperatures[40-45] and, in addition, suppresses oxidation-enhanced diffusion [41,46,47]. iii) The growth rate of OISF is independent of their mutual distance even if their distance is smaller than their length[27].

Basic Formulation of Shrinkage and Growth Kinetics. Within this section we consider the shrinkage or growth of a circular *interstitial-type* dislocation loop of radius r_{SF} containing a stacking fault. The basic formula for the shrinkage or growth kinetics by climb processes via the emission or absorption of point-defects [9, 16, 47] is given by

$$dr_{SF}/dt = -\alpha_{eff}D_IC_I^{eq}[\exp (\Delta G/kT) - C_I/C_I^{eq}]A/\Omega \tag{5}$$

for an interstitial model and by

$$dr_{SF}/dt = -\alpha_{eff}D_VC_V^{eq}[C_V/C_V^{eq} - \exp (-\Delta G/kT)]A/\Omega \tag{6}$$

for a vacancy model. In (5,6) C_I and C_V denote the point-defect concentrations far away from the partial dislocation bordering the OISF, A the area per atom in the stacking fault, and Ω the atomic volume. ΔG, which is positive, is the free enthalpy gained by taking away one self-interstitial from the OISF (or adding one vacancy) and is determined by the line tension, the stacking-fault energy γ, and mechanical stresses acting on the dislocation line. Mechanical stresses are not considered any further in the following, although they play an essential role in the nucleation process of OISF, especially at the wafer surfaces[17,38,49]. The quantity α_{eff} contains all factors related to interaction potentials, e.g., of elastic or electrostatic nature, between the dislocation and the diffusing point-defects as well as a possible reaction barrier opposing climb processes at the dislocation. Without a reaction barrier or long-range interactions α_{eff} for a diffusion-controlled process (denoted by α_d) may be written as [50]

$$\alpha_d = 2\pi/\ell n(8r_{SF}/r_o), \tag{7}$$

where r_o is the distance from the dislocation line at which immediate reaction can be assumed to occur. For attractive interactions $\alpha_{eff} > \alpha_d$ holds and for repulsive interactions or a reaction barrier an $\alpha_{eff} < \alpha_d$ results. It is worth mentioning that interaction potentials and reaction barriers do *not* influence the value of C_I/C_I^{eq} (or C_V/C_V^{eq}) at which the change-over from shrinkage to growth occurs. The requirement for growth, as usually observed under oxidizing conditions, is [16]

$$C_I/C_I^{eq} > \exp (\Delta G/kT) > 1 \tag{8a}$$

or

$$C_V/C_V^{eq} < \exp (-\Delta G/kT) < 1. \tag{8b}$$

Enhanced diffusion of substitutional dopants migrating via a defect mechanism requires a point-defect supersaturation, i.e. $C_I/C_I^{eq} > 1$ or $C_V/C_V^{eq} > 1$. This is schematically shown in Fig. 2 for a simple case without reaction barrier or interaction potentials. As already recognized by Hu[18] in his pioneering work, the experimental results of growth of OISF and of a diffusion enhancement under the same oxidizing conditions allow only one conclusion: The SiO_2-Si interface injects Si self-interstitials into the Si bulk during the oxidation process, thus creating a supersaturation of I. Before we discuss some of the aspects of detailed models proposed for the growth kinetics during oxidation we first turn to the simpler case of OISF shrinkage in an inert atmosphere.

Shrinkage Kinetics in an inert Atmosphere. OISF shrink away in an inert atmosphere (e.g., N_2) at sufficiently high temperatures; roughly above 1050°C [38, 51-56]. The activation energy for this process has been found to lie (with only one gross exception[51]) between 4.2eV and 5.2eV, which falls into the measured range of self-diffusion energies [9,23,24,57-59]. The available experimental data may be averaged and give in terms of the stacking-fault diameter of 'length' $L_{SF} = 2 r_{SF}$ a time-independent shrinkage rate

$$dL_{SF}/dt = -6.6x10^{15} \exp (- 4.3eV/kT)[\mu m/hr]. \tag{9}$$

A typical value is, e.g.,, $13\mu m/hr$ at $1200°C$. The shrinkage kinetics expressed by (5) with $C_I = C_I^{eq}$ or by (6) with $C_V = C_V^{eq}$ is essentially the same for both models since $\Delta G/kT \ll 1$. For the usually considered OISF with diameters of 10 or more μm, ΔG is practically completely determined by the stacking-fault energy γ of about 65 erg/cm^2, or in more convenient units for our purposes $\gamma = 0.026eV/atom$, where we have used $A = 6.38 \times 10^{-16}cm^2$ for the area per atom in the OISF. For a recent review on stacking-fault energies in Si see Alexander et al.[60]. Let us now assume for the moment that the shrinkage is purely diffusion-controlled. Then α_{eff} is given by α_d from (7) and turns out to be ~0.5. From the experimental shrinkage data we may now calculate via (5) $D_I C_I^{eq}$ (or via (6) $D_V C_V^{eq}$ within a vacancy model). The result is shown in Fig. 3 and compared to various direct tracer measurements of the self-diffusion coefficient D^{SD}[23,24,57,58] and to the results indirectly obtained by annealing of perfect dislocation loops in an electron microscope by Sanders and Dobson[9]. For our analysis we have put the correlation factor f_I or f_V equal to about unity. We mention that the absolute values of D^{SD} calculated from the OISF shrinkage data of Sanders and Dobson[20], although showing a much too small activation energy of about 2.9eV, fit well into the range of diffusion data obtained from other experiments on OISF shrinkage.

Both the shrinkage of OISF and of (much smaller) perfect dislocation loops[9] appear to indicate fairly high D^{SD} values compared to those obtained from tracer measurements. The assumption of an attractive interaction between the dislocation and the diffusing point-defects leading to higher values of α_{eff}, would shift the calculated values of D^{SD} into the expected lower range of measured values. We will come back to this consideration later on. We conclude that the shrinkage kinetics of OISF in an inert atmosphere can be reasonably well understood, even quantitatively, by calculating the basic driving force from the well known stacking-fault energy γ.

Growth and Shrinkage Kinetics in an Oxidizing Atmosphere

Discussion of Models. By now it is well established that during oxide growth Si self-interstitials are injected into silicon from the SiO_2-Si interface, basically because of an incomplete adaptation of the two times larger volume of a SiO_2 unit cell compared to the atomic volume Ω of silicon. Any reasonable model should identify the sinks which prevent a continuous accumulation of injected I in a Si wafer during oxidation. Given such a model the remaining task is the calculation of C_I/C_I^{eq} as a function of oxide growth parameters such as time, wafer orientation or oxygen partial pressure.

As long as condition (8a) for C_I/C_I^{eq} is fulfilled existing OISF will grow, for smaller C_I/C_I^{eq} they will shrink ('retrogrowth'). If C_I/C_I^{eq} decreases with time, this change-over may occur during the oxide growth process. Most models on OISF kinetics try to find C_I/C_I^{eq} as a function of the growth rate dy/dt of an oxide layer of thickness y, given (after a short initial time period) by

$$dy/dt \sim Bt^{-1/2} \qquad (10)$$

where B depends again on oxide growth parameters such as wafer orientation or the chlorine content in the oxidizing atmosphere[61]. Since the most reliable time dependence for OISF growth below $1200°C$ appears to be approximately [19,36,39]

$$r_{SF} \sim t^{3/4} \qquad (11)$$

recent efforts [27,37,62,63] are mainly directed towards obtaining this relationship analytically.

In his model Murarka[37] assumes a *time-independent* C_I/C_I^{eq} and maintains that - contrary to the growth law $r_{SF} \sim t$ predicted by (5) for constant C_I/C_I^{eq} - the resulting growth kinetics is given by (11). Since his calculation procedure appears to be based on inapplicable arguments, we will not consider his original model[37] or slightly modified versions[64] any further. In all the other models[18,27,62,63] C_I/C_I^{eq} starts with a value $C_I/C_I^{eq} > 1$ and decreases with time, as indicated in Fig. 4b. All of them assume implicitly or explicitly that the supersaturation of I extends throughout the wafer, which under usual experimental conditions is oxidized on both sides. It is obvious that this requires a comparatively high diffusivity of I and a low density of internal sinks for I within the bulk of the Si wafer. In Fig. 4a it is schematically indicated what is expected if the density of I sinks in the bulk such as vacancies in thermal equilibrium within the corresponding models[17,20,30], OISF themselves or microdefects in the form of 'swirls'[11,15] is high enough to affect the concentration of I. Under such circumstances the oxidation-related supersaturation becomes an effect limited to the surfaces being oxidized[30,65].

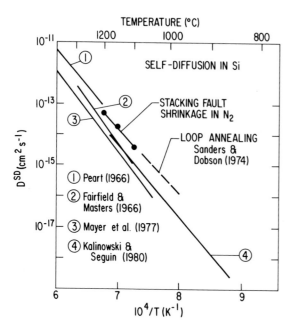

TEMPERATURE (°C)

SELF-DIFFUSION IN Si

—STACKING FAULT
SHRINKAGE IN N_2

—LOOP ANNEALING
Sanders &
Dobson (1974)

① Peart (1966)

② Fairfield &
Masters (1966)

③ Mayer et al. (1977)

④ Kalinowski &
Seguin (1980)

$10^4/T (K^{-1})$

Fig. 3. Self-diffusion coefficient D^{SD} calculated from data on the shrinkage of oxidation-induced stacking faults compared to D^{SD} values obtained by tracer measurements [23,24,57,58] or by loop annealing[9].

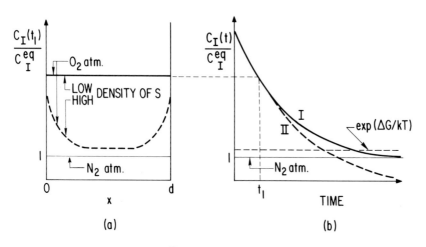

(a)

(b)

Fig. 4. Normalized concentration C_I/C_I^{eq} of self-interstitials in a silicon wafer of thickness d during oxidation from both sides: (a) as a function of depth x for a low and for a high density of sinks S for self-interstitials, (b) as a function of time. Curve I describes a case where C_I approaches C_I^{eq}, curve II a case where $C_I < C_I^{eq}$ for long times.

In the first quantitative model for OISF growth Hu[18] assumes a *linear* relationship between the growth rate (10) of the oxide layer and the excess self-interstitials, which gives

$$r_{SF} \sim t^{1/2} \qquad (12)$$

instead of the experimentally observed relationship (10). By taking surface steps at the SiO_2-Si interface as the major sinks for the I produced by the growing oxide Hu[18] could explain the different degrees of I supersaturations observed for different wafer orientations. For an updated version of Hu's model see ref.[66].

Leroy[27] proposes that the concentration of excess self-interstitials is proportional to some power n of the oxide growth rate dy/dt. Then he chooses n in such a way that the experimental results on OISF growth at temperatures well below 1200°C are described best. His final expression for r_{SF} is

$$r_{SF} = 1640t^{3/4}p^{1/4} \exp (- 2.5eV/kT)F - 3.6 \times 10^{10}t \exp (- 5.02eV/kT) \qquad (13)$$

where r_{SF} is in cm, t in seconds, and p is the oxygen partial pressure in atmospheres. The parameter F depends on the wafer orientation and is given by 1 for (100) and ~ 0.7 for (111) surface orientation. Leroy implicitly assumes that C_I/C_I^{eq} approaches zero at long times, which means that within his model the interface should act as a sink for I, which in turn would imply a retarded diffusivity for substitutional dopants. Lin *et al.*[62] and Fair [63] derive expressions which are similar to (13); also composed of two terms, one of which describes growth, the other one shrinkage of OISF.

Influence of Chlorine. A small percentage of HCl or Cl_2 in the oxidizing atmosphere leads to a much shorter growth period of OISF (reducing to almost zero at high HCl concentration) and to their subsequent shrinkage at temperatures where OISF in dry or wet oxygen grow[40-45]. Under the same conditions the diffusion enhancement of substitutional dopants is much smaller than without these additives in the oxygen atmosphere or even negligible[41,46,47]. Thus we conclude that chlorine reduces the efficiency of the SiO_2-Si interface as a source of I, probably by reacting chemically with I at the interface. As discussed in various extensive articles on this subject[67,68], chlorine-containing oxidizing atmospheres represent a practical possibility to avoid OISF in devices. We concentrate here on the question how fast the shrinkage of already pre-existing OISF can be. Some of the available date indicate that in the presence of chlorine the shrinkage at high temperatures (> 1200°C) is even faster than expected for an inert ambient (see, e.g., Fig. 6). This result implies that the SiO_2-Si interface acts as *sink* for Si self-interstitials, which in turn leads to an *undersaturation* of I,$C_I < C_I^{eq}$, required for the observed higher shrinkage rate. Within a pure interstitial model it follows directly from (5) that the maximal shrinkage rate expected for an interface acting as a perfect sink for I, could be up to 5 times higher than in an inert atmosphere. Unfortunately, there exist hardly any corresponding data on the chlorine influence on the diffusion of substitutional dopants at or above 1200°C. Such diffusion experiments at 1200°C with about 1 percent HCl would be very helpful in distinguishing between vacancy and interstitial models for point-defects in silicon: Based on the results of Fig. 6 an interstitial model predicts a diffusion *retardation* (as compared to an inert atmosphere) of about 50% for substitutional dopants, whereas within a vacancy model[67] a diffusion *enhancement* is expected.

Distance-Independent Growth of OISF. The growth rate of OISF does not depend on their relative distance. Two OISF separated by much less than their radius grow at the same rate as isolated ones[27]. For a purely diffusion-controlled climb reaction the respective diffusion fields around the OISF overlap and consequently a lower growth rate is expected. If there is a reaction barrier to be overcome in the climb process, practically no interference of diffusion fields occurs since then the point-defect concentration is almost constant up to the partial dislocation. Hu[66] elaborates on this concept in his contribution to these proceedings.

In the presence of a reaction barrier, α_{eff} in (5) is appreciably smaller than α_d given by (7). According to our formulation (5) this smaller α_{eff} then also holds for the shrinkage in an inert ambient. Since the D^{SD} values calculated from shrinkage data (Fig. 3) assuming $\alpha_{eff} = \alpha_d$ are higher than the tracer diffusion data, we would rather expect α_{eff} to be larger than α_d, which in turn would require an *attractive interaction* between the dislocation and the diffusing point-defects. In fact, one can construct an attractive interaction potential in such a way that the defect concentration is constant almost up to the dislocation line and then drops sharply. This results in a similar concentration profile as in the case of a reaction barrier and a distance-independent growth of OISF can be explained in such a way as well. The only potential which fulfils the necessary requirements is that generated by *electric charges* on the partial dislocation. There exist quite a number of indications on charges and dangling bonds on various types of dislocations (for a recent review see [69]), but at the moment it seems premature to speculate any further without detailed calculations on the required charge density on the partial dislocation.

Oxidation-Enhanced Diffusion

Oxidation-enhanced diffusion of boron [28-30], phosphorus [29-33], or arsenic [34] is due to a supersaturation of self-interstitials injected from the SiO_2-Si interface into the silicon by the growing oxide. The observed diffusion rates should be proportional to the I supersaturation C_I/C_I^{eq}, which decreases with time, as discussed in previous sections. A diffusion experiment performed up to a certain time t_D renders information on $C_I(t)/C_I^{eq}$ averaged from $t=0$ to $t=t_D$. The diffusion enhancement actually observed does therefore depend on time [18,29,30]. Because of its much lower diffusivity arsenic requires a much longer diffusion time t_D than boron or phosphorus and consequently a smaller diffusion enhancement is observed [18,34], as expected from the decreasing average over $C_I(t)/C_I^{eq}$.

Before results on oxidation-enhanced diffusion can be interpreted quantitatively, the following question has to be answered: Are the excess self-interstitials just added to *self-interstitials* already present under thermal equilibrium conditions [16] or are they added to a point-defect population consisting mainly of *vacancies*? The latter possibility is favored by most of the researchers working in this field mainly because the enhancement factors found for P, B, and As appear to differ from each other. This could be explained by different ratios of the vacancy and interstitialcy diffusion components for these dopants. After checking the data on oxidation-enhanced diffusion available in the literature, the present authors believe that the experiments are just not accurate enough for such a decision. It appears possible within the scatter of these data that diffusion of all substitutional dopants investigated (B, P, As) is governed by self-interstitials. The diffusion coefficient D_S^* under oxidizing conditions, compared to that in an inert atmosphere D_S, is then given by [18]

$$D_S^*/D_S = 1 + \int_0^{t_D}(C_I-C_I^{eq})dt/C_I^{eq}t_D. \tag{14}$$

For conditions in which OISF grow and $C_I >> C_I^{eq}$ we can approximately replace exp $(\Delta G/kT)$ in (5) by unity, thus arriving at [18]

$$D_S^*/D_S = 1 + r_{SF}(t)\Omega/(t_D\alpha_{eff}AD_IC_I^{eq}). \tag{15}$$

Within a pure interstitial model $D_IC_I^{eq}$ can approximately be replaced by the measured tracer self-diffusion coefficient. Eq. (15) is no longer applicable in the 'retrogrowth' temperature regime, where the above mentioned replacement of exp $(\Delta G/kT)$ by 1 would lead to grossly wrong results.

Oxidation-Retarded Diffusion

Francis and Dobson [33] report that oxidation in dry O_2 for long times (>17 hours) at temperatures above 1150°C *retards* phosphorus diffusion. Their results indicate that at a sufficiently high temperature and oxide thickness the SiO_2-Si interface may act as a *sink* for *self-interstitials*. This conclusion should be checked by measurements of the shrinkage rate of OISF in the 'retrogrowth' regime for similarly long oxidation times. Because of the undersaturation of I suggested by the diffusion retardation, a higher shrinkage rate of OISF than in an inert atmosphere is expected. Such a result would also clearly demonstrate that the point-defects governing diffusion in silicon are self-interstitials and not vacancies. In the following section we will discuss experiments which strongly indicate that the diffusion of substitutional dopants such as P or B is actually dominated by Si self-interstitials.

64

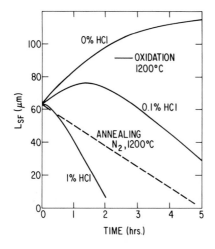

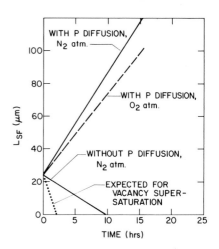

Fig. 5. Length L_{SF} of pre-existing oxidation-induced stacking faults as a function of time for different atmospheres (after Shiraki [41]).

Fig. 6. Length L_{SF} of interstitial-type oxidation-induced stacking faults as a function of time under the influence of indiffusing phosphorus at 1150°C for different atmospheres (after Claeys et al. [75]).

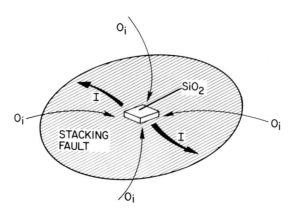

Fig. 7. Schematic of interstitial-type stacking fault surrounding an SiO_2 precipitate. The fluxes of interstitial oxygen O_i and of self-interstititals I during oxygen precipitation are indicated by arrows (partly after Patel et al. [89]).

APPLICATION OF DISLOCATION CLIMB EXPERIMENTS TO ANOMALOUS DIFFUSION PHENOMENA

Description of anomalous diffusion phenomena. Diffusion of group-III and group-V elements in Si shows several unexpected features which have been termed 'anomalous' [2,4,8,70]. According to ref. [16, 70] the three most prominent anomalous diffusion phenomena, which appear related to each other, may briefly be described as follows: (i) The emitter-push effect is the extremely rapid diffusion of the base dopants (e.g., B or Ga) near the emitter region of double diffused npn structures resulting in an enhanced movement of the base-collector boundary below the (usually highly phosphorus-doped) emitter. (ii) Movement of marked layers denotes the abnormally rapid diffusion of dopants (e.g., P or B) in marked or buried layers that is observed if the phosphorus concentration at the surface is high. (iii) A kink-and-tail structure of P-diffusion profiles occurs when such profiles are generated by diffusion of P starting with high surface concentrations. In the tail the diffusion is much higher than expected from isoconcentration studies.

The following features common to the anomalous diffusion phenomena (i)-(iii) are considered established *[4,8,70].(a) The diffusivity enhancement is caused by a supersaturation of intrinsic point-defects (V or I) promoting the diffusion of the dopants. The supersaturation ratios are much higher than in oxidation experiments. (b) The point-defect supersaturation is related to a high surface concentration of phosphorus (or in some cases of boron [71,72]). (c) Mechanisms involving dislocations in an essential way, in particular dislocation climb, are not the origin of the supersaturation.

Until recently it has been commonly assumed that the point-defects present in supersaturation are vacancies [73,74], although this assumption is not based on any specific experimental observations. In recent years the nature of the point-defects in supersaturation has been determined experimentally by several independent groups [75-77]. The basic idea of all these experiments is as follows: The observed diffusion enhancement requires a supersaturation of point-defects and cannot be explained in terms of an undersaturation. This means that isolated dislocations, lying in the region of enhanced diffusion, climb by absorption and not by emission of point-defects. Hence, the direction of climb of a given dislocation is uniquely related to the nature of the point-defects involved (i.e. opposite for V and I).

The Climb Experiments of Claeys et al.

Claeys et al.[75] investigated the growth or shrinkage kinetics of interstitial-type bulk stacking faults in Si under the influence of indiffusing phosphorus in different atmospheres. As the experiments were performed at 1150°C which is a temperature where the anomalous diffusion effects are not very pronounced any more it was especially important to ensure that the effects observed were not due to self-interstitials produced by oxidation processes. For this purpose Claeys et al. [75] performed their experiments in an oxidizing as well as in an inert atmosphere. An example of their results is shown in Fig. 7. The indiffusion of phosphorus in N_2 atmosphere leads to the growth of OISF, which shrink away in N_2 atmosphere without P diffusion. An O_2 atmosphere during the indiffusion of P yields a slightly lower growth rate. The growth of interstitial-type stacking faults demonstrates that self-interstitials are present in supersaturation. A vacancy supersaturation would lead to a higher shrinkage rate (see Fig. 7). Because both shrinkage and growth are linear in time it is possible to estimate that the supersaturation ratio C_I/C_I^{eq} is less than two. Experiments at lower temperatures, where the anomalous diffusion phenomena indicate much higher supersaturations, are therefore desirable and have in fact been performed by Strunk et al.[16,77].

---

*The diffusion retardation sometimes observed for arsenic emitters is caused by electric field effects [4,70].

The Climb Experiments of Strunk et al.
    Strunk et al.[16,77] investigated bipolar transistors with consecutively diffused boron base and phosphorus emitter in a high-voltage electron microscope. These transistors showed the emitter-push effect, i.e. enhanced diffusion of the base dopant. In several transistors isolated helical dislocations could be detected in the emitter-base region, which had been formed by the climb of screw dislocations during the final emitter diffusion at 950°C. The contrast analysis of dipole segments in more irregular helices proved them to be of *interstitial* type. Since the final emitter diffusion was performed in an oxidizing atmosphere it appears appropriate to calculate the supersaturation ratio C_I/C_I^{eq} by the formula [78]

$$\ell n(C_I/C_I^{eq}) = 2\pi n_L Gb^4/kT \qquad (16)$$

where b is the length of the Burgers vector, G the shear modulus, and n_L the number of turns per unit length of the helix. A typical number for n_L is 9 turns/μm, which leads to $C_I/C_I^{eq} \sim 140$. This supersaturation ratio is much higher than that derived from oxidation-enhanced diffusion, which ranges from about 2 to 5, so that we can safely exclude oxidation effects.

Conclusions Drawn from Climb Experiments
    The experimental results obtained by Claeys et al. [75] and Strunk et al.[16,77] as well as related observations [76,79,80] demonstrate consistently that the anomalous diffusion phenomena described above are caused by a supersaturation of Si self-interstitials. This implies that at least phosphours and boron diffusion occurs predominantly via self-interstitials. Experiments on the diffusion enhancement of arsenic in a buried layer - unfortunately not available as yet - would be helpful in deciding whether a dual diffusion mechanism involving I *and* V [12,18,29,3034] exists or not. A similarly large diffusion enhancement for As as for P and B would indicate that all these dopants diffuse via an interstitialcy mechanism only.

INTERSTITIAL-SUBSTITUTIONAL INTERCHANGE MECHANISM

    In the present section we discuss what can be learned on intrinsic point defects in silicon by studying the diffusion behavior of impurities such as Au or Ni which diffuse via an interstitial substitutional interchange mechanism [1,2,22,81-83]. First we concentrate on the case of Au in Si. Au atoms may occupy both substitutional (Au_s) and interstitial (Au_i) sites. The solubility of Au_s is much larger than that of Au_i, and the diffusivity of Au_i exceeds that of Au_s by many orders of magnitude [82,83]. Until recently it has generally been believed that the interchange between Au_s and Au_i occurring during diffusion of Au in Si is controlled by the Frank-Turnbull mechanism [81] (also called dissociative diffusion mechanism) involving vacancies

$$V + Au_i \rightleftarrows Au_s. \qquad (17)$$

Within the frame-work of this mechanism the shape of Au diffusion profiles in Si [82] and the kinetics of the rise of the Au_s concentration in the center of thin silicon wafers during Au diffusion from both surfaces could not be explained. It has been shown [84,85] that the experimental observations may be understood quantitatively if (17) is replaced by the 'kick-out' mechanism

$$Au_i \rightleftarrows Au_s + I \qquad (18)$$

involving Si self-interstitials. For a detailed discussion of (18), see Frank et al. [86] in these proceedings. This result strongly suggests that the point-defects dominating self- and impurity diffusion are indeed self-interstitials, in complete agreement with the conclusions drawn from the climb experiments described in the previous section.
    Let us now look into the case of Ni in Si. The only major difference to Au in Si appears to be that the solubility of interstitial Ni (Ni_i) is much higher than that of substitutional Ni (Ni_s). Yoshida and Saito[22] measured the precipitation of Ni_s in Si and calculated from their data the self-diffusion coefficient D^{SD} in terms of the Frank-Turnbull mechanism. Extrapolation of their D^{SD} values (determined between 700°C and 900°C) to higher temperatures gives reasonable agreement with

available tracer data within a factor of 20 or less. Since the kick-out mechanism predicts a different precipitation kinetics than the Frank-Turnbull mechanism it seems contradictory that D^{SD} can be calculated from Au diffusion data in terms of (18) involving I [84,85] and also from Ni precipitation data in terms of the Frank-Turnbull mechanism involving V. We consider two basically different explanations for the apparent discrepancy. i) The Yoshida-Saito data have been interpreted in terms of the Frank-Turnbull mechanism because no other mechanism was thought of at that time. Knowing about the two mechanisms (17) and (18) a close inspection of the Ni precipitation data could prove the kick-out mechanism to be applicable. If this is the case, we may neglect vacancies under thermal-equilibrium conditions, and self- and impurity diffusion are completely dominated by self-interstitials. ii) There exists *no real discrepancy* and *both results are correct*. This is actually possible in principle if one assumes local equilibrium between I and V according to (4), as, e.g., in Sirtl's model [20]. For such a situation we give the formula governing the concentration C_s of substitutional impurites (Au$_s$ or Ni$_s$) in precipitation experiments as well as in experiments on the up-take of substitutional atoms in the center of a Si wafer. Assuming the sources or sinks for I and V to be dislocations of areal density ρ and taking into account (4) we arrive at

$$dC_s/dt = D_I C_I^{eq}\alpha\rho[(C_s^{eq}/C_s)-1] + D_V C_V^{eq}\alpha\rho(1 - C_s/C_s^{eq}), \qquad (19)$$

where α is a geometrical factor of the order of one and C_s^{eq} the solubility of Au$_s$ or Ni$_s$. If *accidentally* $D_I C_I^{eq} \approx D_V C_V^{eq}$ (within a factor of ten or so), then the indiffusion experiment with $C_s \ll C_s^{eq}$ would actually be governed by the kick-out mechanism, since the vacancy term in (19) can be neglected. A precipitation experiment on the other hand, with $C_s \gg C_s^{eq}$, could be described in terms of the Frank-Turnbull mechanism since now the interstitial term in (19) can be neglected.

Since the validity of (5) and (19) requires the energetically unfavorable thermal creation of vacancy-interstitial pairs in the ideal silicon lattice the present authors feel that the possibility i) is more likely than ii) as long as a careful re-investigation of the Ni precipitation kinetics does not prove the opposite to be true.

OXYGEN PRECIPITATION

Growth Kinetics. When present in supersaturation, interstitially dissolved oxygen in silicon may precipitate, mainly in the form of a SiO_2 phase, provided a nucleation barrier is overcome[87-88]. Because of the volume expansion connected with the formation of SiO_2 precipitates, mechanical stresses are created, which can lead to the ejection of I. These self-interstitials in turn may agglomerate as interstitial-type stacking faults surrounding a central precipitate[87], see, e.g., Fig. 7. The most elaborate experimental and theoretical investigation on the growth of such stacking faults during oxygen precipitation has been presented by Patel *et al.*[87,89]. They assume that the growth of a central SiO_2 precipitate, basically of a near-spherical three-dimensional shape, is rate-limited by the diffusion of I away from the precipitate. If a constant fraction $\delta < 1$ of the I emitted by the growing precipitate is incorporated into the surrounding stacking fault, their model predicts for the growth of the stacking-fault radius r_{SF} approximately

$$r_{SF} = \delta(4A/3\Omega)^{1/2}[(C_i-C_i^{eq})D^{SD}\Omega t/2\pi]^{3/4} \qquad (20)$$

where C_i is the actual concentration of interstitial oxygen present in supersaturation and C_i^{eq} its solubility corresponding to the temperature of the experiment. The other symbols have the same meaning as in previous sections. Although the model is successful in explaining the experimentally observed time and temperature dependence, Patel *et al.*[89] do not compare *experimental* and *calculated* values of r_{SF}. Determining r_{SF} from (20) we find the calculated values of r_{SF} to be at least *four orders of magnitude smaller* than the observed ones, even for the maximally feasible value of $\delta=1$. Re-evaluation of the model of Patel *et al.*[89] under the assumption of local equilibrium between oxygen interstitials and self-interstitials at the precipitate surface leads to

$$r_{SF} = \delta(4A/3\Omega)^{1/2}[2D^{SD}((C_i/C_i^{eq})^2-1)t]^{3/4}. \qquad (21)$$

Eq. (21) gives the right order of magnitude for the absolute values of r_{SF}, but fails to reproduce the measured temperature dependence as nicely as (20), so that (21) cannot be regarded as satisfactory either.

For the derivation of both (20) and (21) a three dimensional, approximately spherical precipitate has been assumed, whereas experimentally the precipitates have been found to be two-dimensional platelets[90,91]. A corresponding derivation for two dimensional precipitates of an effective radius r_p gives $r_{SF} \sim r_p \sim t$ and no longer the 3/4 power law observed. Recently Wada *et al.*[91] measured r_p of the central platelet and got a 3/4 power time dependence for r_p, which is in contradiction to the predicted growth law $r_p \sim t$. It seems fair to say that the quantitative understanding of the growth kinetics of oxygen precipitates and of the corresponding stacking faults is far from being satisfactory.

Nucleation. The nucleation process of SiO_2 precipitates at various heat treatments is even less understood than the growth kinetics. Homogeneous as well as heterogenous nucleation processes have been invoked [92-98]. We devote this last section to some speculations on the latter ones without excluding the possibility of homogeneous nucleation.

Recent experiments on oxygen precipitation show rather clearly that a high carbon concentration can be very helpful in the nucleation process [94-97]. Shimura *et al.*[97] suggest specifically that a critical nucleus could be provided by an agglomerate of carbon atoms, vacancies and self-interstitials, whereas Hu [99] favors small vacancy agglomerates. We recall that 'B-type' swirl defects observed after the growth of silicon single crystals under appropriate conditions are also closely related to the carbon content of the Si crystals[100]. Whereas 'A-type' swirl defects have been established to consist of interstitial-type dislocation loops[11,15], the nature of B-swirls has not yet been identified properly. Within the model put forward by Föll *et al.*[100] B-swirls are agglomerates of carbon atoms and Si self-interstitials, hold together by an entropy gain and therefore stable only at high temperatures in the presence of a supersaturation of I. Below a critical temperature, which depends on the supersaturation of I and the carbon content[100], the agglomerates tend to dissolve. Because of the much lower diffusivity of C compared to I[100], only self-interstitials manage to escape from the agglomerate, thus leaving behind a carbon-rich zone. Since carbon atoms are smaller than silicon atoms such carbon-rich zones would exhibit a vacancy- type strain field [21, 101] just appropriate for the nucleation of oxygen precipitates, which involve a volume expansion. This speculative model for hetergeneous nucleation at carbon-rich zones has attractive features. It predicts the formation of nuclei at temperatures below the stability regime of carbon self-interstitial agglomerates. For annealing at higher temperatures (say, roughly above 1000°C) two competing processes are likely to occur. (i) If self-interstitials are still present in a sufficiently high supersaturation the carbon-rich zones will be decorated again by I and provide no nucleation sites any more. Nucleation sites may be re-introduced by annealing at temperatures below the stability regime (say, e.g., at 800°C). (ii) If self-interstitials are no longer present in a sufficiently high supersaturation, because they have diffused to sinks such as nearby wafer surfaces, then there is no driving force any more for rebuilding carbon self-interstitial agglomerates. On the contrary, the carbon- rich zone will dissolve if the temperature is high enough for an effective migration of carbon atoms. The high temperature treatment will therefore result in an *irreversible* dissolution of the nucleation sites. The nucleation sites cannot be re-introduced by a low temperature annealing as in case (i). Although there are some experimental indications [96,99] that nucleation centers actually behave in a similar way as described above this speculative nucleation model should mainly serve as a guideline for further investigations.

ACKNOWLEDGMENTS

One of the authors (U. Gösele) is very much indebted to Dr. T. Tan for many fruitful and enjoyable discussions.

REFERENCES

1.   A. Seeger and K.P. Chik, Phys. Stat. Sol. *29*, 455 (1968).

2.   S.M. Hu, in: *Diffusion in Semiconductors*, D. Shaw ed. (Plenum Press, London 1973) p. 217.

3.   D. Shaw, Phys. Stat. Sol. (b) *72*, 11 (1975).

4.   A.F.W. Willoughby, Rep. Prog. Phys. *41*, 1665 (1978).

5.   R.O. Simmons and R.W. Balluffi, Phys. Rev. *125*, 862 (1962).

6.   L.C. Kimerling and D.V. Lang, in: *Lattice Defects in Semiconductors 1974* (Inst. Phys. Conf. Ser. *23*, 1974) p. 589.

7.   J.C. Bourgoin and J.W. Corbett, Rad. Effects *36*, 157 (1978).

8.   A. Seeger, W. Frank, and U. Gösele, in: *Defects and Radiation Effects in Semiconductors 1978* (Inst. Phys. Conf. Ser. *46*, 1979) p. 148.

9.   I.R. Sanders and P.S. Dobson, J. Mater. Science *9*, 1987 (1974).

10. B.J. Masters, Sol. State Comm. *9*, 283 (1971).

11. P.M. Petroff and A.J.R. de Kock, J. Cryst. Growth *30*, 117 (1975).

12. R.B. Fair, in: *Semiconductor Silicon 1977*, H.R. Huff and E. Sirtl eds. (The Electrochem. Soc., Princeton 1977) p. 968.

13. J.A. Van Vechten, Phys. Rev. *B17*, 3197 (1978).

14. J.C. Bourgoin and M. Lanoo, Rad. Eff. *46*, 157 (1980).

15. H. Föll and B.O. Kolbesen, Appl. Phys. *8*, 117 (1975).

16. U. Gösele and H. Strunk, Appl. Phys. *20*, 265 (1979).

17. S. Prussin, J. Appl. Phys. *43*, 2850 (1972).

18. S.M. Hu, J. Appl. Phys. *45*, 1567 (1974).

19. S.M. Hu, J. Vac. Sci. Technol. *14*, 17 (1977).

20. E. Sirtl in ref. [12], p. 4.

21. A.J.R. de Kock and W.M. van de Wiggert, J. Cryst. Growth *49*, 718 (1980).

22. M. Yoshida and K. Saito, Jap. J. Appl. Phys. *6*, 573 (1967).

23. R.F. Peart, Phys. Stat. Sol. *15*, K119 (1966).

24. J.M. Fairfield and B.J. Masters, J. Appl. Phys. *38*, 3148 (1967).

25. A. Seeger, W. Frank, and H. Föll, in: *Lattice Defects in Semiconductors 1976* (Inst. Phys. Conf. Ser. *31*, 1977) p. 12.

26. See, e.g.,: M.D. Matthews and S.J. Ashby, Phil. Mag. *27*, 1313 (1973).

27. B. Leroy, J. Appl. Phys. *50*, 7996 (1979).

28. W.G. Allen, Sol. State Electr. *16*, 709 (1973).

29. D.A. Antoniadis, A.G.Gonz,alez and R. Dutton, J. Electrochem. Soc. *125*, 814 (1978).

30. K. Taniguchi, K. Karosawa, and M. Kashiwagi, J. Electrochem. Soc. *127*, 2243 (1980).

31. K. H. Nicholas, Solid State Electr. *9*, 35 (1966).

32. G. Masetti, S. Solmi, and G. Soncini, Solid State Electr. *16*, 1419 (1973); Phil. Mag. *33*, 613 (1976).

33. R. Francis and P.S. Dobson, J. Appl. Phys. *50*, 280 (1979).

34. D.A. Antoniadis, A.M. Lin, and R.W. Dutton, Appl. Phys. Lett. *33*, 1030 (1978).

35. R. Conti, G. Corda, R. Mattecci, and G. Ghezzi, J. Materials Science *10*, 705 (1975).

36. S.P. Murarka and G. Quintana, J. Appl. Phys. *48*, 46 (1977).

37. S.P. Murarka, Phys. Rev. *B16*, 2849 (1977).

38. J.E. Lawrence, J. Appl. Phys. *40*, 360 (1969).

39. S.M. Hu, Appl. Phys. Lett. *27*, 165 (1975).

40. H. Shiraki, Jap. J. Appl. Phys. *14*, 747 (1975).

41. H. Shiraki, Jap. J. Appl. Phys. *15*, 1 (1976).

42. T. Hattori, J. Electrochem. Soc. *123*, 945, (1976).

43. C.L. Claeys, E.E. Laes, G.J. Declerck, and R.J. Van Overstraeten, in ref. [12], p. 773.

44. Y. Hokari and H. Shiraki, Jap. J. Appl. Phys. *16*, 1899 (1977).

45. T. Hattori and T. Suzuki, Appl. Phys. Lett. *33*, 347 (1978).

46. Y. Nabeta, T.Uno, S. Kubo, and H. Tsukamoto, J. Electrochem. Soc. *123*, 1416 (1976).

47. Y. Hokari, Jap. J. Appl Phys. *18*, 873 (1979).

48. J. Friedel, *Dislocations* (Pergamon Press, New York 1964) pp. 104-127.

49. For references see: C.M. Shelvin and L.J. Delmer, Phil. Mag. A*40* 685, (1979).

50. A. Seeger and U. Gösele, Phys. Lett. *61A*, 423 (1977).

51. I.R. Sanders and P.S. Dobson, Phil. Mag. *20*, 881 (1969).

52. H. Hashimoto, H. Shibayama, and H. Ishikawa, Fujitsu Sci. Techn. J., March 73 (1977).

53. Y. Sugita, H. Shimizu, A. Yoshinaka, and T. Aoshima, J. Vac. Sc. Techno. *14*, 44 (1977)

54. W.K. Wu and J. Washburn, J. Appl. Phys. *48*, 3747 (1977).
55. H. Shimizu, A. Yoshinaka, and S. Sugita, Jap. J. Appl. Phys. *17*, 747 (1978).
56. C.L. Claeys, G.J. Declerck, and R.J. Van Overstraeten, Appl. Phys. Lett. *35*, 797 (1979).
57. H.J. Mayer, H. Mehrer, and K. Maier, in ref. [25], p. 186.
58. L. Kalinowski and R. Seguin, Appl. Phys. Lett. *35*, 211 (1979); *36*, 171 (1980).
59. J. Hirvonen and A. Anttila, Appl. Phys. Lett. *35*, 703 (1979).
60. H. Alexander, H. Eppenstein, H. Gottschalk, and S. Wendler, J. Microsc. *118*, 1 (1980).
61. For references see e.g.,: W. Tiller, J. Electrochem. Soc. *127*, 621 (1980).
62. A.M. Lin, D.A. Antoniadis, R.W. Dutton, and W.A. Tiller, Electrochem. Soc. Meeting, Boston, 1978 (unpublished).
63. R.B. Fair, to be published.
64. S.P. Murarka, Phys. Rev. *B21*, 692 (1980).
65. S.J. Ingrey and S. Maniv, Electrochem. Soc. Meeting, Boston, 1978 (unpublished).
66. S.M. Hu, these proceedings.
67. H. Shiraki, in ref. [12], p.546.
68. G.J. DeClerck, 9th Europ. Solid State Device Conf., Munic, 1979, to be published.
69. P.B. Hirsch, J. Microscopy *118*, 3 (1980).
70. A.F.W. Willoughby, J. Phys. D *10*, 455 (1977).
71. J.E. Lawrence, J. Appl. Phys. *37*, 4106 (1966).
72. C.L. Claeys, G.J. DeClerck, and R.J. Van Overstraeten, Rev. Physique Appl. *13*, 797 (1978).
73. R.B. Fair and J.C.C. Tsai, J. Electrochem. Soc. *124*, 1107 (1977).
74. M. Yoshida, Jap. J. Appl. Phys. *18*, 479 (1979).
75. C.L. Claeys, G.J. Declerck, and R.J. Van Overstraeten, in: *Semiconductor Characterization Techniques*, P.A. Barnes and G.A. Rozgonyi eds. (Electrochem. Soc., Princeton 1978) p. 366.
76. A. Armigliato, M. Servidori, S. Solmi, and I. Vecchi, J. Appl. Phys. *48*, 1806 (1977).
77. H. Strunk, U. Gösele, and B.O. Kolbesen, Appl. Phys. Lett. *34*, 530 (1979).
78. J. Weertman, Phys. Rev. *107*, 1259 (1957).
79. H.S. Grienauer and K.R. Mayer, in ref. [6], p. 550.
80. W.F. Tseng, S.S. Lau, and J.W. Mayer, Phys. Lett. *68A*, 93 (1978).
81. F.C. Frank and D. Turnbull, Phys. Rev. *104*, 617 (1956).
82. W.R. Wilcox and T.J. LaChapelle, J. Appl. Phys. *35*, 240 (1964).
83. W.M. Bullis, Solid State Electronics *9*, 143 (1966).
84. U. Gösele, W. Frank, and A. Seeger, Appl. Phys. *23*, 361 (1980); U. Gösele, F. Morehead, W. Frank, and A. Seeger, to be published in Appl. Phys. Lett.
85. A. Seeger, Phys. Stat. Sol. (b), in press.
86. W. Frank *et al.*, these proceedings.
87. For references see: J.R. Patel, in ref. [12], p. 521.
88. K. Tempelhoff *et al.*, Phys. Stat. Sol. (a) *56*, 213 (1979).
89. J.R. Patel, K.A. Jackson, and H. Reiss, J. Appl. Phys. *48*, 5297 (1977).
90. T.Y. Tan and W.K. Tice, Phil. Mag. *34*, 615 (1976).
91. K. Wada, N. Inoue, and K. Kohra, J. Cryst. Growth *49*, 749 (1980).
92. P.E. Freeland *et al.*, Appl. Phys. Lett. *30*, 31 (1977).
93. J. Osaka, N. Inoue and K. Wada, Appl. Phys. Lett. *36*, 289 (1980).
94. S. Kishino *et al.*, Appl. Phys. Lett. *35*, 213 (1979).
95. Y. Matsushita *et al.*, Jap. J. Appl. Phys. *19*, L101 (1980).
96. S. Kishino *et al.*, J. Appl. Phys. *50*, 8240 (1979).
97. F. Shimura *et al.*, Appl. Phys. Lett. *37*, 483 (1980).
98. S. Shirai, Appl. Phys. Lett. *36*, 156 (1980).
99. S.M. Hu, to be published in J. Appl. Phys.
100. H. Föll, U. Gösele, and B.O. Kolbesen, J. Cryst. Growth *40*, 90 (1977).
101. S. Yasuami, J. Harada, and K. Wakamatsu, J. Appl. Phys. *50*, 6860, (1979).

Published 1981 by North-Holland, Inc.
Narayan, and Tan, eds.
Defects in Semiconductors

A CHANNELING STUDY OF DEFECT-BORON COMPLEXES IN Si

M.L. SWANSON, L.M. HOWE, F.W. SARIS* AND A.F. QUENNEVILLE
Atomic Energy of Canada Limited, Chalk River, Ontario, Canada K0J 1J0

ABSTRACT

Si crystals were doped with 0.1-0.2 at% ^{11}B in the near
surface region by ion implantation followed by thermal diffusion at 1373 K or by ruby laser annealing. The position
of the B atoms in the Si lattice was determined by channeling measurements, utilizing both the yield of H$^+$ ions (of
incident energy 0.7 MeV) backscattered from Si atoms and the
yield of alpha particles from the ^{11}B(p, α)^{8}Be nuclear reaction. Initially, 95-99% of the B atoms were substitutional.
Irradiation at 35 K or 293 K with 0.7 MeV H$^+$ displaced B
atoms from lattice sites. The displacement rate was greater
at 293 K than at 35 K, and was greater for diffused samples
than for laser annealed samples. Following 35 K irradiations, a large increase in the fraction f_{dB} of displaced B
atoms occurred during annealing near 240 K. At higher
annealing temperatures, f_{dB} decreased over a broad temperature range from 425-825 K. Angular scans through <110>
channels for the laser annealed samples after 293 K irradiation or after 35 K irradiation plus 293 K annealing showed a
pronounced narrowing of the dip in ^{11}B(p, α)^{8}Be yields compared with the dip in yields from Si, whereas no narrowing
was observed for <100> channels. These results indicate
that B atoms were displaced into specific lattice sites by
the migration of an interstitial B defect (the EPR G28 defect) near 240 K.

INTRODUCTION

From measurements of electron paramagnetic resonance (EPR), infrared absorption and charge transport properties, it has been shown that a multitude of defect complexes can be produced in irradiated semiconductors by the trapping at
solute atoms of irradiation-induced vacancies and self-interstitials [1-11].
For example, EPR data have shown that vacancy-As, vacancy-Sb, vacancy-O and va-
cancy-B pairs are created in irradiated Si during vacancy migration near 170 K,
and that interstitial Al and B centers are created by the trapping of Si self-
interstitials during electron irradiation at temperatures as low as 4 K [3-11].
During the doping of semiconductors by ion implantation, the irradiation conditions are conducive to the formation of such complexes, as well as to the formation of larger clusters of defects in the displacement cascades.

Ion channeling is an especially suitable method of studying defect interactions in ion implanted or diffusion-doped materials in the near-surface region
where the concentrations of solute atoms are relatively high (10^{-4}-10^{-2} atomic
fraction). By the channeling method, the positions of specific solute atoms in
the lattice can be determined [12-14]. Thus the displacement of solute atoms
from lattice sites can be measured as defect complexes are created, and the de-

---

* Permanent Address: FOM Institute for Atomic and Molecular Physics,
  Kruislaan 407, Amsterdam.

fect configurations can be inferred. This technique has been successfully applied to the identification of defect configurations in metals [13].

Several channeling experiments have shown that in Si group III and V solute atoms are displaced from lattice sites by irradiation [15-20]. The displacement can be produced directly by ion implantation, or by particle irradiation of diffusion-doped samples. There is considerable evidence that Coulomb interaction is important in creating defect complexes under these conditions [2,3, 17-20]. Positively charged group V solute atoms trap negatively charged vacancies, while negatively charged group III solute atoms trap positively charged self-interstitials. The configurations of these complexes have not been identified.

Although channeling measurements have been used in several cases [21-23] to determine the lattice position of implanted solute atoms in Si (e.g., Bi, Zr, Yb), no systematic measurement of the configuration of specific defect-solute complexes in Si has been made. In the present paper, we study defect-B complexes in Si which are created by the migration of the G28 EPR defect [6] (interstitial B).

EXPERIMENTAL PROCEDURE

Si crystals were doped with ^{11}B in the near-surface region by two methods: (a) implantation at 100 keV to a fluence of 10^{16} cm^{-2}, followed by thermal diffusion at 1373 K for 1800 s; and (b) implantation at 25 keV to a fluence of 1.5×10^{15} cm^{-2}, followed by ruby laser annealing (1.6 J cm^{-2}). The diffusion depth of the B in the first case [24] was 1 μm and in the second case [25] was 0.25 μm, giving average B concentrations of 0.2 and 0.1 at%, respectively. The crystals were mounted in the low temperature target chamber [26] of the 2.5 MV CRNL Van de Graaff accelerator. The sample temperature could be varied between 35 and 400 K while a cryogenic shield surrounding the sample was kept at approximately 20 K. A sample could also be mounted on a target assembly whose temperature was adjustable from 293-870 K. Both the irradiations and channeling analyses were done with 0.7 MeV H$^+$ ions. For the irradiations, the beam was swept over a 0.04 cm^2 area, in a random crystallographic direction; the channeling analyses were done on a concentric 0.01 cm^2 area. The ion fluence was measured by integrating the current collected by the sample. To suppress secondary electrons, the shield was maintained at a negative bias of 100 V with respect to the sample.

The B atoms were detected by measuring the yield of alpha particles from the nuclear reaction ^{11}B(p,α)^{8}Be, which has a broad resonance near a proton energy of 670 keV [27]. A surface barrier detector at a scattering angle of 150°, covered by a 1 mg cm^{-2} aluminized Mylar film, was used to measure the energies of the alpha particles. Pinholes in the Mylar film permitted simultaneous measurement of the backscattered proton energies.

The position of B atoms in the Si lattice was determined by comparing normalized yields X_B from B atoms and X_{Si} from Si atoms at the same depth near given <lmn> channels [12-14]. The normalized yield is defined as the yield for a given incident angle of the ion beam divided by the yield obtained for a random incident angle. For specific lattice positions of the B atoms, characteristic yield profiles from B atoms would be obtained for angular scans through different axial and planar channels. For example, if the B atoms were in the center of a given channel, a peak in the nuclear reaction yield from B atoms would occur at the aligned direction. If the B atoms were displaced only a small distance into a given channel, the yield profile from B atoms would be narrower than that from Si atoms. Detailed analytical or Monte Carlo calculations of the flux distributions of ions within a channel are required for an accurate lattice site determination [12-13].

In addition to angular scans, measurements of yields for perfect alignment along specific <lmn> channels give quick information on lattice sites [12,13-20]. For example, if $\chi_B^{<lmn>} > 1$, it is apparent that the B atoms would be near the center of that <lmn> channel, whereas if $\chi_B^{<lmn>} \sim \chi_{Si}^{<lmn>}$, the B atoms would lie close to the strings of atoms bordering that channel. A quantity readily measured by channeling is the apparent fraction of B atoms which are displaced from lattice sites into a given <lmn> channel [28],

$$f_{dB}^{<lmn>} = (\chi_B^{<lmn>} - \chi_{Si}^{<lmn>})/(1 - \chi_{Si}^{<lmn>}) \tag{1}$$

Because the ion flux is not uniform in a channel [12-14], the true displaced fraction of B atoms is given by $f_{dB}^{<lmn>}/F_i^{<lmn>}$ where $F_i^{<lmn>}$ is the normalized ion flux at the position of the displaced solute atoms in an <lmn> channel. Near the center of a channel, $F_i^{<lmn>} \sim 2.0$, while $F_i^{<lmn>} < 1$ near the channel edges. [12-14].

For the analysis of B atoms displacements, it is important to measure the normalized yields from B and Si atoms at the same depth in the crystal. For the present thermally diffused samples, previous results [24] indicated that the B was distributed almost uniformly to a depth of ~ 1 μm. Within this depth, $\chi_{Si}^{<lmn>} \sim 0.1$ for the major channels, so that little error arose from dechanneling effects. Nevertheless, an accurate comparison of absolute values of $f_{dB}^{<lmn>}$ for different channels was difficult because of the uncertainties in depth scales. For the present analysis, $\chi_{Si}^{<lmn>}$ was measured for a depth increment of 150-900 nm, and $\chi_B^{<lmn>}$ was determined for alpha particle energies from 3.0-4.5 MeV. For the laser annealed samples, the lack of depth information obtainable from the alpha particle energy was not important, as the dechanneling within the depth 0.25 μm of the B atom distribution was negligible.

RESULTS

Displacement Rates

In the present thermally diffused or laser annealed Si crystals, the B atoms occupied substitutional lattice sites. Prior to irradiation, typically $\chi_{Si}^{<lmn>} = 0.03$ and $\chi_B^{<lmn>} = 0.04$, giving $f_{dB}^{<lmn>} = 0.01$. The B atoms became displaced from lattice sites by irradiation with 0.7 MeV H^+ at 35 K or 293 K, as shown in Fig. 1. In laser annealed samples (as shown previously for thermally diffused samples [20]) the displacement rate produced by 35 K irradiations was much less than that produced by 293 K irradiations. These measurements were taken at the irradiating temperature in each case. We have demonstrated that $f_{dB}^{<lmn>}$ is affected relatively little by the measuring temperature.

For the laser annealed samples, the apparent displaced fraction of B atoms in <100> channels, $f_{dB}^{<100>}$, was much larger than $f_{dB}^{<110>}$. In the thermally diffused sample, this difference was less.

The displacement rate of B atoms $df_{dB}^{<lmn>}/d\phi$ caused by 293 K irradiation was much greater for the thermally diffused crystal than for the laser annealed sample. The value of $(df_{dB}^{<100>}/d\phi)_{\phi=0}$ (extrapolated to zero fluence) was equal to 0.7×10^{16} cm^2 for the thermally diffused sample, as compared with 0.14×10^{16} cm^2 for the laser annealed sample. This difference may be attributed partly to uncertainty in the determination of $\chi_{Si}^{<100>}$, which would arise if the actual diffusion depth of B were greater than the assumed value of 1 μm. However, the main cause of the difference in displacement rates may well be the presence of additional trapping sites, such as oxygen or vacancy-type defects in the laser annealed crystals. The point defect structure produced by laser annealing has not yet been well described, and measurements of the pre-

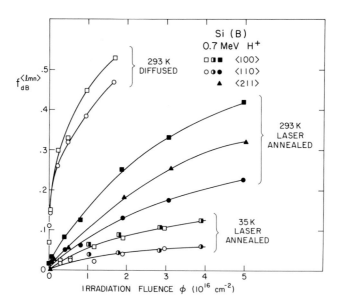

Fig. 1. The apparent displaced fraction of B atoms $f_{dB}^{<lmn>}$ is shown for
<100>, <110> and <211> channels as a function of 0.7 MeV H$^+$ irradiation
fluence ϕ at 35 K and 293 K. Results for 293 K irradiations are shown for both
thermally diffused and laser annealed Si. The equivalent random fluence of
each aligned spectrum, given by the aligned fluence multiplied by $\chi_{Si}^{<lmn>}$ at
the measuring depth, was included in the total fluence [29]. The measurements
were taken at the irradiation temperatures indicated. Two separate samples for
the 35 K irradiations are indicated by different symbols. (Sample 1: O,□;
Sample 2: ◑,◙.)

sent type should be useful in characterizing defect structures caused by the
laser annealing process. This problem will be studied further.

Annealing
    The effect of isochronal annealing on the apparent displaced fractions
$f_{dB}^{<lmn>}$ of B atoms for <100> and <110> channels is shown in Fig. 2 for irradiated laser-annealed samples. One of the samples had been irradiated with 0.7
MeV H to a fluence of 7 x 10^{16} cm^{-2} at 35 K, and the second sample had been
irradiated to a fluence of 4 x 10^{16} cm^{-2} at 300 K. The major displacement of B
atoms occurred during an annealing stage centered at 240 K (as observed previously for a thermally diffused crystal [20]), while the recovery of this B
atom displacement occurred over a broad temperature range from 425–825 K. This
latter recovery behaviour was in marked contrast to that reported [30,31] for
unannealed B-implanted Si, where the displaced fraction of B atoms increased
during annealing from 450–750 K, and then decreased during further annealing
from 750–1250 K. After irradiations at either 35 or 300 K, and throughout most
of the annealing steps for these laser-annealed samples, the apparent displaced

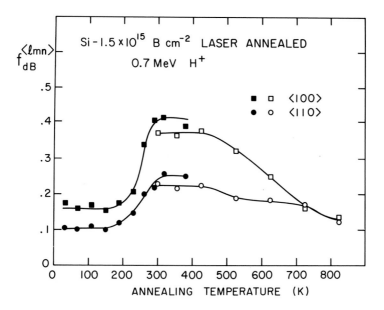

Fig. 2 - Thermal recovery of the irradiation-induced displacement of B atoms from lattice sites in laser annealed Si. The apparent displaced fraction of B atoms $f_{dB}^{<lmn>}$ for <100> and <110> channels is shown as a function of annealing temperature (600 s pulse anneals) for two samples. The first was irradiated with 0.7 MeV H to a fluence of 7 x 10^{16} cm^{-2} at 35 K, and all measurements were at 35 K (●,■). The second was irradiated to a fluence of 4 x 10^{16} cm^{-2} at 300 K and all measurements were at 300 K (O,□).

fractions of B atoms were larger for <100> channels than for <110> channels. Thus preferred sites for the B atoms are indicated. During the annealing steps from 425 K to 725 K, $f_{dB}^{<100>}$ gradually approached $f_{dB}^{<110>}$, indicating that the B atoms which remained displaced from lattice sites at 725 K were accommodated on "random" sites, i.e. associated with larger defect clusters or precipitates.

Angular Scans
    Detailed angular scans near channeling directions were performed to obtain information on the lattice location of the displaced B atoms. In Fig. 3, angular scans through <100>, <110> and <111> axial channels are shown for laser-annealed samples after 35 K irradiation followed by a 293 K anneal. A striking difference is seen between the B yield profiles for <100> and <110> channels. The narrowness of the dip in B yield near the aligned direction for a <110> channel, together with a pronounced shoulder at larger angles, indicates that approximately equal fractions of the B atoms were in substitutional lattice sites and in sites displaced a small distance into <110> channels [12,13,21]. In contrast, the very flat B profile for a <100> channel shows that ∿ 40% of the B atoms were displaced approximately 0.1 nm into this channel [12,13].

Similarly the <111> scan indicates that ∿ 40% of the B atoms were displaced
∿ 0.1 nm into that channel. Similar angular scan data were obtained for crystals which were irradiated at 293 K.

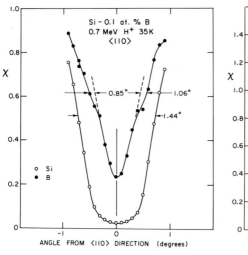

(a)

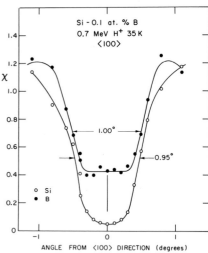

(b)

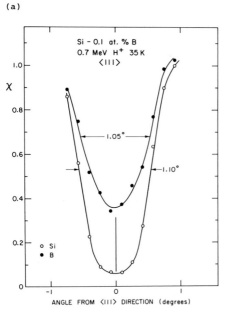

(c)

Fig. 3 - Angular scans for a laser
annealed Si-0.1 at% B crystal after
irradiation at 35 K with 0.7 MeV H$^+$
ions to a fluence of 5 x 10^{16} cm^{-2}
followed by a 20 h anneal at 293 K.
The normalized yields X_{Si} of H$^+$ ions
(incident energy 0.7 MeV)
backscattered from Si atoms and the
normalized yields X_B of alpha
particles from the ^{11}B(p,α)^{8}Be
nuclear reaction are plotted as a
function of the angle from the
indicated axes. The measurements
were taken at 35 K. The lines were
drawn to guide the eye.
(a) <110> channel; (b) <100> channel; (c) <111> channel.

DISCUSSION AND CONCLUSIONS

(1) Relatively little displacement of B atoms occurred during 35 K irradiation of Si crystals (Fig. 1), where the EPR spectrum G28, associated with interstitial B, is observed [6]. Thus the present results indicate that the G28 defect does not involve large displacements of B atoms from lattice sites. That is, interstitial B atom positions close to the hexagonal or tetrahedral sites can be excluded.

(2) During annealing of Si crystals which had been irradiated with H^+ at 35 K, B solute atoms became displaced from lattice sites near 240 K, indicating the formation of a defect-B atom complex (Fig. 2). This complex was an interstitial-type defect, as further irradiation at 35 K (where self-interstitials but not vacancies are mobile) did not decrease the apparent displaced fraction of B atoms. If the defect were of vacancy type, it would be annihilated by the trapping of interstitial Si atoms. (Note that such annihilation has been observed [17] for vacancy-As complexes in Si.)

The annealing temperature (240 K) at which this defect-B complex appeared is consistent with the migration of the EPR defect G28, which has an activation energy for migration of 0.6 eV, and which is attributed to interstitial B atoms or Si-B split interstitials [6]. Taking the jump frequency [6] of defect G28 to be $5 \times 10^{12} \exp(-0.60/kT)s^{-1}$, there are 1.6 jumps per second at 240 K, which is consistent with the present defect concentration of $\sim 5 \times 10^{-4}$ and the annealing time of 600 s.

The concentration of this defect-B complex was almost the same for irradiations at 293 K as for irradiations at 35 K to the same fluence, followed by annealing at 293 K. This was shown by the similar values of $f_{dB}^{<1mn>}$, as well as the similar angular dependence of χ_B in these two cases.

In p-type Si, vacancies are neutral or possibly V^{++} [32], and migrate from 150-180 K [3-5]. However, displacement of B atoms did not occur during annealing below 200 K (Fig. 2), so that the trapping of vacancies by B atoms did not cause appreciable displacement.

(3) The major change in $f_{dB}^{<1mn>}$, which was attributed to the defect-B complex, annealed out gradually from 425-825 K (Fig. 2). The annihilation could have been due in part to divacancy migration [33]. This annealing behaviour is in contrast to that observed for Si which had been implanted with B at room temperature [30,31] (but not subsequently annealed), where the fraction of displaced B atoms increased during annealing from 450-750 K. It can be concluded that the implantation damage produced a different and probably more complicated defect configuration containing B. This is expected because the defect distribution produced in cascades from ~ 50 keV B ion irradiation is more complex than that from 0.7 MeV H^+ irradiation.

(4) The presently observed defect-B complex had a configuration which is characterized by the well-defined structural differences observed for the angular scans across <110>, <100> and <111> axial channels (Fig. 3). In particular, the narrowing of B dips for <110> but not for <100> channels showed that the displaced B atoms were close to <110> atomic strings but about 0.1 nm from <100> strings. These results are consistent with (a) $\sim 40\%$ of the B atoms being near bond-centered positions, and the remainder being on substitutional lattice sites; or (b) 60% of the B atoms being displaced 0.08 nm along <100> directions, and the remainder being on substitutional lattice sites. The relation between these sites and those suggested by Troxell and Watkins [34] for charged interstitial B will be discussed in a subsequent publication. Certainly, the present results show that the B atoms were not near tetrahedral or hexagonal interstitial sites, as these give peaks in yields for <110> channels.

(5) The larger production rates of the defect-B complex for thermally diffused samples than for laser annealed samples was likely due to residual impurities or vacancy-type defects created during the laser annealing.

ACKNOWLEDGEMENTS

The authors appreciate the technical assistance of J. Lori and
G.R. Bellavance, and useful discussions with J.A. Davies, I.V. Mitchell and
S.T. Picraux.

REFERENCES

1. H.J. Stein and F.L. Vook, Rad. Eff. $\underline{1}$, 41 (1969).
2. A.R. Bean, S.R. Morrison, R.C. Newman and R.S. Smith, J. Phys. C5, 379
    (1972).
3. G.D. Watkins, in Proc. Int. Conf. on Lattice Defects in Semiconductors,
    Freiburg, 1974 (Institute of Physics, London 1975), p.1.
4. G.D. Watkins, in Radiation Damage in Semiconductors, P. Baruch ed. (Dunod,
    Paris, 1965) p.97.
5. G.D. Watkins, in Radiation Effects in Semiconductors, F.L. Vook ed.
    (Plenum Press, New York, 1968) p.67.
6. G.D. Watkins, Phys. Rev. B12, 5824 (1975).
7. G.D. Watkins, Phys. Rev. B13, 2511 (1976).
8. E.L. Elkin and G.D. Watkins, Phys. Rev. 174, 881 (1968).
9. G.D. Watkins and J.W. Corbett, Phys. Rev. 134A, 1359 (1964).
10. G.D. Watkins and J.W. Corbett, Phys. Rev. 121, 1001 (1961).
11. K.L. Brower, Phys. Rev. B1, 1908 (1970).
12. J.A. Davies, in Channeling in Solids, D.V. Morgan ed. (J. Wiley and Sons,
    New York, 1973), Ch.11.
13. S.T. Picraux, in New Uses of Ion Accelerators, J.F. Ziegler ed. (Plenum
    Press, New York, 1975) p.229.
14. D.S. Gemmell, Rev. Mod. Phys. 46, 129 (1974).
15. J. Haskell, E. Rimini and J.W Mayer, J. Appl. Phys. 43, 3425 (1972).
16. W.H. Kool, H.E. Roosendaal, L.W. Wiggers and F.W. Saris, Nucl. Instr.
    Meth. 132, 285 (1976).
17. M.L. Swanson, J.A. Davies, A.F. Quenneville, F.W. Saris and L.W. Wiggers,
    Rad. Eff. 35, 51 (1978).
18. L.W. Wiggers and F.W. Saris, Rad. Eff. 41, 149 (1979).
19. L.W. Wiggers and F.W. Saris, Proc. Int. Conf. IBMM, Budapest 1978,
    J. Gyulai et al. eds. p. 583.
20. M.L. Swanson, L.M. Howe, A.F. Quenneville and F.W. Saris, Rad. Eff. Lett.
    50, 139 (1980).
21. S.T. Picraux, W.L. Brown and W.M. Gibson, Phys. Rev. B6, 1382 (1972).
22. B. Domeij, G. Fladda and N.G.E. Johansson, Rad. Eff. 6, 155 (1970).
23. J.U. Andersen, O. Andreasen, J.A. Davies and E. Uggerhøj, Rad. Eff. 7,
    25 (1971).
24. W.K. Hofker, Thesis, University of Amsterdam (1975).
25. R.T. Young, et al., Appl. Phys. Lett. 32, 139 (1978).
26. J. Bøttiger, J.A. Davies, J. Lori and J.L. Whitton, Nucl. Instr. Meth.
    109, 579 (1973).
27. O. Beckman, T. Huus and C. Zupancic, Phys. Rev. 91, 606 (1953).
28. M.L. Swanson, L.M. Howe and A.F. Quenneville, Phys. Stat. Sol. (a), 31,
    675 (1975).
29. M.L. Swanson, P. Offermann and K.H. Ecker, Can. J. Phys. 57, 457 (1979).
30. G. Fladda, K. Björkqvist, L. Eriksson and D. Sigurd, Appl. Phys. Lett.
    16, 313 (1970).
31. J.C. North and W.M. Gibson, Appl. Phys. Lett. 16, 126 (1970).
32. G.D. Watkins and J.R. Troxell, Phys. Rev. Lett. 44, 593 (1980).
33. G.D. Watkins and J.W. Corbett, Phys. Rev. 138, 543, 555 (1965).
34. J.R. Troxell and G.D. Watkins, Phys. Rev. B22, 921 (1980).

Published 1981 by North-Holland, Inc.
Narayan, and Tan, eds.
Defects in Semiconductors

PROCESSING EFFECTS ON THE ELECTRICAL AND OPTICAL PROPERTIES OF SULFUR-RELATED
DEFECT CENTERS IN SILICON AND SIMILARITIES TO THE OXYGEN DONOR*

RICHARD A. FORMAN, ROBERT D. LARRABEE, DAVID R. MYERS, WILLIE E. PHILLIPS AND
W. ROBERT THURBER
Electron Devices Division, National Bureau of Standards, Washington D.C. 20234

ABSTRACT

The properties of sulfur-related defects in silicon are
shown to differ dramatically from those that would have been
expected on the basis of effective mass theory for a simple
substitutional double donor. The ratio of the densities of
the sulfur states as measured by capacitance-voltage techniques has been observed to vary in specimens fabricated
from the same starting resistivity. Optical absorption
studies have shown that the deepest sulfur level has a manifold of ground states which anneal at unequal rates at 550°C.
Deep-level measurements show that the thermal emission rate
at a given temperature and the variety of effects produced
depends on annealing history and total sulfur density. The
variability of properties of samples of sulfur-doped silicon
is similar to those found for the oxygen donors in silicon,
thus suggesting a chemical trend for the column VI impurities
in silicon.

INTRODUCTION

Impurities with energy levels that are much greater than thermal energy
from the nearest band edge (i.e., the deep-level impurity centers) are recognized to be important in the performance of a variety of silicon devices
[1-3]; however, methods for the unambiguous characterization of these centers
have not yet been adequately developed. Specimens of silicon doped with sulfur have provided an excellent test for both the experimental and conceptual
aspects of the problems associated with the development of an adequate method-
ology for deep-level impurity characterization. Previous studies of the sulfur system have shown that the apparent electrical activation energy of the
deeper center varies from sample to sample [4], that the two dominant electrically active centers do not arise from two charge states of the same impurity center [5], and that the optical absorption spectrum near 2 μm is not
assignable to a single center [6]. These results are inconsistent with the
prevailing theoretical model that identifies the two dominant electrically
active centers as arising from an isolated substitutional sulfur atom [7] and
instead suggests that these levels may arise from different sulfur-related
complexes. As defect complexes introduced in silicon by radiation damage are
known to react and evolve under thermal treatment [8], the changes in characteristics of the sulfur-related centers with thermal annealing were examined
in order to investigate the nature of these defects and also to explore the

*The work was conducted as part of the Semiconductor Technology Program at
NBS. Portions of this work were supported by the Division of Electric Energy
Systems, Department of Energy (Task Order A021-EES). Not subject to copy-
right.

causes of specimen-to-specimen variations in electrical properties [4,5]. The
present study includes isothermal transient capacitance (ITCAP) [9], optical
absorption [6], and double boxcar deep-level transient spectroscopy (DLTS)
[10] measurements on sulfur-doped specimens prepared by closed tube diffusions
[6] as well as by ion implantation [11].

PROCESSING EFFECTS ON OPTICAL PROPERTIES

Optical absorption spectra of silicon specimens doped with sulfur exhibit
several series of sharp infrared absorption lines including a prominent group
at 2.11, 2.16, and 2.18 μm, which have been attributed to a number of sulfur-
related defect complexes [6]. To investigate further the origin of these
absorption lines, a silicon ingot doped with sulfur by closed tube diffusion
[6] was repeatedly annealed at 550°C in flowing dry nitrogen and the 9 K optical absorption spectra recorded after total annealing times of 1, 2, 3, 4, 5,
and 23 h. The effects of this annealing on the integrated area of these lines
are shown in Fig. 1. It is apparent that the three transitions do not anneal
at the same rate. These differences in annealing rates are definitive evidence that these transitions arise from different sulfur-related centers.
These results predict that annealing will lead to a variation in the population of these centers that would likely be interpreted as a shift in electrical activation energies or, equivalently, in thermal emission rates at fixed
observation temperatures in an electrical measurement.

PROCESSING EFFECTS ON ELECTRICAL PROPERTIES

Figure 2 shows the DLTS curves of three sulfur-doped silicon specimens, two
for n-type starting material doped by ion implantation of ^{32}S (specimens A and
B) and one for n-type starting material doped by a closed tube diffusion with
naturally occurring isotopic sulfur (i.e., 95% ^{32}S). The test devices for the
^{32}S-implanted specimens are pn junction structures of test pattern NBS-2 [12].
The test devices for the specimen prepared by the closed tube diffusion (specimen C) are pn junctions formed by indium-gallium solder dots with the necessary "ohmic" connection provided by a backside contact of gold-antimony (0.6%
Sb by weight) alloyed at 500°C for 5 min. The three traces of Fig. 2 were
obtained with a double boxcar DLTS system using gate delay times t_1 = 157 μs
and t_2 = 930 μs, yielding an effective emission rate window peaking at 435 μs,
an operating frequency of 20 MHz, and a transient initiation pulse varying
from zero volts (trap filling value) to a reverse bias of 10 V (trap emptying
value). Specimen A $\left(\text{implanted to a } ^{32}S \text{ dose of } 2 \times 10^{14} \text{ cm}^{-2}\right)$ shows only two
well-resolved thermal emission peaks of essentially equal magnitude. However,
both specimen B $\left(\text{implanted to a } ^{32}S \text{ dose of } 7 \times 10^{15} \text{ cm}^{-2}\right)$ and specimen C
(diffused at 1350°C) exhibit at least four emission peaks. The density of the
centers responsible for the low-temperature DLTS emission peaks increases with
fluence faster than that from the centers responsible for the high-temperature
DLTS peak as evidenced by the spectra of Fig. 2. The unequal dependence on
sulfur implant density is consistent with previous implantation studies of the
sulfur system [11]. There appears to be a correspondence of the emission
peaks of the higher dose implanted specimen B and the closed-tube diffused
specimen C. The existence of four electrically active sulfur-related levels
has been observed in measurements of resistivity and Hall effect [13], but has
not previously been reported in transient-capacitance measurements.

Figure 3 shows the effect of annealing on the thermal emission rate from
the shallower sulfur-related center of specimen A. These data were observed
using the ITCAP system which provides an output plot of the capacitance transient as a function of time. For this measurement, the test device was
mounted in a transistor package, measured, removed from that package for

annealing at 550°C for 5 h in an ambient of dry nitrogen, then remounted in the package and measured again. In the leftmost portion of the traces of Fig. 3, steady-state conditions have been established at a reverse bias of about 2 V. This bias was then removed to fill the deep levels and the capacitance increased to a much larger value. When the bias is reestablished at the initial value, the capacitance undershoots the steady-state value because of the fixed charge on the deep levels. The capacitance returns to the initial steady-state value with a time constant inversely proportional to the thermal emission rate of carriers into the conduction band. The time constant before annealing was 9.7 s at -144.2°C for a wide range of bias voltages (up to 20 V), thus showing that there was no detectable dependence on the electric field in the junction depletion region. After the anneal, the time constant decreased to 8.1 s at the same temperature. The traces shown in Fig. 3 have the same amplitude of capacitance transient, but the steady-state reverse bias capacitance, C_f, increased significantly (44%) as a result of the anneal, thus indicating that significant changes in both the shallow- and deep-level concentrations took place. A similar change in the time constant was observed for the deeper sulfur level with annealing, but the change was not as pronounced as that shown in Fig. 3 for the shallower level.

DISCUSSION

The variation of the relative magnitudes of the optical absorption transitions of the deepest sulfur center with annealing is unambiguous evidence that a manifold of ground states is responsible for the optical transitions and not the excitations of a single center, as was previously proposed [14]. Further evidence is obtained from the electrical measurements, which demonstrate significant differences in the relative densities of the various sulfur levels and the values of their characteristic thermal emission rates. Taken together, these results show that sulfur gives rise to a variety of species in silicon, the properties of which depend on sulfur density and on processing. Therefore, the discovery that at any given fixed temperature the thermal emission rate (the measured quantity from which an activation energy is computed) is not a unique characterization parameter for sulfur-doped silicon may seriously affect the conventional application of electrical deep-level measurements for characterization purposes.

In this regard, it also becomes important to examine the behavior of other chemical impurity systems in silicon. Studies of the oxygen donor by Wruck and Gaworzewski [15] have shown a density-dependent ratio of the shallow-to-deep levels of this system, a reduction in the ionization charges of these levels with heat treatment, and a manifold of ground states from which resonant infrared absorption lines arise. The effects of processing have also been shown to be important for manganese [16], and strong evidence has been presented for impurity association phenomena for the deep gold centers in silicon [17]. Examination of the literature shows that there are many impurities in silicon whose characterization by different techniques at different laboratories has yielded significantly different results [18]. Rather than attribute all of these differences to differing experimental techniques, experimental error, or different data analysis techniques, one must also consider the possibility of attributing many of these differences to impurity association phenomena. In Table 1, we list some common impurities which are known to exhibit complex formation. In the first column of the table, we show the probable chemical complexes and in the second column, the values of activation or ionization energies associated with each of these complexes. These data demonstrate that for even one of the most studied semiconductors (i.e., silicon), sufficient experimental results do not yet exist to address properly the problem of the theoretical understanding of deep-level impurities.

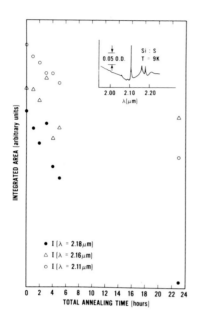

Figure 1. Dependence of the integrated areas of the three prominent peaks of the sulfur deepest level optical spectrum vs time (in hours) of annealing at 550 °C. The inset to the figure shows a typical spectrum of an unannealed specimen. the vertical scale in the insert is linear, the vertical scale of the figure is logarithmic.

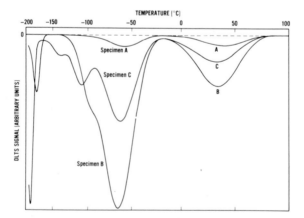

Figure 2. DLTS curves of three silicon specimens doped with sulfur. Specimen A was prepared by ion implantation of ^{32}S to a dose of 2×10^{14} cm^{-2} with subsequent thermal annealing. The preparation of specimen B was similar except that the dose was 7×10^{15} cm^{-2}. Specimen C was doped by closed tube diffusion at 1350 °C. The gate delay times were $t_1 = 157$ μs and $t_2 = 930$ μs. The heating rate was approximately 0.1 °C/s. A negative DLTS signal corresponds to majority carrier (electron) emission.

83

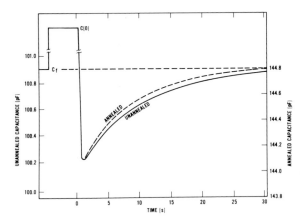

Figure 3. Isothermal transient capacitance of the ^{32}S defect center in a pn junction similar to specimen A before (solid line) and after (dashed line) a 5-h anneal at 550°C.

Table 1

Species	Reported Energy Levels (eV)	Reference
S_2^+	$E_C - 0.37$	a, b, c
Mn_4	$E_C - 0.53$	d
	$E_C - 0.59$	e
Pt	$E_g + (0.21$ to $0.45)$	f
Pt – 0 (?) (Pt (II))	$E_V + 0.45$	g
Pt (Pt (I))	$E_g + 0.34$	
Mg	$E_C - 0.227$	
	$E_C - 0.115$	h
	$E_C - 0.40$	
$[Au-Mn]^+$	unknown	i
divacancy	$E_C - 0.4$	j,k
	$E_V + 0.25$	

a. G. W. Ludwig, Phys. Rev. 137, A1520 (1965).
b. L. C. Kravitz, Ph.D. Thesis, Harvard University (1965).
c. Reference [13] (Camphausen et al.).
d. G. W. Ludwig, H. H. Woodbury, and R. O. Carlson, J. Phys. Chem. Solids 8, 490 (1959); ibid. 8 506 (1959).
e. A. O. Evwaraye, Defect Levels in Silicon Doped with Transition Elements, 19th Electronic Materials Conference, Ithaca, New York, June 29-July 1, 1977.
f. S. Braun, H. G. Grimmeiss, and K. Spann, J. Appl. Phys. 48, 3883 (1977).
g. H. W. Woodbury and G. W. Ludwig, Phys. Rev. 126, 466 (1962).
h. E. Ohta and M. Sakata, Solid State Electronics 22, 677 (1979).
i. H. H. Woodbury and G. W. Ludwig, Phys. Rev. 117, 1287 (1960); ibid. 117, 102 (1960).
j. G. D. Watkins in: Radiation Effects on Semiconductor Components (Jornée D'Electronique, Toulouse, France, 1967), vol. I, paper #A1.
k. H. J. Stein in: Radiation Effects in Semiconductors, Albany, 1970, J. W. Corbett and G. D. Watkins, Eds. (Gordon and Breach Science Publishers, London 1971) p. 125.

84

In previous simple models, deep-level chemical impurities were considered to exist as isolated substitutional atoms which satisfied their atomic bonding by completing the core shells with the excess (or deficit) electrons and thus producing hydrogenic or helium-like centers [19]. This picture is simplistic in assuming that the reactions between impurities in dilute solution in solids are controlled only by the densities of these impurities and their atomic configurations. The prevalence of impurity association phenomena, however, argues that local atomic bonding is likely to be satisfied by chemical associations with other available trace impurities or defects.

The likelihood of complex formation is already well established in studies of radiation-induced defects in silicon [8] where defect complexes are formed between primary defects (such as the isolated vacancy) and contaminants (such as carbon or oxygen) as soon as sufficient thermal energy is supplied to allow the primary defects to become mobile. Further, it is known that long-range forces (Coulombic interactions) control not only the characteristic annealing temperatures for simple radiation-induced defects (such as the isolated vacancy), but also determine which defects will be formed when simpler defects anneal [8]. These considerations suggest that any satisfactory empirical approach to the understanding of deep-level centers in covalent semiconductors will necessarily involve an investigation of impurity-complexing. All deep-level spectroscopic studies should include the effects of processing and background impurity concentrations in those cases where impurity association phenomena are known or suspected so that the correspondence between the deep-level characterization parameters and the local microscopic environment may be attained, as is required for a complete understanding of impurity effects.

ACKNOWLEDGMENT

The authors would like to acknowledge the assistance of Y. M. Liu and L. A. Robinson for device fabrication and D. V. Lang for making a preprint of reference 17 available.

REFERENCES

1. S. K. Ghandi in: Semiconductor Power Devices (John Wiley & Sons, New York 1977) pp. 2-8.
2. P. Migliorato and C. T. Elliott, Solid State Electronics 21, 443 (1978).
3. P. K. Chatterjee, G. W. Taylor, A. F. Tasch, Jr., and H-S. Fu, IEEE Trans. Electron Devices ED-26, 564 (1979).
4. D. R. Myers and W. E. Phillips, Appl. Phys. Lett. 32, 756 (1978).
5. D. R. Myers and W. E. Phillips, J. Elec. Mater. 8, 781 (1979).
6. R. A. Forman, Appl. Phys. Lett. 37 (9), 776 (1980).
7. S. Pantelides and C. T. Sah, Solid State Communications 11, 1713 (1972)
8. See, e.g., J. W. Corbett, Radiat. Eff. 6, 3 (1970).
9. M. G. Buehler and W. E. Phillips, Solid State Electronics 19, 777 (1976).
10. D. V. Lang, J. Appl. Phys. 45, 3023 (1974).
11. D. R. Myers, R. Y. Koyama, and W. E. Phillips, Radiat. Eff. 47, 91 (1980).
12. M. G. Buehler, Semiconductor Measurement Technology: Microelectronic Test Patterns: An Overview, NBS Spec. Publ. 400-6 (August 1974).
13. D. L. Camphausen, H. M. James, and R. D. Sladek, Phys. Rev. B2, 1899 (1970).
14. T. H. Ning and C. T. Sah, Phys. Rev. B4, 3482 (1971).
15. D. Wruck and P. Gaworzewski, phys. stat. sol. (a) 56, 557 (1979).
16. A. O. Evwaraye, J. Appl. Phys. 48, 3813 (1977).
17. D. V. Lang, H. Grimmeiss, E. Meijer, and M. Jaros, Phys. Rev. B, to be published.
18. See, for example, S. Braun, H. G. Grimmeiss, and K. Spann, J. Appl. Phys. 48, 3883 (1977).
19. A. G. Milnes in: Deep Impurities in Semiconductors, Chapter 1 (John Wiley and Sons, Inc., New York 1973).

Published 1981 by North-Holland, Inc.
Narayan, and Tan, eds.
Defects in Semiconductors.

DEFECT CHARACTERIZATION BY JUNCTION SPECTROSCOPY

L. C. KIMERLING

Bell Laboratories, Murray Hill, New Jersey   07974

ABSTRACT

Junction spectroscopic techniques provide a unique tool for the study of imperfections in semiconductor materials. New, fundamental defect processes have been discovered and direct observations of the role of defects in device manufacture and operation have been made. The principles of application of capacitance transient spectroscopy, electroluminescence, and SEM-charge collection microscopy are briefly reviewed. Examples of singular accomplishments of these techniques are given, and directions of future research are outlined.

INTRODUCTION

A new class of material characterization techniques has evolved during the past decade which utilizes semiconductor junctions as the test device. The application of these methods has led to the discovery of new phenomena and provided a unique tool for the understanding of materials processing for semiconductor devices. The test specimens consist of Schottky barrier, $p-n$ junction, or MOS structures — the standard building blocks of solid state circuitry. The characterization techniques are known as capacitance transient spectroscopy, SEM-charge collection microscopy, and electroluminescence. These methods, as schematically illustrated in Figure 1, form a cross correlated set of diagnostics for characterization of the electrical and optical properties of defects in semiconductors. This paper will briefly review the physical principles underlying these techniques and highlight some current materials research applications with both fundamental and applied implications. Each section concludes with directions for future research.

SEMICONDUCTOR JUNCTIONS

The relevant properties of abrupt, semiconductor junctions are outlined below. [The physics of semiconductor junction devices are discussed in books by Sze [1] and Moll [2].] The junction is defined as the boundary across which the majority carrier changes sign and changes density by more than a factor of ten. Figures 1 and 2 and the relations below refer to an n-type substrate.

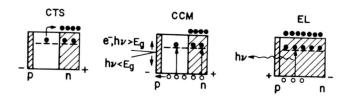

Fig. 1. Three modes of junction spectroscopy: Capacitance transient spectroscopy (CTS), charge collection microscopy (CCM), and electroluminescence (EL).

86

Schottky barriers and p$^+$n junctions. The depletion width W is determined by the applied *reverse* voltage V_R, the barrier height V_B, and the free carrier concentration, N_D.

$$W = \left[\frac{2\epsilon_s(V_R+V_B)}{q\,N_D}\right]^{1/2} \qquad (1)$$

The junction electric field varies linearly through the depletion width in uniformly doped material with a maximum value, E_{max}, at the junction and a zero value at W.

$$E_{max} = [2\,q/\epsilon_s(V_R+V_B)N_D]^{1/2} \qquad (2)$$

The capacitance, C_s, of a junction of area A is given by

$$C_s = \frac{\epsilon_s A}{W} = A\left[\frac{\epsilon_s q N_D}{2(V_R+V_B)}\right]^{1/2} \qquad (3)$$

In equations (1), (2), and (3), ϵ_s refers to the relative dielectric constant of the semiconductor ($\epsilon_s = \epsilon'_s\epsilon_o = \epsilon'_s \cdot 8.86 \times 10^{-14}$F/cm) and q refers to the unit electrical charge ($= 1.6 \times 10^{-19}$ Coul.).

The junction depletion region vanishes under *forward* bias as an exponentially varying distribution of minority carriers is injected into the substrate. The average concentration of minority carriers existing within a diffusion length (L_p) of the junction is given by

$$p = J/q\,(L_p/D_p) \qquad (4)$$

where J is the injected current density and D_p is the minority carrier diffusion constant. Schottky barrier structures allow, however, only majority carrier flow.

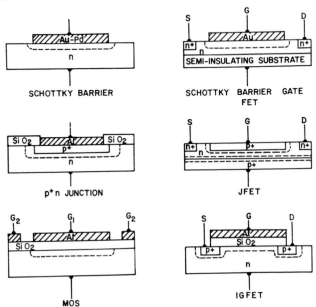

Fig. 2. Measurement structures employed in junction spectroscopy. The depletion region is depicted by dashed lines.

<u>MOS structures</u>. The ideal MOS structure is considered as two capacitances in series

$$C = \frac{1}{C_o} + \frac{1}{C_s} \qquad (5)$$

The oxide capacitance, C_o, is determined by the oxide thickness as

$$C_o = \frac{\epsilon_o A}{t_o} \qquad (6)$$

In equation 5, C_s is associated with the depletion width capacitance generated by the applied gate voltage V_g (see Figure 2) and any trapped charge residing in the oxide or at the interface.

CAPACITANCE TRANSIENT SPECTROSCOPY

Capacitance transient spectroscopy is the most powerful form of a class of measurements which detect the current or capacitance response of a semiconductor junction to various outside stimuli. The methods were first summarized by Sah, et. al. [3] in 1970. Recent reviews which summarize developments in experimental techniques and applications have been written by Miller, et. al. [4], Grimmeis [5], Kimerling [6], Lang [7], and Chen and Milnes [8].

Steady state and 'single-shot' junction capacitance methods have been applied to defect state studies for over fifteen years. The recent introduction of bias pulsing [9] and transient correlation techniques [10,11,12] has created a spectroscopic tool (DLTS-deep level transient spectroscopy) for the observation of the defect states. The method is based on controlling the occupation of a defect state in the depletion region by isolating it from free carriers and subjecting it to external stimuli such as heat or light. The emission or capture process can be accurately monitored as a change in junction capacitance or a junction current. For small defect state concentrations (N_T) the capacitance change is expressed as

$$\Delta C = \frac{N_T}{2N_D} C \qquad (7)$$

Defect state signals are typically observed through the thermal emission process. The temperature dependence of the thermal emission rate yields directly the activation energy for the process. For example, the thermal emission rate for electrons is given by

$$e_n = \sigma_n <v_n> N_c \exp[-E_T/kT] . \qquad (8)$$

The activation energy derived from $d[ln(e_n)]/d[1/T]$ need not be identical to the trap depth E_T (e.g., for a temperature dependent σ_n), and the preexponential factor is typically not as simple function of σ_n (e.g., with electric field or ionization entropy contributions).

Photon stimulated emission processes can be similarly studied by changes in the junction capacitance. Typically, the photocapacitance change is a slowly varying function of the photon energy and fitting parameters are required to determine the threshold energy and optical cross section. However, studies of both optical and thermal emission have proven valuable in revealing properties of defects with multiple states in the gap [5,9]. These measurements have been recently summarized by White [13] and Bois, et. al. [14].

CHARGE COLLECTION MICROSCOPY

Charge collection microscopy is a generalization of junction photocurrent and electron beam-induced current measurements. Recent reviews in this field have been written by Kimerling [6], Miller [22], Varker [23], and Leamy [24]. The techniques rely on the generation of charge by an external excitation [which may be of energy greater than (intrinsic) or less than (extrinsic) the semiconductor band gap] and the collection of this charge by the electric field of the junction structure. Furthermore, scanning of the external excitation over the junction area gives information regarding the spatial distribution of defect states and an image of defect structures at a resolution of, typically, one micron.

The collected current I_c is given by

$$I_c = g\alpha[W + (1/\alpha - W)<\eta>],$$ (9)

where g is the generation flux, α is an effective absorption coefficient, W is the depletion width, and $<\eta>$ is the spatially averaged charge collection efficiency.

Extrinsic excitation is used, primarily, to determine the energy dependence of α by observation of the induced junction current on a near zero background. Broad area, whole diode studies have yielded excellent results [5],but microscopy applications are slow in emerging [15]. Widespread application is expected to flourish with the development of tunable laser sources in the infrared.

Intrinsic excitation probes the variations in $<\eta>$ which relate to the local minority carrier lifetime within the probe volume of the exciting beam. The probe volume is typically a minimum of one micrometer in diameter and is determined by the minimum focal spot diameter for a photon beam or multiple scattering processes in the solid for an electron beam.

Both defect contrast and dopant striation contrast have been obtained with this technique [16]. The relation of defect contrast to Fermi level position [17] and pulsed beam current transients [6,18] have been employed to determine defect state energy positions. The sacrifice in sensitivity required to obtain spatial resolution has slowed developments in the pulsed beam microscopy.

Recent experiments have employed quantitative charge collection measurements to supplement bulk electrical techniques [19,20]. Thin sample junctions have been employed to image both the strain and electrical contrast of dislocations [7]. Such studies demonstrate clearly that a proper investigation of an inhomogeneous material system must combine both a spectroscopy and an appropriate microscopy.

ELECTROLUMINESCENCE

Electroluminescence is a developing practice of the future rather than an accomplished discipline of the present. Typically, measurements on large band gap materials have been reported, reflecting the lack of sensitive photodetectors in the infrared. Recent reports regarding point defects in *GaP* [25] and crystals growth defects in *AlGaAs* laser structures [26] demonstrate the emerging potential of the measurement. Most appealing is the opportunity for direct correlation with transport measurements on the same junction structure (capacitance transient spectroscopy and charge collection microscopy) and optical measurements on similar materials (optical absorption and optically detected magnetic resonance).

The observation of sharp, bound exciton lines with this technique is negated by the occurrence of dopant freeze out at temperatures of 30-40°K leading to a loss of junction properties. The use of MIS structures at 4°K has apparently overcome this difficulty in recent bound exciton studies in silicon [27]. For deep centers, strong lattice coupling can produce broad band luminescence which may be studied, initially, with a 77°K junction measurement and, in more detail, by photoluminescence over a wider range of temperatures.

As the electronics technology moves toward the use of light to carry information rather than electrons, the optical properties of materials will become a necessary and intense subject of investigation.

MATERIALS RESEARCH WITH JUNCTION SPECTROSCOPY

Oxygen in silicon

Oxygen is the dominant impurity in device grade silicon and is typically present in concentrations of 10^{18} cm^{-3}. Oxygen enters the silicon lattice during crystal growth from a melt contained in a quartz crucible or during device fabrication from heat treatment in an oxidizing ambient. The oxygen atom normally occupies a puckered, bond-centered, interstitial position which puts the surrounding lattice under a compressive strain but does *not* produce electrical activity. Upon heat treatment at low temperatures (400-700°C), oxygen present in supersaturated concentrations will cluster and precipitate giving rise to a variety of electrically active structures. The evolution of these defect structures is shown in Figure 3.

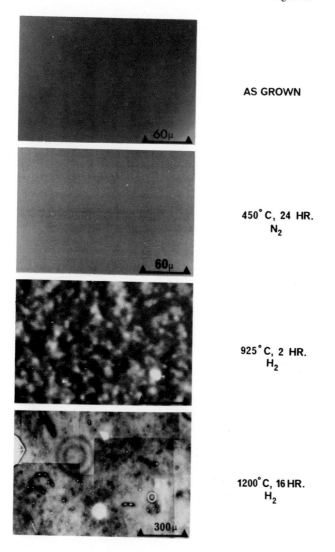

AS GROWN

60μ

450°C, 24 HR.
N$_2$

60μ

925°C, 2 HR.
H$_2$

1200°C, 16 HR.
H$_2$

300μ

Fig. 3. Charge collection micrographs of heat treated, quartz crucible grown silicon. The material is n-type and the measurement structure is an Au−Pd metallized Schottky barrier.

The CCM micrographs show no evidence of macroscopic defect structures in the as-grown silicon or the material which was heat treated at 450°C. CTS spectra taken on the same devices (Fig. 4) show, however, a shallow defect state, $E(0.13)$, which is present after growth and increases in concentration upon annealing at 450°C.

The $E(0.13)$ defect state has been associated with the ubiquitous 'oxygen donor' state and has been employed as a spectroscopic marker to follow the material dependent nature of the donor introduction [29].

Heat treatment of silicon containing oxygen at temperatures above 900°C generates both deep defect states and associated macroscopic defect structures. The cluster-like defects observed by CCM following 925°C anneal are not observable by strain field contrast with X-ray topography or transmission electron microscopy. Following 1200°C heat treatment, a one-to-one correspondence can be shown between the single- and double-fault structures, precipitates, and dislocations observed by CCM and diffraction contrast [16].

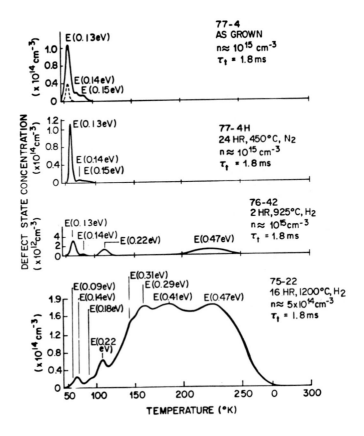

Fig. 4. Capacitance transient spectra taken on the same junctions used in Figure 3.

Junction capacitance measurements have uniquely revealed a strong electric field enhanced emission process for the $E(0.13)$ oxygen donor state (Fig. 5). The magnitude and electric field dependence of the process follows precisely the Poole-Frenkel mechanism [28], indicating that the defect is a simple donor with a size which is small relative to the extent of the point charge potential. Additionally, these studies have demonstrated that: 1) oxygen, carbon, and boron influence the formation of $E(0.13)$; 2) silicon lattice vacancies are not directly involved in the defect structures; and 3) the structure and properties of the donor vary with time and temperature [29].

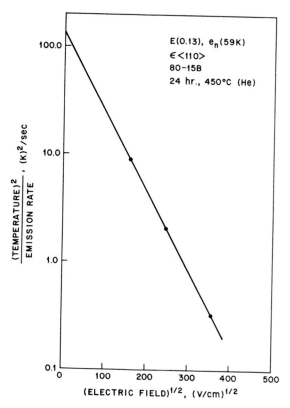

Fig. 5. Electric field enhancement of emission rate from $E(0.13)$ defect state. Both the $(E)^{1/2}$ dependence and the slope fit the Pool-Frenkel mechanism without adjustable parameters [28].

Iron in silicon

The properties of 3d transition metal impurities in silicon have been a subject of study for over twenty years. The electron paramagnetic resonance studies of Ludwig and Woodbury [30] observed the tetrahedral interstitial site for V, Cr, Mn, Fe, Ni following diffusion at 1250°C and quench. Gerson, et al [31] have correlated a CTS defect state, $E(0.45)$ with the EPR signal of (Fe_i). Feichtinger [32] and, recently, Graff and Pieper [33] and Kimerling et al. [34] have observed the reversible decay of $E(0.39)$ with the coordinated production of a shallow state $H(0.1)$ which was attributed to the (Fe_i-B_s) ion pairs. A systematic CTS study of defect states of transition metal elements in silicon has revealed a family of ion pair states for the fast diffusing Mn, Fe, Ni, and Cu impurities [34].

We have directly monitored the ion pairing equilibria and kinetics by CTS for the case of $(Fe_i)^+$ and $(B_s)^-$. Both phenomena display thermal activation and *electronic stimulation*. The association kinetics yield directly the diffusion constant of (Fe_i). Over a limited temperature range near room temperature (300°K) the data yield $D(Fe_i) = 4 \exp[-0.9/kT]$. This expression is to be compared with the high temperature data (1400°K) of Struthers [35] which give $D(Fe) = 6.2 \times 10^{-3} \exp[-0.87/kT]$. The equality of the activation energy suggests that the same species and diffusion path are being studied. The difference in the preexponential factor is indicative of the dissociative diffusion mechanism. The CTS, ion pairing studies measure diffusion of (Fe_i) directly. The lower preexponential factor of the high temperature data suggests that only 10^{-3} of the total Fe concentration at high temperatures is mobile, (Fe_i). This direct correlation is in general agreement with Fe solubility and diffusion studies at intermediate temperatures [36].

The dissociation of (Fe_i-B_s) was observed to be stimulated by white light illumination at 300°K [33]. We have observed both thermal and minority carrier injection stimulated dissociation by CTS in n^+p junctions containing Fe. The variation of the paired fraction of (Fe_i) as a function of temperature confirms the electrostatic pair binding energy of ~0.45 eV. Below 300°K, where the pairs are thermally stable in our samples, near complete dissociation could be effected by forward bias injection. The dissociation rate increases linearly with injection current (Fig. 6) and is weakly temperature dependent with an activation energy of ~ 0.1 eV. The electronically stimulated reaction is effectively quenched at 77°K. A detailed characterization of this system is to be published.

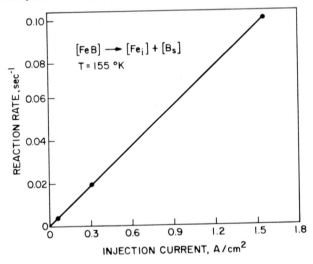

Fig. 6. Dissociation kinetics of (Fe_i-B_s) pairs, $H(0.1)$, stimulated by minority carrier injection.

IMPATT diode reliability

A good example of the application of junction spectroscopy to practical, semiconductor device problems is a recent study of the degradation of $GaAs$ IMPATT diodes [37]. The device is a $Pt{:}GaAs$ Schottky barrier structure which operates by the conductivity modulation established by an avalanche breakdown process. The breakdown voltage is determined by the free carrier concentration in the *GaAs*.

An interfacial reaction between the *Pt* metallization and the *GaAs* occurs at temperatures above 300°C, as shown schematically in Fig. 7, and produces a metallic *Ga* layer on the surface and an excess arsenic condition in the *GaAs* under the *Pt* layer. Accelerated aging tests at high temperatures demonstrated that this reaction would not critically limit device performance when a sufficiently thick *n—GaAs* layer was employed.

When the devices failed under operation at times much shorter than predicted, junction spectroscopy techniques were employed as nondestructive characterization tools. These studies demonstrated clearly that a second degradation mechanism was operative during device operation which limited the device life. Junction capacitance measurements detected compensation of the free carrier concentration by defects which in-diffuse from the interface. CTS measurements determined the electronic properties of these defects and related their accelerated diffusion to an electronically enhanced process which is stimulated under avalanche conditions. These microscopic kinetics were in excellent agreement with device lifetime studies when aging was performed on *operating devices*. The degradation path was quenched when a thin, *Ti* diffusion barrier was placed between the *Pt* and the *GaAs*, and the interfacial reaction and the defect generation process were retarded.

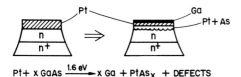

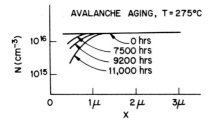

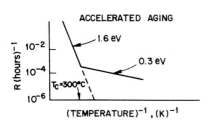

Fig. 7. Schematic presentation of the failure mode of Pt:GaAs IMPATT diodes: interface reaction, free carrier compensation by in-diffusing defects, and low activation (enhanced) failure kinetics under operation.

FUTURE DIRECTIONS

Future developments in junction spectroscopy will probe, primarily, the optical properties of imperfections. This direction evolves for the same reasons as have present techniques. One, *new information* regarding the microscopic atomic and electronic structure of imperfections can be accessed best through the *adiabatic* nature of the optical transition. Two, the

94

optical properties of semiconductor materials are of increasing practical importance as the opto-electronics field matures. Thus, the synergy of the fundamental probe and the device structure will continue to flourish.

Defect structure can be probed by detection of stress (or field) induced dichroism of photoexcitation or luminescent processes. A preliminary experiment of the type was reported by Cheng and Kimerling [4] which indicated a $<111>$ symmetry for the $H(0.21)$ state in electron bombarded silicon-consistent with its assignment to the divacancy defect [38].

The role of *metastable and excited states* in the electrical and physical properties of imperfections is at the initial stages of investigation. Again, photoexcitation must be used to establish a partial occupation of the excited state under study. 'Negative U' systems, in particular, can be studied most conveniently through combined photo and thermal excitation [39].

Quantitative charge collection microscopy promises to revolutionize high resolution transport property measurements. A continuous, spatial representation of conductivity, lifetime, mobility, and luminescence efficiency can precisely evaluate crystal growth and materials processing procedures. *Tuneable, monochromatic photon beams* in the infrared can sensitively detect the presence of specific defect structures.

In summary, junction spectroscopy is developing into a complete set of techniques for the study of imperfections in semiconductor materials. Important results have been obtained in the three primary areas of practical and fundamental concern: *formation* (equilibrium and nonequilibrium), *structure* (atomic and electronic), and *properties* (electrical, physical and chemical). Specifically, these techniques have opened the door to a fundamental inquiry regarding the nature of thermally activated and electronically stimulated processes in covalent crystals.

ACKNOWLEDGEMENT

The contributions of J. L. Benton to all aspects of this research are gratefully acknowledged.

REFERENCES

1. S. M. Sze, *Physics of Semiconductor Devices* (Wiley-Interscience, New York 1969).

2. J. L. Moll, *Physics of Semiconductors* (McGraw Hill, New York 1964).

3. C. T. Sah, L. Forbes, L. L. Rosier, and A. F. Tasch, Jr., Solid-St. Electron. **13**, 759 (1970).

4. G. L. Miller, D. V. Lang, and L. C. Kimerling, Ann. Rev. Mater. Sci. **7**, 377 (1977).

5. H. G. Grimmeiss, Ann. Rev. Mater. Sci. **7**, 341 (1977).

6. L. C. Kimerling in: *Physics of Semiconductors 1978* (Inst. Phys. Conf. Ser. No. 43, 1979) pp. 113-122.

7. D. V. Lang in: *Topics in Applied Physics, Volume 37* (Springer-Verlag, Berlin 1979) pp. 93-133.

8. J. W. Chen and A. G. Milnes, Ann. Rev. Mater. Sci. **10**, 157 (1980).

9. C. H. Henry, H. Kukimoto, G. L. Miller, and F. R. Merritt, Phys. Rev. **B7**, 2486 (1973).

10. D. V. Lang, J. Appl. Phys. **45**, 3023 (1974).

11. G. L. Miller, J. V. Ramirez, and D. A. H. Robinson, J. Appl. Phys. **46**, 2638 (1975).

12. L. C. Kimerling, IEEE Trans. Nucl. Sci. **NS-23**, 1497 (1976).

13. A. M. White in: *Physics of Semiconductors 1978* (Inst. Phys. Conf. Ser. No. 43, 1979) pp. 123-132.

14. D. Bois, A. Chantre, G. Vincent, and A. Nouailhat in: *Physics of Semiconductors 1978* (Inst. Phys. Conf. Ser. No. 43, 1979) pp. 295-298.

15. D. V. Lang and C. H. Henry, Solid St. Electronics **21**, 1519 (1978).

16. L. C. Kimerling, H. J. Leamy, J. L. Benton, S. D. Ferris, P. E. Freeland, and J. J. Rubin in: *Semiconductor Silicon/1977* (Electrochem. Soc., Princeton 1977) pp. 468-480.

17. L. C. Kimerling, H. J. Leamy, and J. R. Patel, Apply Phys. Lett. **30** 217 (1977).

18. P. M. Petroff and D. V. Lang, Appl. Phys. Lett. **31**, 60 (1977).

19. J. Y. Chi and H. C. Gatos, J. Appl. Phys. **50**, 3433 (1978).

20. C. Donolato, Appl. Phys. Lett. **34**, 80 (1979).

21. P. M. Petroff, R. A. Logan, and A. Savage, Phys. Rev. Lett. **44** 287 (1980).

22. G. L. Miller, D. A. H. Robinson, and S. D. Ferris in: *Semiconductor Characterization Methods*, P. A. Barnes and G. A. Roygonyi eds. (Electrochem. Soc., Princeton 1978) pp. 1-31.

23. C. J. Varker in: *Nondestructive Evaluation of Semiconductor Materials and Devices*, J. N. Zemel ed. (Plenum, New York 1979).

24. H. J. Leamy in: *Physical Electron Microscopy*, O. C. Wells, K. F. J. Heinrich, and D. E. Newberry eds. (Van Nostrand Reinhold, New York 1981).

25. C. E. Barnes, J. Electronic Mater. **7** 589 (1978).

26. S. Metz, Appl. Phys. Lett. **33**, 198 (1978).

27. J. Weber, W. Schmid, R. Sauer, and M. H. Pilkuhn, 22nd Electronic Materials Conference, AIME, Cornell, 1980.

28. L. C. Kimerling and J. L. Benton, to be published.

29. L. C. Kimerling and J. L. Benton in: *Extended Abstract of the 158th Electrochemical Society General Meeting*, Holywood, Florida (Electrochem. Soc., Princeton 1980) pp. 415-417.

30. G. W. Ludwig and H. H. Woodbury in: *Solid State Physics*, **13** (Academic Press, New York 1962) p. 223.

31. J. D. Gerson, L. J. Cheng, and J. W. Corbett, J. Appl. Phys. **48** 4821 (1977).

32. H. Feichtinger in: *Defects and Radiation Effects in Semiconductor 1978* (Inst. Phys. Conf. Ser. 46, London 1979) p. 528.

33. K. Graff and H. Pieper, to be published.

34. L. C. Kimerling, J. L. Benton and J. J. Rubin in: *Defects in Semiconductors 1980* (Institute of Physics, London 1981).

35. J. D. Struthers, J. Appl. Phys. **27**, 1560 (1956).

36. C. B. Collins and R. O. Carlson, Phys. Rev. **108**, 1409 (1957).

37. W. C. Ballamy and L. C. Kimerling, IEEE Trans. Electron Dev. **ED-25**, 746 (1978).

38. L. C. Kimerling in: *Defects and Radiation Effects in Semiconductors 1976* (Inst. Phys. Conf. Ser. No. 31, London 1977) pp. 221-230.

39. G. D. Watkins, this volume.

Published 1981 by North-Holland, Inc.
Narayan, and Tan, eds.
Defects in Semiconductors

ION CHANNELING TECHNIQUES FOR DEFECT AND SURFACE STUDIES*

B. R. APPLETON
Solid State Division, Oak Ridge National Laboratory, Oak Ridge, Tennessee 37830

ABSTRACT

The channeling and blocking phenomena are well enough
understood that they can be combined with ion scattering or
ion induced reactions to investigate defects and surface
structures. A brief review is given of the origins of these
phenomena, of associated concepts which are useful for
defect and surface analysis, and of experimental constraints
and requirements. Applications to selected problems are
used to illustrate the advantages and limitations of these
techniques.

INTRODUCTION

This review is one of three companion papers designed to provide an introduction to the use of channeling for defect studies. The primary emphasis of this
paper is on the origins of ion channeling and the associated concepts which make
it useful for defect analysis and surface studies. Particular emphasis is given
to surface and interface analysis. The companion reviews by Chu [1] and Picraux
[2] concentrate more on specific applications. The interested researcher should
be able to assess from these papers whether channeling techniques are suitable
for his particular problems, and find ready reference to the publications
necessary to effect their application. No attempt was made in this paper to
provide a complete or historical accounting of the development of the ideas or
the first applications of channeling. Consequently, the examples used and
references given were chosen for emphasis and do not necessarily represent a
balanced view. There are a number of excellent reviews where such information
can be found [3-7].

ORIGINS OF THE CHANNELING EFFECT

The circumstances responsible for the channeling of energetic ions in single
crystals are represented schematically in Fig. 1 and are considered in numerous
books and articles [1-5]. When ions impinge on a disordered solid or a misaligned single crystal, they encounter the atoms at random and sample a random
distribution of impact parameters. This assumption underlies most descriptions
of ion-solid interaction phenomena. However, if the bombarded sample is a
single crystal aligned so that the ions are incident parallel to an axial or
planar "channel," such as those shown in Fig. 1, then the ions experience successive, correlated collisions with the symmetrically arranged atoms (center
schematic Fig. 1) and acquire oscillatory trajectories in the open regions, or
channels, between the rows or planes of atoms. The minimum distance of approach
of a channeled ion to atoms on normal lattice sites is $\rho_{min} \sim 10^{-9}$ cm. Since many
ion-solid interactions such as Rutherford scattering require much closer impacts

*Research sponsored by the Division of Materials Sciences, U.S. Department of
Energy under contract W-7405-eng-26 with Union Carbide Corporation.

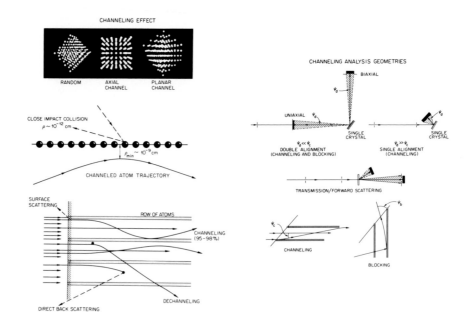

Fig. 1. Origins of the channeling effects.

Fig. 2. Schematic of several experimental geometries useful in channeling applications.

$\rho \sim 10^{-12}$ cm, the reduction in the yield of these phenomena become sensitive monitors of crystal perfection.

Several consequences of channeling such as those shown schematically in the bottom portion of Fig. 1 can be exploited for defect studies in single crystals. When a beam of ions is incident parallel to an axial direction in a single crystal ~97% of the ions impact at a distance $> \rho_{min}$ from the atoms which terminate the rows of atoms at the surface. These ions acquire oscillatory trajectories within the channels whose amplitudes of oscillation are determined by their initial impact parameters (channeling) [4]. The fraction of ions which impact within ρ_{min} (~3%) will either interact directly with the surface atoms (surface scattering) or be deflected through too large an angle to be confined to a channel and assume random trajectories with normal interaction probabilities. This interaction fraction of the beam which impacts within ρ_{min} is the least yield reduction you can hope to achieve for a perfect crystal and is called the minimum yield. In its simplest form it can be expressed

$$\chi_{min} = Nd \ \pi \rho_{min}^2 \qquad (1)$$

where N is the number of atoms/cm^3 of the target, and d is the spacing in cm between atoms along the axial row of atoms. More accurate expressions for χ_{min} are given several places [3,5,26]. Those particles which were initially channeled become sensitive to disorder in the bulk of the crystal either by <u>direct backscattering</u> from displaced atoms, such as depicted in Fig. 1, or they can be <u>dechanneled</u>, by thermally vibrating or displaced atoms, acquire a random trajectory, and interact with the lattice atoms as a random ion. The problem reduces to devising experiments sensitive to these channeling phenomena and interpreting the results properly.

EXPERIMENTAL ARRANGEMENTS

The experimental apparatus required for doing ion channeling is detailed elsewhere [4,6,7] and will not be duplicated here. Numerous experimental geometries have been utilized in channeling studies and various phenomena (ion scattering, nuclear reactions, x-rays, energy loss, etc.) have been exploited as sensors. Four frequent arrangements are shown schematically in Fig. 2 and several of these will be illustrated using measurements of ion scattering or energy loss.

Single Alignment (Channeling)

The most common experimental set up utilized in damage analyses is the single alignment geometry in Fig. 2 in which the acceptance angle of the detector ψ_d is much greater than the critical angle for channeling ψ_c. An example of its use in recent studies to determine the damage and annealing behavior of super-saturated substitutional alloys fabricated by ion implantation doping and pulsed laser annealing is illustrated in Fig. 3 [8]. These spectra show the yields of 2.5 MeV He ions scattering through an angle of approximately 140° from a Si single crystal before (virgin) and after implantation with As, and annealing with a single 15 x 10^{-9} s pulse from a Q-switched ruby laser. The scattered ion distribution from the implanted As is separated in energy from that from Si because of the mass specificity of the Rutherford scattering process. Because Rutherford scattering is quantitative, and since the energy loss processes of He ions is well understood, it is possible to convert the scattering ion distributions in Fig. 3 into quantitative depth concentrations for both the implanted As distribution and the Si host as shown [7,8]. The upper Si spectra in Fig. 3 illustrates the alteration in the Rutherford scattering yield for ions incident on a Si single crystal continously rotated to simulate an amorphous solid (light dashed curve), and the same number of ions incident parallel to the <110> axial direction of a virgin single crystal (dark dashed curve). The overall scattering yield is reduced to ~2.3% as a result of the channeling effect. The pronounced peak at the threshold for scattering from Si at ~1.5 MeV is due to ions scattered from Si surface atoms and will be discussed in detail later. The aligned Si channeling spectrum taken after As implantation shows a scattering yield within ~0.2 μm of the surface which is the same as the random yield. This shows that as a result of As ion implantation damage the near surface region of the single crystal has been turned amorphous. In addition to contributing this direct backscattering yield, the amorphized surface layer also broadens the initially parallel incident ion beam by multiple scattering causing dechanneling in the perfect single crystal behind the surface; this explains the increased yield in Fig. 3 at depths beyond the amorphous layer. The As distribution in the upper portion of Fig. 3 was measured to be the same for the rotating single crystal (random) and aligned measurements confirming that the As is distributed in a structureless region. The lower portion of Fig. 3 shows that as a consequence of pulsed laser annealing: (1) The Si channeling spectrum became nearly identical to the virgin single crystal measurement indicating that the amorphous surface region recrystallized with a high degree of perfection. (2) The As became distributed in depth and lattice site with a spectrum resembling the channeling spectrum of the Si host.

These results suggest the versatility of ion channeling as a nondestructive tool for studying the radiation damage and annealing behavior of semiconductors. In the remainder of this paper and in the companion papers in these proceedings the strengths, weaknesses, and possibilities of this technique will be developed.

The major advantages of the single alignment geometry are that it is simple, compact, easily interpreted and requires very small beam currents (~ 0.1-10×10^{-9} amps).

Forward Scattering

The forward scattering geometry and forward scattering through thin samples has been utilized to advantage in a number of applications [9-16]. Going to forward scattering angles reduced the beam current needed for analysis because of the angular dependence of the Rutherford scattering effect. With the proper detection geometry, knock-on forward recoil scattering can be exploited to depth profile light impurities in heavy hosts [12]. When combined with ion channeling, forward ion scattering through thin single crystals is capable of providing unique insights into defect structures [10-12,14], and interface structures such as SiO_2 on Si [14-16].

It is primarily because of the potential of alternate geometries such as this and the many useful non-Rutherford ion scattering resonances [17] that I prefer the term ion scattering to the more commonly used Rutherford backscattering spectrometry (RBS) to describe the use of ion beams for analysis.

Transmission

Channeling studies of ions transmitted through thin single crystals have provided, and continue to provide, pertinent insights and understandings of the phenomenon which are essential for its use as a damage analysis tool [9-11,18-20]. Since the energy loss rates of channeled ions can be directly related to their trajectories in the channel, measurements of the energy loss and angular distributions of transmitted ions have provided information on the dependencies of stopping powers, impact parameters and potentials for channeled ions [18-20]; the critical half thicknesses and trajectory dependence of dechanneling [9]; the appropriate stopping powers to assume for well-channeled and dechanneled ions [9,11]; and a more detailed understanding of the flux distributions, multiple scattering phenomena, and statistical equilibrium of channeled ions [4]. Transmission measurements are, of course, also directly sensitive to damage [11,18-20].

Double Alignment (Channeling and Blocking)

The double alignment geometry illustrated in Fig. 2 utilizes both channeling and blocking [21]. Blocking is closely related to channeling through the principle of reversibility [3-5]. As shown schematically in Fig. 2, an incident ion which scatters from an atom on a normal lattice site is blocked from emerging along the crystal row or plane which includes this atom by the same correlated collisions responsible for channeling. As illustrated in Fig. 2, in the double alignment geometry an analysis detector with an acceptance angle ψ_d much smaller than the critical angle for blocking ψ_b is aligned with a major axis or plane. Consequently, when the ion beam is incident along a channeling direction only the minimum yield fraction $\sim X_{min}$ interacts with atoms on normal lattice sites because of the channeling effect, and only $\sim X_{min}$ of this fraction (or X_{min}^2) reaches the double alignment detector because of the blocking effect. Thus the "background level" for scattering from a perfect lattice is reduced from $X_{min} \sim 2 \times 10^{-2}$ to $X_{min} \sim 10^{-3}$ by going from the single to double alignment geometry [23], and this greatly improves the sensitive for detecting displaced atoms in defect studies and in lattice location investigations. The double alignment geometry also provides additional sensitivities for surface studies and this will be discussed in a later section.

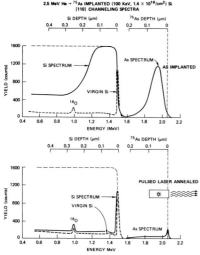

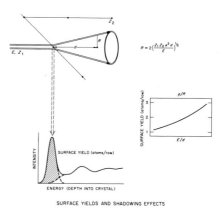

Fig. 3. Ion scattering/channeling analysis of (110) Si implanted with As and pulsed laser annealing.

Fig. 4. Origins of the surface peak in channeling.

CHANNELING CONCEPTS USEFUL IN DAMAGE ANALYSIS

Surface Scattering

Ion channeling provides a special sensitivity to both the quantity and geometrical positions of atoms located at the surfaces of single crystals. In this subsection I will outline the origin of this sensitivity and in the next section, Investigations of Surfaces and Interfaces, give several examples of how to utilize this feature for surface and interface studies. This topic will be a major emphasis in this paper.

The factors which contribute to the surface peak such as the ones observed for the channeling spectra in Fig. 3 are shown schematically in Fig. 4. When a beam of ions is incident parallel to an axial direction as in Fig. 4, the atoms on the surface have a normal scattering probability. The subsurface atoms on the other hand are shadowed by the surface atoms and have a greatly reduced scattering probability. A simple two atom model which assumes a static lattice and Coulomb scattering [14] shows that as a result of this shadowing effect the distance of closest approach to the second atom in the string is given by the simple expression $R = 2(Z_1 Z_2 e^2 d/E)^{1/2}$ in Fig. 4. The backscattering/channeling spectrum in the lower portion of Fig. 4 was obtained for 1.0 MeV He ions scattered from the $\langle 110 \rangle$ axial direction of a Au(100) single crystal [24,25]. In this case $R = 0.164$ Å, and since the thermal vibration amplitude for Au at room temperature is $\rho \lesssim 0.1$ Å it is clear that substantial shadowing of the second atoms by the surface atoms still occurs even in a vibrating lattice. The area under the surface peak in Fig. 4 (shaded area) is a direct measure of the number of atoms exposed to the incident beam in this case and was measured to be ~ 2.2 atoms/row [24,25]. The expected variation of the surface peak yield with incident ion energy, E, lattice spacing, d, and thermal vibration amplitude is shown schematically on the right side of Fig. 4. The simple two atom model does not provide a sufficiently accurate description of the variation of the surface peak yield [24-26,52]. This will be discussed in the next section along with applications to defect studies.

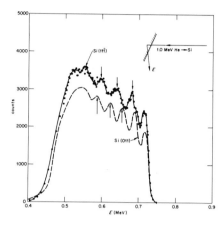

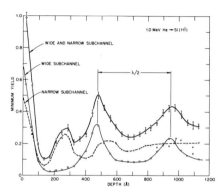

Fig. 5. Planar yield oscillations observed by 1.0 MeV He ion scattering from the (110) and (11$\bar{1}$) planar channels of a 0.25 μm thick Si single crystal.

Fig. 6. Computer simulation calculations of the yield oscillations versus depth for Si(111).

Yield Oscillations

The variations in scattering yield directly behind the surface peak in Fig. 4 are referred to as yield oscillations [26-30]. They occur for ions directed along both axial and planar channels and originate from the variation in ion flux with depth (and angle) induced by the channeling effect. The ions in the incident beam are initially parallel to a channeling direction, and each ion acquires a transverse energy dictated by where it enters the channel in relation to the string or planes of atoms. Those which impact within ~ ρ_{min} of the atoms either scatter from the surface atoms or receive too large a transverse energy to remain channeled as we have discussed. Those ions which impact beyond ρ_{min} of the atoms form the shadow cone discussed in relation to Fig. 4 and acquire oscillatory trajectories within the channel as discussed with Fig. 1. These discrete oscillatory trajectories have wavelengths and amplitudes within the channel determined by their tranverse energies, and it is the distribution of these trajectories which determines the ion flux within the channel.

The yield oscillations are most clearly illustrated by considering planar channeling [29]. Those ions with large amplitudes of oscillation within the planar channel approach closely to the planes of atoms every $\lambda/2$ where λ is the wavelength of oscillations in the channel, and these ions have an increased probability for scattering at that depth. The consequences of this are that the ions scatter selectively at certain depths as is evident from Fig. 5 which shows the yield of 1.0 MeV He ions scattered from a 0.25 μm thick Si single crystal for incidence parallel to the (011) and (11$\bar{1}$) planar channels respectively.

With decreasing scattered ion energy (increasing depth) the discrete yield enhancements observed in Fig. 5 arise from increased scattering of the ions as their trajectories approach the turning point at the planar walls. Although such oscillations have not been utilized extensively for defect studies to date, they can be understood in sufficient detail for such applications. To illustrate this let us pursue the results in Fig. 5 further. Since the scattered energy in Fig. 5 can be converted to depth if the channeled stopping powers are known, the peak positions represent direct measurements of the wavelengths of the channeled trajectories. The converse is also true. In Fig. 6 the yield vs. depth has been calculated specifically for the Si(11$\bar{1}$) case using computer simulation methods [28]. The (111) planar channel is unusual in that it is composed of a set of widely spaced atomic planes (2.35 Å separation) bordered by a narrow set (0.78 Å separation), and as a result there will be superimposed yield oscillations from both the wide and narrow subchannels. The calculations in Fig. 6 show the results expected in this case and the depths where yield maxima can be expected. Comparison to the measurements in Fig. 5 shows that all the yield maxima expected are observed including the small maximum after the surface peak which is due to the narrow subchannel oscillation. The arrows in Fig. 5 represent the depths where the computer simulations predicted identifiable maxima at the $\lambda/2$ trajectory turning points. The (11$\bar{1}$) and (011) results (the (011) is a normal planar channel) show good agreement with calculations. The stopping powers utilized in the depth conversions were determined from correlated transmission measurements [29,30].

These results show that the planar yield oscillations can be characterized with considerable precision and a similarly complete understanding of axial yield oscillations is possible through computer simulations [27]. These oscillations provided enhanced sensitivities to specific (and predictable) lattice sites, and as such they can be used as selective probes of defect configurations and impurity positions. Yield oscillations provide information concerning potentials and stopping powers necessary in defect analysis [29]. These oscillations are also partially responsible for the special sensitivity to displaced surface atoms discussed in Surface Scattering because an incident ion which fails to scatter from the surface atoms must traverse the open channel until it encounters another scattering site and this normally extends hundreds of angstroms into the crystal. Since the detectors used have finite depth resolutions (due to finite energy resolutions) this fortuitous characteristic helps separate the surface peak from the subsurface yields. It will also become evident after subsection Lattice Location that these yield oscillations are precursors of the flux peaking effect which is so useful in lattice location studies and that they may, in fact, be a new complement to lattice site determinations.

Direct Backscattering and Dechanneling

Channeling spectra are most easily interpreted when the direct backscattered ions can be easily distinguished from those which become dechanneled and then backscatter. An example of such a case is shown in Fig. 7. These results were taken from recent studies of supersaturated substitutional alloys of silicon fabricated by ion implantation doping and pulsed laser annealing [8]. The [110] channeling spectrum (solid circles) shows that as a consequence of the radiation damage resulting from As implantation the first ~1500 Å of the single crystal has a scattering yield which has increased from the virgin yield (closed squares) to a value equivalent to the random scattering yield (open circles). Microstructural analysis shows this region has been rendered amorphous. The scattering yield behind this amorphous region has increased due to dechanneling of the incident beam. The origin of this dechanneling is the increased angular divergence of the ion beam, due to multiple scattering in the amorphous overlayer, before entering the crystalline substrate. In this situation the scattering can be unambiguously separated into its direct and dechanneled (or random) components and it is easy and straightforward to extract a quantitative damage distribution as a function of depth for the damaged surface. This and analogous cases have been treated in several excellent review articles [1-7,10,31-35].

Once the nature of the damage becomes structured, interpretation of the ion scattering/channeling data becomes very model dependent. The fraction of the ions which are backscattered directly and the fraction which becomes dechanneled depends on the nature of the defects and their interaction with the channeled

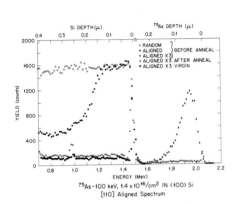

Fig. 7. Ion scattering/channeling analysis of ion implantation damage in a Si single crystal.

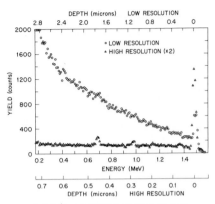

Fig. 8. Utilization of depth resolution for background detection.

ions. The status of channeling as an analysis tool in these situations is reviewed by Chu [1] and Picraux [2] in the companion review papers to this one in these proceedings.

Background Suppression

Another attribute of the channeling effect that is often utilized but sometimes taken for granted is the general reduction in "background" scattering from the host which can be exploited when one wishes to study surface disorder or light impurities. Some of these features are illustrated by the results in Fig. 8 which contrast the energy spectra for 2.5 MeV He ions scattered 144° from a virgin Si crystal, but with the analysis detector located in a grazing angle geometry to improve the depth resolution (high resolution case) [36]. The interest in this case was to correlate the presence of oxygen and carbon with the surface disorder to determine if only an air formed oxide could account for the observed surface peak yield. The peaks corresponding to ion scattering from oxygen (Energy ~ 0.98 MeV) and carbon (Energy ~ 0.68 MeV) are clearly visible in the high resolution case because the overall background scattering from silicon has been reduced by the channeling effect, and enhanced by the grazing angle geometry [37]. Neither peak could be analyzed for ion scattering from an amorphous or randomly oriented sample. In the upcoming section on the use of channeling for the analysis of surfaces and interfaces it will become evident that this consequence results in a powerful analysis tool.

Lattice Location

Among the most significant advantages to ion channeling is its ability to determine directly the lattice positions of impurities and defect structures. Impressive examples of this are given in the accompanying reviews by Chu [1] and Picraux [2] and in several previous reviews [3,4,6,7]. Particularly noteworthy are applications of lattice location techniques to study the interactions between irradiation produced defects and impurities [38-44]. These studies have provided direct understanding of a variety of radiation damage mechanisms, defect configurations, annealing mechanisms, cluster formation, etc. [1,2,38-44]. In all such applications the position of the defect or impurity atom is deduced

105

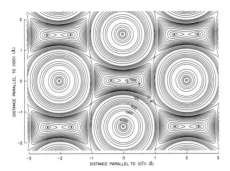

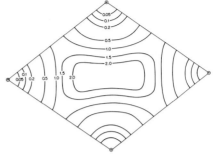

Fig. 9. Calculated continuum
potential energy contours for the
⟨110⟩ axial direction of an fcc
lattice.

Fig. 10. Equiflux Contours Relative to
a uniform distribution for 1 MeV He in
Al[011] at 40 K, averaged over the depth
range 50-150 nm.

by interpreting the interaction of the channeled component of the ion beam with
the displaced atom. The probability that the channeled beam will interact with
the displaced atom depends on: (a) the magnitude of the displacement (i.e., it
must be $\gtrsim \rho_{min}$) and the crystallographic orientation of the displacement relative to the incident beam; (b) the flux distribution of the channeled beam at
the position of the displaced atom; and (c) the interaction cross section.

Flux Peaking and Statistical Equilibrium. Since channeled particles are
excluded from the area $\pi \rho_{min}^2$ around a string, the flux of particles per unit
area in the channel, and thus the probability of interacting with an interstitial
atom, is higher than for a randomly directed beam. This is the basis of the
flux peaking effect which has proved so useful in atom location studies [45,46].
The origin of this effect can be understood by considering how the spatial
distribution of a uniform, parallel beam of ions is altered upon entering a
single crystal parallel to an axial direction. Figure 9 shows a set of calculated potential energy contours constructed from the continuum potentials for
the rows of atoms (R) of the [011] channel of an fcc single crystal [20]. When
an ion enters the crystal into this channel it acquires a transverse
energy of motion $E_+ = \phi(r)$ where $\phi(r)$ is the potential energy of the contour at
the position r where the ion enters the channel. Thus, ions entering on the
contour labeled 50 would acquire that transverse energy and their trajectories
(neglecting multiple scatterings, energy loss, etc.) would be confined within
the area enclosed by that contour. After penetrating a sufficient depth into
the crystal, the probability of finding a given ion at a given position is independent of the position within the area accessible to it and zero outside that
area; this condition is referred to as statistical equilibrium [5]. Many analytical calculations of the flux profiles of ions within the channel assume that
statistical equilibrium is valid. Such calculations can be used to interpret
channeling measurements provided they are applied at sufficiently large depths
or are suitably averaged over large enough depth intervals to approximate the
conditions of statistical equilibrium. As we have seen in the case of planar
channeling (Fig. 5 and 6) the ion flux initally exhibits pronounced yield
oscillations with depth in the channel, and the same is true for axial
channeling. (Note, for example, the yield oscillations behind the surface peak
in Fig. 4.) Thus, care must be taken to avoid these effects, or as will be
suggested later, to exploit them purposely as more sensitive probes of defect
sites. If they are to be utilized, however, more accurate flux profiles must be
calculated. This can be accomplished from computer simulation calculations
which follow and record individual ion trajectories [26,43,44]. An example of

the ion flux distributions averaged over a depth of 50-150 nm is given in Fig.
10 for 1 MeV He ions incident parallel to [011] of Al [47]. This figure
illustrates how crucial an accurate flux distribution is in interpreting the
position and concentration of displaced impurities.

Triangulation. The flux distribution in the channel also varies with ion
energy and the angle of incidence of the ions relative to the channeling
direction. These variations provide additional sensitivity to the type and
location of lattice impurities [1-7, 38-46]. Another simple concept which helps
identify defect sites is triangulation. A comparison of the interaction yields
for the host lattice relative to impurity measured along different intersecting
crystallographic directions can be utilized to pinpoint defect sites with great
accuracy. Excellent examples of this are given in the comparison papers by Chu
[1] and Picraux [2] and the references therein.

INVESTIGATIONS OF SURFACES AND INTERFACES

As indicated in the preliminary discussion in the preceding section, channeling provides a special sensitivity to the order (disorder) of single crystal
surfaces. In a perfect single crystal the shadowing effect and yield oscillations insure that the area under the surface peak is directly proportional to
the number of near-surface atoms exposed to the incident beam. Similarly, if
surface disorder increases the surface peak yield increases. The yield of scattered ions results from direct backscattering events before any dechanneling can
occur and is thus easily interpreted. The model descriptions necessary for analysis of the results are based on simple, well-established, classical ion-atom
potentials. In this section a number of channeling concepts will be cited to
illustrate the usefulness of this effect for studying surfaces and interface
structures.

Surface Peak Yields: Reconstructed Au(001)
Use of the surface peak yield as a surface probe can be illustrated by
investigations of the Au (001) surface [24,25]. The results of this investigation are summarized in Table 1. This was a coordinated study comparing low
energy electron diffraction (LEED), positive ion crystallography of surfaces
(PICS), and model computer simulation calculations to determine the nature of
the Au (001) single crystal surface. The "normal" Au(001) surface (Table I)
could be prepared by annealing in air to 600°C. This resulted in the retention
of approximately one monolayer of impurities which stabilized the surface so
that it had the same (1x1) termination as the bulk lattice. Furthermore, PICS
measurements of the surface peak yield, L, and minimum yield behind the surface
peak, X_{min}, were in excellent agreement with the model simulations for a normal
surface. As will be shown in the subsection, Angular Dependence of the Surface
Peak Yield: Au(001) and Au (011) the angular variation of the surface peak
yield was also in good agreement with expectations for a normal, unreconstructed
surface. When this surface was sputter cleaned and annealed in ultra high
vacuum, a reordered (001) surface resulted with a complex (5x20) LEED pattern and
very different surface peak yields as shown in Table I. Combined analyses
showed [24,25] that this reordered surface consisted of a single hexagonal (111)
layer contracted 5% and superimposed on (001) bulk structure. The increased
surface peak yields in Table I, when compared with the model calculations, were
a direct verification of this.

Energy Dependence of Shadow Cone: GaAs (111)
Since the radius of the shadow cone created by the surface atoms increases
with decreasing energy (see Fig. 4), it is possible to utilize ion beams of
various energies for structural determinations of surfaces. One possibility for
studying relaxed surfaces by this method is represented schematically in Fig. 11.

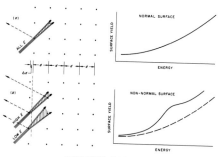

ENERGY EFFECTS: RELAXED SURFACES

Fig. 11. Schematic representation of the use of energy dependence of the shadow cone for studies of relaxed surfaces.

Part (a) shows that the surface peak yield for a normal surface can be expected to increase with increasing energy ; this is because the shadow cone radius decreases and allows the incident ions to come closer to the thermally vibrating subsurface atoms. Part (b) hypothesizes a surface whose outermost layer is relaxed outward an amount Δd compared to the bulk spacing. In this case, some discontinuity can be expected as the energy is increased, depending on the nature and magnitude of the relaxation, and comparison to model calculations can

TABLE I
PICS Investigations: Au(001) Surface

	PICS Measurements			Model Simulations				Comments
	$\langle 100 \rangle_{0°}$ χ_{min}	L	$\langle 110 \rangle_{45°}$ χ_{min}	L	$\langle 100 \rangle_{0°}$ χ_{min}	L	$\langle 110 \rangle_{45°}$ χ_{min}	L	
Normal (001) Annealed in Air 600°C									Surface peak yield, angular scans for $\langle 100 \rangle$ and $\langle 110 \rangle$ not conclusive
LEED → (1x1)	0.030	1.95	0.021	2.33	0.032	1.89	0.023	2.17	
AES → C,O,S									
Reordered (001) Sputter cleaned 500 eV Ar									Reordered surface
350°C Anneal in UHV	0.036	2.48	0.025	3.58	0.034	2.49	0.025	3.43	1 reordered layer
LEED → (5x20)						3.09		4.69	2 reordered layers
AES → None						3.69		5.95	3 reordered layers

Summary: 1. (001) appears most "normal" surface for Au.
          2. Reordered surface consistent with model in which only the top monolayer reorders into a hexagonal, contracted (5%) (111) layer.

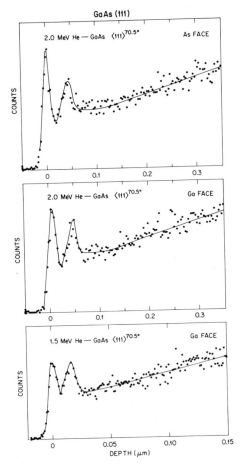

Fig. 12. Ion scattering/channeling analysis of the faces of a GaAs (111) single crystal.

be used to identify this relaxation. This discontinuity occurs at an ion energy where the shadow cone cast by the relaxed surface atoms no longer shields the bulk atoms from the incident ion beam. As the ion energies are reduced from high (MeV) to medium (10^2 KeV) [48] to low (< KeV) [49], the analysis techniques become much more surface specific. There are advantages and complications in all energy ranges and one should tailor the energy range to the application.

An example of the use of the shadow cone effect is given in Fig. 12. Recent measurements at ORNL [50] have shown that pulsed laser annealing of Si, Ge and GaAs single crystals in ultra high vacuums can lead to metastable surface structures as a result of rapid melting and recrystallization. LEED studies on a GaAs (111) single crystal showed that pulsed laser annealing of one crystal face led to a clean (1x1) diffraction pattern which resulted from a quenched, metastable surface structure with the same atomic arrangement as the bulk. (Normally, clean GaAs has a reconstructed surface.) However, pulsed laser annealing of the opposite face of the crystal led to a poorly defined surface. Since the opposite faces of GaAs (111) single crystals are known to terminate

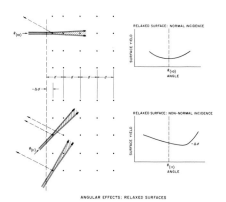

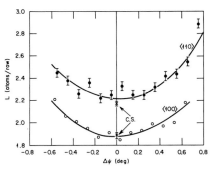

1 MeV He →Au(001) Contaminated Surface.

Fig. 13. Schematic representation of the use of the angular variation of the surface peak yield for studies of contracted surfaces.

Fig. 14. Surface peak yield versus angle of beam incidence to ⟨100⟩ and ⟨110⟩ of a contaminated Au(001) single crystal surface.

with all Ga atoms on the surface of one face and all As on the other, ion scattering/channeling was used to identify the faces and to correlate with the pulsed laser annealing results. The top and middle spectra in Fig. 12 compare the scattering of 2.0 MeV He ions channeled along ⟨111⟩ axes at 70.5° to the surface. In the top figure analysis shows that As atoms reside at the end of each ⟨111⟩ row and shadow the underlying Ga atoms, and in the middle spectrum the opposite is true. This accounts for the relative heights of the As (first peak) and Ga signals in the two cases. In the lower spectrum of Fig. 12 the incident energy was reduced to 1.5 Mev to increase the shadow cone radius. When incident on the Ga face the expected enhancement of Ga is observed. This analysis revealed that it was the As face which led to a well-defined LEED pattern following pulsed laser annealing and further results from this study will be reported elsewhere. The objective here is to illustrate the use of the shadow cone for surface characterization.

Angular Dependence of the Surface Peak Yield: Au(001) and Au(011)

Just as the angular variation of the yield of ions scattered from defects in the bulk can be used to determine their lattice sites, so the angular variation of the surface peak yield can be utilized to determine the geometrical arrangement of surface atoms relative to the bulk. Figure 13 illustrates this principle for a surface whose top layer is contracted an amount $-\Delta d$ from the normal bulk spacing d. As the angle of incidence is varied relative to the normal direction ⟨10⟩ (upper illustration) a symmetric variation of the surface peak yield is expected because the contraction is along this direction and alters the shadowing effect only slightly. However, when the beam is incident near a channeling direction away from normal, such as ⟨11⟩ in the lower illustration, the contracted surface atoms shadow the underlying atoms at an angle off ⟨11⟩ by an amount which is proportional to the displacement $-\Delta d$. If the surface is relaxed outward the minimum will occur at an angle on the opposite side of ⟨11⟩. There is also an overall increase in the magnitude of the surface peak yield for both cases. Thus, the magnitude of the yield increase, and the magnitude and sign of the angular minimum provide an accurate characterization of relaxed surfaces.

Normal Surface: Contaminated Au (001). It was demonstrated in the Surface Scattering section that the contaminated Au(001) was a normal (unreconstructed) surface. This is confirmed by the angular variation of the surface peak yields

TABLE II
PICS Investigations: Au(011) Surface

SURFACE PREPARATION AND CHARACTERIZATION	PICS Measurements				Model Simulations			
	$\langle 100 \rangle_{0°}$		$\langle 110 \rangle_{60°}$		$\langle 100 \rangle_{0°}$		$\langle 110 \rangle_{60°}$	
	X_{min}	L	X_{min}	L	X_{min}	L	X_{min}	L

Contaminated (011)

Annealed in Air 600°C

| LEED → (1x1) | 0.018 | 1.98 | 0.020 | 2.84 | 0.023 | 2.17 | 0.023 | 2.17 |

AES → C,0,S

Surface Peak Yield, Angular scans of $\langle 110 \rangle_{0°}$ and $\langle 110 \rangle_{60°}$ shows features like relaxed surface (-Δd)

Clean-Reordered (011)

| 500 eV Ar | 0.021 | 1.98 | 0.017 | 2.68 | 2.67 | | 3.17 | |
| | | | | Paired Row Model | | | | |

350°C in UHV

LEED → (1x2)					2.17		2.17	
				Missing Row Model ($\Theta_\perp = \Theta_\parallel = \Theta_B = 182°K$)				
AES → None					2.17		2.34	
				Missing Row Model ($\Theta_\perp = \Theta_B/2$, $\Theta_\parallel = \Theta_B$)				

Temperature
Transition ~600°C

Summary: 1. Contaminated and clean (011) show correlated lattice vibration effects.
2. Contaminated (011) surface is relaxed inward.
3. Paired row model is ruled out.
4. Missing row model is not confirmed, even with anisotropic thermal vibrations.

(L, in atoms/row) in Fig. 14 measured at various angles (Δψ) from the $\langle 100 \rangle$ axis normal to the single crystal surface, and the $\langle 110 \rangle$ at 45° to the crystal normal. Both show symmetric variations about the channel center, and the points labeled C.S. show that the magnitudes of the surface peak yields are in good agreement with model computer simulations for a normal surface. The model calculations assumed that the surface atoms had uncorrelated thermal vibrations the same magnitude as bulk Au atoms.

Contracted Surface: Contaminated Au(011). Investigations of the Au(011) surface have shown that when this surface is clean it reorders, and when it is contaminated with trace amounts of impurities it contracts [51]. The results for this surface are summarized in Table II and Fig. 15. For the present discussion emphasis will be on the contaminated Au(011) surface to demonstrate the utility of channeling for interface studies.

When the Au(011) surface is annealed to 600°C in air it acquires trace amounts ~1 monolayer of C, O or S (top row, Table II). LEED analysis indicates a normal (1x1) surface structure which is the same symmetry as the bulk.

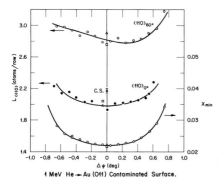

1 MeV He → Au (011) Contaminated Surface.

Fig. 15. Surface peak yield and minimum yields versus angle of beam incidence to <110> axes at 0° and 60° of a contaminated Au(011) single crystal surface.

However, PICS analysis such as that in Fig. 15 shows that the surface is not normal. The lower curve in Fig. 15 shows the minimum yield fraction as a function of the angle of beam incidence. Since X_{min} is determined by the channeling phenomenon deep within the crystal it can be used, as here, to locate the true channeling direction. The middle curve (closed circles) shows the measured surface peak yield versus $\Delta\psi$ for the <110> axes normal to the crystal surface. Although the yield variation is symmetric about $\Delta\psi=0$ it has a yield <u>less</u> than that calculated by computer simulations (point marked C.S.). Furthermore, when the surface peak yield is measured for the <110> axes at 60° to the crystal normal (uppermost curve), the magnitude of the yield is larger than predicted by C.S. and the minimum at $\Delta\psi\sim0.3$ is consistent with the behavior expected for a contracted surface. A detailed analysis of this contaminated surface suggests two main conclusions. First, the measured surface peak yields for both the contaminated and clean (reordered) surfaces (Table II) are less than calculated. This is believed to be due to correlated thermal vibrations [51]. The role of correlated lattice vibrations has been discussed in a recent paper [52]. Second, the behavior of the contracted surface can be understood if several monolayers participate in the process. Although the clean-reordered (011) is not consistent with either of the models suggested from LEED measurements, its reordered structure as well as that of Au(001) and Au(111) seem to follow a consistent trend to attempt to stabilize in a modified (111) surface [53].

The results from the contaminated Au(011) measurements highlight an important feature of PICS studies of surfaces. Because of the nature of ion scattering and channeling, it is possible, as we have seen, to examine in detail the changes in surface structure <u>at the interface</u> that results from a surface deposit. This consequence has been applied recently to study SiO_2, Si interfaces, $NiSi_2$ epitaxial structures on Si, and numerous other interface problems [14]. Related examples are also given in the companion articles by Chu [1] and Picraux [2].

CONCLUDING REMARKS

It is particularly appropriate that a review of ion channeling be included in a proceedings on analysis of defects in semiconductors because the strengths and capabilities of ion scattering/channeling are ideally matched to many of the problems in semiconductors. Ion implantation and laser annealing have joined thermal diffusion to become dominant processing techniques in semiconductor technology. These techniques lead to near-surface modifications of the material. Ion implantation and diffusion are usually employed for high dose doping with the important questions concerning the dopant concentration, depth profile, lattice location, and residual damage. Laser and thermal annealing are

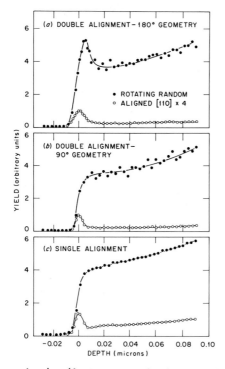

Fig. 16. Comparison of channeled (open circles) spectra for 1.0 MeV He channeled in Au(001). a) Uniaxial double alignment geometry ⟨110⟩-⟨110⟩, b) biaxial double alignment geometry ⟨110⟩-⟨110⟩ and c) single alignment geometry ⟨110⟩.

used primarily to remove the damage and redistribute the dopants. As we have seen, ion scattering/channeling is most useful for near surface analysis. It can provide detailed information on the lattice positions of impurities, the quantity and depth distributions of near-surface damage, the kinetics of defect-impurity interactions, the effectiveness of annealing treatments, near-surface stoichiometries etc. It is particularly sensitive to studies of surface alterations, interfaces, and epitaxial growth. The technique is quantitative and non-destructive. Semiconductors are even ideal materials because ion and laser induced damage in the bulk is often in the form of amorphous regions where channeling interpretation is fairly unambiguous, and surface damage usually leads to reconstructed layers suitable for PICS analysis. As stressed in all the review papers, however, specific defect analysis is usually model dependent and is most powerful when combined with complementary analysis techniques.

Ion channeling is a developing technique and it is probably worth mentioning several areas where future improvements can be expected. The flux peaking effect provides greatly improved sensitivities for precise lattice location studies. However, as we have seen, the yield oscillations which exist can often complicate interpretation of model calculations which assume statistical equilibrium. Computer simulation calcuations are available which can calculate ion fluxes precisely. Thus it is possible to utilize nonstatistical equilibrium or the yield oscillations to provide increased sensitivities to defect positions in the near-surface region. In general any lattice location measurement can be made with sufficient accuracy that such computer simulation methods should be used to interpret the results. Recent measurements have shown that nonequilibrium effects exist for 1.0 MeV He ions in Si⟨110⟩ to depths as great as 3000 Å [54].

Double alignment techniques were discussed in Section III as a means of achieving increased sensitivity to bulk defects. These techniques also provide increased sensitivity for lattice location measurements as well as for determining surface structures [14,48]. The potential advantages of double alignment over single alignment are demonstrated by the spectra in Fig. 16 which contrast double alignment measurements for 180° and 90° scattering/channeling with a single alignment spectrum. The double alignment geometry substantially reduces the scattering from bulk atoms on normal lattice sites and thus increases the sensitivity of this technique for displaced atoms in the bulk. It was recently discovered [55] that an enhanced scattering yield exists for the 180° double alignment geometry; this explains the pronounced peak on the random spectrum in Fig. 16. This enhanced yield also provides increased sensitivity for displaced atoms in the first few monolayers of the surface in this geometry [56]. Consequently, one might exploit this effect for near surface damage studies.

REFERENCES

1. W. K. Chu, "Application of Channeling to Defect Studies in Crystals," these proceedings.

2. S. T. Picraux, "Ion Channeling Analysis," these proceedings.

3. D. S. Gemmell, Rev. Mod. Phys. 46, N.1, 129 (1974) and references therein.

4. Channeling, ed. by D. V. Morgan, John Wiley and Sons, New York, 1973.

5. J. Lindhard, Kgl. Danske Videnskab. Selskab. Mat.-Fys. Medd. 34, No. 14 (1965).

6. New Uses of Ion Accelerators, ed. by. J. F. Zielger, Plenum Press, New York, 1975.

7. W. K. Chu, J. W. Mayer and M. A. Nicolet, Backscattering Spectrometry, Academic Press, New York, 1978.

8. C. W. White, S. R. Wilson, B. R. Appleton and F. W. Young, Jr., J. Appl. Phys. 51, 738 (1980).

9. L. C. Feldman and B. R. Appleton, Phys. Rev. B 8, 935 (1973).

10. Atomic Collisions in Solids, Vol. 1 and Vol. 2, ed. by S. Datz, B. R. Appleton and C. D. Moak, Plenum Press, New York, 1975.

11. F. H. Eisen and J. Bottiger, Ref. 10, p. 910.

12. R. Berliner and J. S. Koehler, Rad. Effects, 28, 141 (1976).

13. J. L'Ecuyer, C. Brassard, C. Cardinal and B. Terreault, Nucl. Instrum. and Methods 149, 271 (1978) and references therein.

14. L. C. Feldman, ISISS 1979 Surface Science, Recent Progress and Perspective, CRC Press, Inc., Cleveland, Ohio, 1980 and references therein.

15. L. C. Feldman, I. Stensgaard, P. J. Silverman and T. E. Jackman, The Physics of SiO2 and Its Interfaces, ed. by S. T. Pantelides, Pergamon Press, New York, p. 344, 1978.

114

16. L. C. Feldman, P. J. Silverman, J. S. Williams, T. E. Jackman and I. Stensgaard, Phys. Rev. Lett. 41, 1396 (1978).

17. Ion Beam Handbook for Material Analysis, ed. by J. W. Mayer and E. Rimini, Academic Press, New York, p. 112, 1977.

18. S. Datz, B. R. Appleton and C. D. Moak, p. 153 in Ref. 4 and references therein.

19. B. R. Appleton, S. Datz, C. D. Moak and M. T. Robinson, Phys. Rev. B 4, 1452 (1971) and M. T. Robinson, Phys. Rev. B 4, 1461 (1971).

20. B. R. Appleton, C. D. Moak, T. S. Noggle and J. H. Barrett, Phys. Rev. Lett. 28, 1307 (1972); and B. R. Appleton, J. H. Barrett, T. S. Noggle and C. D. Moak, Rad. Eff. 13, 171 (1972).

21. V. S. Kulikaushas, M. M. Maloo and A. F. Tulinov, Soviet Phys. JETP 26, 321 (1968); Zh. eksp. teor. Fiz. 53, 487 (1967).

22. E. Bogh, Can. J. Phys. 46, 653 (1968).

23. B. R. Appleton and L. C. Feldman, Atomic Collisions in Solids, ed. by D. W. Palmer, M. W. Thompson and P. D. Townsend, North Holland, Amsterdam, p. 417, 1970.

24. B. R. Appleton, T. S. Noggle, J. W. Miller, O. E. Schow III, D. M. Zehner, L. H. Jenkins and J. H. Barrett, Proceedings of the Third Conference on Applications of Small Accelerators, ed. by J. L. Duggan and I. L. Morgan, ERDA Conf. 74-1040-PK (1974).

25. B. R. Appleton, D. M. Zehner, T. S. Noggle, J. W. Miller, O. E. Schow III, L. H. Jenkins and J. H. Barrett, Ion Beam Surface Layer Analysis, Vol. 2, ed. by. O. Meyer, G. Linker and F. Kappeler, Plenum Press, New York, p. 607, 1976.

26. J. H. Barrett, Phys. Rev. B 3, 1527 (1971).

27. J. H. Barrett, Phys. Rev. Lett. 31, 1542 (1973).

28. J. H. Barrett as in Ref. 26 and 27.

29. J. H. Barrett, Phys. Rev. B 20, 3535 (1979) and references therein.

30. O. W. Holland and B. R. Appleton, Scientific and Industrial Applications of Small Accelerators, ed. by J. L. Duggan, I. L. Morgan and J. A. Martin, IEEE Proceedings 76CH 1175 9NPS, p. 585, 1976.

31. F. Grasso, Chapter 7, Ref. 4, and references therein.

32. F. H. Eisen, Chapter 14, Ref. 4, and references therein.

33. J. W. Mayer, Chapter 16, Ref. 4, and references therein.

34. E. Rimini, Materials Characterization Using Ion Beams, ed. by J. P. Thomas and A. Cachard, Plenum Press, New York, p. 455, 1978.

35. Y. Quere, Rad. Eff. 28, 253 (1976).

36. C. W. White, P. P. Pronko, S. R. Wilson, B. R. Appleton, J. Narayan and R. T. Young, J. Appl. Phys. 50(5), 3261 (1979) and references therein.

37. J. S. Williams, Nucl. Instrum. and Methods 149, 207 (1978).

38. M. L. Swanson and F. Maury, Can. J. Phys. 53, 1117 (1975).

39. M. L. Swanson, L. M. Howe and A. F. Quenneville, J. Nucl. Mater. 69/70, 744 (1978).

40. M. L. Swanson, L. M. Howe and A. F. Quenneville, Nucl. Instrum. and Methods 170, 427 (1980) and references therein.

41. M. L. Swanson, L. M. Howe, A. F. Quenneville, P. Offermann and K. H. Echer, J. Phys. Metal Phys. 10, 599 (1980) and references therein.

42. L. W. Wiggers and F. W. Saris, Rad. Eff. 41, 149 (1979) and references therein.

43. N. Matsunami, M. L. Swanson and L. M. Howe, Can. J. Phys. 56, 1057 (1978).

44. J. H. Barrett, Proceedings of Scientific and Industrial Applications of Small Accelerators, IEEE 176CH 1175-9NPS, p. 571, 1976.

45. J. U. Anderson, O. Andreasen, J. A. Davies and E. Uggerhoj. Rad. Eff. 7, 25 (1971).

46. D. Van Vliet, Rad. Eff. 10, 137 (1971).

47. J. H. Barrett, private communications.

48. R. G. Smeenk, R. M. Tromp, J. F. Van der Veen and F. W. Saris, Surf. Sci. 95, 156 (1980) and reference therein.

49. E. Taglauer and W. Heiland, Appl. Phys. 9, 261 (1976) and references therein.

50. D. M. Zehner, J. R. Noonan, H. L. Davis and C. W. White, J. Vac. Sci. Technol., March-April 1981 (in press).

51. B. R. Appleton, VII Inter. Conference on Atomic Collisions in Solids, September 19-23, 1977, Moscow State University, USSR, unpublished data.

52. J. H. Barrett and D. P. Jackson, Nucl. Inst. and Methods 170, 115 (1980).

53. B. R. Appleton, J. H. Barrett and D. M. Zehner, unpublished data.

54. O. W. Holland and B. R. Appleton, submitted to Phys. Rev.

55. P. P. Pronko, B. R. Appleton, O. W. Holland and S. R. Wilson, Phys. Rev. Lett. 43, 779 (1979).

56. J. H. Barrett, B. R. Appleton and O. W. Holland, Phys. Rev. B 22, 4180 (1980).

Published 1981 by North-Holland, Inc.
Narayan, and Tan, eds.
Defects in Semiconductors

APPLICATION OF CHANNELING TO DEFECT STUDIES IN CRYSTALS

W. K. Chu
IBM General Technology Division, East Fishkill,
Hopewell Junction, New York 12533

ABSTRACT

Channeling of fast, light ions in crystals has been
widely used as a tool for studying crystal defects. Backscattering yield measurement on ions incident along major
axial or planar crystalline directions provides information
on the depth distribution of the structural defects in the
first few microns. The channeling technique in defect
detection is not as sensitive as Transmission Electron
Spectroscopy, nor is it accurate in measuring the absolute
numbers of defect density. Channeling measurements can give
only an indication of the degree of lattice disorder. It is
possible to distinguish one type of defect from another by
carefully studying the energy dependence of the dechanneling.
The dechanneling interpretation is not always unique, and in
practice it is difficult to obtain structure information
through that method. Despite these negative qualities,
channeling is an attractive and unique method in certain
defect studies. For example, it is sensitive for studying
the lattice location of impurity atoms at substitutional or
interstitial sites. Clustering of substitutional impurity
atoms will show a displacement of the impurity atoms from
lattice sites due to the change of bond distance. Channeling
is sensitive for measuring impurity displacement as small
as 0.1Å. This has been demonstrated in the study of arsenic
clustering formation in Si. Interfacial relaxation and
contraction in a multi-layered structure made by molecular
beam epitaxy has been detected by dechanneling along various
axial directions. Channeling study on surface and interface
structures has developed over the past few years. In this
paper, I will use examples to illustrate the unique features
of the channeling technique and its application to defect
studies in single crystals.

INTRODUCTION

Channeling phenomenon was first discussed and predicted in 1912 by Stark[1].
His paper on the subject was ignored and forgotten, however, perhaps due to a
lack of instrumentation and the birth of atomic and nuclear physics. Half a
century later, the phenomenon was rediscovered by Robinson and Oen[2] during a
computer simulation on the directional dependence of the penetration depth of
charged particle in single-crystal solids. Experimental evidence for channeling
was obtained almost simultaneously in different labs during 1963.

In 1965 Lindhard[3] gave a theoretical description of channeling phenomenon. HE said that charged particles penetrating the single crystal along or
near a major axis are experiencing a collective string potential produced by
the rows of atoms along that axis. If the incident direction of the ion beam

is nearly parallel with the string of atoms, the string potential will steer the charge particle beam forward. A critical angle is defined as the limiting angle between the incident direction and row of atoms such that the steering effect exists. When charged particles are incident in a direction exceeding the critical angle, those particles have transverse kinetic energy exceeding the collective string potential. The collective steering effect subsequently disappears. An excellent review on the theory of channeling and the experimental study of channeling effects was given by Gemmell[4].

A comprehensive discussion of channeling phenomenon theory, its observation and applications was collected into a book by Morgan[5] in 1973. Many chapters are devoted to channeling applications. They are applications to radiation damage[6,7], surface studies[8], and to semiconductor technology[9]. Excellent discussions on lattice location of foreign atoms[10] are also included. Many other comprehensive reviews with varying emphasis on channeling have been written over the past 15 years. Review chapters and papers are very useful references for people working in the ion beam channeling field. The purpose of this paper is to discuss the application of channeling in defect studies for people who are not familiar with the process. The paper is designed to help them get a feel for the channeling technique, its principle of defect detection, and its sensitivity in comparison with other techniques. I will emphasize its uniqueness with examples.

APPLICATION OF CHANNELING TO DEFECT STUDIES

Ion channeling in conjunction with ion backscattering measurements has been used extensively for the study of defects in single crystal near the surface. Most of the defect studies by channeling involve backscattering of light particles such as protons or helium ions with energies of a few hundred keV to a few MeV. A particle accelerator plus ion beam focusing, steering, and analyzing magnets are needed to produce a monoenergetic ion beam. A goneometer is required for target manipulation and crystal alignment. A surface barrier detector and signals processing electronics are needed to detect and analyze the energy of the backscattered particles. This equipment is standard and readily available in most of the low-energy atomic and nuclear physics laboratories, or in any of the material analytical labs that specialize in the ion beam material studies.

I will illustrate the channeling technique with examples of defect detection and identification.

Depth Distribution of Lattice Disorders

Application of ion beam discipline includes ion beam micro-analysis as well as ion implantation, which by itself is a mature and well-studied subject. It is very natural to study the ion-implanted samples by ion beam backscattering and channeling. One of the salient capabilities of ion beam backscattering is that the depth scale can be established by the analysis of the energy of the backscattered ions. Ions backscattered from the buried depth of the sample will have less energy than those backscattered from the surface. This is due to energy loss of the ion traveling into and out of the sample before and after scattering. The subject of backscattering has been summarized elsewhere[11,12]. Figure 1 illustrates the depth capabilities of backscattering and channeling. This figure is extracted from Csepregi, Mayer, and Sigmon's paper[13], which deals with the regrowth of Si crystal from amorphous layers created by Si self implantation into samples of various orientations. The top portion of Fig. 1

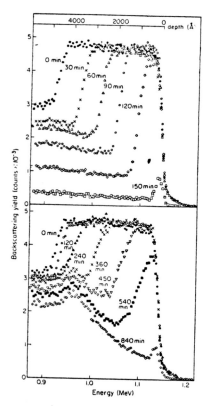

Fig. 1. Channeled backscattering spectra for 2 MeV ^{4}He ions incident on silicon samples implanted at liquid nitrogen temperature, pre-annealed at 400°C for 60 minutes and annealed at 550°C. Upper portion shows the epitaxial regrowth of <100> Si sample. Lower portions show the regrowth of <111> Si sample. The regrowth rate is much slower, and the interface is not as sharp as that of <100> regrowth and there is a residual damage region near the original interface (from Ref. 13).

shows six channeling spectra of 2 MeV He ions backscattered from <100> Si samples ion implanted and regrown at 550°C for 30 to 150 minutes. The energy scale is translated into a depth scale shown on top of the figure. This figure indicates that high-dose Si self ion implantation at liquid nitrogen temperature renders the top 500nm of Si into amorphous Si which is regrown at 550°C linearly with time. The lower portion of Fig. 1 shows the regrowth on <111> samples. The growth rate is considerably slower and the smoothness of the growth point is not as good as that of <100>. A residual defect region with peak concentration at 400 to 500 nm is also observed on <111> regrown samples. The smoothness of the interface, depth and nature of the defect were also studied[13] by etching, Scanning Electron Microscopy (SEM), and Transmission Electron Microscopy (TEM).

Conversion from channeling spectrum to depth distribution of defect is illustrated in Fig. 2, which shows channeled backscattering spectra of 2.4 MeV He$^+$ ions from single-crystal Si pre-bombarded with high dose H$^+$ at various energies[14]. Figure 3 shows a direct translation from energy spectra to the buried defect distribution. Note that the defect concentration given in Fig. 3 is on a relative scale. If one assumes that the peaks in Fig. 2 are produced

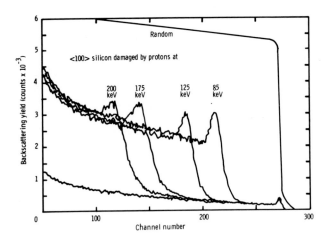

Fig. 2. Channeled backscattering spectra of 2.4 MeV [4]He ions from Si bombarded with 4 x 10[16] protons/cm[2] at various energies. Channeled and random spectra for an umbombarded Si are also shown as the background (from Ref. 14).

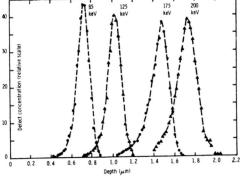

Fig. 3. Defect distribution, extracted from Fig. 2 (from Ref. 14).

by the scattering of Si atoms randomly distributed in the defect region, the relative concentration scale can be considered to be the percentage of Si atoms randomly displaced from the lattice site. Foti, et al.[15] have developed a method to obtain the distribution of defects in a plane normal to a low-index axis direction by channeling along and near the axial direction. Therefore, if the Si atoms are not randomly displaced from the lattice, their distribution with respect to the low-index direction can be attained.

The discussion so far involves scattering centers of individual atoms displaced from the lattice site. It has been pointed out, for example by Quéré[16], that unless one has clear evidence that the irradiation has produced exclusive amorphization of the surface layer, one should give up the term of "displaced atoms" and only should use scattering and dechanneling as useful indications of the degree of disorder or damage. They should not be used as measurements of the number of disordered atoms or defects in the crystal.

Backscattering is good for determining the depth distribution of the disorder, but TEM is better for defining the nature of the defect. It is possible to observe the energy dependence of the dechanneling to obtain information on the nature of the defects. We will elaborate on this in the next few sections.

Dislocation, Stacking Fault, and Twin.
The above section indicates that defects cause and increase scattering and dechanneling of the ion beam. Therefore, an accurate interpretation of the dechanneling data requires knowing the nature of the defect. This defeats the purpose of using channeling to analyze that defect. It has been demonstrated often that channeling is quite helpless by itself in the definitive characterization of crystal defects. It is much more powerful when combined with other techniques or a prior knowledge of the sample. One of the approaches in the development of channeling application on defect characterization is to study defective samples which are well-characterized by independent measurements such as TEM, and to see how channeling measurement responds to a known defect and its sensitivity.

Regarding dislocations: Foti, Picraux, and Campisano[17] have observed by TEM that Zn-implanted Al single crystals have large density of dislocations. When they studied the same sample by channeling, they observed that the amount of dechanneling is proportional to the square root of the energy of the analytical ion beam. Figure 4 shows their dechanneling measurement for 1, 2, and 3 MeV He ions. In all cases, the minimum yield of the unimplanted Al crystal was less than 5%. For implanted sample, the minimum yields are 5 to 8% near the surface region and increase appreciably to a depth of about twice the projected range (R_p), which is about 79 nm and indicated on the figure by arrows. The

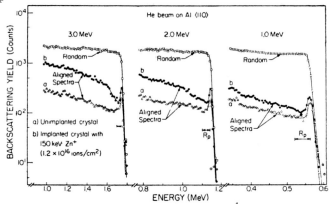

Fig. 4. Backscattering spectra of 3, 2 and 1 MeV [4]He ions incident along <110> direction of Al single crystal with and without 150 keV 1.2 x 10^{16} Zn/cm^2 implantation (from Ref. 17).

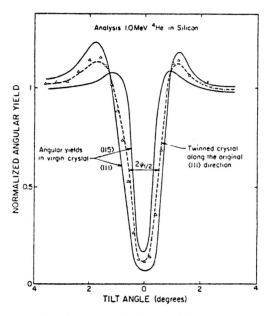

Fig. 5. Angular yield profiles for 1 MeV [4]He ions incident along <111> and <115> in virgin crystal (solid lines) and along the <111> direction in a highly twinned sample (triangles). The dashed line represents an average of <111> and <115> angular yield profiles (two solid lines) (from Ref. 17).

energy dependence of the dechanneling measurement of Fig. 4 is in good agreement with a distortional steering model given by Quéré[18]. A minimum dislocation density for detection by channeling is estimated to be 10^9 - 10^{10} cm length of line/cm^3.

Regarding stacking faults: Campisano, Foti, Rimini, and Picraux[19] have shown that the amount of dechanneling is independent of ion beam energy. Their study was made on crystal Si grown epitaxially on sapphire. TEM micrographs indicate that Si grown on sapphire is loaded with stacking faults plus twin lamellac. They conclude that a stacking defect density to 10^{15} displaced rows/cm^2 is required for the detection by single-alignment channeling. This corresponds to approximately one projected atomic plan normal to the beam direction and would give rise to an increase in the channel yield of 3%. In contrast, this sensitivity limit for channeling analysis represents an approximate upper limit for quantitative TEM analysis, because an effective area coverage of greater than one monolayer will result in appreciable defect image overlap.

Regarding Twins: it has been demonstrated by Foti, et al.[20,21], in the study of regrowth of self-implanted (111) Si crystals, that after annealing and recrystallization there is a high concentration of twinned regions lying on the three (111) planes inclined to the (111) surface. For particles channeled along a <111> axial direction in the host crystal, the inclined twins have a lattice with <115> orientation and consequently the ions will have a dechanneling rate which is much higher than along the original <111> direction. Therefore, the critical angle for the twinned crystal will be narrower. The amount of critical angle narrowing and dechanneling rate increment can be interpolated between that measured for <111> and <115> single-crystal Si to give an indication of the fractional concentration of the twinned and untwinned region. For example, Fig. 5 (extracted from Ref. 21) shows the angular scan of the backscattered He ions incident along <111> and <115> direction on perfect Si crystals and along <111> host crystal in a twinned sample (ploted as triangles). The dashed curve is an average of the two solid curves which represent equal contributions from the <111> and <115> angular yield profiles. The close agreement between the triangles and dashed curve indicates that about 50% of the samples consist of twinned regions.

It is obvious that channeling technique is not very sensitive to structure defects such as dislocations, stacking faults, and twins. From energy dependence of dechanneling or angular scan, channeling is capable of distinguishing the nature of the defect in a marginal manner. It is common practice to use channeling to identify the depth of the defect region and then use TEM to perform detail analysis on the nature and extent of the defect. Comparison study of TEM and channeling has helped those using the latter to understand its capabilities on defect study. Samples with well-defined defect structure will help in the development of channeling application on defect study, and will shed light on some high-order effects on channeling per se.

Foreign Atom Location, Trapping, and Clustering.
One of the unique features of channeling is its capacity for revealing the lattice location of the solute atoms in single-crystal solid solution. In general, a foreign atom species desolved in a single crystal could take one of the following positions: (1) A substitutional site; (2) A well-defined interstitial site; (3) A well-defined displacement from a substitutional site; and (4) A random distribution within the lattice.

The channeling technique is capable of showing the foreign atom location.

On certain occasions, foreign atoms can take more than one of these positions. Superposition of the different locations gives rise to superposition of various channeling results. It then becomes difficult to provide a unique answer on the lattice locations.

Excellent reviews on channeling application of foreign atom location are available. In this section we briefly illustrate.

Substitutional Case. From the analytical point of view, detectable foreign atoms located at the substitutional site are the simplest case in channeling characterization. The yield attenuation of the foreign atoms along all axial and planar channeling directions should be identical to that of the host lattice. Alexander and Poate[22] have shown that Au atoms in single-crystal Cu could assume the substitutional site perfectly. Figure 6 shows their results on the angular dependence of the normalized backscattering yield of 1.2 MeV He ions from Au and from Cu atoms in a single-crystal Cu sample continuing 2 at. % Au. The identical half angle on the angular scan for impurity and first atoms indicates that the Au is substantial and the gold atoms are completely shadowed by the Cu atoms. This 100% substitutional case is rarely observed. Usually a small percentage of impurity will occupy a non-substitutional site, which will produce a higher minimum yield for the backscattering signals from the impurity atoms.

Angular scan along more than one channeling direction is often required to deduce the lattice location. Planar and axial channeling are helpful in the triangulation analysis. This is especially true when foreign atoms take

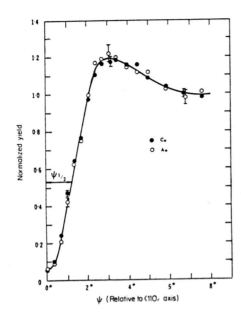

Fig. 6. Angular yield profiles of 1.2 MeV ^{4}He ions backscattering from Au and Cu atoms in single-crystal Cu containing 2 at % Au (from Ref. 22).

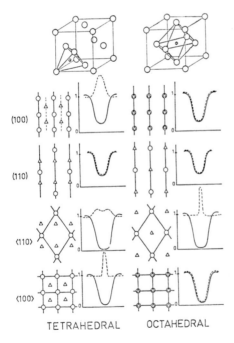

Fig. 7. Impurity atom taking tetrahedral and actahedral inter-stitial sites for FCC crystals. The schematic of the lattice planes and strings and their corresponding form of the angular yield profiles (from Ref. 23).

interstitial locations.

Interstitial Case. For a simple cubic crystal, an interstitial site is located
at the center of the cubic. For b.c.c., f.c.c. or diamond structures, there
are more than one type of interstitial sites. Figure 7 illustrates[23] the
tetrahedral interstitial and octahedral interstital site of a f.c.c. single
crystal. For (100) planar direction, the tetrahedral interstitials are located
between the (100) planes, and the angular scans indicate a peak for the foreign
atom at the well-channeled direction. This is not the case for octahedral
interstitials which are hidden behind the host atoms in (100) and (110) directions. For planar channeling cases, the ion beam is incident parallel to the
lattice planes, which are defined by the atoms on the plane. For axial channeling cases, the ion beam is incident parallel to the major axial direction
(normal to the drawing). Flux peak is shown in <100> tetrahedral site and <110>
octahedral site, and good shadowing in <110> tetrahedral interstitial. The
foreign atoms are not located in the center of the channel but off centered by
a well-defined distance. A double peak can be expected. Through careful
triangulation in various channeling directions, a well-defined lattice location
can be determined. Examples of hydrogen locations in f.c.c. crystal[23] and
b.c.c. crystal[24,25] are the classic demonstrations of this type.

Small Displacement from Substitutional Site. There are many cases indicating
that good channeling conditions for impurity atoms are observed. However,
sometimes the width of the angular scan from the impurities is much narrower
than that of the host atoms. The amount of half-angle narrowing is much more
than indicated in Fig. 7, where a slight offset of dashed and solid curves is
needed for clarity to indicate the two identical curves. The angular width
narrowing is an indication of small displacement from a substitutional site.
There are many reasons why foreign atoms will displace from a lattice site.
For example, foreign atom and vacancy pair could move the foreign atom off the
lattice site (toward the vacancy). Small foreign-atom clusters could have
slightly different bond lengths between them that make the foreign atoms off
from the lattice site. It does not matter what causes the displacement. The
amount of the displacement can be measured by observing the amount of half-
angle narrowing versus that of the angular scan of the host atoms.

This concept can be illustrated by showing that when Si crystals are
heavily doped with As, As atoms tend to form small clusters and are displaced
from the lattice by ∿ 0.015 nm. When the sample is subjected to laser irradiation, As clusters disolve in Si solution and As atoms take an exact substitutional site. The channeling study of As displacement[26,27] and electrical
measurement[28] has been shown. We will use the As cluster study as an example
to illustrate the capability of channeling for measuring small displacement of
foreign atoms from a lattice site.

Figure 8 shows a schematic drawing for a silicon sample that has high
concentration of As with all As atoms on a substitutional site. This condition
is achieved by laser annealing which disolves the As clusters and quenches them
in lattice sites. Tilting the crystal with respect to the ion beam will produce an angular scan such that the half angle for the As and Si is nearly
identical. The angular scan of such a sample along <110> direction is given in
Fig. 9. The slightly higher minimum yield of As signal than that of the Si is
an indication that not all of the As atoms are on a substitutional site.

126

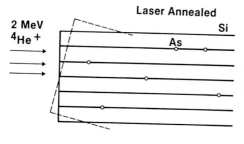

Laser Annealed

2 MeV
⁴He⁺

Fig. 8. Schematic drawing for a laser annealed Si sample where high concentration of As atoms are on lattice site. Rocking of the crystal (dashed line) with respect to the ion beam produces angular yield profiles given in Fig. 9.

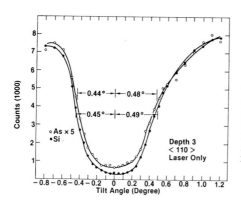

Fig. 9. Angular yield profiles for 2 MeV ⁴He ion backscattered from Si (As) which is irradiated by a pulsed laser. Nearly all As atoms are on lattice site and no half-angle narrowing for As scan is observed (from Ref. 27).

When the sample is subjected to 950°C, one-hour heat treatment, the moderate annealing allows As atoms to diffuse in Si for a distance sufficient to agglomerate into small clusters. This subsequently alters the bond length between (or among) the As atoms. This is illustrated in Fig. 10. Although Fig. 10 schematically shows two As atoms located at the second nearest neighboring site with a bond length contraction, we do not imply that the drawing is the configuration of an As cluster. Figure 10 is useful to illustrate only that the bond length is altered due to the clustering of two or several As atoms. This small displacement will define a stringent condition for the

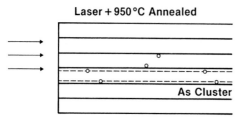

Laser + 950°C Annealed

As Cluster

Fig. 10. Schematic drawing for a Si sample with high concentration of As forming small clusters. This thermal dynnamic equilibrium condition is achieved by 950°C annealing on samples with or without laser annealing. The cluster formation displaces the As atoms from lattice sites and that defines a stringent channeling condition and narrower half angle.

channeling beam such as the dashed line given in Fig. 10. A larger probability of ion beam scattering from As atoms under tilting conditions and a narrower half angle will be observed when compared with that of Si. If the As-As bond expands rather than contracts, this expansion also will cause the displacement of As atoms and the narrowing of the half angle on the As angular scan. An

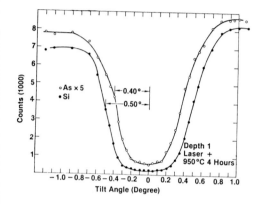

Fig. 11. Angular yield profiles for 2 MeV [4]He ion backscattering from Si loaded with As clusters. The displacement of As atoms from lattice site can be seen from the half-angle narrowing of the As angular scan (from Ref. 27).

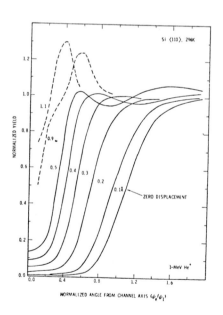

Fig. 12. Angular yield calculation as a function of equilibrium displacement displacement distance from Si lattice row for 1 MeV [4]He ions backscattering along <110> axial direction (from Ref. 30).

angular scan measurement on a sample being heat treated is given in Fig. 11
where a 0.1° halfangle narrowing is observed.

The narrowing of the angular scan is a measurement of the projected displacement to a plane normal to the channeling direction. The 0.1° half-angle
narrowing is equivalent to $\sim$ 0.015 nm. The translation is nonlinear, and the
interpretation is made from the channeling study of ion beam-induced Si L X-ray
production[29], and from a calculation by Picraux, et al.[30]. In the channeling
ion-induced, X-ray study, Lurio, et al.[29] observe that Si K X-ray production
is identical to the backscattering angular scan, while Si L X-ray yield has a
half angle considerably narrower than that of the backscattering. This is
because the L-electrons are sticking out, occupying a space away from the
lattice and providing a stringent condition for channeling beam in L X-ray
production. (Imagine that the distance between the dash line and the solid
line of Fig. 10 is the average radius of the L orbital electrons of Si atom.)
Angular scan calculation base on flux distribution derived from a string potential has been made by Picraux, et al.[30]. Their calculation can be illustrated
in a frequently quoted figure. Figure 12 is a reproduction of their calculation
on single-alignment angular distributions as a function of equilibrium displacement distance from the row for 1 MeV He$^+$ beam along <110> axis in Si at 296°K.
The calculation has been motivated by their lattice location study of Bi implanted in Si. Their calculation and measurement have shown a case where $\sim$ 50%
of the Bi atoms displaced by 0.45Å from the Si lattice sites and the remaining
Bi atoms are on a substitutional site.

Channeling measurement is capable of detecting and measuring the amount of
displacement of impurity from the lattice site. The cause of such a displacement has to be determined or conjectured by other means. There are two major
reasons for impurity to displace from the lattice sites. One is the impurity-
impurity clustering already discussed. The other is impurity-vacancy or impurity-
interstitial trapping. The subject has been discussed in great detail[31,32].
From the channeling point of view, clustering and trapping give a measurable
displacement of impurity from the lattice site. Channeling cannot indicate the
difference.

Surface and Interface of Single Crystal.
Using backscattering channeling to study the surface structures has been
suggested by Bøgh and Uggerhoj[33,8]. A recent review by Feldman[34] covered
the principle and applications of backscattering channeling to surface structure determination. The main concept involved is the reduction of the scattering events from the second, third, and other layers of atoms due to the shadowing effect of the first layer on the surface of major axial channeling directions. Any surface relaxation or reordering will change the amount of shadowing effect and expose the inner layer(s). Therefore, by measuring the number
of apparent surface atoms per unit area exposed to the ion beam at various
experimental conditions (beam energies and incident and scattering angles), it
is possible to deduce the surface structure by computer simulations to match
the observation. For example, the reconstruction and reordering of W (001)
with hydrogen coverage has been studied by Stensgaard, et al.[35]. Surface
relaxation of Pt (111) has been studied by Davies, et al.[36], Bøgh and
Stensgaard[37], and by Van der Veen, et al.[38]. Surface reconstruction of Au
(100) has been studied by Appleton, et al.[39] and the subject will be reviewed

by Appleton[40] during the symposium.

Application of channeling to interface structure is a very recent topic. Since the ion beam is capable of probing into the solid, it is natural to study the interface structure by channeling. Feldman, et al.[41,42] have studied the interface of SiO_2 on single-crystal silicon. A thin self-supported, single-crystal Si foil was used in their experiment. This enabled them to study front and back sides of the SiO_2/Si interfaces. The back side of the interface was exposed to the channeled ion beam which has flux distribution peaked at the center of the Si channel. The front side of the interface was exposed to ion beam with uniform flux. The difference in ion beam flux distribution enabled them to extract the information of the stoichiometry of the thin oxide and the structure of the underlying Si layer at the interface.

Using self-supported Si foils as channeling samples has been extended to the interface study between silicide and Si. For example Cheung, et al.[43] have studied the reactivity and interface structure of Ni film in Si (111). The interface structure of $NiSi_2$ on <111> Si has been studied by Chiu, et al.[42].

Interface relaxation and contraction of InAs-GaSb superlattices have been studied by Saris, et al.[45]. The superlattice samples were made with a Molecular Beam Epitaxy (MBE) apparatus by periodically opening the In and As or Ga and Sb beam from the four elemental sources used[46]. Figure 13 shows their backscattering and <100> channeling study of 20 periods of 41 nm InAs and 41 nm GaSb on GaSb substrate. The oscillatory nature of the random and channeled spectra is caused by the overlapping of the backscattering signals from different masses at different depths as indicated in the insert. The dashed curves

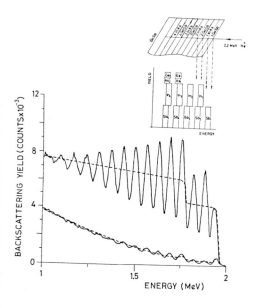

Fig. 13. Backscattering spectra for 2.225 MeV [4]He ions incident along a <100> channeling or random directions in an InAs-GaSb superlattice with 20 periods of 41 nm InAs and 41 nm GaSb. For comparison, backscattering spectra from a GaSb single crystal are also given. The insert gives a schematic representation of the origin of the oscillatory backscattering yield. The low <100> channeling spectrum indicate that the superlattice is of good quality when measured along the growth direction (from Ref. 45).

130

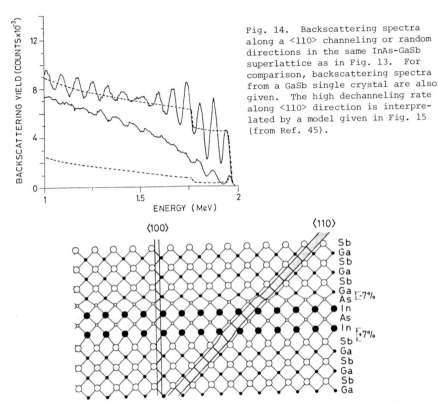

Fig. 14. Backscattering spectra along a <110> channeling or random directions in the same InAs-GaSb superlattice as in Fig. 13. For comparison, backscattering spectra from a GaSb single crystal are also given. The high dechanneling rate along <110> direction is interpreted by a model given in Fig. 15 (from Ref. 45).

Fig. 15. Schematic diagram of a model for the InAs-GaSb interfaces showing a side view of a thin InAs-layer sandwiched between GaSb layers. An ion beam aligned along the <100> surface normal will not observe any obstructions at the interfaces but for alignment along <110> there will be a kink in the atomic string at each interface (from Ref. 45).

are random and <100> channeling spectra of single-crystal GaSb. The good agreement between the two channeled spectra is an indication that the superlattice has been grown in perfect epitxial relation with its GaSb substrate. On the other hand, when channeling study is made on the same sample but along <110> direction (see Fig. 14), a very high dechanneling rate is observed for the superlattice sample. They attribute the high dechanneling rate observed along <110> direction but not along the growth direction <100> to interface relaxation and contraction. This is illustrated in Fig. 15. Although the lattice matching between GaSb and InAs is excellent, the interface relaxation of In-Sb bond and contraction of Ga-As bond will produce kinks at each interface and cause high dechanneling along <100> direction.

CONCLUSION

In addition to this review talk, there are two review talks by Appleton[40]

and Picraux[47] during the symposium. Although all three speakers address the
topic of channeling applications, different authors stress different areas.
Limited effort was made in drafting the outline of our papers. Channeling [47]
analysis vs. TEM on epitaxially grown layers and defect trapping of gas atoms in
single-crystal metals are emphasized in Picraux's paper[47]. The analysis of
surface structure and channeling fundamentals is emphasized in Appleton's
paper[40]. My talk touches several cases with emphasis on the relaxation. All
our examples indicate that the channeling technique is highly specialized and
unique in certain cases. For example, its capability for depth perception and
on foreign atom locations and surface interface structures is quite different
from the traditional microscopic techniques. On the other hand its lack of
lateral resolving power limits its sensitivity to "see" some defects, such as
those routinely found by TEM. In general, backscattering and channeling measure the collective effect of defects or foreign atoms over a sample area of
~ 1 mm^2. It is not sensitive in detecting or determining the nature of defects.
The channeling technique is unique in its ability to measure the depth scale.
the foreign atom location, and surface and interface structures.

ACKNOWLEDGMENT

I am grateful to my many colleagues who allowed me to use their works as
examples for this talk, and I want to thank Bill Appleton and Tom Picraux in
their collective effort in preparing our papers.

REFERENCES

1. J. Stark, Physik. Z. 13, 973 (1912).

2. M. T. Robinson and O. S. Oen, Phys. Rev. 132, 2385 (1963).

3. J. Lindhard, Kgl. Danske Videnskab. Selskab. Mat.-Fys. Medd. 34
    No. 14 (1965).

4. D. S. Gemmell, Rev. Mod Phys. 46, N. 1, 129 (1974).

5. Channeling, Ed. D. V. Morgan, John Wiley and Sons, NY (1973).

6. F. H. Eisen, Chapter 14 of Reference 5.

7. R. S. Nielson, Chapter 9 of Reference 5.

8. E. Bøgh, Chapter 15 of Reference 5.

9. J. W. Mayer, Chapter 16 of Reference 5.

10. J. A. Davies, Chapter 13 of Reference 5.

11. W. K. Chu, J. W. Mayer and M. A. Nicolet, Backscattering Spectrometry,
     Academic Press, New York (1978).

12. W. K. Chu, J. W. Mayer, M. A. Nicolet, T. M. Buck, G. Amsel and
     F. Eisen, Thin Solid Films 17, 1 (1973).

13. L. Csepregi, J. W. Mayer and T. W. Sigmon, Appl. Phys. Lett. 29,
     92 (1976).

14. W. K. Chu, R. H. Kastl, R. F. Lever, S. Mader and B. J. Masters, Phys., Rev. B16, 3851 (1977).

15. G. Foti, P. Baeri, E. Rimini and S. U. Campisano, J. Appl. Phys. 47, 5206 (1976).

16. Y. Quérè, Radiation Effects 28, 253 (1976).

17. G. Foti, S. T. Picraux and S. U. Campisano, in Ion Implantation in Semiconductor 1976, Ed by F. Chernow, J. A. Borders and D. K. Brice, Plenum Press, p. 247 (1977).

18. Y. Quérè Phys. Stat. Sol. 30 713 (1968).

19. S. U. Campisano, G. Foti, E. Rimini and S. T. Picraux Nucl. Inst. and Methods 149, 371 (1978).

20. G. Foti, L. Csepregi, E. F. Kennedy, J. W. Mayer, P. P. Pronko and M. D. Rechtin, Phil. Mag. 37, 591 (1977).

21. G. Foti, L. Csepregi, E. F. Kennedy, P. Pronko, and J. W. Mayer, Phys. Letter 64A, 265 (1977).

22. R. B. Alexander and J. M. Poate, Rad Effects 12, 211 (1972).

23. J. P. Bugeat, A. C. Chami and E. Ligeon, Phys. Letter 58A, 127 (1976).

24. H. D. Carstangen and R. Sizmann, Phys. Letter 40A, 93 (1972).

25. S. T. Picraux and F. L. Vook, Phys. Rev. Letter 33, 1216 (1974).

26. W. K. Chu and B. J. Masters in Laser-Solid Interactions and Laser Processing - 1978, edited by S. D. Ferris, J. H. Leamy and J. M. Poate AIP Conf. Proc. 50, 305 (1979).

27. W. K. Chu in Laser and Electron Beam processing of Electronic Materials Edited by C. L. Anderson, G. K. Celler and G. A. Rozgonyi, The Electrochemical Soc. Proc. 801, 361 (1980).

28. W. K. Chu, Appl. Phys. Letter 36, 273 (1980).

29. A. Lurio, J. Keller and W. K. Chu, Nucl. Ins. and Methods, 149, 387, (1978).

30. S. T. Picraux, W. L. Brown and W. M. Gibson, Phys. Rev. B6, 1382 (1972).

31. M. L. Swanson (this conference).

32. M. L. Swanson, J. A. Davies, A. F. Quenneville, F. W. Saris, and L. W. Wiggers, Rad. Effects, 35, 51 (1978).

33. E. Bøgh and E. Uggerhoj, Nucl. Inst. Methods 38, 216 (1965).

34. L. C. Feldman, ISISS 1979 Surface Science. Recent Progress and Perspectives, CRC Press Inc., Cleveland, Ohio (1980).

35. I. Stensgaard, L. C. Feldman, and P. J. Silverman, Phys. Rev. Letter 42, 247 (1979).

36. J. A. Davies, D. P. Jackson, N. Matsunami, P. R. Norton and J. U. Anderson, Surface Sci. 78, 274 (1978).

37. E. Bøgh and I. Stensgaard, Phys. Lett. 65A, 357 (1978).

38. J. Van der Veen,, R. G. Smeenk, and F. W. Saris, Surf. Sci. 79, 219 (1979).

39. B. R. Appleton, D. M. Zehner, T. S. Noggle, J. W. Miller, O. E. Schoww, III, L. H. Jenkins and J. H. Barrett in Ion Beam Surface Layer Analysis, Volume 2, edited by O. Meyer, G. Linker and F. Kappeler, p. 607, Plenum, New York (1976).

40. B. R. Appleton, this symposium, this proceeding.

41. L. C. Feldman, P. J. Silverman, J. S. Williams, T. E. Jackman and I. Stensgaard, Phys, Rev. Letter 41, 1396 (1978).

42. N. W. Cheung, L. C. Feldman, P. J. Silverman and I. Stensgaard, Appl. Phys. Letter 35, 859 (1979).

43. N. W. Cheung, R. J. Culbertson, L. C. Feldman, P. J. Silverman, K. W. West and J. W. Mayer, Phys. Rev. Letter 45, 120 (1980).

44. K. C. R. Chiu, J. M. Poate, L. C. Feldman and C. J. Doherty, Appl. Phys. Letter 36, 544 (1980).

45. F. W. Saris, W. K. Chu, C. A. Chang, R. Ludeke and L. Esaki, Appl. Phys. Letter (to be published).

46. C. A. Chang, R. Ludeke, L. L. Chang and L. Esaki, Appl. Phys. Letter 31, 759 (1977).

47. S. T. Picraux, this symposium, this proceeding.

Published 1981 by North-Holland, Inc.
Narayan, and Tan, eds.
Defects in Semiconductors

ION CHANNELING ANALYSIS OF DISORDER*

S. T. PICRAUX
Sandia National Laboratories[†], Albuquerque, New Mexico, 87185, USA

Abstract

The use of ion channeling to characterize disorder in semiconductors
is briefly reviewed. At high defect densities the ion channeling/back-
scattering technique can give the depth distribution of defects with a
resolution ~ 10 nm. Quantitative analysis of the defect depth profile
requires that a single type of defect dominates the scattering of particles out of channeling trajectories. This scattering process is characterized by two defect-specific quantities: the direct scattering
factor and the dechanneling cross-section. For impurity-associated
defects the lattice location of the impurity atoms can be determined
to within ~ 0.1 Å by simultaneously measuring the impurity and host
atom signals as a function of tilt angle about channeling directions.
The channeling technique can detect a wide range of intrinsic defects
including interstitials, dislocations, stacking faults, microtwins,
and amorphous clusters. Examples of the application of channeling to
study defects in silicon will be given for the cases: 1) hydrogen
trapping at defects; 2) ion implantation doping; and 3) epitaxial
growth of layers.

INTRODUCTION

The study of disorder by the ion channeling technique [1] is based
on the principle that channeled ions detect displacements from regular
lattice sites. Thus, ion channeling is sensitive to interstitial host
atoms or changes in the host periodicity, e.g., stacking faults, but is
insensitive to isolated vacancies, except for a very small contribution
in the case of lattice relaxations ≥ 0.1 Å around the vacancy. Some
examples of defects which have been detected by the channeling effect
are given in Table I. The technique is versatile but relatively insensitive. However, this can be an advantage in studies which involve high
densities of defects.
Ion channeling is a direct lattice technique. It contrasts to dif-
fraction-based, reciprocal lattice techniques in that the interpretation
of channeling data is very geometrical and qualitative aspects are subject to little uncertainty. However channeling is a spectroscopy only
in the sense that by ion beam techniques impurity and host atom signals
can be separated and resolved in depth. Therefore if a variety of defects involving a single impurity or the host atoms are present simultaneously, their respective signals will be summed together complicating
interpretation. For this reason transmission electron microscopy is an
important complementary technique to ion channeling studies of disorder.

*This work was supported by the Department of Energy, Division of Basic
  Energy Sciences, under contract DE-AC04-76-DP00789.
†A U. S. Department of Energy facility.

TABLE I

Some defects which have been studied by the ion channeling technique.

Defect	Typical Concentrations Needed	Representative Study		Reference
Substitutional impurity	0.1 at.%	Si(As)	laser annealing	2
Near substitutional impurity	0.1 at.%	Si(Bi)	defect association	3
Interstitial impurity	0.1 at.%	Si(D)	implanted hydrogen	4
Interstitial host atoms and clusters	10 at.%	Si(B)	implantation disorder profiles	5
Amorphous zones	10 at.%	Si(Te)	cascade effects	6
Amorphous layers	> monolayer	Si(Si)	solid phase epitaxy	7
Dislocations	10^9–10^{10}/cm^2	Si(P)	anneal-induced network	8
Twins	10%	Si(Si)	<111> laser annealed	9
Stacking faults	1 monolayer*	Si-on-sapphire annealing		10
Voids/gas bubbles	1 monolayer*	Al(He)	thermal evolution	11
Gunier-Preston zones	1–5 at.%	Al(Cu)	aging	12

*Projected total surface area over depth < 100 nm.

A primary advantage of the ion channeling technique is that in favorable cases the depth profile of the disorder is obtained with a typical resolution of ~ 10 nm[13]. The range of probing depths is ~ 1 to 10 μm. The depth determination is due to the energy lost by the ion beam as it penetrates the solid and is nondestructive in that layer-removal techniques are not required. Nondestructive measurements are particularly convenient for studies of the kinetics of defect generation or annihilation, as during thermal annealing.

A second important advantage lies in the ability to determine the crystallographic position of impurity atoms[14]. This provides understanding of the structure and nature of the defect center being studied. Finally, under favorable conditions quantitative determination of the defect concentration vs. depth can be obtained. Quantitative analysis requires that a single type of defect contribute to the channeled spectrum and that the dechanneling and direct scattering factors be reasonably well-known.

Defect studies by the ion channeling technique can take one of two directions: either the impurity atom or the host atom displacements can be studied. In both cases the measurement depends on the fact that during channeling there is a high density of channeled particles near the center of the channel and a low density near the rows (Fig. 1). Thus close encounter

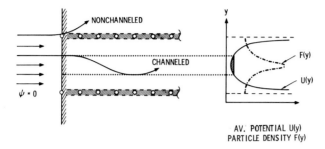

Fig. 1. Schematic of channeling process.

events, such as ion backscattering, are reduced by up to two orders of
magnitude in a perfect crystal containing no displaced atoms and are
enhanced by a factor ~ 2 for atoms near the center of channels. In
the case of impurity-related defects, defect structural information is
usually sought in channeling studies by impurity site determinations.
In intrinsic bulk defect studies the channeling method is usually
restricted to use as a monitor of the defect depth profile rather than
providing any new structural information.

  The present review is restricted to applications of channeling to
the study of disorder within the crystal bulk. Other reviews in this
symposium proceedings discuss channeling applications to surface, interface and modulated layer structures[15,16]. In what follows we give a
qualitative overview of the channeling technique for disorder analysis
and give examples for Si from three areas: 1) lattice location studies of
hydrogen-associated defects; 2) implantation disorder production; and
3) epitaxial layer growth and resulting defect generation and annealing.

CHANNELING TECHNIQUE

  Impurity Location. In the case of impurity or solute atoms the
crystallographic site of the impurity can be obtained for reasonably high
symmetry sites by tilting the beam along high symmetry crystal directions[14].
The change in the impurity signal relative to the host atom signal allows the
projected position of an impurity along a particular direction to be determined. Measurements along various crystallographic directions then can
isolate the impurity site in three dimensions. In the case of a single
high symmetry site, determination of the impurity position can be made,
sometimes to within 0.1 Å within the unit cell. When more than two sites
are involved, unique site determinations are usually impossible.

  Interstitial impurity location studies require that the channeled
particle spatial density across the channel be calculated for comparison
to experiment. However, even without calculation, the major high symmetry
interstitial sites usually can be distinguished by inspection for cubic
crystals. Furthermore, large qualitative differences are seen between
interstitial, near substitutional and substitutional sites (Fig. 2). For
the substitutional case the impurity and host angular scans are both dips
and they coincide. For near substitutional impurities the impurity dip
is narrower than the host lattice dip. Whereas for interstitial impurities
the impurity signal increases to give a narrow "flux peak" along directions
where the impurity is in the center of the channel.

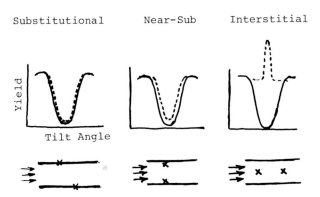

Fig. 2. Angular yield curve for host (solid line) and impurity (dashed line) as function of impurity position in the channel.

Intrinsic Defects. Displaced host atoms influence the channeled beam in two ways. First, they greately enhance the dechanneling over that found in a perfect crystal due to deflections of the channeled particles to angles greater than the critical angle for channeling. Secondly, they may directly scatter the beam into the detector, giving rise to a direct scattering peak in the backscattering spectrum at the depth of the disorder. As shown in Fig. 3, this direct scattering contribution is significant for interstitial atoms, clusters and amorphous zones, but is usually negligible in the case of dislocations, stacking faults or twins where the dechanneling contribution dominates the observed spectrum.

For a given defect the dechanneling and direct scattering factors can be calculated for quantitative analysis of the disorder profile[17]. The defect dechanneling factor has quite different dependences on channeling parameters according to the type of defect. For example, consider the energy of the channeling beam. For defects characterized primarily by distortion of the crystal rows, such as for dislocations, the dechanneling increases with increasing beam energy. For defects characterized by new internal surfaces, as for stacking faults along directions other than the stacking direction, the dechanneling is nearly independent of energy. For point scattering centers, as for interstitials, interstitial atom clusters, amorphous zones or amorphous layers, the dechanneling decreases with increasing beam energy. Differing functional dependences such as these can be exploited to enhance the sensitivity to the type of defect studied. However, if more than one type of defect gives comparable contributions to the dechanneling, a quantitative separation of the defects present cannot generally be obtained. In defect studies electron microscopy can often be used to identify the microscopic nature of the defects. Channeling measurements are used to determine the number of defects and their depth profile.

APPLICATIONS TO DISORDER STUDIES

Hydrogen Lattice Location. One of the most definitive uses of channeling has been in the lattice site location of impurities in association with defects in crystals. Channeling studies of hydrogen have been particularly fruitful because of the difficulty of directly probing hydrogen by other de-

fect-sensitive techniques. Ion beam studies of hydrogen in metals have given
the first evidence for the identification, structure and binding energy of the
hydrogen-vacancy center[18]. Studies of hydrogen implanted in crystalline Si
have given less well-resolved channeling angular scans, but have suggested well-
defined locations for the hydrogen associated with defects[4, 19].

As seen in Fig. 4, D implanted into crystalline Si at room temperature
exhibits a depth profile as determined by $D(^3He, p)^4He$ ion-induced nuclear
reaction analysis consistent with the ion range distribution. Since the
untrapped D is mobile in Si at room temperature, this suggests an excess
of lattice defect traps produced by the D ions as they slow down, consistent
with theoretical estimates of about 20 vacancy-interstitial primary pairs
produced per each 13 keV D ion.

Channeling angular scans for the D signal relative to the Si signal
along various axial and planar directions indicate the D is located along
tetrahedral coordination directions about 1.6Å from Si lattice sites, a
distance only slightly larger than typically found for the Si-D bond in
silane molecules. As seen by the <110> angular scan in Fig. 5, channeling
measurements indicate the D is located primarily in the "anti-bond" direction
away from a nearest neighbor Si and in the direction of the tetrahedral
interstitial site (see Fig. 6).

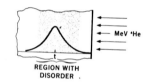

a) AMORPHOUS ZONES

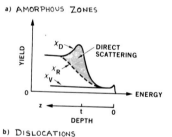

b) DISLOCATIONS

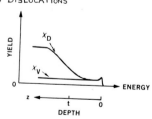

Fig. 3. Influence of a disorder profile on channeling energy spectra
for a) amorphous zones and b) dislocations, where χ_R is the dechanneled
component and χ_D and χ_V are the observed signals for a disordered
and for a virgin crystal, respectively.

140

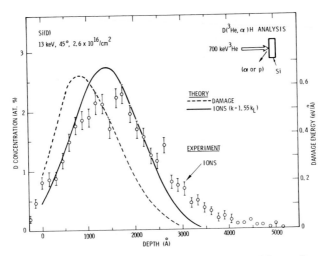

Fig. 4. Depth profile of D implanted in Si determined by nuclear reaction analysis along with theoretical profiles for D damage and ion range distributions[4].

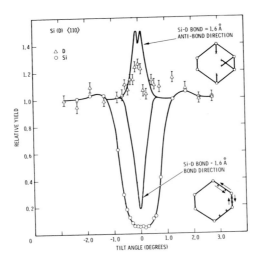

Fig. 5. The ⟨110⟩ angular scans for D-implanted Si and calculated distributions for all the D located 1.6 Å from a Si site along the ⟨111⟩ in the anti-bond or in the bond directions from the Si site[4].

While occupation of this more open region of the Si lattice by other impurities has been suggested by electron spin resonance studies (Fig. 6), no location information for hydrogen has been obtained by other techniques. The detailed nature of the Si defect center with which the hydrogen is associated cannot be specified at present, since channeling location measurements do not allow the location of the host atoms surrounding the impurity to be determined. However, it is believed that either Si vacancies or interstitials are present to provide for the hydrogen bonding to Si, as suggested by the observed infrared absorption bands associated with local mode vibrations.

Ion Implantation Disorder

A major use of the channeling technique for disorder studies has been the study of disorder production in Si during ion implantation doping and the subsequent removal of defects by furnace or laser annealing. In Fig. 7, an example of channeling measurement of the disorder profile in the case of B implantation into Si at -150°C is shown[5]. The direct scattering peak is clearly seen in the <110> aligned spectrum due to interstitial clusters and amorphous zones. The increased yield at deeper depths (below channel 120) in the aligned spectrum indicates the increased dechanneling in the defected layer over that for an unimplanted crystal. An iterative procedure is used to separate the direct scattering and the dechanneling contributions to the spectrum. This allows the density of displaced Si scattering centers to be determined as a function of depth. Comparison in the lower part of Fig. 7 of the normalized defect density to the theoretical calculations of the energy into atomic damage processes [20] shows good agreement. Thus the defect production approximately scales with the primary energy deposited into displacement processes.

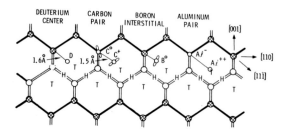

IMPURITY CENTERS IN SILICON

|110| PLANE OF SILICON

⊗ Si ATOMS IN PLANE
○ Si ATOMS ABOVE/BELOW PLANE
o IMPURITY ATOMS

Fig. 6. Schematic of various impurity sites in Si determined by channeling (D) and by EPR (C, B, Al)[4].

142

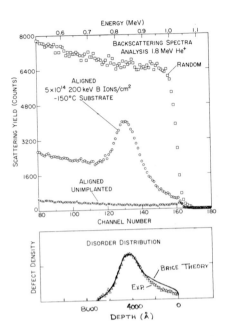

Fig. 7. Channeling energy spectrum and the analyzed disorder profile for
B-implanted Si[5].

In general for low flux implants in Si at room temperature or below,
scaling between the defect profile and the energy into atomic processes
has been observed. However scaling of the absolute number of displaced
Si with ion mass is not observed. With increasing mass a decreasing
amount of energy is required per displaced Si atom[21]. Except for very
light ions (e.g., H, He) an appreciably smaller energy per displaced Si
is required than the values (~ 15-30eV) predicted by collision theory for
isolated displacements. For heavy ions at low keV energies, the values
are less than 1eV. This effect is due to the dense energy cascades created
by keV ions. Collision cascade density effects have been directly demonstrated by showing that molecular implants, where collision cascades can
instantaneously overlap (Fig. 8), produce more displacements per atom than
atomic implants of the same particles (e.g., Te_2^+ vs Te^+)[21]. A continuous
transition in this displacement efficiency from light to heavy ions is
observed and attributed to inter-cascade and intra-cascade interactions
at high energy densities[21].
Furnace or laser annealing has been found to effectively remove the
implantation disorder to allow electrical activation of the implanted ions.
Ion channeling disorder measurements have been useful in helping to identify
the correct processing conditions. Under certain conditions secondary
defects can be produced which are difficult to remove by subsequent
processing. Often these are extended defects such as dislocations. An
example of the influence of a dense dislocation network on the channeling
spectra is given in Fig. 9 for the case of P-implanted Si to $1\times10^{16}/cm^2$

143

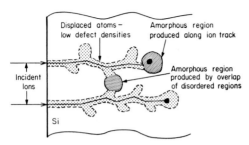

Fig. 8. Schematic of molecular ion implantation disorder tracks.

followed by an annealing sequence to 550°C, 900°C and 1050°C[22]. The
lattice contraction induced by the high P concentration has resulted in a
regular array of misfit edge dislocation as seen by TEM at a precise depth
at the tail of the P concentration. The <111> axial and (110) planar
channeling spectra both exhibit well-defined dechanneling steps in the
spectra but no direct scattering (as discussed previously). From the
measured step height and the dechanneling width (~40-350Å for
axial and planar directions) around the dislocations for edge dislocations,
the total projected length of dislocations in the layer can be determined.

Epitaxial Layers
    Thin crystal growth on crystalline substrates has had great technologi-
cal impact on semiconductor electronics, and investigations of novel or
improved methods to form epitaxial layers continue to be important. Ion
backscattering and channeling is an ideal technique for the depth resolution
and thickness range of interest. This technique allows determination of the
epitaxial layer thickness and rate of growth, the presence and depth profile
of growth defects (e.g., stacking faults and twins) when present at high
densities, the mosaic spread of epitaxial layers and the influence of thermal
treatments on the growth defects.

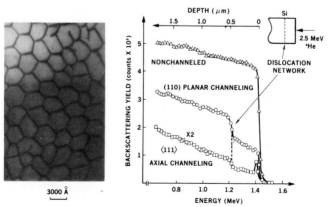

Fig. 9. TEM and channeling spectra for a dislocation network at a fixed
depth below the surface of Si[22].

Solid phase epitaxial growth of amorphous Si has been studied in detail by ion channeling[7]. High purity Si amorphous layers without interface contamination are formed on cyrstalline Si by Si implantation. Channeling measurements allow nondestructive determination of the amorphous layer thickness and the layer growth is observed to proceed by uniform movement of the amorphous-crystalline interface. At a given temperature the interface velocity is constant for (100) growth in the solid phase. Regrowth for (111) orientations is a factor ~ 4 slower (see Fig. 10). For (100) Si the channeling approaches that for a virgin defect-free crystral, whereas for (111) Si appreciable dechanneling remains even after complete layer regrowth (840 min. anneal spectrum in Fig. 10); TEM analysis shows that ~ 30-40% of the volume contains twinned material. The slower (111) growth rate and the presence of defects has been proposed to be due to growth along inclined (111) planes. For (111) growth rotation around the single bonds extending to the next plane is possible whereas the two bonds across the (100) planes do not allow the bond rotation, consistent with the uniform, defect-free growth for the (100) orientation. The orientation dependence is summarzied in Fig. 11[17].

The temperature dependence of the regrowth rate is seen in Fig. 12 to be an activated process with a characteristic energy E_a = 2.35 eV for Si[7]. Since regrowth is observed at the order of one-half the melting point for Si, the solid phase regrowth process cannot involve long range atomic motion. It has been inferred that E_a represents the activation energy for bond breaking to allow reordering at the interface. The addition of certain impurities at concentrations greater than 0.1 at.% alters the regrowth rate (see Fig. 13). For example, O, C, and N slow the regrowth, presumably because these impurities tie up broken bonds. In contrast, P, B, and As increase the regrowth rate and it is believed that this is because these impurities tend to weaken bonds and hence increase the probability of bond breaking.

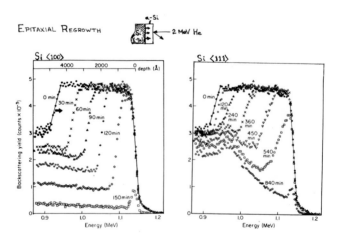

Fig. 10. Channeling spectra for Si amorphous layers on Si formed by Si implantation as a function of regrowth time for annealing at 525C for the <100> orientation and 550C for the <111> orientation[7].

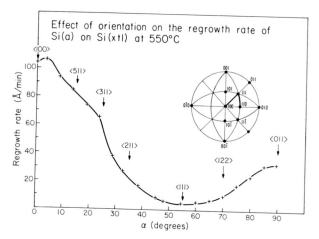

Fig. 11. Orientation dependence of solid phase amorphous Si epitaxial regrowth rate[7].

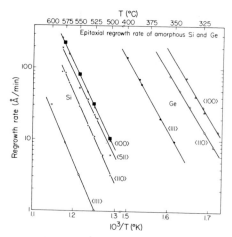

Fig. 12. Temperature dependence of solid phase amorphous Si and Ge growth rates[7].

Epitaxial layers of silicon grown on sapphire substrates typically have a high density of growth defects. If crystalline layers of high quality could be obtained on insulating substrates they would have distinct advantages such as higher speed and lower power dissipation in certain microelectronic circuit applications.

Figure 14 shows the channeling spectra for a typical 500 nm layer of <100> Si on <1102> sapphire [10]. The high yield for the channeled spectrum relative to bulk Si indicates a large amount of dechanneling due to defects. The inset shows a transmission electron micrograph identifying the presence

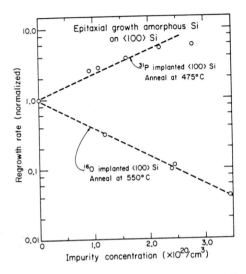

of stacking faults and microtwin lamallae at high densities. The slope of the spectrum increases rapidly with decreasing energy as the Si/sapphire interface is reached (at ≈ 0.55 MeV in Fig. 14). This corresponds here to increased dechanneling and indicates a higher density of defects near the interface as is typical for Si on sapphire.

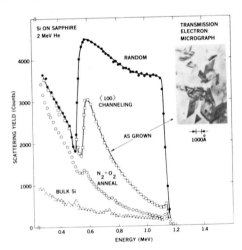

Fig. 14. Channeling spectra for <100> Si epitaxial layers as grown on <1102> sapphire and after 1150°C furnace anneal in $N_2 + O_2$ gas[10].

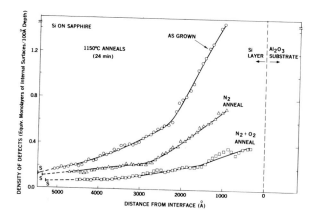

Fig. 15. Defect depth profiles in Si epitaxial layers on sapphire obtained from channeling spectra like those in Fig. 14 for 1150°C anneals[10].

Application of channeling analysis allows the depth profile of the defects to be determined, as shown in Fig. 15. The projected area of twin or stacking fault surface encountered in an increment of depth is expressed in units of equivalent monolayers. Studies of furnace heat treatments of these Si heteroepitaxial layers have shown that annealing in flowing mixtures of N_2 and O_2 gas can result in a large reduction in the defects present (see Fig. 15)[10]. It is believed that the presence of O_2 enhances the rate of vacancy generation at the SiO_2-Si interface near the Si surface. Enhancements in the flow of vacancies and/or interstitials over equilibrium values can enhance annihilation of growth defects such as stacking faults and microtwin lamallae.

Changes in the defect depth profile such as shown in Fig. 15 are difficult to infer from electrical measurements alone. In the case of these high temperature treatments some undesirable migration of Al from the sapphire into the Si accompanies the defect annealing. At higher anneal temperatures (1200°C) this Al contamination appears to be responsible for the observed inhibiting of the defect annealing.

SUMMARY

The channeling technique is particularly well suited for studies involving high densities of defects where the depth profile of the disorder within the first several micrometers of the sample surface is of interest. For intrinsic defects the channeling technique primarily provides a monitor of the density of defects present. The non-destructive nature of the technique makes it well suited for investigation of mechanisms and kinetics associated with defect generation, evolution and annihilation. Quantitative profile measurements require that a single type of defect be present and often necessitate complementary, defect-specific techniques such as transmission electron microscopy. For impurity-associated defects the crystallographic site of the impurity atom can often be determined by angular scans along channeling directions. This provides valuable information on the nature and structure of such defects.

148

REFERENCES

1. D. S. Gemmell, Rev. Mod. Phys. $\underline{46}$, 129 (1974).

2. C. W. White, S. R. Wilson, B. R. Appleton, and F. W. Young, J. Appl. Phys. $\underline{51}$, 738 (1980).

3. S. T. Picraux, W. L. Brown, and W. M. Gibson, Phys. Rev. $\underline{B6}$, 1382 (1972).

4. S. T. Picraux and F. L. Vook, Phys. Rev. $\underline{B18}$, 2066 (1978).

5. J. E. Westmoreland, J. W. Mayer, F. H. Eisen, and B. Welch, Radiation Effects $\underline{6}$, 161 (1970); J. E. Westmoreland, PhD Thesis, Caltech, 1970.

6. D. A. Tompson, R. S. Walker, and J. A. Davies, Radiation Effects $\underline{32}$ 135 (1977).

7. L. Csepregi, J. W. Mayer, and T. W. Sigmon, Appl. Phys. Lett. $\underline{29}$, 92 (1976); S. S. Lau, J. W. Mayer, and W. F. Tseng in "Handbook on Semiconductors" (S. P. Keller, ed.) Vol. 3, Chap. 7 (1980).

8. S. T. Picraux, D. M. Follstaedt, P. Baeri, S. U. Campisano, G. Foti, and E. Rimini, Radiation Effects $\underline{49}$, 75 (1980).

9. G. Foti, Nucl. Inst. and Methods (in press).

10. S. T. Picraux and P. Rai-Choudhury, in Semiconductor Characterization Techniques, Ed. by P. A. Barnes and G. A. Rozgoniji (Electrochemical Society, Princeton, NJ, (1978) p. 447.

11. D. Ronikier-Polonsky, G. Desarmot, N. Housseau, and Y. Quere, Radiation Effects $\underline{27}$, 81 (1975).

12. D. Desarmot and Y. Quere, Acta Metallurica (in press).

13. W. K. Chu, J. W. Mayer, and M. A. Nicolet, Backscattering Spectrometry (Academic Press, NY, 1978).

14. S. T. Picraux in New Uses of Ion Accelerators, Ed. by J. F. Ziegler (Plenum Press, NY, 1975) p. 229.

15. B. R. Appleton (these proceedings).

16. W. K. Chu (these proceedings).

17. L. C. Feldman, J. W. Mayer and S. T. Picraux, Channeling Analysis of Materials, (Academic Press, NY, to be published).

18. S. T. Picraux, Nucl. Inst. and Methods (in press).

19. S. T. Picraux, F. L. Vook, and H. J. Stein, Inst. Phys. Conf. Series No. 46, (Institute of Physics, London, 1979), p. 31.

20. D. K. Brice, Ion Implantation Range and Energy Deposition, (Plenum Press, NY 1975).

21. D. A. Thompson, R. S. Walker and J. A. Davies, Radiation Effects 32, 135 (1977).

22. S. T. Picraux, J. A. Knapp and E. Rimini (private communication).

Published 1981 by North-Holland, Inc.
Narayan, and Tan, eds.
Defects in Semiconductors

DIFFUSE X-RAY SCATTERING FOR THE STUDY OF DEFECTS IN SILICON*

B. C. LARSON AND J. F. BARHORST
Solid State Division, Oak Ridge National Laboratory, Oak Ridge, TN 37830, USA

ABSTRACT

X-ray diffuse scattering can be used to study the size, concentration, and nature of lattice defects and defect clusters in crystalline materials. The availability of analytical models and detailed numerical calculations for the "Huang" diffuse scattering close to Bragg reflections and the "asymptotic" diffuse scattering at somewhat larger distances from Bragg reflections has made it possible to carry out detailed analyses of the intensity and angular distribution of the diffuse scattering in terms of specific defect parameters. Accordingly, diffuse scattering has been used to study point defects, dislocation loops, and precipitates in metals, semiconductors, and insulators and has provided information on defect geometries, size distributions, and concentrations. In this paper, the theoretical framework necessary for the interpretation of the "Huang" and "asymptotic" diffuse scattering from defect clusters is presented and examples of diffuse scattering calculations for defects in silicon are given. Diffuse scattering from clustered defects in silicon is discussed in terms of the type of information available from this scattering and the relative merits of diffuse scattering as an investigative tool for defect clusters. Diffuse scattering analysis techniques for the determination of separate size distributions for vacancy and interstitial dislocation loops in silicon are presented and applied to diffuse scattering measurements on neutron irradiated silicon. Dislocation loop densities and sizes determined in the as-irradiated state and after thermal anneals are compared with electron microscopy results.

INTRODUCTION

X-ray diffuse scattering has been used as an investigative probe in a variety of lattice defect studies in recent years. The lattice distortions around crystal defects such as single interstitials and vacancies, vacancy and interstitial loops, and solute precipitates and inclusions, give rise to diffuse scattering in the vicinity of Bragg reflections. This scattering can be analysed to obtain information about the number, size, and characteristic shape of the defects by comparison with theoretical scattering models [1,2], and a number of studies of point defects in metals [3,4] and insulator [5] crystals, dislocation loops in irradiated materials [6,7], and solute precipitation in dilute alloys [8] have been carried out using this technique.

*Research sponsored by the Division of Materials Sciences, U. S. Department of Energy under contract W-7405-eng-26 with Union Carbide Corporation.

152

Defect clusters in neutron irradiated silicon have been investigated in
some detail using electron microscopy [9,10], anomalous x-ray transmission [11],
and integral x-ray diffuse scattering [12]. Although some earlier reports have
indicated the presence of very large defect clusters in neutron irradiated silicon, recent work [12] by both integral x-ray diffuse scattering and transmission
electron microscopy have found no evidence for such large (>25 nm radius)
defect clusters. Rather, high concentrations of small dislocation loops, that
were found to thermally anneal at about 700°C, were identified.

In the present paper, we will discuss the application of differential diffuse scattering to the study of defect clusters in silicon and use measurements
on neutron irradiation induced loops in silicon to illustrate this application.
The theoretical diffuse scattering framework will include the Huang scattering
close to Bragg reflections and will discuss the possibility of differentiating
between planar and spherical defect aggregates. Similarly, the asymptotic
diffuse scattering found at somewhat larger distances from the Bragg reflections, will be discussed in terms of distinguishing between vacancy and
interstitial loops as well as the determination of their size distributions and
concentrations.

Theoretical scattering models will be demonstrated through the use of
detailed numerical calculations, and these calculations will, in turn, be used
to perform the loop concentration and size distribution analysis on the interstitial and vacancy loops. In addition, the effects of thermal annealing on
these loops will be considered and the results will be compared with transmission electron microscopy measurements.

THEORY

The theoretical framework for the scattering of x-rays from defect clusters
has been discussed in a number of papers [1, 2, 13, 14]. The differential
diffuse scattering cross-section is given by

$$\frac{d\sigma(\vec{K})}{d\Omega} = (r_e F)^2 \left| A(\vec{K}) \right|^2 \tag{1}$$

where $A(\vec{K})$ is the scattering amplitude, $\vec{K}$ is the scattering vector (reciprocal
space position of the scattering), $r_e F$ is the product of the classical electron
radius, r_e, and the atomic scattering factor, F, including thermal and polariza-
tion effects. The scattering amplitude for dislocation loops is given by

$$A(\vec{K}) = \sum_j e^{i\vec{K}\cdot\vec{r}_j^L} + \sum_i e^{i\vec{q}\cdot\vec{r}_i} (e^{i\vec{K}\cdot\vec{S}_i} -1) \tag{2}$$

where the first sum is over the atoms comprising the dislocation loop and the
second sum is over the atoms surrounding the loops. The scattering vector $\vec{K}$
is comprised of a reciprocal lattice vector $\vec{h}$ and a vector $\vec{q}$ relative to $\vec{h}$.
$\vec{S}_i$ is the displacement of the ith atom as a result of the displacement field
of the dislocation loop.

$A(\vec{K})$ can be written [14] as the sum of three terms

$$A(\vec{K}) = A_1(\vec{K}) + A_2(\vec{K}) + A_3(\vec{K}) \tag{3}$$

where

$$A_1(\vec{K}) = \frac{b\pi R^2}{V_c} \frac{2J_1(QR)}{QR} e^{i\Delta_h^L}$$ (4)

and

$$A_2(\vec{K}) = i\vec{K}\cdot\vec{S}(\vec{q})$$ (5)

and

$$A_3(\vec{K}) = \frac{1}{V_c} \int \text{Cos}(\vec{q}\cdot\vec{r}) \left[\text{Cos}(\vec{K}\cdot\vec{S}(\vec{r}))-1 \right] d\vec{r}$$

$$- \frac{1}{V_c} \int \text{Sin}(\vec{q}\cdot\vec{r}) \left[\text{Sin}(\vec{K}\cdot\vec{S}(\vec{r})) -i\vec{K}\cdot\vec{S}(\vec{r}) \right] d\vec{r}$$ (6)

In these expressions, b is the Burgers vector of the loop, R is the loop radius, $J_1(QR)$ is the first order Bessel function, Q is the projection of $\vec{q}$ onto the dislocation loop plane, V_c is the cell size, $e^{i\Delta_h^L}$ is the phase factor of the atoms in the loop, and $\vec{S}(\vec{q})$ is the Fourier transform of the displacement field, $\vec{S}(\vec{r})$. Eq. (4) is the Laue, or direct scattering from the loops, Eq. (5) is the so-called Huang scattering, which results from the long range part of the strain field from the defects, and Eq. (6) represents the scattering from the highly distorted $\vec{K}\cdot\vec{S}(r) > 1$) region near the loops. The Laue and Huang scattering terms can be calculated in closed form [14]; however, the scattering in Eq. (6) must be calculated through numerical integration, making use of numerically calculated displacement fields, $\vec{S}(\vec{r})$.

At small q $(q \ll 1/R)$ the scattering can be given by

$$\frac{d\sigma(\vec{K})}{d\Omega} = (r_e F)^2 \left|A_2(\vec{K})\right|^2$$ (7)

and for somewhat larger q $(q \lesssim 1/R)$, the scattering can be approximated by

$$\frac{d\sigma(\vec{K})}{d\Omega} = (r_e F)^2 \left[\left|A_2(\vec{K})\right|^2 + 2 A_2(\vec{K}) A_3^s(\vec{K}) \right]$$ (8)

where $A_3^s(\vec{K})$ represents the first term in Eq. (6). Whereas $A_2(\vec{K})$ is anti-symmetric with respect to q, Eq. (7) is, of course, symmetric. $A_3^s(\vec{K})$ is symmetric, so Eq. (8) has an asymmetric form. The nature of the asymmetry associated with Eq. (8) is such that, for defects with positive misfit volumes (interstitial type), the scattering is larger for q parallel to h than for q anti-parallel to h, and vice-versa for defects with negative misfit volumes.

The interference effects associated with the large distortions near the dislocation loops lead to a transition from the $1/q^2$ scattering law as indicated in Eq. (7) to a $1/q^4$ falloff in the scattering for $q > 2/R$. This larger q region is referred to as the asymptotic scattering region and the scattering in this region is roughly proportional to R^2/q^4. The scattering is therefore proportional to the total area of the loops, independent of their size or size distribution.

154

Because of the complexity of the scattering and the large amount of interference in the scattering for larger q's, all of the terms in Eq. (3) must
be included in calculating the scattering for the asymptotic region.
For experimental measurements carried out on samples with distributions
of defect sizes, the scattering power is given by

$$I(\vec{K}) = \frac{I_o}{2\mu_o} \sum_i C_i \frac{d\sigma_i}{d\Omega} (\vec{K}) \Delta\Omega \tag{9}$$

in symmetric scattering geometry. In this expression, μ_o is the linear absorption coefficient, C_i is the volume concentration of defects with scattering
cross-section of $d\sigma_i/d\Omega$ and $\Delta\Omega$ is the solid angle subtended by the detector.

EXPERIMENT

The silicon investigated in this study was irradiated with 1×10^{20} n/cm^2
(E > 1 MeV) at ~50°C in the ORR reactor as part of an investigation [12,16] of
neutron transmutation doping in silicon. The samples examined were cut from
ingots in the form of 111 platelets and chemically polished to obtain damage
free surfaces. The x-ray diffuse scattering measurements were carried out in
the reflection geometry as shown schematically in Fig. (1). The x-rays were
generated by a 12 KW rotating anode generator, monochromated with a flat
graphite monochromator to provide CuKα x-rays, and the scattered x-rays were
detected with a linear position sensitive detector. The measurements were
carried out at 10 K in order to minimize thermal diffuse scattering, and the
Compton and thermal diffuse scattering were separated from the defect scattering
by making measurements on unirradiated samples.
The scattering geometry, shown as a composite of real and reciprocal space
in Fig. (1), indicates the use of the position sensitive detector and shows the
corresponding reciprocal space measuring position. Measurements were made
radially outward from the reciprocal lattice point by orienting the crystal
very near to the Bragg reflection, and by setting the crystal at regular
intervals off the Bragg angle, measurements were carried out along arcs not
intersecting the reciprocal lattice point.

CALCULATIONS

The availability of numerical displacement field data for dislocation loops
in elastically anisotropic materials [17] makes it possible to carry out
accurate scattering calculations using Eqs. (3-6). In this study, calculations
were performed assuming an elastic continuum [14,18], and for silicon, the
summations in Eq. (2) ranged over unit cells containing two atoms, so that F
represented the structure factor of the two atom unit cell [13]. (111) edge
type dislocation loops lying on {111} planes in silicon were considered, with
radii varying from 0.375 to 2.0 nm. As an illustration, Fig. (2) shows
calculations for 0.75 nm interstitial and vacancy loops in silicon. These
calculations were performed for q's near the 111 reciprocal lattice point,
along the trace of the Ewald sphere for CuKα x-rays and in the plane defined
by the [111] and [2̄11] directions. Making use of the experimental geometry
in Fig. (1), the Ewald sphere trace makes an angle of 14.22 degrees with the
[111] direction (this trace is shown directly in Fig. (8a)). The calculated
cross sections in Fig. (2) are multiplied by q^4 in order to emphasize the
structure in the asymptotic region. The important aspects of the scattering

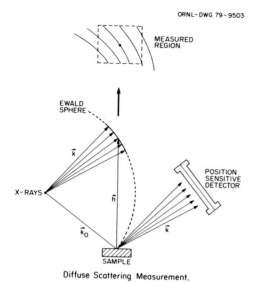

ORNL-DWG 79-9503

MEASURED REGION

EWALD SPHERE

$\vec{k}$

POSITION SENSITIVE DETECTOR

X-RAYS

$\vec{h}$

$\vec{k}_o$

$\vec{k}$

SAMPLE

Diffuse Scattering Measurement.

Fig. 1. Schematic view of the diffuse scattering geometry. This is a composite of real and reciprocal space, indicating the incoming wave vector, $\vec{k}_o$, the scattered wave vector, $\vec{k}$, and the role played by the Ewald sphere in determining the measuring position in reciprocal space.

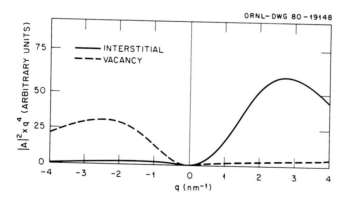

Fig. 2. Numerically calculated scattering cross-sections for 0.75 nm vacancy and interstitial loops on {111} planes of silicon. The calculations pertain to the 111 reflection using CuKα radiation.

are that the intensity is confined largely to the positive q range for interstitial loops and to the negative q range for vacancy loops, and that there is a well defined peak in the scattering when scaled by q^4. These peaks in the

scattering are related to Bragg-like scattering from the highly strained region
near the loops [14,18], and therefore, the reversal in the position of this
peak for vacancy as compared to interstitial loops is expected because of the
reversal in the sign of the strain. The strain in the vicinity of dislocation
loops has been shown to be ~-b/4R, and although detailed calculations are
needed to determine the precise nature of the scattering, the calculated peak
in the scattering has been found to occur at q/h~-b/4R. This means that the
size of loops as well as the type and number can be determined from diffuse
scattering measurements.

Similar calculations for spherical precipitates are shown in Fig. 3.
These calculations were made for .75 nm inclusions with misfit volumes equal
to that of .75 nm loops. In these calculations we see significant intensity
on both sides of the Bragg reflection. Such behavior is explained by the fact
that the volume inside of the sphere has an expanded lattice parameter while
the volume immediately outside the sphere is compressed as a result of the
expanded sphere; thus leading to Bragg like scattering from both compressed
and expanded regions [15].

Whereas the scattering in the asymptotic region from spherical precipitates or inclusions would be difficult to distinguish from the scattering from
a combination of both vacancy and interstitial loops, it has been well

ORNL-DWG 80-19434

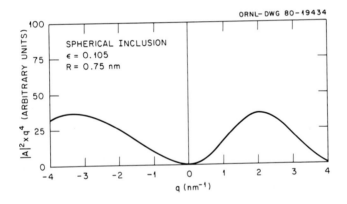

Fig. 3. Numer-
ically calcu-
lated diffuse
scattering
cross-sections
for 0.75 nm
spherical in-
clusions in
silicon. The
strain inside
the inclusion
was $\varepsilon = 0.105$.

established [8] that the scattering in the Huang region is quite different for
the two cases. Loops scatter more or less isotropically, while the scattering
from spherical defects contains a nodal plane normal to the reciprocal lattice
vector. This aspect will be discussed in more detail later.

RESULTS

The diffuse scattering measured on neutron irradiated silicon is presented
in Fig. (4). This figure shows the scattering from the as irradiated silicon
and the scattering after annealing for 30 min. at 550°C and 650°C. The
intensity plotted in Fig. (4) represents the measured diffuse scattering after
scaling by q^4 and removing the geometrical parameters and the structure factors
in Eq. (9). The scattering in Figs. (4) can, therefore, be compared directly
with the calculations in Fig. (2). Accordingly, we see the measured intensity

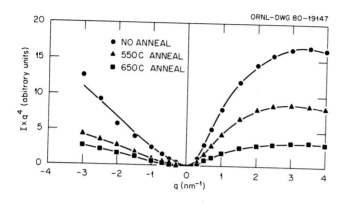

Fig. 4. Diffuse scattering measurements made on 1×10^{20} n/cm^2 irradiated silicon. The measurements have been scaled by q^4 so that they can be compared directly to Figs. 2, 3. Measurements were made in the as-irradiated state and after annealing at 550°C and 650°C.

at small q increases parabolically, as expected for the Huang region under a q^4 weighting, and the intensity at the larger q's is qualitatively similar to that at the larger q's in Fig. (2). The results for the 550°C anneal and the 650°C anneal indicate substantial decreases in the defect concentrations after these anneals.

The shape of the calculated data in Fig. (2) indicates that 0.75 nm loops would not be adequate to explain the measured results, but, recalling that the position of the peaks in the calculated data vary as q $\sim$ -bh/4R, it can be anticipated that the inclusion of loops with varying sizes would provide an adequate fit to the data. Assuming that only dislocation loops are present, it can be inferred that substantial numbers of both vacancy and interstitial loops are represented.

The solid lines in Fig. (4) are fits to the measured data using vacancy and interstitial loops of 0.375, 0.75, 1.0, 1.5, and 2.0 nm radii, where the concentrations of loops of each size and type were used as independent parameters. Although there is some discrepancy for the non-annealed case at negative q, the fit is in general quite good. The results of these fits are shown in Figs. (5-7). Fig. (5) shows the size distribution corresponding to the fit to the non-annealed case. Separate size distributions are shown for the vacancy and interstitial type loops, as well as for the composite size distribution. The histograms were generated by distributing the 0.375 and 0.75 nm loop concentrations over ± 0.375 nm, and by distributing the 1.0, 1.5, and 2.0 nm loops over ±0.5 nm and converting the concentrations to loops/cm^3/nm. The area of the graph represents the total loop concentration in this case.

The loops were found to be quite small in size. The interstitial loops have the large majority of their number in the size range of less than 1.0 nm, and the vacancy loops are represented soley by the 0.375 nm size, which is distributed over ±0.375 nm. In the composite size distribution, a rather exponential like falloff in concentration with increasing size is evident.

Similar results are shown for the 550°C and 650°C anneals in Fig. (6) and Fig. (7). The size distributions for the annealed cases are remarkably similar to the non-annealed size distribution, the only substantial difference being in the defect concentrations. Table I shows the point defect concentrations contained in these loop distributions along with the average sizes of the loops and the ratio of vacancies to interstitials in the distributions.

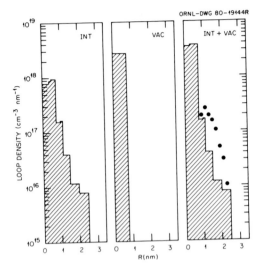

Fig. 5. Size distributions of vacancy and interstitial dislocation loops in neutron irradiated silicon. The concentrations are plotted in the form of loop density per cm^3 per nm. Points: TEM measurements [9] scaled to 1 x 10^{20} n cm^{-2}.

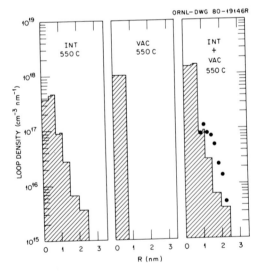

Fig. 6. Size distribution of vacancy and interstitial loops in neutron irradiated silicon after annealing at $550°C$ for 30 minutes. Points: TEM measurements [9] scaled to 1 x 10^{20} n cm^{-2}.

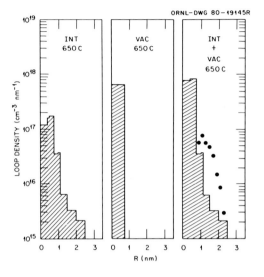

Fig. 7. Size distribution
of vacancy and interstitial
loops in neutron irradiated
silicon after annealing at
650°C for 30 minutes.
Points: TEM measurements
[9] scaled to 1×10^{20} n cm^{-2}.

In order to test the validity of the interpretation of the scattering in
terms of loops, the Huang scattering near the 111 reflection was measured for
the non-annealed case. The raw scattering data and the symmetric part of this
scattering are shown in Fig. (8a, b). These contours are to be compared
with the scattering calculations in Fig. (8c, d), which are based on the
measured size distribution in Fig. (4). The calculated contours have the same
general shape as the measured contours, so that the interpretation in terms of
loops seems to be reasonable; however, the calculated intensity for q's normal
to h =[111] seem to contain somewhat more intensity than found in the measurements. Calculations for spherical precipitates, as shown in Fig. (8e, f),
contain a nodal plane perpendicular to the [111] direction so that dislocation
loops seem to be the dominant defect, however.

DISCUSSION

The size distributions of the dislocation loops found in this study indicate most of the defects to be clustered into rather small loops. Although the
average sizes of these loops are smaller than those reported by electron
microscopy [9,12] (as shown in Figs. (5-7)), it should be noted that the
smallest loops reported here would likely be below the resolution limit of
the electron microscopy studies. Therefore, the smaller size loops would
not be expected in the microscopy results. The reasonable agreement between
the x-ray and microscopy measurements in the larger size ranges indicates
consistency between the two methods. On the other hand, studies involving
techniques such as magnetic resonance and optical measurements [19] directed
toward the study of much smaller (1 - 6 atom) aggregates in irradiated silicon,
have reported high concentrations of defect clusters with even smaller sizes
than those reported in Figs. (5-7). To an extent then, the x-ray diffuse
scattering technique provides a bridge between the studies of the small,
atomistic-like defects and the microscopy studies of the larger, extended
defects.

160

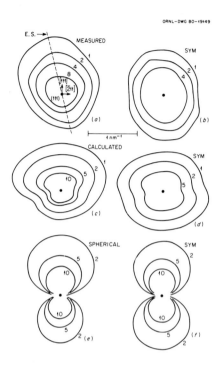

ORNL-DWG 80-19149

E.S.

MEASURED

SYM

(a)

(b)

1 nm⁻¹

CALCULATED

SYM

(c)

(d)

SPHERICAL

SYM

(e)

(f)

Fig. 8. Huang diffuse scattering intensity contours (small q region) near the (111) Bragg reflection of silicon. (a) Measured scattering from 1×10^{20} n/cm^2 irradiated silicon; (b) symmetric part of (a); (c) Huang scattering calculated for dislocation loops with the size distribution in Fig. 6; (d) symmetric part of (c); (e) Huang diffuse scattering calculated for 0.75 nm spherical inclusions in silicon; (f) symmetric part of (e).

The numbers of small defects, such as di-vacancies as determined by infra-red measurements and reported in ref. (12) for neutron irradiated silicon, should not be ignored when considering the numbers of larger defects produced. For the dose of 1×10^{20} n/cm^2, a di-vacancy concentration of 3.5×10^{20}/cm^3 was inferred through the use of the 1.8μm infrared absorption band. Comparing this to the x-ray value shown in Table I of 2.4×10^{19} point defects per cm^3, we see that the x-ray method has not included these defects. On the other hand, the differences between the measured and fitted data in Fig. 4 for the as-irradiated state would be consistent with the presence of vacancy-like clusters of very small sizes. Furthermore, this difference does not appear after annealing to 550°C, by which temperature di-vacancies anneal. This aspect needs further investigation in the x-ray diffuse scattering.

With regard to the rather small size of the dislocation loops found in this study, it should be remarked that the diffuse scattering techniques employed here make use of experience gained on radiation damage studies in metals [6,14,18]. While the measurement and analysis techniques are similar, the smaller size of the clusters found in the silicon case raises the question of the applicability of the continuum approximations used in the diffuse scattering calculations. It would seem likely, for instance, that calculations involving the smallest sizes (0.375 nm) would benefit from atomistic calculations, such as lattice statics methods [4]. The calculations for the spherical inclusions show, however, that for diffuse scattering along (or nearly along) the recipro-cal lattice vector, the scattering power depends to a large extent on the

TABLE I
Parameters obtained from the dislocation loop size distributions in Figs. 5-7.
In this table, V and I denote point defects and VL and VI denote vacancy and
interstitial loops, respectively.

	AS-IRRAD.	550°C	650°C
N_{PD} (10^{19} cm^{-3})	2.4	1.1	0.5
N_V/N_I	57/43	48/53	62/38
$<R>_{VL}$ (nm)	0.375	0.375	0.375
$<R>_{IL}$ (nm)	0.48	0.49	0.51

misfit volume and the size of the clusters, and not so sensitively on the shape
of the clusters. To be sure, there are significant differences between the
scattering from loops (Fig. (2)) and inclusions (Fig. (3)), but, in a broader
sense, the scattering from 0.75 nm vacancy and interstitials loops is qualitatively (and within a factor of 2-3, quantitatively) the same as that from
0.75 nm inclusions with the same misfit volumes. This observation should not be
taken to imply that calculations for one type of cluster may be substituted for
another, but rather to emphasize that the scattering in this scattering region
(along the reciprocal lattice vector) is more sensitive to the strength and
number of the clusters than to the detailed nature of the clusters.

As alluded to earlier, and as shown in Fig. (8), the Huang scattering is
quite sensitive to the shape of the defects; however, because of the quadradic
dependence of the Huang scattering on the misfit volume, it is not as well
suited for size determinations when a distribution of sizes are present. This
is due to the R^4 and R^6 weighting of the scattering from loops and spherical
clusters, respectively. The symmetric part of the Huang scattering, in
particular, contains the information on the characteristic shape of the
clusters, and Fig. (8) shows the symmetric part of the measured scattering as
well as calculations of the symmetric scattering for edge type loops and
spherical clusters. Clearly, the dislocation loop calculations fit the measured
data better, although the agreement could be improved. It is conceivable, for
instance, that a small admixture of scattering from spherical clusters would
improve the agreement between the experimental measurements in Fig. (8b) and
the calculations in Fig. (8d). Recall that the calculations in Fig. (8c,d)
were made using the size distribution obtained in the fit to the data in Fig.
(4). The consideration of more variables in the fitting procedure should,
however, await additional measurements, say normal to the reciprocal lattice
vector or around other reciprocal lattice points.

CONCLUSIONS

X-ray diffuse scattering can be used for the study of the size and concentration of small clusters in silicon. Information is available on the nature
of the clusters, as to planar or three dimensional, and as to the vacancy or
interstitial nature of dislocation loops. In the diffuse scattering application
reported here, the clusters were found to be small dislocation loops of both
vacancy and interstitial nature. Although reasonable agreement was found
between the x-ray results reported here and TEM results on similar systems; the
x-ray diffuse scattering measurements indicated a significant number of defects

with sizes below the TEM resolution. Correlations of these small defect clusters should be made with the high concentrations of even smaller defect clusters
inferred from magnetic reasonance and optical measurements.

REFERENCES

1. P. H. Dederichs, J. Phys. F $\underline{3}$, 471 (1973).

2. H. Trinkaus, Phys. Status Solidi (b) $\underline{54}$, 209 (1972).

3. P. Ehrhart, H. -G. Haubold, and W. Schilling, Advances in Solid State
    Research, $\underline{14}$, 84 (1974).

4. H. -G. Haubold and D. Martinsen, J. Nucl. Mat. $\underline{69-70}$, 644 (1978).

5. H. Peisl, J. Appl. Cryst. $\underline{8}$, 643 (1975).

6. B. C. Larson, in Fundamental Aspects of Radiation Damage, ed. by M. T.
    Robinson and F. W. Young, Jr, ERDA CONF-75 1006-P1, Oak Ridge, TN, 1976,
    p. 820.

7. B. C. Larson and F. W. Young, Jr., J. Appl. Phys. $\underline{48}$, 880 (1977).

8. B. C. Larson and W. Schmatz, Phys. Rev. $\underline{B10}$, 2307 (1974).

9. J. M. Pankratz, J. A. Sprague, and M. L. Rudee, J. Appl. Phys. $\underline{39}$, 101
    (1968).

10. P. L. F. Hemment and E. M. Gunnerson, J. Appl. Phys. $\underline{37}$, 2912 (1966).

11. T. O. Baldwin and J. E. Thomas, J. Appl. Phys. $\underline{39}$, 4391 (1968).

12. B. C. Larson, R. T. Young, and J. Narayan, in Neutron Transmutation
    Doping in Semiconductors, ed. by J. M. Meese (Plenum Press, New York 1979)
    p. 281.

13. H. Trinkaus, Phys. Status Solidi $\underline{51}$, 307 (1972).

14. B. C. Larson and W. Schmatz, Phys. Status Solidi $\underline{99}$, 267 (1980).

15. H. Trinkaus, Z. Angew, Phys. $\underline{31}$, 229 (1971).

16. J. W. Cleland, P. H. Fleming, R. D. Westbrook, R. F. Wood, and R. T. Young,
    reference 12, p. 261.

17. S. M. Ohr, Phys. Status Solidi (b) $\underline{64}$, 317 (1974).

18. P. Ehrhart, B. C. Larson, and H. Trinkaus (to be published).

19. J. W. Corbet, J. C. Bourgoin, L. J. Cheng, J. C. Correlli, Y. H. Lee,
    P. M. Mooney, and C. Weigel, Radiation Damage and Defects in Semiconductors
    (Inst. Phys. Conf. Ser. 31) p. 1.

DISLOCATION NUCLEATION MODELS FROM POINT DEFECT CONDENSATIONS IN SILICON AND GERMANIUM

T.Y. Tan
IBM T.J. Watson Research Center, Yorktown Heights, N.Y. 10598

ABSTRACT

The process of dislocation nucleation from point defect condensations in Si(Ge) is discussed. Based on the assumption that during the dislocation nucleation stage, the dominant factor in the configurational energy is the number of dangling bonds per point defect incorporated, rather than the more commonly recognized factor of strain energy, it is possible to model the dislocation nucleation process. In order to minimize the number of dangling bonds, point defects would condense into row configurations elongated in <110>, called intermediate defects (IDC), and then the IDCs would evolute into undissociated 90° edge -, 60°, and Frank partial dislocations.

INTRODUCTION

For many reasons the process of dislocation nucleation from condensations of point defects in Si(Ge) need to be studied. A most fundamental reason is that, being covalent materials, it is reasonable to expect the bonding nature of the atoms will influence the dislocation nucleation process to a significant extent so that it may be quite different from that in metals. There are a number of practical reasons too. To name a few important ones: (1). In electron, neutron and low dose ion damaged Si(Ge), simple point defects are usually produced. These are detectable by EPR, Raman or IR Spectroscopies because of their electrical activities.[1] After an annealing, however, such electrical activities cease and the defects are generally said to have been annealed out. Some such defects may have just disappeared due to vacancy-interstitial annihilation or migration to the surface, but it is very improbable that some of them did not aggregate together to form some sort of extended defect. Presently, it is not known just what such extended defects are but we could expect that some must be dislocations. (2). In high energy electron damaged thin Si(Ge) samples, and in thermally recrystallized, ion-implantation induced amorphous Si, there exists a peculiar class of defects not found in metals: the so-called rod-like defects (which, in our opinion, should also included {113} stacking faults).[2-13] The mystery of this class of defects is so far unresolved. It is possible that they are dislocations in an early nucleating stage or defects with structures closely associated with the dislocation nucleation process. (3). In high dose rate implantation, the production rate of point defects is high and the probability of point defect coalasing into dislocations would also be high. Such coalasing processes need obviously to be studied. In this paper we present atomic models describing the manner by which point defects may condense and generate dislocations. The morphology, energetic and kinetic factors implied by the models are discussed.

164

MODELING

Assumptions. In crystalline materials, the free energy associated with a point defect is high, and in the presence of a large number of them in supersaturation; they must eventually collapse into some form of extended defects, usually dislocations to reduce the total energy. For metals, because the outer shell electrons are free, we need only to compare the strain energy and the surface energy (line tension) of the dislocation loop to the free energy of the total number of point defects annihilated to form the loop.

For Si(Ge), the atoms form covalent bonds, and each dangling bond amounts to ~1eV of free energy. Thus, in these semiconducting materials the dislocation nucleation problem differs from that in metals in the sense that dangling bonds must be considered. Ideally, we need to consider both dangling bonds and strain. This, however, cannot be easily done, and, in constructing the models, our first assumption is that the dominant factor of the configurational energy of the defects is the number of dangling bonds per point defect incorporated rather than the more usually recognized strain energy. This assumption is actually not too surprising, thoug it may at first appear to be at odds with the traditional wisdom the dislocation community is familiar with, i.e., the strain energy. There are at least two factors indicating it may be so. First, vacancy aggregates (up to 4 vacancies) adopted configurations for which the number of dangling bonds for each involved atom are minimized, and interstitials are thought to behave similarly.[1] Second, dislocations are supposed to have reconstruct-ed cores with a minimum number of dangling bonds for the core atoms.[14] Our second assumption is, to begin with, vacancies and Si(Ge) self-interstitials are treated, but the effect of impurity atoms or interstitials is ignored. There are experimental evidences showing that the formation of rod-like defects (hence, in our view, dislocation nucleations) is dependent on the presence of certain impurites, e.g., B, C, and O.[8,11] This would lead one to think that such defects may be impurity precipitates. Unfortunately, the work 'precipitates' may immediatedly imply a phase or compound with a definitive structure which may not have covalent bonding. It thus seems reasonable that, in view of the fact that these atoms form covalent bonds with Si (Ge) atoms, at least at the beginning let us not consider their effect so as not to confuse the issue. Some effect of these impurities will be discussed later on.

Modeling Procedure. We construct the defect models by inspection which, at this stage, is about all one can rely upon. Such a procedure would yeield the possible low energy configurations. In two equivalent ways this may be carried out. In the first, defect configurations were obtained by defining some suitable point defect aggregates and inserting them into proper cuts made in the crystal lattice and reform bonds. In the second, the same defect configurations were obtained by generalizing some experimentally known or proposed point defect aggregates by rearranging bonds. The defect configurations are developed in two states: the point defects are first condensed into some intermedi-ate defect configurations (IDCs) and then the IDCs evolute into dislocations.[15]

The Defect Models. We consider the defect models of the 90°-edge the 60°-, and the Frank-partial dislocations for both intrinsic (accommodating vacancy) and extrinsic (accommodating interstitial) cases. The 90°-edge dislocation will be fully described according to the modeling process and the results for all other dislocations are presented.

We first define a point defect aggregate in Fig. 1. This corresponds to a <110> row of Si atoms in the crystal. A <110> row type of aggregates is the best choice because a reconstructed dislocation core with a minimum number of dangling bonds per atom has line directions in the <110>.

For the intrinsic case, the cut is shown in Fig. 2(a). After removal of defect aggregate (Fig. 1), the bond of the atoms surrounding the cut is reconstructed to arrive at Fig. 2(b). Fig. 2(b) actually shows the way the four-vacancy chain bonds, as detected by EPR, reconstructed. With some bond rearrangement in Fig. 2(b), we obtain Fig. 2(c), the IDC, which consists of two conjugate pairs of 5-7 membered atomic rings. If the bonds in Fig. 2(c) are further rearranged, we arrive at Fig. 2(d) which is clearly recognized as a dislocation dipole by recognizing the core configuration of the two disloca-tions. The bond rearrangement from Fig. 2(b) to 2(d) needs a shear stress on (001) plane in [110] direction. A dipole may also result from climb of the IDC, Fig. 2(e). We have shown here how to condense vacancy to IDC to dislocation dipole by the lattice cutting method as well as by deriving it from the known vacancy chain configuration.

For the extrinsic case, the cut is shown in Fig. 3(a). After inserting the aggregate and reform bonds we obtain Fig. 3(b), the IDC. This IDC may also be obtained form an already proposed interstitial aggregate configuration. Fig. 3(c) shows the necessary bond rearrangement from two adjacent <110> chain of interstitials (each occupying a split-<100> position) to arrive at Fig. 3(b). With bond rearrangement in Fig. 3(b) we obtain the dipole shown in Fig. 3(d) via glide of the IDC. Fig. 3(e) shows that obtained by climb of the IDC.

Similarly, we obtain the IDCs and dipoles for the other dislocations. Fig. 4 shows that for 60° intrinsic case, Fig. 5 for the Frank partial intrinsic case, Fig. 6 for the 60° extrinsic case and Fig. 7 for the Frank-partial extrinsic case. We have presented the 90°-edge dislocation of the glide set because two dangling bonds per core atom are present for the shuffle set (and hence less energetically favorable). We have presented the 60° dislocation of the shuffle set because the bond rearrangement is more complex for the glide set (hence kinetically less favorable).

DISCUSSIONS

Nature of the IDCs. The IDCs are the key to the problem of dislocation nucleation by point defect condensations in Si(Ge). Their further evolution produces dislocation configurations and themselves may be derived from an evolution of known point defect chain configurations. Thus, formation of IDCs represents a natural, low energy means (per bonding argument) of condensing a large number of point defects into chain type aggregates in Si(Ge). Both IDCs and the dislocation dipoles are rod-like (elongated in <110>). This is different from the case in metals for which the dislocations would be small circular loops. Also, the formation of the IDCs has no direct equivalence in metals. The generation of dislocation dipoles via the formation of IDCs shows that the dislocation nucleation process in Si(Ge) can be a gradual, step by step process and its activation energy need not be unreasonably high. Geometrically, we now recognize that the IDCs are but two dislocation core configurations placed right next to each other in the lattice in an appropriate manner. This is necessarily so, for only then they are low energy configurations from which dislocation core configurations may be derived.

Energetic and Kinetic Implications of the Models. It is obvious that the configurational and motional energies of the defect models need now to be calculated. However, before obtaining such information, a few implications may already be obtained by inspection of the models. If we now take the number of dangling bonds as an indication of configurational energy, then an inspection of the models would reveal that the 90° edge (and its IDC) is the lowest in energy, followed by the Frank partial and then by the 60° dislocations. If we now take the complexities of bond rearrangement (bond breaking and reforming) as an indication for activation energies in forming the defects then we see the 60° dislocation is the easiest to nucleate, followed by the 90° edge and then by the Frank partial dislocations.

We also note that the models are for undissociated dislocations. This need not to be regarded as strange, although most dislocations encountered daily all seem to have been dissociated. The reason is simply that before the undissociated dislocation dipoles are generated, creation of fault ribbons in the matrix crystal cannot proceed simultaneously with point defect condensations. A process which would nucleate dissociated dislocations from point defect condensations directly would amount to the creation of two fault ribbons in the perfect crystal (dissociation) and simultaneously these two fault ribbons need to climb past each other (condensation of point defects). Such a process is so complicated that, obviously, it cannot be (kinetically) favorable. we note, however, that once a dipole is generated from the IDC, dissocation may proceed if the temperature is high enough. We also note that a dissociated dislocation dipole would not accommodate any more point defects than an undissociated one and its climb motion would also be more difficult.

The Role of Impurities. The present models were constructed by the incorporation of Si(Ge) self-interstitials or vacancies, to emphasize that the dislocation nucleation process is primarily dictated by one of the materials intrinsic properties, namely, the bonds, rather than some properties of the impurity atoms which might confuse the issue if they were considered from the beginning. This, however, does not mean that impurity atoms do not play a role in the dislocation nucleation process. In fact, since a few important impurity atoms, i.e., B, C, and O do form covalent bonds with Si(Ge) atoms, their role may be as important as Si(Ge) self-interstitials. In Fig. 8 we show a possible way these impurity atoms may be participating in the formation of the $90°$ edge dislocation IDC. Here 50% of the Si(Ge) self-interstitials is replaced by C (or B) atom, and some O interstitials are occupying some bond-centered positions. C atoms may also replace some Si atoms and occupying positions such as 1, i.e., regular lattice atoms positions. It is noted that the ways illustrated for incorporating impurities is more like the normal segregation process. The presence of these impurities do not effect the evolution process of IDCs into dislocation dipoles. The manner these atoms are incorporated is totally compatible with the bonding argument. Thus, these models may have provided IDCs for impurity atom 'precipitation', but the impurities do not form any known stable compounds or phases. The presently discussed possibilities are consistent with experimental observations that: (1) B, C and O are important in the formation of rod-like defects,[8,11] (2). that the rod-like defects are meta-stable, they evolute into dislocations.[12,13] We further note that the incorporation of these impurities in the IDCs may lower the strain energy of the IDCs and may thus promote their formation.

CONCLUDING REMARKS

The factors that influence the process of dislocation nucleations appeared to be numerous, and it should not be expected that any simple proposal at this stage could have taken into account all the considerations. It is for the sake of convenience that in the present approach we assumed that during the dislocation nucleation stage strain energy is relatively unimportant and that there is a homogeneous presence of point defects and their diffusion is isotropic throughout the crystal. We notice that, however, the generation of dipoles from IDCs are dependent upon stresses, the strain energy of the IDCs are dependent upon impurity atom incorporations (which may also serve to 'pin' the IDCs or dipoles). As the IDCs develop, their self-stress field may bias the diffusion of point defects, hence at some stage it becomes anisotropic. During irradiation, the point defect distribution will not be homogeneous. Furthermore, the constituents of the IDC or dipole core region may be charged, which are dependent upon the impurities incorporated as well as the type (p or n) of the matrix material. All these factors would influence the stability of the IDCs and dipoles and hence the sizes of the rod-like defects, whether they cluster together, their temperature dependence of formations etc.. It is obvious that the effects of these factors need to be systematically investigated. Irrespective of these complications, however, it is believed the basic manner the point defects would condense to nucleate dislocations has been described, and it is hoped that models of the present type would provide a good basis for gaining a more complete understanding of the dislocation nucleation process from point defect condensations in Si(Ge), and of the nature of the rod-like defects.

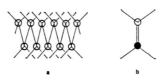

Fig. 1. The best choice of a point defect chain. (a). Oblique view, and (b). End view.

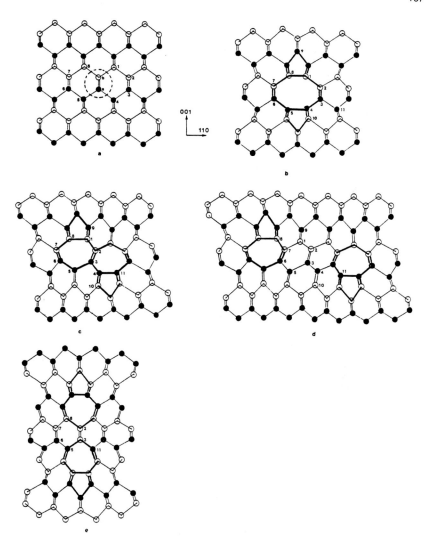

Fig. 2. Nucleation model of intrinsic 90° edge dislocation dipole.
(a). The cut, (b). Bond configuration for incorporating the vacancy
chain, (c). The IDC obtained by rearranging bonds in (b): breaking
5-4, joining 3-5 and 4-11, (d). The 90° edge dislocation dipole
due to glide of the IDC. The bond rearrangement is similar to that
obtaining (c) from (b), and (e). A dipole due to climb of the IDC.

168

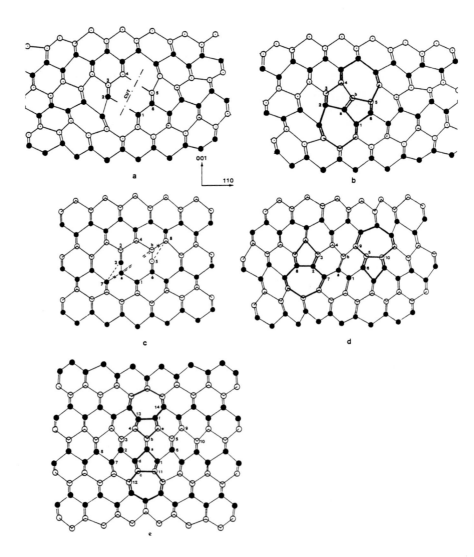

Fig. 3. Nucleation model of extrinsic 90° edge dislocation dipole.
(a). The cut, (b). The IDC obtained by inserting the point defect
chain into the cut and rearranging bonds, (c). Bond rearrangement
leading to the IDC from two adjacent rows of 100 -split interstitials, (d). The dipole due to glide of the IDC, and (e). The
dipole due to climb of the IDC.

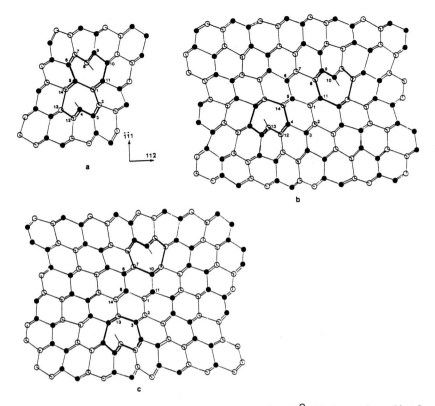

Fig. 4. Nucleation model of the intrinsic 60° dislocation dipole.
(a). The IDC, (b). The dipole due to glide of the IDC, and (c).
The dipole due to climb of the IDC.

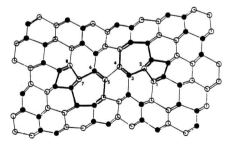

Fig. 5. Nucleation model of the intrinsic Frank partial dislocation
dipole.

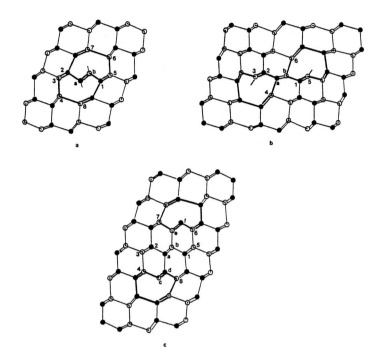

Fig. 6. Nucleation model of the extrinsic 60° dislocation dipole.
(a). The IDC, (b). The dipole due to glide of the IDC, and (c).
The dipole due to climb of the IDC.

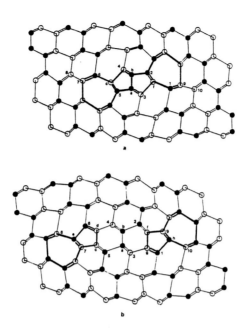

Fig. 7. Nucleation model of the extrinsic Frank partial dislocation dipole.

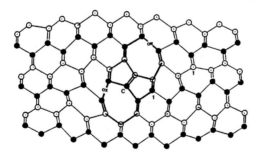

Fig. 8. Possible manner of incorporating impurity atoms (C, B, O) in an extrinsic IDC.

REFERENCES

1. J.W. Corbert and J.C. Bourgoin, in "Point Defects in Solids", eds. J.H. Crowford, L.M. Slifkin, (Plenum, N.Y., 1975) vol. 2, p.1.

2. D.J. Mazey, R.S. Nelson and R.S. Barnes, in "Proc. 6th International Conf. on Electron Microscopy", ed. R. Uyeda (Maruzen Co. L 1966) p. 363.

3. R.W. Bicknell, Proc. R. Soc. A31, 75 (1969).

4. S.M. Davidson and G.R. Booker, Rad. Eff. 6, 33 (1970).

5. L.T. Chadderton and F.H. Eisen, Rad. Eff. 7, 129 (1971).

6. P.K. Maden and S.M. Davidson, Rad. Eff. 14, 271 (1972).

7. M.D. Matthews and S.J. Ashby, Phil. Mag. 27, 1313 (1973).

8. W.K. Wu and J. Washburn, J. Appl. Phys. 48, 3742 (1977).

9. K. Seshan and J. Washburn, Rad. Eff. 14, 267 (1972).

10. E. Nes and J. Washburn, J. Appl. Phys. 42, 3559 (1972).

11. A.L. Aseev, V.V. Bolotov, L.S. Smirnov and S.T. Stenin, Sov. Phys. Semicond. 13, 764 (1979).

12. I.G. Salisbury and M.H. Loretto, Phil. Mag. A39, 317 (1979).

13. C.A. Ferreira Lima and A. Howie, Phil Mag. 34, 1057 (1976).

14. J. Hornstra, J. Phys. Chem. Solids 5, 129 (1958).

15. T.Y. Tan, Phil. Mag. to be published.

UNDISSOCIATED DISLOCATIONS AND INTERMEDIATE DEFECTS IN As+ ION DAMAGED SILICON

H. FOELL, T. Y. TAN AND W. KRAKOW
IBM Thomas J. Watson Research Center, P. O. Box 218, Yorktown Heights, New York 10598, USA

ABSTRACT

Undissociated dislocation dipoles and intermediate defects are detected in As+ ion damaged Si using the lattice resolution technique of transmission electron microscopy. The present results are consistent with our proposed dislocation nucleation models.

INTRODUCTION

Defect creation in Si due to radiation damage has been extensively studied. From EPR measurements, the configuration of a number of vacancy related defects has been well established although much less is known of the Si self-interstitials.[1] EPR experiments rely on the detection of the hyperfine (HF) spectrum produced by the dangling bonds, and they are interpreted by comparison to calculated electronic states of the defects. If the number of point defects in a cluster exceeds a few, HF interactions may produce a band and the spectra become uninterpretable. EPR, therefore, is not suited to investigate defects formed if the damage is extensive. Here more extended defects (dislocations) are produced. When the size of dislocations exceeds a few hundred Å they may be readily analyzed using TEM by observing their strain contrast, but if the dislocation size is within a few hundred Å conventional TEM will cease to provide useful information. There is no easy experimental way to bridge the gap between the observation of point defects by EPR and the analysis of larger sized dislocations by TEM. Perhaps for this reason the manner by which point defects condense into dislocations has not been extensively discussed, although there exists a large body of literature wherein the experimental results may be interpreted as observations of early stages of dislocation generations (see references in [3]). To address this subject, we have now modeled the point defect condensation process of dislocation nucleations.[2,3] Because of dangling bond considerations, it was reasoned that dislocations nucleate in the form of $<110>$ elongated, undissociated narrow dipoles as opposed to small circular loops in the case of metals. Before dislocation dipoles are generated, the point defects must first condense into some linear configuration called "Intermediate Defects" (IDC), a step which has no direct parallel in metals. In this paper we report a high resolution TEM study of As+ ion implanted Si which showed that our approach of modeling the dislocation nucleation process is essentially correct in that undissociated dislocation dipoles as well as IDC's were observed.

EXPERIMENTAL

For our experiments, P-type, 12 Ωcm, $<001>$ oriented Si wafers were implanted with As+ ions at 80keV to a dose of $8 \times 10^{15}/cm^2$ at $100\mu A$ ion current. This dose of implantation created an amorphous Si layer of about 800Å thick. In the crystalline Si beneath the α-Si many defects formed in situ during implantation due to the high dose rate. $<110>$ oriented cross-sectional thin foils, about 200Å thick, were prepared from these wafers and examined in a Siemens 102 electron microscope at 125kV. Defects were examined by the lattice imaging technique with seven Bragg beams, 000, ± $\bar{1}11$, ± 111 and ± 002, included in the objective lens aperture. The beams were tilted away from the optical axis by $\sim\frac{1}{4} |g_{002}|$.

RESULTS

As a result of this study, we have detected over 30 dislocation type defects, as well as a number of IDC type defects. Except for one case, all dislocations are undissociated. The majority of them are the 60° type with only two of the 90° edge type, and no Frank-partials. All IDC's are, however, of the 90° edge type. Four pairs of the 60° dislocations are spaced close enough to be identified as dipoles. Three dipoles have incorporated interstitials. The other incorporated vacancies and is shown in Fig. 1 together with drawings showing the possible structure. The core-core separation of this dipole is only 12.6Å. Figure 2 shows an IDC of the 90° edge case incorporated vacancies, and Fig. 3 shows an IDC of the 90° edge case incorporated interstitials.

DISCUSSION

We have used somewhat tilted beams to obtain the present micrographs. This was necessary for two reasons: (1) Since the thin part of the foil is always deviated a little from an ideal <110> orientation one can set up by using thicker parts of the foil, it was necessary to make small corrections of the beam incident direction. (2) More importantly, however, is that the microscope is having a 3.0Å point to point resolution, which is not sufficient for resolve the small localized configurational changes of the IDC's. However, they are resolvable with the beams tilted properly. We have carried out image calculations and our calculated micrographs confirms this point, Fig. 4.

The detection of dislocation dipoles and IDC's are sufficient proof that our proposal on dislocation nucleation models[2,3] is essentially correct, i.e. point defect condense into IDC's and evolve into <110> elongated dislocation dipoles. It is significant to note that the present results also contain the following details that are consistent with the models: (1) that the dislocations are primarily undissociated, just as it was proposed. (2) That the IDC of the 90° edge dislocation is very stable, so they are observed, the 60° dislocations (and their IDC's) are easy to form, but the IDC's are unstable, so we observed many 60° dislocations, but no IDC. (3) That the Frank partial should be hard to form, and we have not observed any.

CONCLUSION

The observation of dislocation dipoles and IDC's provided sufficient evidence of the correctness our proposed dislocation nucleation process from point defect condensations. These results also provided a basis for a better understanding of the nature of rod-like defects, namely, they can be point defect rows which generate dislocation dipoles.

175

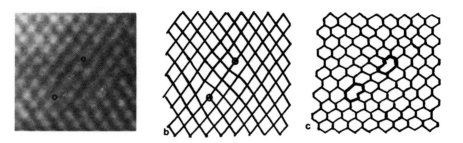

Fig. 1. (a). Lattice image of a 60° dislocation dipole, (b).
Schematic lattice fringes, and (c). Possible structure. The dipole
has a 12.6A core to core separation and the dislocation is not
dissociated.

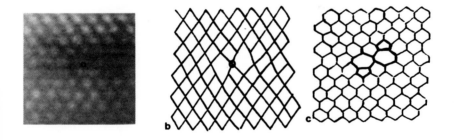

Fig. 2. (a). Lattice image of an IDC of the 90° intrinsic case,
(b). Schematic lattice fringes, and (c). The possible structure.

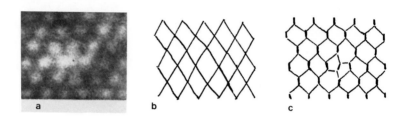

Fig. 3. (a). Lattice image of an IDC of the 90° extrinsic case, (b). Schematic lattice fringes, and (c). The possible structure.

Fig. 4. Calculated micrographs illustrating that small topographical changes in the lattice can not be resolved using axial illumination in a microscope with 3A resolution, (a), but it can be resolved using tilted beam illumination, (b). The model is that of an extrinsic IDC.

REFERENCES

1. J. W. Corbet and J. C. Bourgoin, in "Point Defects in Solids", eds. J. H. Crawford and L. M. Slifkin (Plenum, New York, 1975), Vol. 2, pp. 1.

2. T. Y. Tan, Phil. Mag, to be published.

3. T. Y. Tan, this proceeding.

A TENTATIVE IDENTIFICATION OF THE NATURE OF {113} STACKING FAULTS IN Si - MODEL AND EXPERIMENT

T.Y. Tan, H. Foell, S. Mader* and W. Krakow
IBM T.J. Watson Research Center, Yorktown Heights, N.Y. 10598

*IBM Data Systems Division, Hopewell Junction, N.Y. 12533

ABSTRACT

An atomic model with fully bonded interstitial atoms is proposed for the {113} stacking faults (SF). It is shown that the development of the {113} habit plane is a result of minimizing bond length changes. The fault may be represented by a $\frac{1}{11}<113>$ partial dislocation, and a $\frac{3}{22}<332>$ partial will facilitate an unfaulting reaction to result in a $\frac{1}{2}<110>$ $90°$ edge dislocation. Lattice images were obtained for a {113} SF which showed that the model is essentially correct.

INTRODUCTION

In high energy (1000keV) electon damaged thin samples of Si(Ge),[1-4] and in thermally annealed, ion-damaged Si crystals,[5,6] there exists a peculiar class of defects not found in metals: the so-called rod-like defects and {113} stacking faults (SF). In our previous papers,[7,8] it was shown that a better understanding of the nature of rod-like defects has now been obtained. The purpose of this paper is to show that for {113}SF it is also possible to obtain a model with all four coordinated Si(Ge) atoms and our preliminary experimental results tends to support this model.

THE MODEL

From observation of electron irradiated Si(Ge) thin samples in situ in high voltage electron microscopes, it is established that the {113}SF are characterized by: (1) a prominent growth in a $<110>$ direction and a minor growth in a $<332>$ direction, so that the habit plane of the defect is a {113}.[1-4] (2) The SF unfaults by a shear, which results in a unique 1/2 $<110>$ interstitial loop.[3] (3) The SF is extrinsic,[1-4] and (4) the matrix crystal is displaced by $\sim \frac{1}{25}<116>$.[2] The only reported model of this defect invokes the use of interstitial atoms each with three dangling bonds and the bonds of matrix crystal atoms on either side of the fault plane are stretched by 35-48%.[3] We view this model as highly unlikely because it implies a high energy situation. Here we propose an atomic model for the {113} SF with all 4-coordinated atoms satisfying the above listed properties. Fig. 1(a) shows the first step of the modeling: two neighboring rows of [1̄1̄1] bonds of a row of 6-membered rings are cut and the cell is pulled apart so as to insert a row of interstitial atoms. The formation of new bonds results in two conjugate pairs of 5-7 membered rings - this is just the IDC of the $90°$ edge dislocation.[7] There is no dangling bond for the inserted atoms not at the two ends of the defect. The two end atoms each have a dangling bond and the defect is easy to grow in the [11̄0] direction. The defect can now go into climb motion in a direction lying in the (11̄0) plane. As a result of not allowing the presence of any dangling bonds and minimizing atomic stresses, i.e., bond stretches, only two configurations may be derived, one is the $90°$ edge dipole due to climb,[7] the other is the one shown in Fig. 1b, for which the climb direction is $\pm$ [332] and the habit plane of the defect is $(1\overline{1}3)$. The (minor) growth in <332> direction is due to climb and a minimization of bond stretches, and in Fig. 1(b) every bond length is not changed by more than 5% of the normal Si atom bond length. Because the total number of interstitial atoms in the fault layer equals that of a monolayer of {113} plane, a Burgers vector of $\frac{1}{11}[11\overline{3}]$ may be defined for its bounding partial dislocation, and unfaulting reaction may be accomplished by the nucleation of a $\frac{3}{22}[332]$ partial. The 3/22 [322] partial sweeps through the fault, with bounding rearrangement shown in Fig. 1(c), to arrive at a $\frac{1}{2}[110]$ 90° edge dislocation dipole, see Fig. 1(d). Thus this $(1\overline{1}3)$SF satisfies the above discussed characteristics (1) to (3). However, the matrix crystal displacement is now $\sim \frac{1}{5}[\overline{1}11]$ instead of $\sim \frac{1}{25}<116>$. To rectify this situation, we consider the segregation of C and O interstitials to the defect as Aseev et al has observed.[16] If 50% of Si self-interstitials is replaced by C, and an equal number of O interstitials is segregated to the bond centered positions of the 7-membered rings, we obtain Fig. 2 which has a matrix crystal displacement of $\sim\frac{1}{25}[11\overline{6}]$. Notice that the covalent radii of Si, C and O are respectively 1.17Å, 0.77Å and 0.66Å, we see no severe bond length changes are involved. The incorporation of C and O do not effect the unfaulting reaction, but after unfaulting C and O will tend to de-segregate and they may form small ppt.-like defects. Salisbury and Loretto[3] and Aseev et al[4] have observed such phenomena.

EXPERIMENT

For our experiment, p-type20Ωcm, <001> oriented CZ Si wafers were implanted by 50keV H^+ ions to a dose of $4 \times 10^{16}/cm^2$ at a substrate temperature of 575°C. Weak beam imaging of these samples showed that {113} SF 40Å wide and up to 1μ long were produced in a depth of 2500Å to 5000Å below the sample surface.[6] <110> cross-sectional sample were made from these wafers and imaged in a Siemens 102 electron microscope in the lattice imaging mode by including seven beams, 000, $\pm\overline{1}11, \pm111, \pm002$ in the objective aperture.

Fig. 3(a) shows the lattice image of a {113} F. Two things are immediatedly clear: (1). The habit plane is $(1\overline{1}3)$, (2). When measured against a perfect Si crystal lattice drawn to scale, Fig. 3(b), it is found that the matrix crystal displacement is $\sim \frac{1}{25}[11\overline{6}]$. This is the first direct measurement of the displacement and the second accurate measurement using any method.[2]

The images of the atomic columns on the fault plane are generally hard to see. However, at the lower left end it is possible to analyze the finge patterns, and draw the atomic structures, Fig. 4(a). For this tiny patch it is seen that this is consistent with our model. Fig. 4(b) shows a calculated micrograph for comparison purposes. The micrograph contains regions where a direct comparison with our model is not possible probably because the fault is jogged.

CONCLUSION

A model that meets all the essential characteristics of the {113} SF is obtained with fully bonded interstitial atoms, and on a preliminary basis, is in good agreement with the experimental results obtained.

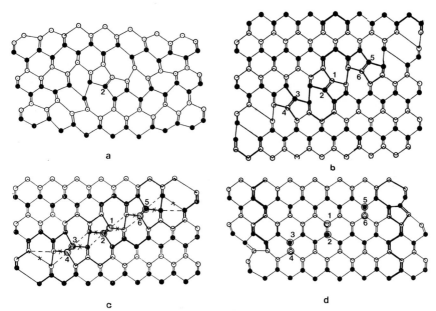

Fig. 1. The ($\bar{1}\bar{1}3$) SF model. (a). The extrinsic 90° edge dislocation IDC, (b). The ($\bar{1}\bar{1}3$) SF with a matrix displacement of 1/5[$\bar{1}\bar{1}1$], (c). Bond rearrangement during unfaulting by a 3/22[332] partial dislocation, and (d). The resultant extrinsic 90° edge dislocation dipole.

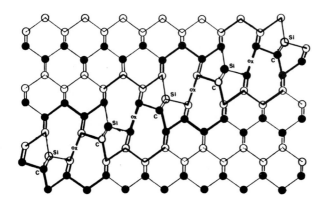

Fig. 2. The ($\bar{1}\bar{1}3$) SF model incorporated C and O to yield a matrix crystal displacement of 1/25[11$\bar{6}$].

182

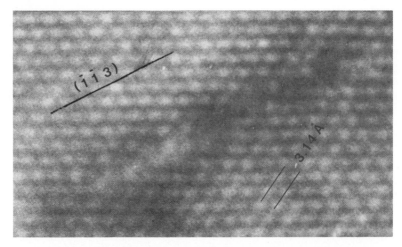

a

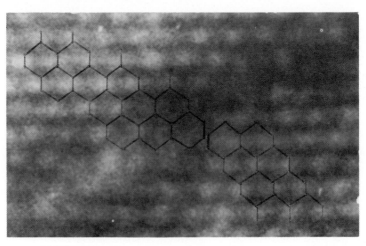

b

Fig. 3. An experimental (113) SF. (a). The lattice image, (b). An enlarged portion of (a) with two lattices drawn in on either side of the SF. The two lattices has a relative displacemant of 1/25 [11$\bar{6}$]. This shows that the matrix displacemant is ~1/25 [11$\bar{6}$].

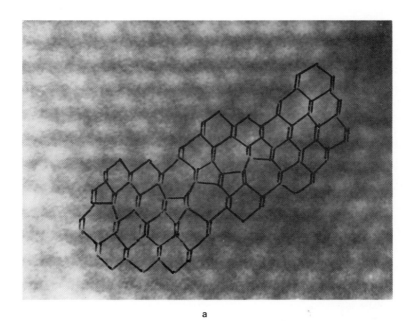

a

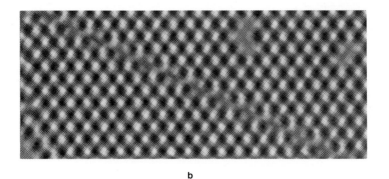

b

Fig. 4. (a). An enlarged view of the lower left part of the SF in figure 3(a) with possible structure drawn in. This shows that at least some part of the SF is consistent with our proposed model, (b). Calculated image of the SF with CTF=1.

REFERENCES

1. M.D. Matthews and S.J. Ashby, Phil. Mag. *27*, 1313 (1973).

2. C.A. Ferreira Lima and A. Howie, Phil Mag. *34*, 1057 (1976).

3. I.G. Salisbury and M.H. Loretto, Phil. Mag. A*39*, 317 (1979).

4. A.L. Aseev, V.V. Bolotov, L.S. Smirnov and S.T. Stenin, Sov. Phys. Semicond. *13*, 764 (1979).

5. W.K. Wu and J. Washburn, J. Appl. Phys. *48*, 3742 (1977).

6. W.K. Chu, R.H. Kastl, R.F. Lever, S. Mader and B.J. Masters, in "Ion Implantation in Semiconductors, 1976", Ed. F. Chernow, J.A. Borders and D.K. Brice (Plenum, N.Y., 1977) p.483.

7. T.Y. Tan, this proceeding.

8. H. Foell, T.Y. Tan and W. Krakow, this proceeding.

DETECTION OF POINT DEFECT CHAINS IN ION IRRADIATED SILICON BY HIGH RESOLUTION ELECTRON MICROSCOPY

W. Krakow, T.Y. Tan and H. Foell

IBM T.J. Watson Research Center, Yorktown Heights, N.Y. 10598 U.S.A.

ABSTRACT

In a lattice imaging study of As^+ ion damaged Si, we have detected $<110>$ chain type defects which are not associated with any significant strain or configurational changes. By image matching of the experimental and calculated micrographs of vacancies and interstitials, it is established that about 100% more interstitial atoms may incorporate into a defective chain. A structure model of this defect is proposed wherein a di-interstitial, occupying the $<100>$ split position, is incorporated into every available site along a $<110>$ chain.

INTRODUCTION

For the past twenty years defect creation in semiconductors due to radiation damage has been extensively studied using electron paramagnetic resonance where defect complexes consisting of a few point defects and impurity atoms have been considered[1]. Alternatively, transmission electron microscopy (TEM) has been used primarily to observe extended defects strain fields which are a minimum of a few hundred Å in extent by diffraction contrast. For smaller defect configurations there has been a minimal effort, since it constitutes a more difficult problem to obtain and interpret the data. It is for this reason the manner by which point defects condense into extended defects has not been discussed extensively in the literature. In this paper we demonstrate that the proper use of high resolution electron microscopy, at the atomic resolution level constitutes a suitable method for studying point defect chains where diffraction contrast would not be sufficient. Whereas dislocation dipoles and other intermediate defect configurations in Si, when viewed along a $<110>$, are easily identified because of significant configurational changes in the lattice images, an interstitial or vacancy chain has almost no lattice distortion and hence requires the precise adjustment of electron microscope imaging conditions to optimize image contrast.

EXPERIMENTAL METHODS AND RESULTS

For our experiments, P-type, 12Ωcm, $<001>$ oriented silicon wafers were implanted with As^+ ions at 80keV to a dose of $8 \times 10^{15}/cm^2$ at 100μA ion current. From these samples $<110>$ oriented cross-sectioned thin foils, about 200Å thick were prepared and examined in a Siemens 102 microscope at 125keV. Defects were then studied by bright-field axial illumination imaging with the foil oriented in a $<110>$ and with seven Bragg beams, (000), $\pm(1\bar{1}1), \pm(111)$ and $\pm(002)$, included in the objective lens apperture. Fig. 1a shows a micrograph containing a linear defect which was image processed to eliminate photographic noise. The micrograph was taken in an underfocussed setting of the objective lens, for which the white dots are interpreted as a chain of two unresolved columns of Si atoms lying approximately 1.36Å apart. Fig. 1b shows the power spectrum of the image demonstrating the presence of the seven beams. It is seen in Fig. 1a that the white dots exhibit a relatively uniform brightness except for the one at the defect center which shows diminished brightness, while its immediate surrounding is a cross-shaped area somewhat darker than the other dark regions between white dots. There is no detectable strain or configurational change associated with this defect and this contrast behavior can only arise from the fact that the scattering ability of the constituents of the atomic chain of the defect are different from the normal Si atom chains. Thus this chain may have incorporated a large number of vacancies, self interstitials or impurity atoms.

COMPUTER MODELING OF IMAGES AND DIFFRACTION PATTERNS

To interpret experimental results it was necessary to obtain models of the defect structure, calculate their electron scattering distributions and electron microscope images to confirm the nature of the observed image contrast. First, atomic position models of the object were constructed using an interactive graphics computer terminal to create, move, or destroy atomic sites in a perfect Si lattice which was one repeat distance thick along the $[1\bar{1}0]$, i.e. $\sim 3.8\text{Å}$. Based upon the model of a stable single di-interstitial defect of Lee et. al [2], similar to that shown in Fig. 2a, a variety of vacancy and interstitial defects was constructed. Fig. 2b shows a number of defect configurations (di-vacancy, di-interstitial, di-vacancy-interstitial pairs) where strain field displacements were ignored since the computed lattice strain of a di-interstitial is a small fraction of the Si lattice parameter.

The image computations of defect chains are then accomplished in two stages by first performing a multislice dynamical diffraction calculation of the scattering distribution emerging from the bottom of a crystal [3]. The computer program for this purpose was capable of calculating diffraction patterns of 256x256 diffraction points in reciprocal space for any given film thickness. Electron micrographs are then by including the appropriate electron microscope parameters (electron wavelength, beam tilt, objective lens spherical abberation, etc.) in a computer program which accepts the diffraction data. Fig. 3a shows the result of a bright-field image computation for the previous model with a film thickness of $\sim 200\text{Å}$ and Fig. 3b is the image power spectrum. Note, there is an almost exact match between the computer generated image of a di-interstitial chain and the real experimental micrograph. Here for {111} diffraction spots the images of the atomic sites of the perfect crystal Si atom pairs appear white. These spots are observed in the power spectrum of the image as shown in fig. 3b to occur in a contrast zone which has a phase shift which produces bright contrast above background. The major contribution to the scattering of interstitial atoms, however, comes from the central zone where the contrast is dark or below background. It has been shown in other publications [4,5], that a free atom and interstitial have the same scattering sign and a vacancy has the opposite sign from a free atom. In this case, a vacancy will produce a whitened or brighter image area. These points are further demonstrated by removing the lattice image contribution of Fig. 3a. by filtering the higher zones from the power spectrum as shown in Fig. 4. Here the image contrast for a vacancy is above background while free atoms which, were put in as fiducial marks and located outside the model field at the top and bottom of the image, appear dark as is the case for an interstitial. Therefore, the modification of the lattice image intensity by vacancies and interstitials can be readily explained by considering the phase contrast transfer function of the microscope, i.e. the image power spectrum. Our analysis has also included the effect of varying the film thickness and changing the defocus which can significantly alter image contrast. It has not proven difficult to vary these parameters and obtain close image matching to experiment. The effect of adding heavy atom impurities to the surface is not significant for crystals over $\sim 15\text{Å}$ thickness where the lattice images overpower their image contrast. It can also be shown when concentration of interstitials in a column is reduced by a factor of two that the image contrast of the interstitial columns is almost negligible. From this we must conclude that the defective chain must have incorporated into it 100% or near 100% more Si atoms.

Other vacancy and interstitial clusters have also been observed in experimental micrographs as well as intermediate defect configurations.[6,7] To show that vacancies can be imaged, fig 5 is an experimental micrograph of what appears to be three columns of adjacent vacancy chains and Fig. 5b is the matching computer generated result.

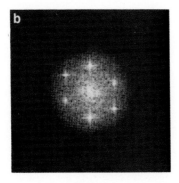

Fig. 1.   a) Bright-field micrograph of a chain defect. (b) Power spectrum of the image a).

188

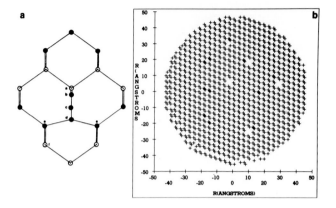

Fig. 2.   a) Atomic model of a split di-interstitial, b) models of various vacancy and di-interstitial configurations for a [110] orientation.

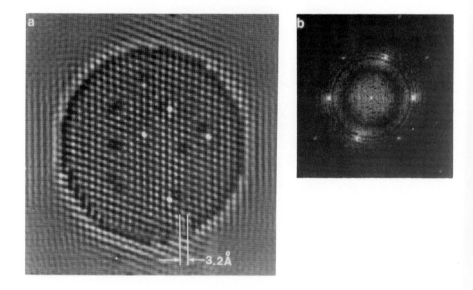

Fig. 3.   a) Computer generated micrograph of the atomic model of Fig. 2b for a ~200Å film thickness. b) power spectrum of image (a) showing phase contrast zones.

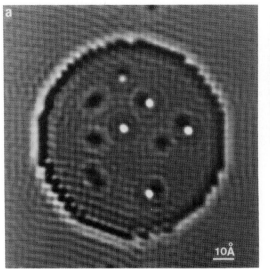
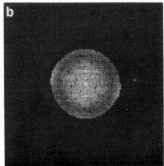

Fig. 4.   a)Filtered image showing defects and free atoms without the lattice image contribution, b)power spectrum of a).

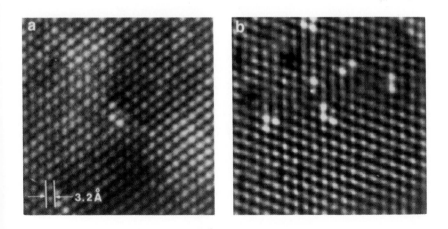

Fig. 5.   a) Experimental image of three adjacent columns of vacanies, b) computed micrograph corresponding to a).

REFERENCES

1. J.W. Corbett and J.C. Burgoin, in "Point Defects in Solids", *2*, p.15, (Plenum, N.Y., edited by J.H. Crawford and L.M Slifkin, 1976).

2. Y.H. Lee, N.N. Gerasemenko and J.W. Corbett, Phys. Rev. *B14*, 4506 (1976).

3. W. Krakow, IBM J. of Res. and Devel., (1980) in press.

4. W. Krakow, Ultramicrosc., *1*, 203 (1976).

5. W. Krakow, A.L. J. Chang and S.L. Sass, Phil. Mag., *35*, 575, (1977).

6. H. Foell, T.Y. Tan and W. Krakow, this proceedings.

7. T.Y. Tan, H. Foell and W. Krakow, this proceedings.

Published 1981 by North-Holland, Inc.
Narayan, and Tan, eds.
Defects in Semiconductors.

RADIATION DAMAGE AND ITS ANNEALING IN SEMICONDUCTORS

J. NARAYAN and J. FLETCHER
Solid State Division, Oak Ridge National Laboratory, Oak Ridge, TN 37830

ABSTRACT

Residual damage in the form of point defect clusters and
amorphous regions has been investigated in ion and neutron
irradiated silicon specimens. Annealing of this damage
during conventional heating, flame annealing, and pulsed
laser irradiation has been studied by plan-view and
cross-section electron microscopy techniques. These results
provide detailed information on annealing mechanisms, and
emphasize the characterization of damage in the as-irradiated
state.

INTRODUCTION

Ion implantation and neutron irradiation (via transmutation) have proved
to be very useful techniques of doping semiconductors [1,2]. However, the displacement damage created by energetic ions must be removed in order to achieve
electrical recovery of dopants. Conventional techniques of removing displacement damage involved thermal heating in a furnace. Recently other methods such
as laser- and electron-beam annealing, although limited to near-surface ($< 1\mu m$)
layers, have provided viable methods of removing displacement damage [3]. There
is a good phenomenological understanding of doping effects produced by various
techniques, however, our basic understanding of the nature of displacement
damage and its annealing is rather limited [4].
In this paper we present our recent work on the characterization of radiation damage and the annealing processes (in thermal, flame, and laser annealing)
associated with neutron irradiated, and B^+ and As^+ implanted silicon. The
residual damage in the neutron irradiated silicon is in the form of vacancy and
interstitial dislocation loops with no amorphous cascades present, the impurity
concentrations in these specimens were rather small. The damage in B^+ implanted
silicon is similar to the neutron damage except for the presence of boron as
an extra atom. The low dose As^+ ion damage contains amorphous regions as well
as dislocation loops, and the high dose As^+ damage consists of an amorphous
layer with a band of dislocation loops underneath this layer. We have studied
annealing processes associated with conventional thermal heating in a furnace,
with flame annealing using an oxyacetelene torch, and with pulsed (ruby) laser
irradiation.

NEUTRON DAMAGE AND ANNEALING PROCESSES

Silicon specimens used in this study were float zone refined (~ 700 Ω-cm
n-type) and were neutron irradiated in the Oak Ridge Research Reactor (ORR) at
temperatures $<100°C$ to a fluence of 1.0×10^{20} n cm^{-2} (E > 1.0 MeV). Thermal-
ized neutrons via ^{30}Si (n, γ, β) ^{31}P reactions produce a phosphorous concentration of $\sim 5.0 \times 10^{16}$ cm^{-3}. The displacement damage is produced primarily by the

*Research sponsored by the Division of Materials Sciences, U.S. Department of
Energy under contract W-7405-eng-26 with Union Carbide Corporation.

fast component of the reactor neutron spectrum. The displacement damage produced by reactor neutrons is unique in the sense that the recoil spectrum is close to that associated with ion irradiations without the introduction of an extra atom. As a result of this, neutron damage and its annealing processes should reflect the intrinsic annealing behavior.

Fig. 1a is an electron micrograph from a neutron irradiated specimen, which shows displacement damage in the form of dislocation loops. The vacancy or interstitial nature of dislocation loops was determined from black-white contrast and location of the images with respect to the free surface. More than 90% of the loops were found to be interstitial type and the rest were vacancy type. The average (second moment of the diameter) size, including both interstitial and vacancy types, of the dislocation loops was determined to be 28 Å. X-ray diffuse scattering [5] on these specimens indicated the average size of the vacancy loops to be about 7 Å, from that it would appear that most of the vacancy loops are below the microscope resolution and only the tail end of the vacancy loop-size distribution is observable in the microscope. Fig. 1 also shows defect clusters after annealing these specimens at 250°C, 400°C, 600°C,

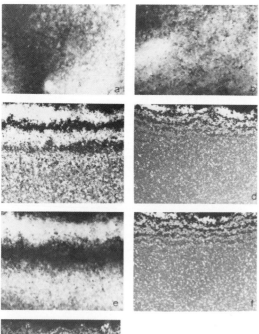

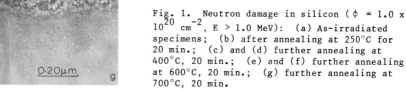

Fig. 1. Neutron damage in silicon ($\phi = 1.0 \times 10^{20}$ cm^{-2}, E > 1.0 MeV): (a) As-irradiated specimens; (b) after annealing at 250°C for 20 min.; (c) and (d) further annealing at 400°C, 20 min.; (e) and (f) further annealing at 600°C, 20 min.; (g) further annealing at 700°C, 20 min.

and 700°C for 20 minutes. Fig. 2 shows size distributions of dislocation loops after annealing at 400°C and 600°C. The differences in size distributions between as-irradiated specimens and those annealed at 400°C, were very small. The results in Fig. 2 indicate that the agreement between size distributions determined by TEM and X-ray [5] is fairly good for larger sizes (≥ 20 Å), however, for loop sizes below 20 Å X-ray measurements consistently indicated higher number densities of dislocation loops. The average size increases only slightly (from 28 to 32 Å) after annealing at 600°C, however, the total number of point defects contained in the dislocation loops remains largely unaffected up to 600°C, and then decrease rather rapidly. Since infrared measurements indicate the presence of high concentrations of divancies [6], an inescapable conclusion in view of these results, is that interstitials and divacancies, which were undetected by TEM and X-ray techniques, annihilate each other below 600°C. The main result from the neutron-damage annealing studies is that most of the intrinsic damage anneals out by 750°C, and during this annealing glide and climb of the dislocations, leading to coarsening of the loops, is not significant.

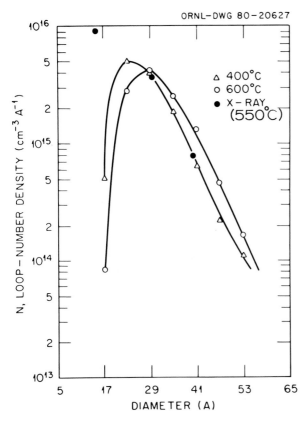

ORNL-DWG 80-20627

Δ 400°C
o 600°C
● X - RAY (550°C)

Fig. 2. Size distribution of dislocation loops in neutron-irradiated silicon after annealing at 400°C and 600°C (as determined by TEM), and at 550°C (as determined by X-ray).

194

One of the interesting observations in the neutron irradiated specimens was the presence of short interstitial dipoles (encircled in Fig. 3) in <110> directions. The number density of these dipoles was small (<10%). The formation of the <110> dipoles may result from clustering of <100> split interstitials [7].

Fig. 3. Dislocation dipoles in neutron irradiated silicon.

BORON ION DAMAGE

In this investigation <100> silicon (2-6 Ω-cm) specimens were irradiated with 35 keV boron ions ($^{11}B^+$) to a dose of 2.0×10^{15} ions cm^{-2} at room temperature. The residual damage in as-implanted specimens is shown in Fig. 4, where the damage (in the form of interstitial and vacancy dislocation loops) is similar to that produced by reactor neutrons. Thermal annealing up to 600°C leads to only small changes in the loop-size distributions and only a small decrease in the number of point defects contained in the dislocation loops. Annealing at 700°C leads to a slight decrease in the number of point defects contained in the loops (Fig. 4b), but 750°C annealing leads to the formation of large dislocation loops and dipoles as shown in Fig. 4c. The micrograph shown as Fig. 4c also contains examples in which faulted a/3<111> loops (indicated as 1 and 2) are converted into a/2<110> perfect loops by nucleation of a Shockley partial: (a/3<111> + a/6<11$\bar{2}$> —·a/2<110>).Fig. 5 shows x-section micrographs from companion specimens after annealing at 600°C and 750°C. As shown in Fig. 5b, the loops are distributed around the projected range $\bar{R}p$ (1200 ± 450 Å). The nature of these (Fig. 5b) dislocation loops and dipoles was determined to be interstitial type, and the number of interstitials contained in these loops was approximately equal to the number of boron atoms in the substitutional sites as determined by Hall measurements. These results imply that most of the intrinsic damage anneals out by 750°C (similar to the

Fig. 5. Cross-section electron micrographs
of specimens similar to Fig. 4. (a) After
annealing at 600°C; (b) after annealing
at 700°C. R̄p denotes the mean projected
range of the ions.

Fig. 4. $^{11}B^+$ (35 keV) damage in
silicon (φ = 2.0 x 10^{15} cm^{-2})
(a) After annealing as-irradiated
specimen at 600°C for 20 min.;
(b) further annealing at 700°C,
20 min.; (c) further annealing
at 750°C, 20 min.

annealing of neutron damage) and the dopants occupy substitutional sites and
create self interstitials by a "kick out mechanism" [8], which then cluster to
form interstitial loops with the peak at R̄p. Annealing at still higher temperatures causes coarsening of the loops via glide and conservative climb processes.
Thus, the microstructures at high temperatures are primarily determined by the
dopant concentrations, and the size distributions of the resulting interstitial
loops, which control the glide and climb of the loops. In order to illustrate
the point that the distribution of residual damage after 750°C annealing in
boron implanted silicon is centered around the projected range, we implanted a
silicon specimen with 50 and 180 keV, $^{11}B^+$ ions resulting in the boron distribution profile shown in Fig. 6. After annealing at 750°C two damage peaks are
observed in Fig. 7, which correspond to the two boron concentration peaks shown
in Fig. 6.

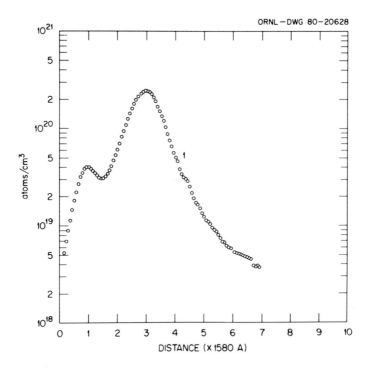

Fig. 6. Boron concentration profile after implantation with 50 and 180 keV $^{11}B^+$ ions (ϕ = 6.25 x 10^{15} cm^{-2}).

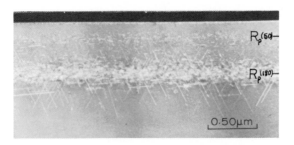

Fig. 7. X-section micrograph (after annealing at 750°C) showing two damage peaks corresponding to the boron concentration peaks in Fig. 6.

As$^+$ DAMAGE AND THE ANNEALING PROCESSES

The displacement damage produced by 100 keV As$^+$ ions was studied in silicon specimens (2-6 Ω-cm) implanted to a dose of 1.0×10^{14} ions cm^{-2} at room temperature. Fig. 8a is a weak-beam, dark-field micrograph where dislocation loops as well as amorphous regions are imaged. Fig. 8b is a bright-field micrograph taken under dynamical diffraction conditions in which dislocation loops exhibit characteristic black-white contrast. From a comparison between Fig. 8a and Fig. 8b it was found that about 85% of the observed images are amorphous regions and the rest dislocation loops.

When these specimens are thermally annealed up to 400°C, the loop number-density increases slightly; otherwise there is very little change in the observed microstructure (Fig. 9a). After 500°C annealing there is a large increase in the number density of loops, and relatively small changes then occur up to 600°C (Fig. 9b). Above 600°C the intrinsic damage starts

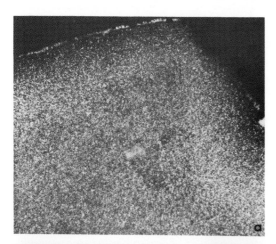

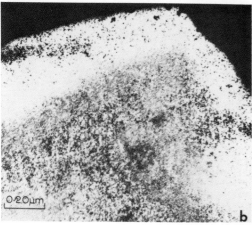

Fig. 8. ^{75}As$^+$ (100 keV) damage in silicon ($\phi = 1.0 \times 10^{14}$ cm^{-2}): (a) As-implanted, weak-beam micrograph; (b) bright-field micrograph (under two-beam dynamical diffraction conditions) of the same area.

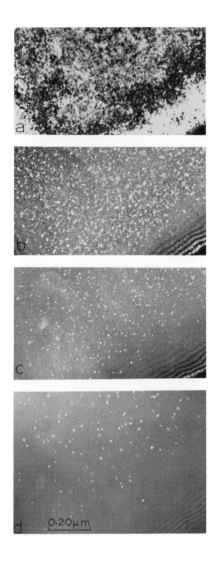

Fig. 9. $^{75}\text{As}^+$ (100 keV) damage in silicon ($\phi = 1.0 \times 10^{14}\ \text{cm}^{-2}$): (a) After annealing at 400°C for 20 min.; (b) further annealing at 600°C, 20 min.; (c) further annealing at 700°C, 20 min.; (d) further annealing at 800°C, 20 min.

annealing out, and above 800°C interstitial loops, with a distribution centered around the projected range of the ions, are observed. The number of interstitials contained in these loops was found to be approximately equal to the dopant concentration, implying a "kickout mechanism" for the creation of self interstitials which cluster to give the observed dislocation loop concentrations.

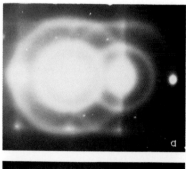

Fig. 10. Diffraction patterns from ion implanted (100 keV, $^{75}As^{+}$, 1.0 x 10^{14} cm^{-2}) silicon: (a) as-implanted; (b) further annealing at 250°C for 20 min., (c) further annealing at 400°C for 20 min.

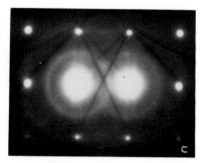

Fig. 10 shows diffraction studies on companion arsenic-implanted specimens. In the as-implanted specimens (Fig. 10a) two diffuse rings, along with the diffraction spots corresponding to crystalline silicon, are observed. The first ring represents disorder in the first nearest neighbor distance, and gives information about the truly amorphous zones. The second ring represents the disorder in the second nearest-neighbors [9,10], i.e., disorder in the tetrahedra. Upon annealing at 250°C (Fig. 10b) the intensity diminishes primarily in the second ring, indicating the repair of disorder in the tetrahedra. By 400°C (Fig. 10c) the repair in the disorder of the tetrahedra is complete and only the first ring is observed. After annealing at 500°C the first ring disappears which is attended by a sudden increase in the number of dislocation loops.

An tentative annealing model which is consistent with the observations in Figs. 8-11 is as follows. The as-implanted state is primarily composed of displacement cascades, divacancies, interstitials, and dislocation loops which are mostly interstitial type (see schematic in (Fig. 12). The displacement cascades consist of amorphous core regions which are surrounded by interstitial rich transition regions which contain some order and are characterized by disorder mainly in the tetrahedra. When this state passes through the temperature region 250 - 400°C where divacancies are mobile, repair mainly in the disorder of tetrahedra (outer regions of the cascades) occurs, leading to a preferential decrease in the intensity of the second diffraction ring. By 400°C only the

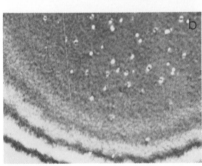

Fig. 11. $^{75}As^+$ (100 keV, ϕ = 1.0 x 10^{14} cm^{-2}) damage: (a) after annealing at 650°C for 20 min., (b) further annealing the same area at 750°C, 20 min.

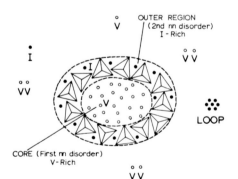

Fig. 12. Schematic model for annealing ion damage in silicon: a typical displacement cascade consists of an amorphous core containing mainly first-nearest neighbor disorder, which is surrounded by the interstitial-rich region containing second-nearest neighbor (tetrahedra) disorder. V stands for vacancies and I for interstitials.

core regions of the cascades (representing the first nearest-neighbor disorder, corresponding to the first ring) and the interstitial dislocation loops survive. Above about 500°C crystalline order in the core regions is restored by solid phase epitaxy and a sudden increase in the loop density is observed due to the clustering of the remaining interstitials and vacancies. No significant changes in loop densities occur up to 600°C, and then the intrinsic damage anneals out by 750°C. The dislocation loops which survive after annealing at 750°C contain interstitial silicon atoms equal to the number of implanted arsenic ions. These interstitial loops, depending upon their size distribution, can coalesce via glide and/or climb processes to form large loops, and thus the high-temperature microstructure is determined by these dislocation loops.

SOLID PHASE EPITAXIAL GROWTH

Epitaxial growth of amorphous layers in silicon has been studied extensively using channeling and backscattering techniques. Although these techniques provide detailed information about the lattice location of dopants, they are rather insensitive to the details of damage microstructures [11-13]. We have investigated (using TEM x-section techniques) the solid phase epitaxial growth in As^+ implanted silicon specimens. The implants, which are similar to those used in transistor devices, were done at room temperature, or at liquid nitrogen temperature, to a dose of 1.0×10^{16} ions cm^{-2}. Fig. 13 shows a composite of TEM x-sections for the case of an [111] silicon specimen, implanted at liquid nitrogen temperature, after annealing at 600°C for 30 (Fig. 13a), 60 (Fig. 13b, c) and 90 min (Fig. 13d). The thicknesses of regrown layers are indicated as 1, 2, and 3 in the As^+ concentration profile (Fig. 14) for 30, 60, and 90 min. annealing respectively. The thickness of the original amorphous layer is indicated as 0. The epitaxial growth rates are nonlinear and the initial growth rate is much faster than the subsequent rates. For <111>-oriented silicon, these regrown layers contain a high density of twins, but <100> and <110> specimens were found to be relatively defect-free.

A band of damage in the form of dislocation loops was found to be present underneath the amorphous layers as shown in Fig. 13c. The damage density in this layer was found to increase with the temperature of ion implantation. Fig. 15b is a cross-section micrograph showing this band of damage in a sample implanted at room temperature and then annealed at 600°C for 20 min., where the damage density is much higher than that for the sample implanted at liquid nitrogen temperature (Fig. 13). Fig. 15a is a plan-view micrograph corresponding to the x-section micrograph, where the damage is clearly observed in the form of dislocation loops and tangles. The dislocation loops in this layer remain unaffected up to 600°C annealing, but higher - temperature annealing (at 1000°C for 20 min.) frequently leads to the formation of large loops and dislocation tangles by glide and climb processes, as shown in Fig. 16. It should be remarked that these dislocation loops and tangles form a microstructure that is very stable against high-temperature thermal annealing, but is easily removed by pulsed laser melting to a depth greater than the outer boundaries of these dislocations [14].

FLAME ANNEALING OF As^+ IMPLANTED SILICON

Silicon specimens (<100>, 2-6 Ω-cm) implanted with As^+ ions (dose = 2.0×10^{15} cm^{-2}) were treated with an oxyacetelene flame for times of 10-20 seconds. The temperature of the hot zone was estimated to be 1150 ± 20°C. After flame annealing no residual damage (Fig. 17) in the form of dislocation loops and dislocations is observed. Some of the flame-annealed specimens were subsequently

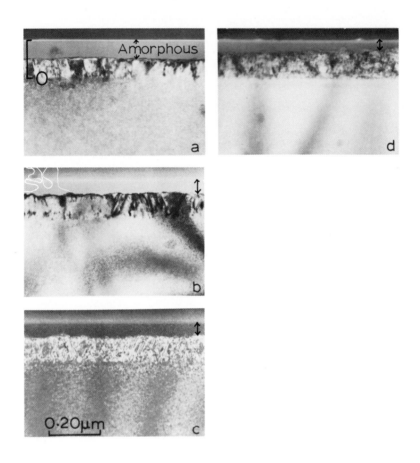

Fig. 13. X-section micrographs showing SPE growth in ^{75}As$^+$ (100 keV, $\phi = 1.0$ x 10^{16} cm^{-2} at liquid nitrogen temperature) implanted silicon: (a) (111) Si annealed at 600°C for 30 min.; (b) and (c) after annealing at 600°C for 60 min.; and (d) after annealing at 600°C for 90 min.

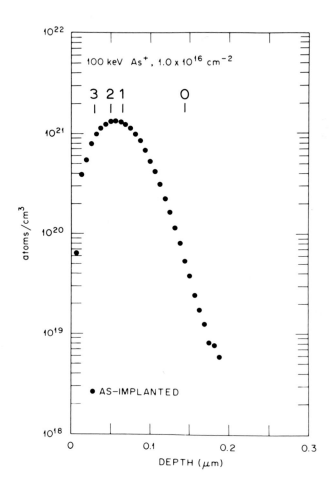

Fig. 14. Arsenic distribution profile after ion implantation, 1, 2 and 3 indicate the thicknesses of SPE grown layers. 0 denotes the thickness of the original amorphous layer.

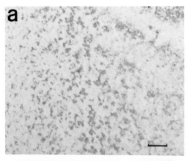

Fig. 15. ^{75}As$^+$ (100 keV, ϕ = 1.0 x 10^{16} cm^{-2} at room temperature) damage annealing: (a) plan-view electron micrograph after annealing at 600°C, 20 min.; (b) corresponding x-section micrograph.

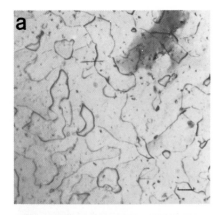

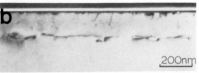

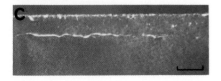

Fig. 16. ^{75}As$^+$ damage annealing of the specimen similar to that in Fig. 15. (a) Plan-view micrograph after annealing at 1000°C, 20 min.; (b) corresponding x-section (bright-field) micrograph; (c) dark-field micrograph.

annealed in a furnace for longer times to investigate the clustering of residual defect clusters which may be below the resolution limit of the microscope. No defect clusters were observed after furnace annealing of the flame annealed specimens. Hall measurements in the flame annealed specimens indicated complete electrical activation of the implanted ions [15].

LASER ANNEALING OF ION IMPLANTED SEMICONDUCTORS

Annealing methods discussed so far involved thermally activated motion of defects in the solid state. Recently it has been shown that nanosecond laser pulses can deliver enough energy to the lattice such that thin surface layers

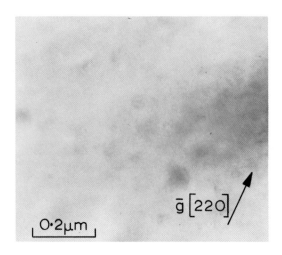

Fig. 17. $^{75}As^+$ implanted (100 keV, $\phi = 2.0 \times 10^{15}$ cm^{-2}) silicon after flame annealing: the electron micrograph shows complete annealing of the displacement damage at 1150°C with an interaction time of 10-20 s.

(<1μm) can be melted and recrystallized with growth rates of meters per second [3]. By controlling the depth of melting (by manipulating the laser parameters) it is possible to grow thin layers with the same perfection as the substrate. Fig. 18 shows annealing of displacement damage in a boron implanted (similar to the specimen shown in Fig. 6) specimen in which the residual damage was in the form of dislocation loops distributed from the surface to a depth of about 0.60μm. The annealed regions were found to be separated from the unannealed regions by sharp boundaries. The dislocation loop size distribution below these boundaries remain unaffected after the laser treatment suggesting that ionization effects are not important in the removal of dislocation loops. The observed annealing characteristics can be explained by a melting model where the depths of annealed regions correspond to the depths of melting. Whenever the melt-front intersects a dislocation loop, two segments of dislocations grow back to the surface (Fig. 17f). In other cases where the damaged layers are completely amorphous, laser melting partway through the damaged layers leads to the formation of polycrystalline regions. Therefore, to obtain a defect-free annealed region it was necessary to exceed the thickness of the damaged layer so that the underlying defect-free substrate could act as a seed for crystal growth.

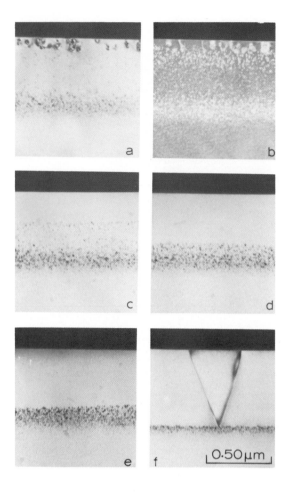

Fig. 18. Laser annealing of $^{11}B^+$ (180 keV, ϕ = 6.25 x 10^{15} cm^{-2}) damage in silicon using 15 ns pulse of Q-switched ruby laser (λ = 0.694μm): (a) and (b) 0.8 J cm^{-2}; (c) 1.0; (d) 1.3; (e) 1.5; (f) 2.0 J cm^{-2}.

SUMMARY

In summary, displacement damage in neutron-irradiated silicon was observed in the form of interstitial and vacancy loops. On subsequent thermal annealing relatively small changes were observed up to 600°C, but most of the damage annealed out by 750°C. The results on B$^+$ implanted specimens show that most of the intrinsic damage anneals out by 750°C, and the dislocation loops observed above 750°C are presumably due to clustering of self interstitials generated as a result of dopant substitution and a "kick-out mechanism". Annealing studies in As$^+$ implanted silicon suggest that divacancy migration in the temperature range 200 - 400°C repairs the disorder associated with tetrahedra in the outer regions of the cascades. In the temperature range ~500°C, core regions of the cascades anneal out by solid phase epitaxial growth. At still higher temperatures dislocation loop annealing occurs similar to that observed in neutron and B^+ implanted silicon. The results on laser annealing are consistent with the melting model.

Finally, it should be remarked that in this paper we have attempted to present a unified picture on the nature and annealing of displacement damage created by energetic ions. Some conclusions in the thermal annealing model are based upon arguments which may be of speculative nature. Further work in this direction is needed to establish these ideas.

REFERENCES

1. Ion Implantation of Semiconductors, G. Carter and W. A. Grant, Arnold (1976); Ion Implantation in Semiconductors, J. W. Mayer, L. Eriksson, and J. A. Davies, Academic Press (1970).

2. Neutron Transmutation Doping in Semiconductors, ed. by J. M. Meese, Plenum Press (1979).

3. J. Narayan, J. Metals. 32 (6), 15 (1980); C. W. White, J. Narayan, R. T. Young, Science 204, 461 (1979).

4. Lattice Defects in Semiconductors, The Institute of Physics, Vol. 23 (London) 1974).

5. B. C. Larson and J. F. Barhorst, these proceedings; B. C. Larson, J. Narayan, and R. T. Young, p. 281 in Neutron Transmutation Doping in Semiconductors, Plenum Press (1979).

6. H. Stein, p. 229 in Neutron Transmutation Doping in Semiconductors, Plenum Press (1979).

7. T. Y. Tan, these proceedings.

8. W. Frank, these proceedings.

9. C. Cemboli, L. Dori, R. Gallioni, M. Servidori, and F. Zignani, Rad. Eff. 36, 111 (1978).

10. J. Narayan (unpublished data).

11. L. Csepregi, J. W. Mayer, and T. W. Sigmon, Appl. Phys. Lett. 29, 92 (1976).

12. L. Csepregi, E. F. Kennedy, T. J. Gallagher, J. W. Mayer, and T. W. Sigmon, J. Appl. Phys. 48, 4234 (1977).

13. S. T. Picraux, these proceedings.

14. J. Narayan, J. Electrochem. Soc. 80-1, 294 (1980).

15. J. Narayan and R. T. Young (unpublished).

FORMATION AND EFFECTS OF SECONDARY DEFECTS IN ION IMPLANTED SILICON

Jack Washburn
Department of Engineering and Lawrence Berkeley Laboratory,
University of California, Berkeley, California 94720

ABSTRACT

The clustering of isolated interstitial silicon, implanted atoms,
and vacant lattice sites produced by low temperature and room temperature ion implantation during subsequent annealing is reviewed. An
electron microscope method for studying the kinetics of the amorphous
to crystalline transformation in silicon is described. The technique
is applied to measurement of the activation energy for interface migration and the formation of microtwins for different growth directions.
A very brief review of the effects of laser annealing after ion
implantation is included.

INTRODUCTION

Because of its flexibility in providing desired dopant distributions and
because undesirable impurity diffusion accompanying high temperature treatments can be minimized, ion implantation has become increasingly utilized for
fabrication of solid state devices. Ion implanatation is also interesting from
a basic science viewpoint. The concomitant radiation damage is severe and
recovers to give a fascinating diversity of secondary defects. These secondary
defects can adversely affect electrical propeties. Consequently, ion implantation conditions and post implantation annealing procedures which avoid
formation of secondary defects are desirable.

When a foreign atom is injected into a silicon crystal with an energy of
~100 KeV, it generally comes to rest in an interstitial space or as part of an
interstitial complex with one or more of the far more numerous displaced
silicon atoms. As long as it remains in such a special site it does not give
rise to the desired donor or acceptor level: it is electrically inactive. Post
implantation annealing decreases the density of special sites gradually
increasing the number of implanted atoms that occupy regular substitational
sites. A recent alternative to post implantation annealing is so called laser
annealing which in reality is not annealing but rather liquid phase epitaxial
regrowth of the surface layers of the wafer in an extremely sharp temperature
gradient. The technique has the advantage that heating of the wafer is almost
entirely confined to the surface layers, thus avoiding diffusion of impurities
in the underlying part of the wafer.

This paper is a brief review of the nature of some commonly observed
secondary defects in silicon and the mechanisms of their formation. Because
other papers in this volume focus major attention on laser "annealing," effects
very little will be included here. Space limitations also make a complete
cataloging of all known kinds of secondary defects impossible. The discussion
will be limited to implantation of boron and phosphoros in silicon.

Growth and Effects of Secondary Defects

During implantation and in the low temperature post-implantation annealing range below 600°C silicon interstitials and the implanted atoms which are also in interstitial sites form clusters or precipitates. In the earliest stages the nature of the complexes is unknown. As the annealing temperature is increased to about 700°C defects large enough to study by transmission electron microscopy begin to appear. Figure 1 shows defects in silicon implanted with $2 \times 10^{14} cm^2$ boron ions at 100 KV at room temperature and then annealed at 700°C [1,2]. These rod defects are of two types: 1) interstitial Frank dipoles (elongated loops on {111} planes in ⟨110⟩ directions), and 2) rod shaped precipitates containing an appreciable fraction of boron atoms lying on {100} with a ⟨100⟩ displacement vector. Both kinds probably grow at the expense of smaller interstitial clusters and they are both interstitial type. Rod defects are most numerous in boron implanted silicon but are also found after implantation with phosphorous. Implantation to higher doses of the order of 10^{16} ions/cm^2 leads to growth of an amorphous zone which is at first buried but at still higher doses or lower implantation temperatures extends to the surface [3]. Figure 2 is a series of transmission electron microscope pictures made in cross-section showing stages in the development of an amorphous layer [4]. When an amorphos layer is formed, secondary defects arise during annealing not only from coarsening of the initially formed interstitial and vacancy clusters, but also from migration of the amorphous/crystalline interface. this will be considered in more detail in a following section.

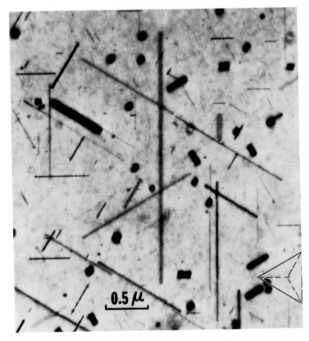

Figure 1. Rod defects in silicon—100 KV room temperature implant—annealed at 700°C—all six sets in ⟨110⟩ directions are in contrast.

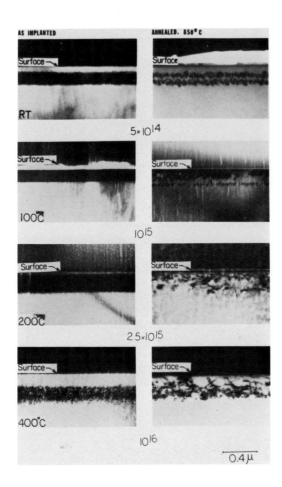

AS IMPLANTED **ANNEALED. 850°C**

Surface Surface

RT

5×10^{14}

Surface Surface

100C

10^{15}

Surface Surface

200C

2.5×10^{15}

Surface Surface

400°C

10^{16}

0.4 µ

Figure 2. Effect of ion dose on amorphous layer formation—left, unannealed right, annealed. amorphous zone is first buried but extends to the surface as dose increases—note complex double layers of defects resulting from regrowth of amorphous layer during annealing.

The type of secondary defects found is also sensitive to orientation of the implanted surface particularly if the ion dose and implantation temperature are such as to produce an amorphous layer [5-8]. Figure 3 shows the effect of implanting $10^{16}/$ p^+ ions/cm^2 at 100 KV at room temperature followed by annealing at 800°C for 1/2 hour for a), {111}, b {110} and, c {100} surfaces. Figure 3d is an unannealed specimen showing an apparently featureless crystal containing an amorphous layer. For the {111} orientation numerous microtwins

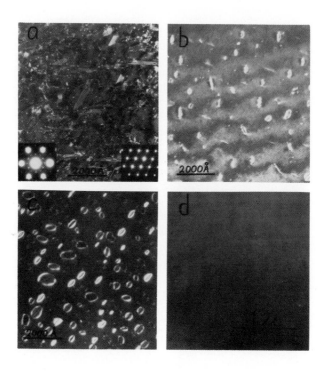

Figure 3. Effect of orientation of implanted surface on secondary defects in
room temperature implanted silicon. Dark field electron micrographs a) 111 ,
b) 110 , and c) 100 wafers all implanted to 10^{16} p$^+$/cm^2 and annealed
at 800°C.

are a prominant part of the secondary defect population while for the other two
orientations only interstitial dislocation loops are seen. The implantation
temperature also plays an important role in determining the type of secondary
defects that form. Figure 4 shows the result of implanting 10^{15} p$^+$/cm^2
at 100 Kv at 100°C into {111} and {100} wafers, (a and b) or 10^{16} p$^+$/cm^2
at 100 KV at 100°C into {111} and {100} wafers, (c and d). At this implantation temperature an amorphous layer was probably not formed in a and b and
probably was formed in c and d. At both dose levels the type of defects formed
was similar for the two orientations: interstitial loops both faulted and
perfect along with rod defects for the lower dose and loops and a dislocation
network for the higher dose. Implants which are nominally made at room
temperature can easily be heated to 100°C or higher during implantation if poor
thermal contact exists between the wafer and the holder.

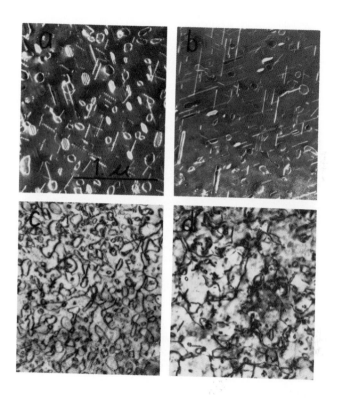

Figure 4. Effect of elevated temperature implantation at 100°C on the formation of microtwins. {100} wafers (left), and {111} wafers (right); 10^{15} p$^+$/cm^2 (top) and 10^{16} p$^+$/cm^2 (bottom). No microtwins were formed in either {111} specimen. The orientation dependence on defect type is absent for elevated temperature implants.

After annealing an implanted wafer in which no amorphous zone has been formed at 800°C or above the total area of the remaining interstitial loops is found to be equal to the area that would be expected if it is assumed that all the extra atoms that have been introduced were to form planar clusters [9].

Therefore, nearly all the vacancies formed during implantation have recombined with a silicon or foreign interstitial. However, at lower annealing temperatures some vacancy type defects are also observed [4]. Figure 5 shows stacking fault tetrahedra in silicon implanted with 10^{16} p$^+$/cm^2 at room temperature and annealed at 1050°C. These small vacancy clusters are easily missed because of their small size, particularly for lower annealing temperatures. As expected, a few vacancy clusters that do grow large enough to be observed are in a layer near the implanted surface.

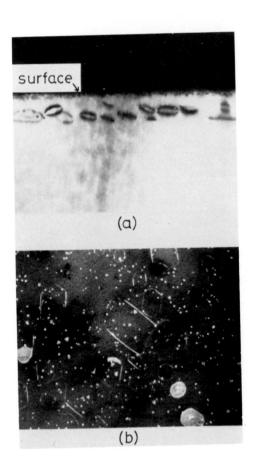

Figure 5. Vacancy defects in implanted silicon stacking fault tetrahedra.

Electrical activity continues to increase to quite high annealing temperatures. Some rod shaped defects have been shown to contain appreciable numbers of the implanted atoms. On heating to temperatures above about 750°C they start to shrink at the ends while near by interstitial loops grow, Fig. 6. The temperature dependence of the shrinkage rate was found to be the same as that for boron diffusion in silicon [10]. As small interstitial clusters and rod defects anneal out the number of special (inactive) sites for implanted atoms rapidly decreases. However, even at the stage where the predominant secondary defect is large interstitial loops there is still segregation of the dopant element to the dislocations. This has been demonstrated by comparing the depth distribution of secondary defects as revealed by cross-sectional TEM with the depth distribution of phosphorous as measured by secondary ion mass spectroscopy [11]. This correlation is shown in Figure 7 where both kinds of measurements were made on parts of the same specimen. In order to make the comparison more convincing a specimen was used which contained two layers of high dislocation density at different depths. Also included in the figure are measurements of carrier concentration and mobility as a function of depth. It can be seen that the phosphorous concentration has peaks that coincide with the

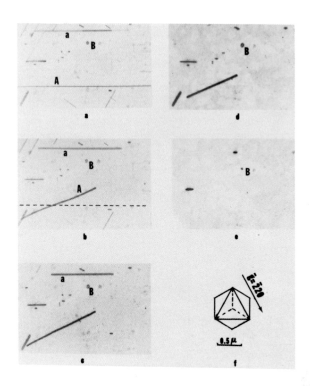

Figure 6. Annealing out of secondary defects. Rod defects are the least stable of the "visible" secondary defects. The sequence shows that they shrink at the ends by absorbing vacancies or giving off intertitial atoms starting at about 750°C.

regions of high dislocation density and that changes in carrier concentration and mobility also correlate with the double buried damage layers. Also shown is a channelled back scattering spectrum from the same specimen. Scattering peaks also reveal the presence of the two high dislocation density regions. Since back scattering is a nondestructive technique it provides a quick, convenient method for revealing subsurface structural damage. Figure 2 shows additional examples of the complex depth distribution of secondary defects in phosphorous implanted silicon.

The Amorphous to Crystalline Transformation

In most practical cases the implantation dose produces an amorphous zone. It has been found that at a given temperature the rate of migration of the

216

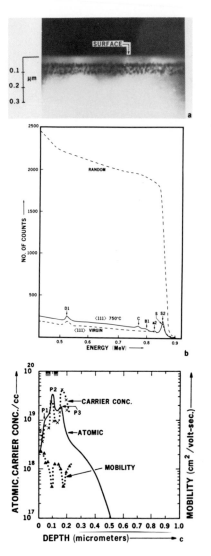

crystalline/amorphous interface is more than an order of magnitude faster for a {100} interface than it is for a {111} interface [12]. The rate is also affected by the pre‑s nce of the implanted ion, being speeded up by boron or phosphorous for example. In order to study the kinetics of the transformation and the formation of microtwins an electron microscope technique has been used which permits monitoring of inteface migration in different crystallo‑graphic directions in the same electron microscope foil [8]. This techique is described in Figure 8. A wafer implanted so as to contain an amorphous layer extending to the surface is thinned from the back side before annealing. The thinnest parts of the electron microscope foil are therefore entirely amorphous but in the thicker part away from the edge there is still crystalline material on top of the amorphous layer. On heating this type of specimen the c/α interface as it moves into the thinner part of the foil is seen edge on. By observing growth sequences at dif‑ferent positions around the edge of the hole at the specimen center it is possible to follow growth in ⟨100⟩, ⟨110⟩ and ⟨111⟩ directions. By measuring interface migration rate vs. temperature it was possible to find the activation energy involved in the transformation process (2.9 ± 0.1 eV). The relative growth rates were in the ratio 1, 6.5 and 15 for the ⟨111⟩, ⟨110⟩, and ⟨100⟩ growth directions, respectively. Figure 9 shows typical growth sequences for ⟨100⟩ and ⟨111⟩ interfaces. For the ⟨111⟩ growth direction microtwins were nucleated as soon as the interface started to move from its original position but for ⟨100⟩ no twins were formed.

These results suggest that when an amorphous layer is in contact with a {100} face a thermal fluctuation which reorients a small group of atoms at the interface is much more likely to put one of the atoms into a position where it can become part of the crystal than is the case for {111}. Figure 10 shows the atomic arrange‑

Figure 7. Segregation of phosphorous at regions of high dislocation density.

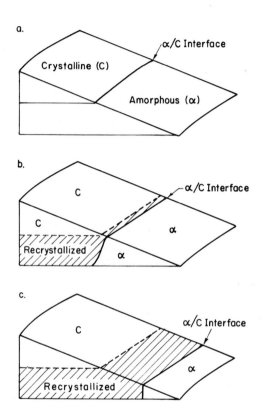

Figure 8. Description of transmission electron microscopy method for measuring amorphous→crystal interface velocity as a function of temperature and growth direction.

ment on {100}, {110}, and {111} faces for the diamond cubic structure. The atomic structure on smooth faces of these three orientations is very different with respect to addition of a new atom to the crystal. If it is assumed that any atom to be considered part of the crystal rather than the adjacent amorphous material must have formed at least two regular bonds to the lattice, then a single atom can add to a {100} face anywhere. For {110} two adjacent atoms must first nucleate a growth step by attaching to the crystal simultaneously. For {111}, nucleation of a growth step requires simultaneous attach-

218

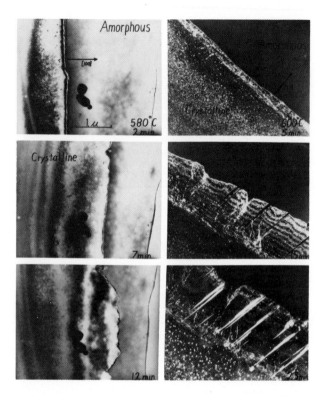

Figure 9. Crystal growth sequences showing motion of {100} (left) and {111} (right) crystal--amorphous interfaces. Note twin nucleation associated with {111} growth.

ment of three adjacent atoms. The microtwins associated with {111} growth are frequently nucleated on a {111} face inclined to the main growth front. Also, no twins are formed during regrowth of α layers produced during high temperature implants (Fig. 4). It seems likely that a growth mistake combined with high internal stress is necessary for twin nucleation.

Figure 11 presents electron diffraction evidence that the C/α interface directly after implantation is rough with a transition region within which some volumes are amorphous but crystalline regions still remain. As the temperature of implantation is increased the scale of the roughness appears to increase. At least at low implantation temperatures the increase in volume associated with the crystal to amorphous transformation is expected to result in high internal stresses at the interface. In a {111} interface with protrusions of crystal into the amorphous material there will be fast growth directions on the sides of the protrusions that are inclined to the main growth front as shown schematically in Figure 12. Thus, the protrusions may rapidly grow together laterally resulting in the meeting of two oppositely moving growth fronts. If

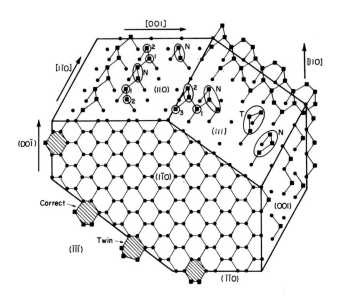

Figure 10. Atomic arrangement on {100}, {110} and {111} surfaces of a diamond cubic crystal.

there are also high internal stresses at the interface the two parts of the crystal may be out of registry where they meet which could nucleate a stacking fault and develop into a microtwin on an inclined plane. Microtwins on one of the inclined habit planes appear to facilitate growth in the {111} case. Where a twin boundary intersects the growth front, special sites are present for nucleation of growth ledges. On one side of the twin at the boundary only two atoms instead of three must be correctly positioned to nucleate a growth step and the step can propagate away from the twin by addition of single atoms along the step. On the other side of the twin growth step propagation is less favorable because single atoms can not add directly to a step; there must be repeated nucleation of kinds which require two atoms. The expected growth morphology near a twin is illustated schematically in Figure 13 and shown in the transmission electromicrograph (Fig. 14). On the fast growing side the side surface of the twin is never exposed because the twin and the main crystal grow at the same rate. The direction of easy ledge motion is away from the boundary both in the twin and in the main crystal. On the other side there is less enhancement of the growth of the main crystal so that side of the twin becomes exposed. It therefore thickens only from that side. This thickening is in turn accelerated by secondary twins.

220

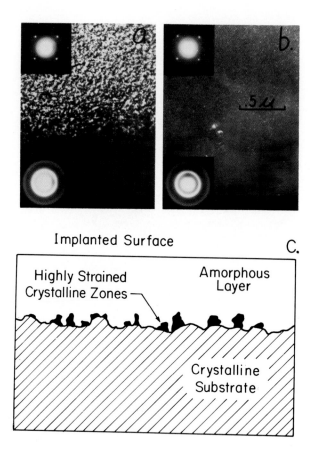

Figure 11. Roughness of the as implanted crystal—amorphous interface.
Electron diffraction contrast producing alternate bright dark regions suggests
that there is a transition zone where strained crystalline regions are partly
surrounded by amorphous material.

Laser Annealing

Laser annealing after ion implantation has the advantage that only the
implanted region of the wafer is appreciably heated. It is also possible under
the right conditions to regrow the crystal almost defect free. When an
amorphous zone is laser melted, so that the depth of melting is less than the
thickness of the amorphous layer, then solidification results in polycrystalline matieral (Fig. 14Bb). If the depth of melting exceeds the thickness of
the amorphous zone solidification is epitaxial and the perfection of the
regrown part of the crystal can be very good (Fig. 14Cc).

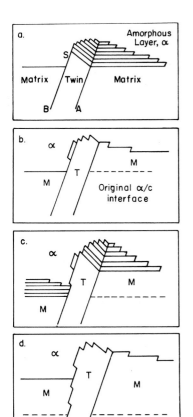

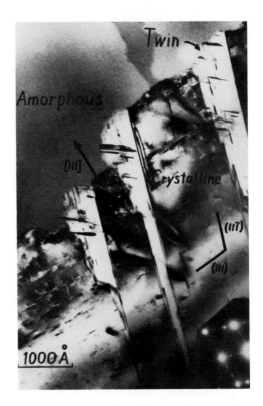

Figure 12. Effect of a twin on subsequent growth at a {111} interface.

Figure 13. Electron microscope
evidence for the effect of twins
on {111} growth.

When more complex damage structure consisting of buried layers of damage
are laser annealed the results are less well understood (Fig. 14Cb and Cc).
Multiple reflection from subsurface interfaces and changes in light absorption
with depth can produce anomolous effects. It becomes very difficult to get
reproducible results. These effects have not been adequately studied and are
beyond the scope of this review.

222

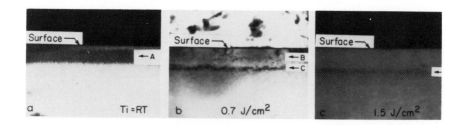

A. 5×10^{14} p$^+$/cm^2

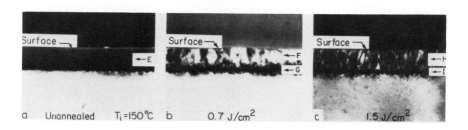

B. 10^{15} p$^+$/cm^2

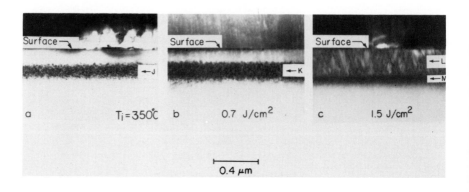

0.4 μm

C. 10^{16} p$^+$/cm^2

Figure 14. Defects resulting from laser annealing.

ACKNOLWEDGEMENT

Much of the experimental work and conclusions reviewed here are due to former grduate students Krishna Seshan, Wei Ku Wu and Bob Drosd and to post-doctoral associate Devendra Sadana. It has been a great pleasure to work with them and it is an honor to acknowledge their many contributions.

This work has been supported by the Department of Energy through the Materials and Molecular Research Division of the Lawrence Berkeley Laboratory, Contract No. W-7405-ENG-48. This support is much appreciated.

REFERENCES

1. K. Seshan and J. Washburn, Phys. Stat. Sol. (a) 26, 345 (1974).

2. P. K. Madden and S. M. Davidson, Rad. Effects 14, 271 (1972).

3. J. Mayer et al., Can. Journ. of Phys. 46, 663 (1968).

4. D. Sadana, unpublished data.

5. H. Muller et al., App. Phys. Lett. 26, 292 (1975).

6. L. Csepregi et al., App. Phys. Lett. 29, 92 (1976).

7. Radiation Effects in Semiconductors, P. Vook, ed., (Plenum Press, 1968).

8. R. Drosd and J. Washburn, J. Appl. Phys. 51, 4106 (1980).

9. K. Seshan and J. Washburn, Rad. Eff. 26, 31 (1975).

10. Wei-Kuo Wu and J. Washburn, J. Appl. Phys. 48, 3742 (1977).

11. D. Sadana et al., Appl. Phys. Lett. 37, 615 (1980).

12. L. Csepregi et al., J. Appl. Phys. 48, 4234 (1977).

A REVIEW OF NTD-INDUCED DEFECTS IN SILICON

J. M. MEESE
University of Missouri Research Reactor, Columbia, Missouri   65211

ABSTRACT

Neutron transmutation doping (NTD) has become the preferred method of phosphorus doping of float zone silicon for
the Si power device industry. This doping process is accomplished by thermal neutron irradiation in a nuclear reactor.
One of the stable isotopes, ^{30}Si, is transmuted to the desired dopant, ^{31}P, by thermal neutron capture and beta decay. This transmuted product becomes electrically active
only after suitable annealing of the accompanying radiation
damage which is the result of a number of different displacement mechanisms. This paper will present a review of
our present understanding of the production and annealing of
these displacement defects.

INTRODUCTION

Neutron transmutation doping of semiconductors was discovered 30 years ago
by Oak Ridge National Laboratory and Purdue University during neutron radiation
damage experiments on Ge [1,2]. The transmutation of Ge isotopes to the dopants
Ga and As by thermal neutron capture was observed and was used to determine, to
high precision, new values of the thermal neutron capture cross sections in
Ge [1,2]. The first systematic transmutation doping studies in Si were performed at Bell Labs in 1961 by Tanenbaum and Mills [3]. They considered the
many possible nuclear reactions which are produced in Si and its common impurities and concluded that the main dopant effect was the production of phosphorus
by the thermal neutron capture reaction

$$^{30}Si(n,\gamma)^{31}Si \rightarrow {}^{31}P + \beta^- \quad . \tag{1}$$

The phosphorus production rate for this process is calculated from

$$[^{31}P] = N_T \sigma_c (^{30}Si)\ \Phi \quad , \tag{2}$$

where N_T = (5 x 10^{22} Si/cm^3 x 0.031), $\sigma_c \cong 0.11$ barns, and Φ is the thermal
neutron fluence. The phosphorus production rate is then found to be about
3.36 ppb per 10^{18} n/cm^2 or 1.68 x 10^{-4} ^{31}P/cm^3 per n/cm^2.
    The major advantages of the NTD process, high doping accuracy and homogene-
ity, were not immediately realized commercially; however, attempts were made to
utilize the excellent doping control inherent in the process to fabricate high
resistivity detectors and integrated circuits as early as 1964 [4,5]. The excellent doping homogeneity of NTD-Si was finally demonstrated in 1971 by Kharchenko and Solov'ev who investigated radial spreading resistance for the first
time [6]. Their work was followed by extensive industrial research [7-15]
which led to the first commercial introduction of NTD float zone for power rectifiers in 1974.
    Over the past several years, NTD-Si has been found to be useful for a number
of device applications including power rectifiers, thyristors, power transistors, extrinsic Si IR detectors, particle and x-ray detectors, avalanche diodes,

thermistors and N.B.S. standard-resistivity wafers for probe calibration [14-15]. We estimate that over 50% of all commercial float zone is now being doped by the NTD process and that the production of NTD Si will exceed 60 metric tons in 1980. Figure 1 illustrates the primary advantages which have made NTD Si a commercial success, namely, highly accurate control of dopant concentration and a dramatic increase in resistivity homogeneity. These characteristics of NTD Si provide unique materials properties which cannot be obtained by any other process today.

The commercial success of NTD-Si has stimulated a renewed interest in neutron induced radiation damage processes in Si which will be the subject of this paper. NTD doping techniques and an analysis of NTD improvement in resistivity microstructure have been reviewed elsewhere [16].

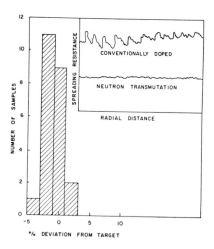

Fig. 1. Advantages of NTD process. Histogram of irradiation accuracy for a commercial sample lot irradiated at MURR. Insert compares spreading resistance traces across wafer diameter for conventionally doped and NTD-Si.

NTD MINORITY CARRIER LIFETIMES

Although the NTD benefits of homogeneity and concentration control provide advantages for many device applications, these benefits are not realized without an accompanying degradation in the minority carrier lifetime after annealing to values in the 100-500 μsec range [17]. To date, the reduction of NTD minority carrier lifetime has not been correlated with a specific radiation damage recombination level with known properties. Schulz and Lefèvre have observed DDLTS spectra in a large number of NTD samples from a variety of NTD manufacturers and find that the trap levels observed, for the most part, appear to be due to impurities introduced during annealing rather than to specific radiation induced levels [18]. In fact, the only DDLTS level they could attribute to a radiation induced defect was always lower in concentration for those samples irradiated in reactor spectra containing a higher fraction of fast neutrons [18]. This rather surprising result has been confirmed by the observations of both Malmros and the Oak Ridge group [17,19]. These authors find that minority carrier lifetime tends to be higher in those samples irradiated in a "hard" neutron spectrum of low Cd-ratio. For unknown reasons, the fast neutron damage appears to improve the quality of the material.

The lifetime and DDLTS experiments in NTD-Si irradiated in different reactor spectra should be compared with the out diffusion of radioactive tracer impurities from amorphous implanted layers [20]. In these implantation experiments it has been found that low dose out diffusion occurs at the temperature of self-diffusion. At higher doses, however, impurity out diffusion occurs at a much lower temperature corresponding to amorphous layer reordering. Although these out diffusion experiments suggest a damage dependent mechanism in which lifetime might be improved in low Cd-ratio neutron irradiated Si, Stein has shown that the formation of truly amorphous regions in neutron irradiated Si is unlikely at fluences typically encountered in neutron doping (50 Ω-cm, Φ_{th} = 6 x 10^{17} n/cm^2) [21].

At present, we cannot explain why different reactors yield differing minority carrier lifetimes in NTD Si. The concentrations of defects in annealed NTD Si are, in fact, too low to be detected by most of the microscopic defect identification tools available. We can describe, however, the damage mechanisms as well as the types and concentrations of defects which we have observed immediately following neutron irradiation in the hope that this information can serve as a clue to further studies.

DAMAGE MECHANISMS

Several damage mechanisms contribute to displacement events during NTD irradiation. These are:
1. Fast neutron knock-on displacements
2. Gamma flux induced displacements
3. Gamma recoil displacements
4. Beta recoil displacements
5. Charged particle knock-on displacements from (n,p), (n,α), etc. nuclear reactions.

Although all of the above mechanisms contribute to the total amount of radiation damage, fast neutron knock-ons and gamma recoils produced, by far, the most displacements [14]. The recoil energy imparted to a silicon isotope during gamma emission is given by

$$E = \tfrac{1}{2}(\hbar\omega)^2/Mc^2$$

where $\hbar\omega$ is the emitted gamma energy and Mc^2 is the recoiling isotope mass in energy units. Attempts to average over all Si isotopes and their gamma spectra have yielded values ranging from 474 eV [22] to 780 eV [23] for the mean recoil energy, $\overline{E}$, imparted to the lattice per thermal neutron capture. The concentration of displacements per unit time, i.e., the damage rate, is then given by

$$R_{th} = (dN_D/dt)_{th} = N_T\sigma_c(Si)\phi_{th}\,\overline{E}/2E_d \tag{3}$$

where N_T = 5 x 10^{22} cm^{-3}, σ_c = 0.16 barns, ϕ_{th} is the total thermal neutron flux and E_d the displacement threshold.

The displacement rate for fast neutrons varies considerably with the neutron energy spectrum. For light water reactors of relatively low Cd-ratio, fast neutron displacements dominate; however, in heavy water reactors of high Cd-ratio, the displacement rate from thermal neutron capture recoils is roughly the same order of magnitude as the fast displacement rate. The displacement rate induced by fast neutrons is given by

$$R_f = (dN_D/dt)_f = N_T \int_{E_d/\lambda}^{\infty} \sigma_d(E_n)\phi(E_n)dE_n \tag{4}$$

where N_T = 5 x 10^{22} cm^{-3} is the concentration of Si target atoms, $\phi(E_n)dE_n$ is the flux of neutrons in dE_n at neutron energy E_n and E_d/λ is the neutron energy corresponding to the threshold recoil energy E_d. The constant λ =

$4M_nM_{Si}/(M_n+M_{Si})^2 = 0.133$ relates the Si recoil energy E to the neutron energy E_n by the expression

$$E_n = \lambda E \sin^2 \frac{\theta}{2}$$

where θ is the neutron scattering angle in the center of mass system.

The displacement cross section is given by an integration over all possible recoil energies, E, by

$$\sigma_d(E_n) = \int_{E_d}^{\lambda E_n} \frac{d\sigma}{dE} \nu(E) dE \tag{5}$$

where $d\sigma/dE$ is the differential neutron scattering cross section and $\nu(E)$ represents the number of displacements in a cascade for a primary knock-on of recoil energy E. Since $\sigma_d(E_n)$ is only material dependent, this parameter can be calculated independent of the reactor spectrum. Several codes are available to obtain $\sigma_d(E_n)$ which include elastic scattering, inelastic resolved scattering, (n,2n) scattering, non-elastic scattering and high energy scattering [24,25]. These programs include the Robinson approximation to Lindhard's theory for ionization energy loss corrections to $\nu(E)$ [26].

The displacement cross section can also be calculated approximately by the Kinchin-Pease model [27] with an ionization threshold correction and isotropic scattering cross section [28]. Under these assumptions, the displacement cross section is given by [29]

$$\sigma_d(E_n) = \begin{cases} \dfrac{\sigma_e \lambda E_n}{4E_d}, & E_n \leq E_i/\lambda \\[2ex] \dfrac{\sigma_e E_i}{4E_d} \left(2 - \dfrac{E_i}{\lambda E_n}\right), & E_n > E_i/\lambda \end{cases} \tag{6}$$

where $E_i = M_{Si}(keV) = 28$ keV and $\sigma_e \cong 3$ barns is the elastic neutron scattering cross section. It should be noted that the value of this displacement cross section saturates above 0.8 MeV due to ionization effects as $E_i/\lambda E_n$ becomes negligible. This same characteristic is also seen in the more exact calculations discussed above. Since σ_d in Si approaches a constant value at high neutron energies, Eq. (4) suggests that only the total number of neutrons, $\int \phi(E_n)dE_n$, above 0.8 MeV need be considered independent of the detailed nature of the neutron energy spectrum. This is a useful property since the total number of fast neutrons can be estimated using a threshold activation foil [30] such as $^{58}Ni(n,p)^{58}Co$ to a reasonable approximation.

For energies below 0.8 MeV down to thermal energies, most thermal reactor spectra follow approximately an inverse energy relationship given by

$$\phi(E_n) = K/E_n \tag{7}$$

The magnitude of the flux constant, K, can be determined experimentally by measuring the Cd-ratio provided that the resonance flux follows the relationship given by Eq. (5). Under these conditions [31]

$$R_{cd} = 1 + \frac{\phi_{th} \sigma_o}{K I_R} \tag{8}$$

where ϕ_{th} is the thermal neutron flux, σ_o the thermal neutron capture cross section and I_R is the resonance integral given by

TABLE I
Resonance integrals and thermal capture cross sections for foils commonly used
to determine Cd-ratio [32]

Isotope	I_R (barns)	σ_0 (barns)	σ_0/I_R
^{197}Au	1550	98.8	6.37×10^{-2}
^{59}Co	75.0	37.5	0.50
^{63}Cu	4.4	5.0	1.14

$$I_R = \int_{0.4 \text{ eV}}^{\infty} \sigma_a(E_n) \frac{dE_n}{E_n} \tag{9}$$

Values of σ_0 and I_R have been determined experimentally for a number of materials and are listed in Table I.

It should be noted from Eq. (8) that the experimentally determined Cd-ratio depends on the foil used for measurement. This is illustrated below in Table II which compares Cd-ratios obtained with Au to those expected for Co and Cu for two extreme reactor types.

TABLE II
Cd-ratios expected for different reactor types using different activation foil
materials

Reactor Type	R_{Cd} (Au)	R_{Cd} (Co)	R_{Cd} (Cu)	ϕ_{th}/K
Hardest Light Water	5	32.4	72.6	62.8
Softest Heavy Water	1000	7840	17900	15680

Because the thermal reactor spectrum is dominated by Eq. (7) for neutron energies above the Cd edge, and because the damage rate in Si is independent of spectral detail at high neutron energies, it is possible to assume that Eq. (7) holds for the entire fast neutron spectrum up to a maximum cut off energy E_f which is then adjusted to agree with high energy threshold foil activation data. We have found for a row-2 reflector position in the University of Missouri Research Reactor, that $E_f = 3.27$ MeV is a good choice for this parameter in order to agree with the reactor design data and threshold foil measurements.

Under the simplified assumptions concerning the reactor spectrum and using the Kinchin-Pease cross section, the damage rate is found from Eq. (4) to be given by

$$R_f = N_T \sigma_e \ K\left(\frac{\lambda E_f}{4E_d}\right) f(\beta) \tag{10}$$

where $f(\beta) = \beta^2 + 2\beta \ \ln(1/\beta)$ is an ionization loss correction term with $\beta = E_i/\lambda E_f$. Taking the ratio of Eqs. (10) to Eq. (3), we then obtain an expression for the ratio of damage rates for fast and thermal neutrons,

$$R_f/R_{th} = \frac{\sigma_e}{\sigma_c} \frac{\lambda E_f}{2\bar{E}} \frac{f(\beta)}{} \left(\frac{K}{\phi_{th}}\right) \tag{11}$$

This expression is not very sensitive to the value E_f selected for the cut off energy. Varying E_f between 1 MeV to 10 MeV only varies the result of Eq. (11) by ± 40%. Eq. (11) can be expressed also in terms of the Cd-ratio defined by Eq. (8). Using the following parameters, $\sigma_e = 3$ barns, $\sigma_c = 0.16$ barns, $\lambda =$

0.133, $E_f = 3.27$ MeV, $E_i = 28$ keV, $\beta = 6.43 \times 10^{-2}$, $f(\beta) = 0.357$, $E_d = 25$ eV and $\bar{E} = 474$ eV, Eq. (11) becomes

$$R_f/R_{th} \cong 3000 \ K/\phi_{th} = \frac{3000 \ I_R/\sigma_o}{(R_{Cd} - 1)} \quad . \tag{12}$$

Although highly simplified, the above model allows the silicon manufacturer, who is faced with the problem of obtaining neutron irradiations from a variety of reactors with poorly characterized neutron energy spectra, to estimate the order of magnitude importance of the dominant displacement rates as a function of Cd-ratio using the cross sections and resonance integrals given in Table I. From Eq. (1), the rate, R_p, at which phosphorus is produced can also be obtained. A simple calculation using Eqs. (1) and (3) yields the ratio

$$R_{th}/R_p \cong 450 \quad ; \tag{13}$$

therefore for $R_{Cd}(Au) = 5$, we find that $R_p:R_{th}:R_f = 1:450:2.15\times10^4$ while for $R_{Cd}(Au) = 1000$, $R_p:R_{th}:R_f = 1:450:86$. The light water reactor damage is dominated by fast neutron displacements while the recoil damage is of equal significance for heavy water reactor irradiations.

We have compared the simple Kinchin-Pease model above to calculations using the displacement cross section data for Si obtained from Argonne codes [33] and a reactor spectrum which approximates the design data and threshold foil data for a row-2 position at MURR,

$$\phi(E_n) = \begin{cases} 6.55\times10^{11}/E_n & , \ E_n < 2 \text{ MeV} \\ 1.0\times10^{12} \ \sqrt{E_n} \ e^{-0.776E_n} & , \ E_n > 2 \text{ Mev} \end{cases} \tag{14}$$

consisting of a down scattered spectrum below 2 MeV and a fission spectrum above 2 MeV. The results are shown in Fig. 2 which is a plot of $\sigma_d(E_n)\phi(E_n)$. The integral of this function is proportional to the number of displacements. The solid line represents a Kinchin-Pease model for the spectrum given by Eq. (14) while the curve with data points is the same spectrum but using the improved values of $\sigma_d(E_n)$ obtained from the Argonne code. The dotted line represents the simple cut off procedure discussed in the model leading to Eq. (12). All three methods yield the same fast neutron displacement rate to $\pm$ 10%.

Fig. 2. Displacement rate per Si atom per unit energy range. Dotted line and solid line for a typical light water reactor spectrum described by Eq. (14). Dotted line represents simple cut off procedure described in text.

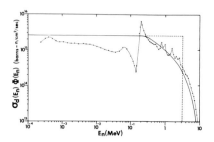

EXPERIMENTAL DATA

In this section we shall review some of the NTD radiation damage results which have recently been observed at the University of Missouri using electrical measurements, DLTS, optical absorption, Raman scattering and EPR [16,34,35].

These will be compared to some extent with the calculations of the previous section.

Figure 3 shows isochronal annealing of resistivity [34] for three float zone samples which were 5250 Ω-cm, p-type before irradiation. Thermal neutron fluences were sufficient to transmutation dope the material to 1 Ω-cm, 100 Ω-cm and near intrinsic. The Cd-ratio on Co for these irradiations was 25-30 as is typical of a light water reactor. All anneals were 15 mins. in argon.

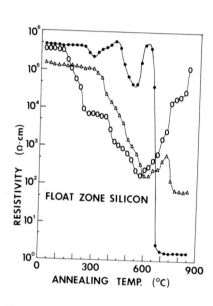

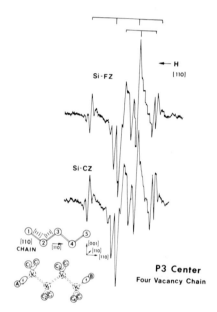

Fig. 3. Dose dependence of isochronal annealing of NTD float zone Si. Initial resistivity was 5250 Ω-cm, p-type. Thermal neutron fluences and the temperature of type conversion (n → p) were as follows: (o) = 1.67 x 10^16 n/cm^2, mixed type at 800°C; (Δ = 3.47 x 10^17 n/cm^2, 725°C; (●) = 1.57 x 10^19 n/cm^2, 425°C.

Fig. 4. EPR spectrum of P-3 four vacancy chain, the dominant defect concentration in room temperature NTD silicon irradiated to ϕ_{th} = 1.34 x 10^19 n/cm^2 in both CZ and FZ. Spectra taken at room temperature at 9.10 GHz.

All samples were slightly p-type immediately following irradiation and nearly intrinsic. This is the case independent of the type of starting material. The transition from p-type to n-type decreases in annealing temperature as the concentration of transmutation produced phosphorus increases as indicated in the caption to Fig. 1. For the highest fluence sample, this type conversion occurs as low as 425°C suggesting that appreciable phosphorus is becoming substitutional at this or even lower temperatures.

A number of recovery stages are observed including a very prominent annealing stage between 500 to 600°C. During this stage acceptors are produced (or donors destroyed) in a concentration exceeding 6 x 10^13 cm^-3. This is apparent in the 1 Ω-cm sample by the reverse recovery peak at 600°C after this sample had converted to n-type at a previous anneal temperature. This same defect is observed as a dip in resistivity in the other two samples which are still p-type at 600°C. The temperature at which this defect is first observed appears to be fluence dependent. The recovery stage occurs at lower temperatures for more lightly irradiated material, however, the concentration of this defect appears to be independent of fluence suggesting a defect interaction with a fixed concentration of impurities.

Some annealing structure is evident at temperatures exceeding 600°C. Similar samples were irradiated to different fast neutron fluences while shielded by boron in order to eliminate transmutations [34]. The annealing of these samples was similar to the 100 Ω-cm sample shown in Fig. 1, however, there was a general tendency for the annealing curves to shift to higher temperatures above 500°C as the fast neutron fluence was increased. This suggests that different annealing procedures might be required depending upon the reactor spectrum used for transmutation doping. In all cases, however, electrical activation of the phosphorus was complete by 850°C.

Isochronal anneals of Czochralski Si neutron doped to 1 Ω-cm were similar to the float zone 1 Ω-cm sample shown in Fig. 1 [34]. Cleland, et al. [19] have shown, however, from isothermal anneals of NTD Czochralski that the process is complicated by oxygen donor activation which must be accounted for if accurate resistivity targeting is to be achieved.

Microscopic information on the defects produced during transmutation doping was obtained by optical absorption and EPR [16,35]. Both Czochralski (CZ) and float zone (FZ) samples were investigated. Optical absorption and EPR data were obtained on the same samples. Both the CZ and FZ samples had initial resistivities of 250 Ω-cm and were n-type. The first experiments were performed on samples irradiated in a light water reactor (MURR) in a Cd-ratio on Co of 25-30 to a thermal neutron fluence of 1.35×10^{19} n/cm^2. This fluence should produce an added phosphorus concentration of 2.2×10^{15} cm^{-3} corresponding to a final resistivity after annealing of about 2 Ω-cm. This is about an order of magnitude greater fluence than is normally encountered in typical NTD material. The concentration of displacements was estimated to be about 4 to 6 x 10^{19} disp/cm^3 according to the simple Kinchin-Pease model discussed in the previous section. All of the defects observed in these samples have been previously reported, however, emphasis is placed here on comparing the relative concentrations of the defects observed.

The dominant defects observed by optical absorption were the divacancy [36] (1.8 µm band) and, in the CZ sample, the A-center [37] (vacancy-oxygen). The concentration of A-centers was estimated from a decrease in the 9 µm oxygen interstitial band and also by assuming that its oscillator strength was similar to the 9 µm band. The A-center concentration was therefore estimated to be about 3×10^{17} cm^{-3} in the CZ sample which contained 6×10^{17} cm^{-3} oxygen interstitials before irradiation. The concentration of divacancies is less certain. The concentration of this defect was estimated to be about 3 to 4 x 10^{17} cm^{-3} based on the fast neutron fluence and previous experiments by ORNL [38].

The E-center [39] (vacancy-phosphorus) EPR resonance is somewhat difficult to observe in NTD-Si but is seen occasionally at lower temperatures in FZ. This is not surprising since the concentration of NTD produced phosphorus is only 2×10^{15} cm^{-3} for this irradiation and several orders of magnitude lower than other defect concentrations present in the sample. The dominant EPR defect is the four-vacancy P-3 center [40]. It appears in a concentration of 3.5 x 10^{17} cm^{-3} in this experiment. The EPR spectra for this defect are shown in Fig. 4 as well as its atomic configuration. The surprising result of this experiment is that the concentration of P-3 centers is identical to better than 1% in both CZ and FZ irradiated to the same fluence. Since the concentration of oxygen and also A-centers in the CZ sample is sufficient to have depleted the P-3 concentration, if this defect had been formed from a vacancy gas by condensation, this experiment strongly suggests that the P-3 defect is a cascade remnant. Finally, the five-vacancy P-1 center [41] is observed in low concentrations ($< 3 \times 10^{16}$ cm^{-3}) following irradiation. Following annealing of the divacancies between 200-300°C, however, the concentration of this defect increases dramatically as the P-3 concentration decreases [42] suggesting that four-vacancy centers become five-vacancies by picking up an extra vacancy supplied by the divacancy annihilation.

The results of this light water reactor irradiation experiment are summarized in Table III. It is apparent from Table III that only a small percentage of the total calculated displacements are observed as recognizable point defects.

TABLE III
Defect bookkeeping of observed NTD vacancy defect concentrations for light water reactor irradiations immediately following irradiation

CALCULATED DISPLACEMENTS = 4×10^{19} cm^{-3}
(1.34×10^{19} n_{th}/cm^2, R_{Cd}(Co) = 25)

DEFECT	CONCENTRATION OBSERVED (cm^{-3})	VACANCY CONCENTRATION (cm^{-3})
Divacancy (V-2)	4.4×10^{17}	0.88×10^{18}
4-vacancy (P-3)	3.5×10^{17}	1.4×10^{18}
5-vacancy (P-3)	$< 3 \times 10^{16}$	$< 0.14 \times 10^{18}$
Vacancy-Phos. (E)	$< 2 \times 10^{15}$	$< 0.002 \times 10^{18}$
Vacancy-Oxy. (A)	3×10^{17}	0.3×10^{18}
	TOTAL	2.8×10^{18}

OBSERVED/CALCULATED = 0.068 (6.8%)

In addition to the vacancy defects listed in Table III, the P-6 center [43] has been observed in concentrations equal to the P-3 center concentration. The P-6 defect is thought to be a di-interstitial in a positive charge state and represents the dominant interstitial concentration after irradiation.

Additional EPR experiments have been performed on samples irradiated at two facilities in Argonne's CP-5 reactor. One facility had a Cd-ratio on Au of 939, a much softer spectrum than previous experiments, while the other facility contained a uranium converter plate and produced a somewhat harder spectrum [45] than the light water reactor irradiations. Figure 5 compares the EPR spectra for two identical FZ samples irradiated in the Argonne facilities. It can be seen that the ratio of P-3 to P-6 concentration is nearly independent of the type of neutron spectrum suggesting that P-6 is also a cascade remnant. In addition, a very broad resonance containing an appreciable spin concentration is apparent in the high Cd-ratio irradiation.

The mean number of displacements per cascade, from the simple Kinchin-Pease model, is given by [15]

$$\bar{\nu} = \frac{\lambda E_f \ f(\beta)}{4E_d \ \ell n \ \frac{\lambda E_f}{E_d}} \tag{15}$$

while the mean number of cascades can be calculated from the relationship, [cascade concentration] = $N_D/\bar{\nu}$. For the light water reactor experiments it is found that the cascade concentration was about 3.8×10^{17} cm^{-3}. Comparison with Table III suggests that there was about one P-3 and one P-6 defect per

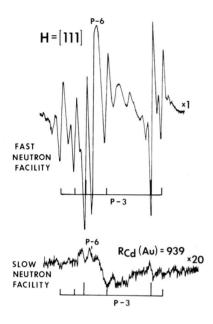

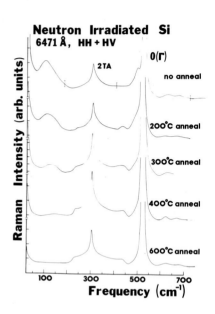

Fig. 5. Room temperature (9.1 GHz) EPR spectra for samples irradiated a) in Argonne's CP-5 fast facility (top trace) and b) in Cd-ratio of 939 on Au. The P-3 and P-6 centers are indicated.

Fig. 6. The Raman spectrum of NTD sample after successive 15 min. isochronal anneals. Sample was irradiated to a thermal neutron fluence of 3.5×10^{19} n/cm^2 in a Cd-ratio on Co of 25 to 30.

cascade. Similar calculations showed that this was approximately true for both Argonne irradiations also and that the P-3 and P-6 concentrations for samples irradiated in the high Cd-ratio were caused by fast neutron contamination.

We believe that these experiments suggest that P-3 and P-6 are produced as a result of a Brinkman displacement spike [46] which is created when the primary knock-on's energy becomes low enough that the mean free path between displacements is equal to the distance between atoms. The primary knock-on energy below which this occurs is 937 eV [15] compared to an energy of 200 to 400 eV required for the production of a four-vacancy. Allowing for some recombination, this model appears to account for the concentration of cascade remnants observed.

Results of these standard microscopic experiments show that recognizable point defects account for only a small fraction of the calculated displacements. Although some recombination of defects is expected for room temperature irradiations, the possibility still exists that large amounts of damage are unobservable due to the proximity of the defects within the cascade. The sample irradiated in the very high Cd-ratio (lower spectrum of Fig. 5) should be dominated by recoil damage, and yet, the only concentrations of defects which are observed can be adequately accounted for on the basis of the spike mechanism and fast neutron contamination. Only the broad resonance in this spectrum, which obviously contains a large fraction of the total spin concentration, gives any hint of the recoil damage.

A second experiment suggests that large amounts of disorder exist in the lattice. Figure 6 shows the Raman scattering spectrum and isochronal annealing of a sample irradiated at MURR (light water, Cd-ratio on Co of 25-30) to a thermal neutron fluence of 7×10^{19} n/cm^2. Two peaks at 100-150 cm^{-1} and at 490 cm^{-1} are produced by irradiation [15]. These peaks are very similar to those seen in the Raman spectra of amorphous Si [47] or ion implanted Si which is nearly amorphous.

If these Raman peaks are due to amorphous tracks, then it is clear that a very large fraction of the displacements is required to produce these peaks. Stein [21] has estimated that the displacement energy density for a 22 keV recoil (the mean recoil energy for this sample irradiation) is about 10^{21} eV/cm^3. The ratio of this energy density to that required to produce totally amorphous material is about 0.5% [21]. We have calculated from the Kinchin-Pease model

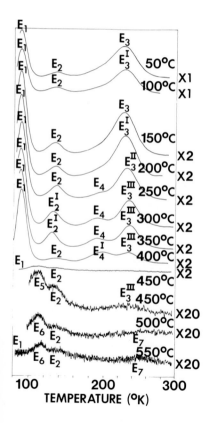

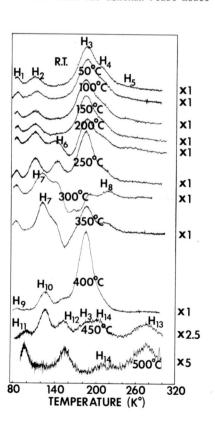

Fig. 7. DLTS electron traps in n-type neutron irradiated Si as a function of isochronal annealing (100 Hz, $\phi_{th} = 1.8 \times 10^{15}$ n/cm^2).

Fig. 8. 10 Hz DLTS hole traps in p-type neutron irradiated Si as a function of isochronal anneals ($\phi_{th} = 1.8 \times 10^{16}$ n/cm^2).

that the cascade density in this sample is about 2×10^{18} cm^{-3} while the concentration of displacements is around 3.6×10^{20} cm^{-3}. This corresponds to 0.7% of the total crystal volume. Comparing these two percentages, one would conclude that if most of the cascade volume were amorphous this could account for most of the calculated displacements. Although the isochronal annealing of the Raman spectra is suggestive of amorphous crystal regrowth, the characteristic EPR signal for amorphous Si has not been observed in these samples. We must conclude, therefore, that the regions observed by Raman scattering might not be amorphous but, rather, highly disordered and probably contain a large fraction of the expected defect concentration.

We wish finally to discuss briefly preliminary DLTS data obtained on p-n junction diodes in CZ NTD material [15]. Previous DLTS data has been reported by Guldberg [48] for n-type FZ annealed above 425°C and by Baliga and Evwaraye [49] for both p- and n-type FZ annealed above 425°C. We have purposely limited the fluence and radiation damage to low values in order to preserve diode operation so that data could be obtained for anneals from room temperature up to 500°C. Spectra for the electron traps in n-type material as a function of isochronal annealing are shown in Fig. 7 while hole trap spectra in p-type material are shown in Fig. 8.

A total of seven electron traps and fifteen hole traps have been observed as are shown in Tables IV and V. Although some of these levels have been observed previously in electron irradiated material, many have not. Identification of the defects associated with these levels is difficult. Reviews of known defect annealing stages in irradiated Si by Guldberg [48], Corbett [50] and Kimerling [51] are useful in guessing at the defect structures, however, considerably more work needs to be done in this area in order to make the identification positive. Table VI lists tentative suggestions as to defect identification for future research purposes.

The E_1 and E_3 levels have been identified previously from electron irradiation data [51]. The E_3 level is observed to shift slightly as a function of annealing suggesting that several different defects might possess levels in this range. The $E_3(III)$, E_2 and H_2 levels are possible candidates for divacancy levels. The H_4 level was tentatively identified as being associated with the di-interstitial P-6 center based on annealing data and relative intensity. This H_4 energy level agrees with a P-6 level assignment suggested by Lee, Gerasimenko and Corbett [43]. The H_3 level has been assigned to the P-3 center mainly on the basis of concentration and annealing. It should be noted that this level disappears at around 300°C, however, an identical level re-appears and dominates the spectrum at 400°C. We have tentatively identified this second occurrence with the appearance of the P-4 center, however, this behavior could also possi-

TABLE IV

Energy levels ($E_C - E_T$) and carrier capture cross sections for electron traps in n-type material

Level	$(E_C - E_T)$ (eV)	σ (cm^2)
E_1	0.16	4.8×10^{-15}
E_2	0.24	4.6×10^{-15}
E_3	0.42	3.7×10^{-15}
E_4	0.34	2.5×10^{-15}
E_5	0.27	7.5×10^{-12}
E_6	0.21	1.0×10^{-14}
E_7	0.52	5.8×10^{-14}

bly be due to Fermi level shifts. The defect levels which occur from 450 to 500°C are likely to be associated with various oxygen complexes. We also have not allowed for the possibility of carbon complexes, however, several such defects are thought to exist. Considerably more DLTS work needs to be done on neutron irradiated Si, particularly annealing above 500°C, in order to establish which lattice defects, if any, contribute to the reduction in minority carrier lifetime in NTD material.

TABLE V
Energy levels ($E_T - E_V$) and carrier capture cross sections for hole traps in p-type material

Level	$E_T - E_V$ (eV)	σ (cm^2)
H_1	0.16	6.4×10^{-15}
H_2	0.18	1.6×10^{-16}
H_3	0.35	2.8×10^{-15}
H_4	0.39	9.7×10^{-16}
H_5	0.61	4.1×10^{-13}
H_6	0.38	6.2×10^{-15}
H_7	0.22	1.2×10^{-15}
H_8	0.62	1.2×10^{-10}
H_9	0.19	8.8×10^{-14}
H_{10}	0.25	1.0×10^{-15}
H_{11}	0.20	3.0×10^{-15}
H_{12}	0.37	6.4×10^{-13}
H_{13}	0.45	2.4×10^{-17}
H_{14}	0.51	7.9×10^{-12}
H_{15}	0.19	5.1×10^{-11}

TABLE VI
Tentative identification of DLTS levels with known defect structures in Si

DLTS Level	Energy (eV)	Defect	Structure
E_7	0.52	Surface	?
E_1	0.16	A-Center	vacancy-oxygen
E_3 (< 150°C)	0.42	E-Center	vacancy-phosphorus
E_3(III)	0.45		
E_2	0.24	V-V	divacancy
H_2	0.18		
E_4	0.34	P-1	five-vacancy
H_3 (300-500°C)	0.35	P-4	trivacancy-oxygen
E_5	0.27	P-2	divacancy-dioxygen
H_3 (R.T. - 300°C)	0.35	P-3	four-vacancy
H_4	0.39	P-6	di-interstitial

SUMMARY AND CONCLUSIONS

A simple Kinchin-Pease model has been presented to illustrate the effects of different reactor spectra on the overall displacement rate. It was argued that most thermal reactor spectra can be approximated by a K/E_n flux distribution and that a measurement of the Cd-ratio can yield order of magnitude correct displacement rates in Si in terms of this parameter if more detailed spectral information is not readily available.

A number of point defects have been observed with EPR and optical absorption in NTD-Si. It has been argued that the four-vacancy (P-3) and di-interstitial (P-6) centers are most likely to be remnants of displacement spikes since there is excellent correlation with the number of centers and the number of cascades regardless of wide variations in the reactor spectrum used for irradiation and since the spike energy corresponds within factors of 2 or 3 to the energy required to make these defects.

Raman scattering and EPR both suggest that highly disordered regions also exist which contain large numbers of unresolved defects. In addition, a vast array of defect energy levels have been observed by DLTS. Tentative defect assignments have been suggested for the more prominent levels, however, considerably more experimental evidence will be required to make these assignments definite.

Finally, the question of NTD-Si minority carrier lifetime remains unanswered. Much of the problem undoubtedly lies with contamination during annealing, however, there appears to be an additional mechanism by which a large amount of fast neutron damage tends to improve the situation somewhat. Both the lifetime measurements of Malmros and the Oak Ridge group as well as the DDLTS measurements of Schultz and Lefèvre agree on this point. We have suggested that enhanced out diffusion of impurities during annealing might provide an explanation, however, no experimental evidence exists to support or reject this hypothesis at present.

ACKNOWLEDGMENTS

We wish to acknowledge the contributions of D. L. Cowan, Meera Chandrasekhar and H. R. Chandrasekhar of the Department of Physics of the University of Missouri-Columbia in obtaining the EPR and optical data and is formulating interpretations of these data. We also wish to thank T. Blewitt of Argonne National Laboratory for providing those irradiations performed at CP-5 reactor.

REFERENCES

1. J. W. Cleland, K. Lark-Horowitz and J. C. Pigg, Phys. Rev. 78, 814 (1950).

2. K. Lark-Horowitz, "Nuclear-bombarded Semiconductors" in Semiconductor Materials, (Butterworths, London, 1951), p. 47.

3. M. Tanenbaum and A. D. Mills, J. Electrochem. Soc. 108, 171 (1961).

4. J. Messier, Y. le Coroller and J. M. Flores, I.E.E.E. Trans. Nuc. Sci. NS-11, 276 (1964).

5. C. M. Klahr and M. S. Cohen, Nucleonics 22, 62 (1964).

6. V. A. Kharchenko and S. P. Solov'ev, Izv. Akad. Nauk USSR, Neorgan. Mat. 7, 2137 (1971).

7. M. S. Schnöller, I.E.E.E. Trans. Electron. Devices ED-21, 313 (1974).

8. K. Platzoder and K. Lock, I.E.E.E. Trans. Electron. Devices ED-23, 805 (1976).

9. M. J. Hill, P. M. Van Iseghem, W. Zimmerman, I.E.E.E. Trans. Electron. Devices ED-23, 809 (1976).

10. J. M. Janus and O. M. Malmros, I.E.E.E. Trans. Electron. Devices ED-23, 797 (1976).

11. H. A. Herrmann and H. Herzer, J. Electrochem. Soc. 122, 1568 (1975).

12. W. E. Hass and M. S. Schnöller, J. Electronic Mat. 5, 57 (1976).

13. Semiconductor Silicon 1977, ed. by H. R. Huff and E. Sirtl (The Electrochemical Society Proceedings, vol. 77-2, Princeton, 1977), Chapter 3.

14. Neutron Transmutation Doping in Semiconductors, ed. by J. M. Meese (Plenum Press, N.Y., 1979).

15. Proceedings of the Third International Conference on NTD-Si, ed. by J. Guldberg (Plenum Press, N.Y., in press).

16. J. M. Meese, D. L. Cowan and M. Chandrasekhar, I.E.E.E. Trans. Nuc. Sci. NS-26, 4858 (1979).

17. O. Malmros, Ref. 14, p. 249.

18. M. Schulz and H. Lefèvre, Ref. 13. p. 142.

19. J. W. Cleland, P. H. Fleming, R. D. Westbrook, R. F. Wood and R. T. Young, Ref. 14, p. 261.

20. J. W. Meyer, L. Erikson and J. A. Davies, Ion Implantation in Semiconductors, (Academic Press, New York, 1970) pp. 116-118.

21. H. J. Stein, Ref. 14, p. 229.

22. R. R. Coltman, Jr., C. E. Klabunde, D. L. McDonald, J. K. Redman, J. Appl. Phys. 33, 3509 (1962); R. R. Coltman, C. E. Klabunde, J. K. Redman, Phys. Rev. 156, 715 (1967).

23. M. V. Chukichev and V. S. Vavilov, Sov. Phys. Solid State 3, 1103 (1961).

24. G. R. Odette and D. R. Dorian, Nucl. Technol. 29, 346 (1976).

25. D. M. Parkin and A. N. Goland, Rad. Effects 28, 31 (1976).

26. M. T. Robinson, in Nuclear Fusion Reactors (British Nuclear Energy Society, London, 1970), p. 364.

27. G. H. Kinchin and R. S. Pease, Rep. Prog. Phys. 18, 1 (1955).

28. D. R. Olander, Fundamental Aspects of Nuclear Reactor Fuel Elements, TID-26711-P1, (Technical Information Center, ERDA, Washington, D.C., 1976) Chapter 17.

29. D. R. Harries, P. J. Barton and S. B. Wright, in Symposium on Radiation Effects on Metals and Neutron Dosimetry, A.S.T.M. Special Tech. Pub. No. 341 (A.S.T.M., Philadelphia), p. 276.

30. Measuring Neutron Flux by Radioactivation Techniques, A.S.T.M. Designation E261-70 (A.S.T.M., Philadelphia).

31. F. A. Valente, A Manual of Experiments in Reactor Physics, (Macmillan Co., N.Y., 1963) p. 69.

32. Handbook on Nuclear Activation Cross-sections, Tech. Reports Series 156 (I.A.E.A., Vienna, 1974), p. 17.

33. The displacement cross section for Si as a function of neutron energy was obtained from L. R. Greenwood, Chemical Engineering Div., Argonne National Laboratory and is based on the methods of Ref. 24.

34. P. J. Glairon and J. M. Meese, Ref. 14, p. 291.

35. J. M. Meese, M. Chandrasekhar, D. L. Cowan, S. L. Chang, H. Yousif, H. R. Chandrasekhar and P. McGrail, Ref. 15.

36. L. J. Cheng, J. C. Corelli, J. W. Corbett, G. D. Watkins, Phys. Rev. 152, 761 (1966).

37. J. W. Corbett, G. D. Watkins, R. M. Chrenko, R. S. McDonald, Phys. Rev. 121, 1015 (1961).

38. B. C. Larson, R. T. Young and J. Narayan, Ref. 14, p. 281.

39. G. D. Watkins and J. W. Corbett, Phys. Rev. 134, A1359 (1964).

40. Young-Hoon Lee and J. W. Corbett, Phys. Rev. B8, 2810 (1973).

41. W. Jung and G. S, Newell, Phys. Rev. 132, 648 (1963).

42. L. Katz and E. B. Hale, Ref. 14, p. 307.

43. Y. H. Lee, N. N. Gerasimenko and J. W. Corbett, Phys. Rev. B14, 4506 (1976).

44. Y. H. Lee, Y. M. Kim and J. W. Corbett, Rad. Effects 15, 77 (1972).

45. M. A. Kirk and L. R. Greenwood, Ref. 14, p. 143.

46. J. A. Brinkman, J. Appl. Phys. 25, 961 (1954).

47. H. M. Brodsky, Light Scattering in Solids, ed. by M. Cardona, (Springer-Verlag, 1975) p. 208 and references therein.

48. J. Guldberg, J. Phys. D11, 2043 (1978).

49. B. J. Baliga and A. O. Evwaraye, Ref. 14, p. 317.

50. J. W. Corbett, J. C. Bourgoin, L. C. Chang, J. C. Corelli, Y. H. Lee, P. M. Mooney and C. Weigel, Radiation Effects in Semiconductors 1976, ed. by N. B. Urli and J. W. Corbett, (Institute of Physics, Bristol, 1977), p. 1.

51. L. C. Kimerling, Ibid., p. 221.

Published 1981 by North-Holland, Inc.
Narayan, and Tan, eds.
Defects in Semiconductors

TRANSIENT CAPACITANCE STUDIES OF A LOW-LYING ELECTRON TRAP IN n-TYPE SILICON

G. E. Jellison, Jr., J. W. Cleland, and R. T. Young
Solid State Division, Oak Ridge National Laboratory,[*]
Oak Ridge, Tennessee 37830

ABSTRACT

A new electron trap has been observed in electron-irradiated n-type silicon at $E_c - E_T = 0.105$ eV using transient capacitance techniques. It is found that the maximum transient capacitance response is observed only when the majority carrier pulse width is much smaller than some characteristic time constant, and when the time between pulses is much larger than another characteristic time constant. It is shown that this defect is related to an electron trap at $E_c-E_T = 0.172$ eV (probably the oxygen-vacancy or A-center); it is believed that this trap is induced by the electric field found in the depletion region of a p-n junction.

INTRODUCTION

Electron irradiation-induced defects in n-type silicon have been studied for many years. The work of Watkins, Corbett, and others has resulted in the identification of the oxygen-vacancy defect (A-center) [1,2], the divacancy defect [3], and the phosphorus-vacancy defect (E-center) [4]. The first transient capacitance study of electron-irradiated n-type silicon was performed by Walker and Sah [5]. They found 6 traps in n-type Si, including the A-center at 0.174 eV, the E-center at 0.471 eV, and an unidentified trap at 0.112 eV. Recently, the technique of deep level transient spectroscopy (DLTS) [6] has allowed workers to spectroscopically determine the energy level, capture cross section, and concentration of defects in semiconductors. Kimerling has applied the DLTS technique, in conjunction with annealing studies, to phosphorus-doped silicon, associating specific DLTS peaks to particular traps; in particular, he associated the A-center to a peak at $E_c-E_T = 0.18$ eV [7], and the E-center to a peak at $E_c-E_T = 0.44$ eV [7]. Evwaraye and Sun [8] have shown that two traps at $E_c-E_T = 0.23$ eV and 0.39 eV could be associated with two different charge states of the divacancy. This paper shows that there is another trap in phosphorus-doped silicon at $E_c-E_T = 0.105$ eV which is probably related to the A-center. It will be shown that this new trap disappears under certain pulsing schemes, and is replaced by a trap at $E_c-E_T = 0.172$ eV. Evidence will also be presented showing that this trap is probably activated by the electric field found naturally in the depletion region of the sample diode.

BACKGROUND AND EXPERIMENTAL CONSIDERATIONS

Transient capacitance techniques have been used by many workers for the characterization of defects found in the depletion region of junction devices

[*]Operated by Union Carbide Corporation for the U.S. Department of Energy under contract W-7405-eng-26.

242

ORNL—DWG 80—11437R

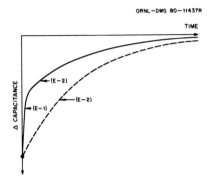

ORNL—DWG 80—13612

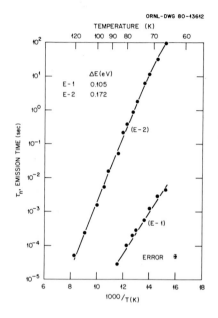

Fig. 1. A schematic of the observed transient capacitance from a majority carrier pulse measured at 77 K of a p^+-n diode irradiated with 1.5 MeV electrons. The solid line is normally observed when the time between pulses is long and the pulse width is short. If either of these conditions are violated, then E-1 will be replaced with E-2. For example, if the pulse width is long, the response will look like the dotted line.

Fig. 2. Emission times for E-1 and E-2 vs inverse temperature.

or Schottky barrier diodes. The reader is referred to one of the several good reviews which are available (for example, see refs. [5], [10], and [11]).

Transient capacitance measurements were made using a variable frequency capacitance bridge similar to that described by Lang (see ref. [11]). Samples were fabricated from float-zoned silicon substrates doped with 3.2×10^{14} P/cm^3 (Sample A) and 8.9×10^{14} P/cm^3 (Sample B), and from a Czochralski-grown substrate doped with 1.4×10^{14} P/cm^3 (Sample C). All substrates were purchased from Monsanto. The back contacts were degenerately doped with phosphorus before applying the aluminum contacts to ensure ohmic contact. The emitter of the diode was degenerately doped with boron by heating a BN source to 900°C for 30 min. Mesa diodes were then made and the resulting samples were mounted in integrated circuit holders by using a room-temperature curing silver epoxy on the back contact and wire bonding the front contacts.

Electron irradiations were carried out on the finished mesa diodes, using 1.5 MeV electrons. Typical currents were less than 5 μA/cm^2, and the sample was cooled by blowing air across the sample surface during irradiation. This results in less than a 25°C rise in sample temperature during irradiation.

RESULTS

The solid line in Fig. 1 shows a schematic diagram of the observed transient capacitance after a short majority carrier pulse (pulse width $t_w \ll \tau_w$), which occurred a long time after the previous pulse (time between pulses $t_p \gg \tau_p$), where the quantities τ_w and τ_p are characteristic time constants for the pulse width and time between pulses, respectively. Two traps are observed at 77 K; E-1, with an emission time τ_{n1} = 250 μs, and E-2, with an emission time τ_{n2} = 1.5 sec. If the pulse width is much greater than τ_w, the observed transient capacitance response is shown schematically by the dotted line in Fig. 1; that is, the response due to trap E-1 is decreased, while the response due to trap E-2 is increased.

Figure 2 shows the emission time versus inverse temperature data for E-1 and E-2. The emission time was taken from the transient capacitance measurements assuming that

$$\Delta C = -h_1(1-\exp(-t/\tau_{n1}))-h_2(1-\exp(-t/\tau_{n2})) \tag{1}$$

where τ_{n1} and τ_{n2} are the emission times for the E-1 and E-2 traps, respectively, t is the time after the majority carrier pulse, and h_1 and h_2 are the heights of the responses of E-1 and E-2, respectively. Within experimental error, it was found that all transient capacitances were exponential. The emission times are given by

$$\tau_n = \frac{g}{\sigma_n \langle v_n \rangle N_c} \ \exp \ (\Delta E/kT) \tag{2}$$

where σ_n is the electron capture cross section, $\langle v_n \rangle$ is the average electron thermal velocity, N_c is the effective density of states at the conduction band edge, g is the degeneracy factor, and ΔE is the activation energy. The emission times were then fit to Eq. (2) using a least squares procedure assuming that $N_c \propto T^{3/2}$ and $\langle v_n \rangle \propto T^{1/2}$; the resulting fit is shown by the solid lines of Fig. 2. The activation energies determined by this method were ΔE(E-1) = 0.105 ± 0.005 eV and ΔE(E-2) = 0.172 ± 0.002 eV.

It has been experimentally observed that the height of the response from E-1 can be expressed as

$$h_1 = h_1^o \ (1-\exp(-t_p/\tau_p))\exp(-t_w/\tau_w) \tag{3}$$

where h_1^o is the maximum possible signal from E-1. Furthermore, it is observed that h_1 + h_2 = constant, but that h_2 is never zero; that is, the creation of an E-1 requires the destruction of an E-2, but that all the E-2's can never be destroyed. In order to see the maximum response from E-1, the time between pulses, t_p, must be longer than a characteristic time constant, τ_p, and the pulse width, t_w, must be shorter than another characteristic time constant, τ_w. Note that h_1 decreases with increased pulse width; this is opposite to what is normally observed. Furthermore, h_1 and h_2 were constant with pulse width for all pulse widths greater than 100 ns, but less than ∿ $t_w/3$, indicating that $\sigma_n \geq 10^{-14}$ cm^{-2} for both traps.

It has been found that both τ_p and τ_w are thermally activated quantities, and can be expressed as

$$\tau_p = \tau_p^o \ \exp \ (\Delta E_p/kT) \tag{4a}$$

and $$\tau_w = \tau_w^o \ \exp \ (\Delta E_w/kT) \ , \tag{4b}$$

where $$\tau_p^o = 7.5 \times 10^{-13} \ s \ and \ \tau_w^o = 7.3 \times 10^{-13} \ s \ , \tag{4c}$$

$$\Delta E_p = 0.174 \ eV \quad and \ \Delta E_w = 0.145 \ eV \tag{4d}$$

Furthermore, within experimental error, it was determined that $\tau_p = \tau_{n2}$; this requires that an E-2 trap must be empty in order that it be converted to an E-1 trap. If the time between pulses is much smaller than τ_p (= τ_{n2}) then all the

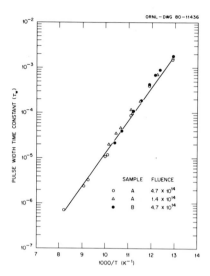

Fig. 3. The pulse width time
constant (τ_w) vs inverse temperature.

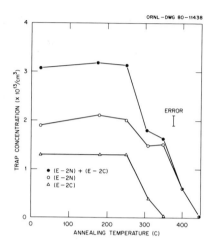

Fig. 4. Trap concentrations after
20 min isochronal anneals at various temperatures of sample A irradiated with 1.4 x 10^{14} e^-/cm^2 at
1.5 MeV. The filled-in circles ($\bullet$)
represent the total E-2C + E-2N
trap concentration, while the open
circles (o) represent the E-2N
traps, and the triangles (Δ) represent the E-2C traps.

E-2 traps will not have relaxed to their prepulsed state, and the number of
observed E-1 traps (which is proportional to h_1) will be less than maximum.

Figure 3 shows the pulse width time constant (τ_w) versus inverse temperature
for samples A and B irradiated with 4.7 x 10^{14} e^-/cm^2 at 1.5 MeV, and another
sample A irradiated with 1.4 x 10^{14} e^-/cm^2 at 1.5 MeV. Clearly, τ_w does not
depend upon either the number of phosphorus atoms present, or the electron
fluence.

It is clear that two different traps are responsible for the E-2 response.
One of these traps (E-2C) can be converted to an E-1 under the proper conditions ($t_p \gg \tau_p$, $t_w \ll \tau_w$), while the other (E-2N) cannot. This can be seen
from the observation that h_2 is never zero, and from the annealing data presented in Fig. 4. The E-2C traps anneal out at 350°C, while the E-2N traps
anneal out at 450°C. Clearly, there must be some structural difference between
the two types of E-2 traps. However, this difference is <u>not</u> manifested in the
emission times determined from transient capacitance measurements, since the
emission times of E-2N and E-2C are equal at all observed temperatures.

Table 1 presents trap concentration data for several samples. By comparing
samples A and A-1, one can see that the trap concentrations scale roughly with
electron fluence; this simply means that the number of traps created is proportional to the electron fluence. By comparing samples A, B, and C, we see that
there does not appear to be any correlation of trap density with phosphorus
concentration. It will be noted that the trap E-1 occurred in all samples,

TABLE I

Trap concentrations of E-1 and E-2N for several different samples. The concentrations of E-1 traps tend to decrease by 30-40% closer to the junction.

Sample	Growth	Carrier concen. ($\times 10^{14}/cm^3$)	Electron fluence ($\times 10^{14}/cm^2$)	Trap concentrations ($\times 10^{13}/cm^3$) N(E-1)	N(E-2N)
A	Fz	3.2	4.7	3.8	5.3
A1	Fz	3.2	1.4	1.3	1.9
B	Fz	8.9	4.7	3.3	5.9
C	Cz	1.4	4.7	0.2	39.0

though the concentration in the Czochralski-grown sample was well below the concentration in the float-zoned samples. Also, profiling measurements indicate that the concentration of E-1 centers decreased closer to the junction, though this decrease was no more than 30-40%.

DISCUSSION

The activation energy ($\Delta E = 0.172$ eV), capture cross section ($\sigma_n > 10^{-14}$ cm^2), and anneal-out temperature of trap E-2N indicate that E-2N is just the well-known A-center, or oxygen-vacancy defect. Note that this trap is seen in Fz material as well as Cz material. This is not surprising when one considers that the diodes were fabricated using diffused junctions, which commonly introduces oxygen into the depletion region [5,9].

The traps E-1 and E-2C apparently are just different manifestations of the same center. Experimentally, it is found that a trap E-1 can be created if the time between pulses (t_p) is long, and the pulse width (t_w) is short. From the t_p dependence, it was concluded that a trap E-1 could be formed only if E-2C were not filled. The pulse width dependence, on the other hand, indicates that a trap E-1 may be created from an E-2C if the trap is placed in an electric field. In the depletion region of a junction diode, there is an electric field of varying strength, being maximum at the junction, and zero at the edge of the depletion region. At normal reverse bias, all traps in the depletion region will experience this electric field. However, when the junction bias is reduced (after the beginning of a majority carrier pulse), the edge of the depletion region is moved closer to the junction (distance x_1 from the junction). At the conclusion of the pulse, the edge of the depletion region returns to its normal reverse-biased position (distance x_2 from the junction). During the time that the pulse is on, the traps located between x_1 and x_2 will experience zero electric field. If the pulse duration is long enough, then the E-1's will relax back to E-2C's. One would expect such a process to be thermally activated, which is observed: the pulse width time constant, τ_w, has an activation energy $\Delta E_w = 0.145$ eV (see Fig. 3). One would also expect such a process to be independent of dopant concentration and electron fluence, which is also observed. Clearly, the creation of an E-1 from an E-2C upon application of the electric field will have a very low threshold field (the minimum field required to cause the transformation), since profiling measurements indicate the existence of E-1 centers very close to the edge of the depletion region, where the junction electric field will be the smallest.

A trap with nearly the same energy has been seen by Walker and Sah [5] at 0.112 eV. Within experimental error, the emission times of the two traps are equal at temperatures above 65 K. However, at low temperatures, we found that it was impossible to measure an emission time, since the emission time of the

A-center (which had to be empty to create an E-1) became prohibitively long; Walker and Sah did not have this problem with their trap. Also Walker and Sah state that their low-lying trap was not oxygen-related in that it was formed only in Schottky barrier diodes fabricated on Fz material. For these reasons we must conclude that the two traps are different.

The fact that the E-2C trap is equivalent to the A-center in transient capacitance response indicates that the E-2C trap may be closely related to the A-center. One possible explanation is that E-2C is an oxygen-vacancy with some impurity nearby. Under zero field, the impurity and the E-2C trap do not interact, resulting in E-2C behaving electrically as an A-center. Under an applied field, however, the impurity and the oxygen-vacancy couple together, resulting in an E-1 trap. Such a scheme could also result in the (E-1) – (E-2C) combination annealing out at a lower temperature.

ACKNOWLEDGMENTS

The authors would like to thank D. H. Lowndes, R. D. Westbrook, R. F. Wood, N. Wilsey, P. Mooney, J. W. Corbett, and C. T. Sah for helpful discussions.

REFERENCES

1. J. W. Corbett, G. D. Watkins, R. M. Chronko, and R. S. McDonald, Phys. Rev. 121, 1015 (1961).

2. G. D. Watkins and J. W. Corbett, Phys. Rev. 121, 1001 (1961).

3. G. D. Watkins and J. W. Corbett, Phys. Rev. 134, A1359 (1964).

4. G. D. Watkins and J. W. Corbett, Phys. Rev. 138, A543 (1965).

5. J. W. Walker and C T. Sah, Phys. Rev. B7, 4587 (1973).

6. D. V. Lang, J. Appl. Phys. 45, 3023 (1974).

7. L. C. Kimerling, "Radiation Effects in Semiconductors," ed. by N. B. Urli and J. W. Corbett (Inst. of Physics, London and Bristol, 1976), p. 221.

8. L. C. Kimerling, H. N. DeAngelis and J. W. Diebold, Solid State Commun. 16, 171 (1975).

9. A. O. Evwaraye and Edmund Sun, J. Appl. Phys. 47, 3776 (1976).

10. G. L. Miller, D. V. Lang, and L. C. Kimerling, Ann. Rev. Mater. Sci. 7, 377 (1977).

11. C. T. Sah, L. Forbes, L. L. Rosier and A. F. Tasch, Jr., Solid State Electronics 13, 759 (1970).

12. D. V. Lang, J. Appl. Phys. 45, 3014 (1974).

Narayan, and Tan, eds.
Defects in Semiconductors

REDUCTION OF α-PARTICLE SENSITIVITY IN DYNAMIC SEMICONDUCTOR MEMORIES
(16K d-RAMs) BY NEUTRON IRRADIATION

CHARLES E. THOMPSON[+] AND J. M. MEESE[*]
[+]Burroughs Corporation, 16701 W. Bernardo Dr., San Diego, California   92127
[*]University of Missouri Research Reactor, Columbia, Missouri   65211

ABSTRACT

Trace radioactive impurities found in all semiconductor devices induce soft errors in semiconductor memories by α-particle emission. Data taken on 16K d-RAMs which have been fast neutron irradiated to fluences from 10^{13} to 10^{16} n/cm^2 show that soft errors in these devices can be significantly reduced while maintaining acceptable device operation. Reduction in soft error rate by factors as high as 80 are observed following irradiation and thermal annealing. The effect on device parameters is discussed as well as the defects responsible for this beneficial radiation processing. Estimates of the soft error rate improvement to be expected on higher density memory devices (64K and 256K d-RAMs) will also be presented.

INTRODUCTION

Economic growth through price reduction in the computer industry has been truly remarkable, due primarily to the application of very large scale integrated (VLSI) memories. Prospects for continued economic growth by an extension of this technology to higher device densities will be limited unless α-particle induced soft errors, which presently plague this technology, are brought under control. Although the industry has made large capital investments in high resolution technology, the devices presently being produced do not meet the demands of large computer systems due to this soft error problem. The implications for the semiconductor industry and the economy as a whole are obvious.

The α-particle soft error problem has been recognized only as recently as 1978 [1-3]. Briefly stated, the problem is as follows. The amount of charge stored in a one bit gate of a 16K device is approximately 10^6 electrons while one bit in a 64K device represents a charge of only 10^4 electrons [4]. Trace impurities of uranium and thorium found in both silicon and aluminum emit α-particles with energies in the 5 to 10 MeV range. Ionization in a single α-track is therefore several times 10^6 electron-hole pairs which is sufficient to flip a single bit in 16K devices or many bits in 64K devices if this ionization is allowed to diffuse to the gates. This effect leads to soft error rates of about 280 bits/10^6 hrs in the 16K d-RAM devices discussed in this paper. Much higher soft error rates are observed for 64K devices currently produced. Although this rate appears negligible for a single chip, a semiconductor memory composed of a megabyte with 16 bit words will contain 1000 such chips and will experience a soft error rate of 280 bits/10^3 hrs ($\approx$ 6 weeks) or about 47 soft errors per week! Totally unacceptable error rates occur for device densities higher than 16K. Estimates have also been made of cosmic ray induced soft error rates [4] which add to this problem for terrestrial applications and suggest that space applications using large computer memories are impossible.

We have investigated one technique which appears promising for the reduction of these soft errors, namely, the introduction of recombination centers by fast neutron irradiation. This idea has been tested experimentally by J. R. Adams and R. J. Sokel [5] on static RAMs, however, the experiments in this paper represent the first attempt to apply this technique to lower-cost dynamic memories. To be successful, sufficient recombination centers must be introduced to provide significant soft error rate reduction without sacrificing device performance through detrimental radiation effects. The results presented here appear encouraging. Furthermore, this application of neutron radiation damage constitutes a case of "defect engineering" for beneficial purposes as recently envisioned by Jim Corbett [6] and therefore constitutes a new stimulus for better understanding radiation produced defects in semiconductors.

EXPERIMENTAL

The devices used in these experiments were 16K dynamic RAMs processed using double poly for word and capacitor-to-transfer-gate lines and Al busing. Bit lines were n^+ diffused into p-type 3" dia. Monsanto CZ starting wafers of 10-16 Ω-cm with $\langle 100 \rangle$ orientation. Chip size was 120 x 225 mils^2 while cell size was 14 x 36 microns^2. Cell capacitance was 76 femtofarads. Devices were operated with a drain voltage V_{DD} = 12 V $\pm$ 10%, a substrait bias V_{BB} = -5.0 V $\pm$ 10% and a charging voltage V_{CC} = 5 V $\pm$ 10%. Typical soft error rates were 280 per 10^6 hrs.

A special irradiation facility was fabricated consisting of a double wall irradiation tube which contained about 1" of boron carbide between the walls to absorb thermal neutrons in order to reduce radioactivity induced in the wafers. Wafers containing about 200 devices per wafer were loaded into standard polypropylene wafer cassettes and irradiated along with the cassettes. Fast neutron flux above 1 MeV was about 2 x 10^{10} n/cm^2/sec as determined by gold activation from the reaction ^{197}Au(n,p)^{196}Au assuming a fission spectrum above 1 MeV. This flux estimate agreed well with estimates of the fast flux from reactor design data of 2.6 x 10^{10} n/cm^2/sec above 0.821 MeV.

Wafers were completely processed through the final Au back contact evaporation. Some thermally induced ^{197}Au(n,γ)^{198}Au activity was detected indicating that the cassettes had produced some thermalization of the fast neutrons. Exempt radioactive limits of 2 x 10^{-3} μCi/gm for gold activity were obtained after 210, 350 and 530 hrs respectively for fast fluences of 10^{13}. 10^{14} and 10^{15} n/cm^2. Bare silicon wafers and wafers which were completely processed except for the final Au evaporation were found to have nearly identical activity which was below exempt limits in most cases only two or three days after irradiation. It is envisioned that production runs will be irradiated prior to Au evaporation and that no radioactivity problems would result from this procedure.

Split lots were used in these tests. All wafers were tested prior to lot splitting. After irradiation and decay to exempt limits, wafers were cleaned, annealed and retested. A total of 14 split lots were tested with approximately 21 wafers per lot. Fast neutron fluences of 10^{11} to 10^{16} n/cm^2 (E > 1 MeV) were used in these experiments.

RESULTS

Isochronal anneals of 5 min. at successively higher temperatures were performed on all wafers until all device performance parameters for reading, writing and storing recovered to within original specifications. This required about 200°C annealing for fluences below 10^{13} n/cm^2, however, annealing to just over 400°C was required for the highest fluence of 10^{16} n/cm^2 as shown in Fig. 1. Following the anneal temperatures shown in this figure, various device parameters were measured on all devices for all fluences and compared to the control devices.

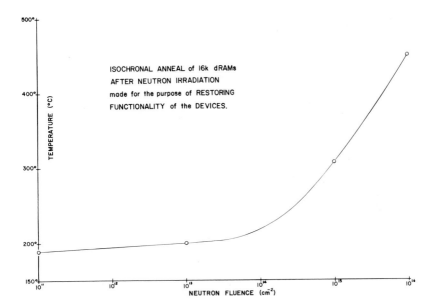

Fig. 1. Final 5 min. anneal temperature required to restore read, write and store functions to within device specifications.

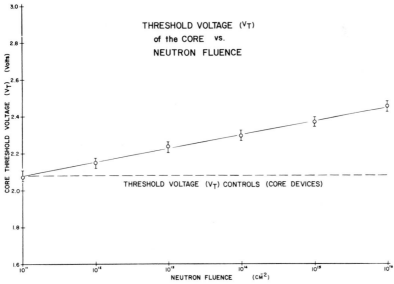

Fig. 2. Change in memory core threshold voltage vs. fast neutron fluence (E > 1 MeV) after annealing.

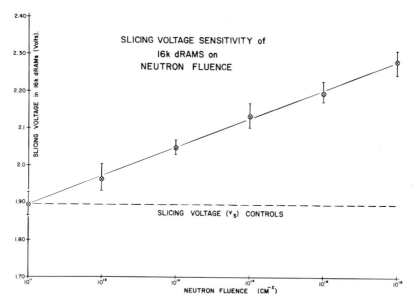

Fig. 3.   Change in slicing voltage vs. fluence after annealing.

Several test capacitors on each wafer were used to measure threshold voltage shift. The capacitance and size of these devices was similar to the core capacitors used in the actual memory cells. An increase in threshold voltage was observed with increasing fluence as shown in Fig. 2. An increase in this voltage is desirable since it represents a decreased sensitivity to random noise. The slicing voltage, or the cell voltage difference, which must be stored in order for the sense amplifier to discriminate between a 0 or 1 bit, is also increased with fluence as shown in Fig. 3. The improvement in bit to sense voltage ΔV_{BS} due to neutron irradiation can be expressed as [7]

$$\Delta V_{BS} = \frac{\pm (V_1 - V_0)}{2(C_{BS}/C_S + 1)} \qquad (1)$$

where ΔV_{BS} is the offset signal across latch nodes of the memory as a result of neutron bombardment, V_1 and V_0 are the stored voltage levels in the memory cell for a "1" or "0" bit while C_{BS} is the bit/sense line capacitance and C_S is the storage cell capacitance after neutron irradiation. For the highest fluence of 10^{16} n/cm^2, Eq. (1) yields a ΔV_{BS} induced by irradiation of 34 mV or almost a 30% increase.

It is also expected that the memory cell storage capacitance should be increased by irradiation which is desirable since a greater charge can be stored as information. The cell capacitance after irradiation is given by [8]

$$C_D = \left[\frac{\varepsilon (N_i - N_D)N_D}{2(|\phi_{Fp}| + |\phi_{Fn}|)N_i} \right]^{\frac{1}{2}} \qquad (2)$$

where N_D is the concentration of defects which has been estimated for a fluence of 1×10^{16} n/cm^2 to be 6×10^{14} cm^{-3} from the observed threshold voltage shift,

$\phi_{Fn(p)}$ are the quasi-Fermi potentials, N_i the number of doping impurities and ϵ the dielectric constant of Si. Eq. (2) yields $C_D \cong 1.4 \times 10^{-9}$ f/cm^2 which for our 16K d-RAM cell is equivalent to an increase of 7×10^{-15} farads or about a 10% increase in cell capacitance.

These benefits are not obtained without some degradation in access time; however, for the highest fluence used, this parameter is observed to increase by only about 3% as shown in Fig. 4 and to be acceptable as compared to the reduction in soft error rate observed in the next figure. Figure 5 shows the reduction in soft error rate obtained on the 16K d-RAMS tested as a function of fluence. These soft error rates were obtained by an accelerated testing procedure in which the wafers were exposed to α-particle bombardment from either ^{230}Th or ^{226}Ra planar sources placed closer than 5 mils to the top surface of the wafers. The source disintegration rate was 80,000 sec^{-1}. The number of soft errors induced in 1 min. was measured for a checkerboard bit pattern of 0's and 1's. The soft error rates for both high and low bits was approximately the same. The mean improvement in soft error rate for a fluence of 10^{16} n/cm^2 was observed to be about 45 while maximum improvement in some devices approached a factor of 100. It appears from Figs. 4 and 5 that further improvements in d-RAM soft error rates can be obtained at the expense of minor reduction in memory speed by increasing the irradiations to still higher fluences.

In order to understand which defect levels are contributing to this soft error improvement, DLTS spectra were obtained using p-n junctions in base material doped to about 10^{16} cm^{-3}. To obtain hole traps, n+p junctions were used while p+n junctions were used to monitor electron traps. The p+n diodes were irradiated to a fast fluence of about 2×10^{14} n/cm^2 while the n+p samples were irradiated to about 2×10^{15} n/cm^2. Isochronal annealing for 15 min. of the DLTS

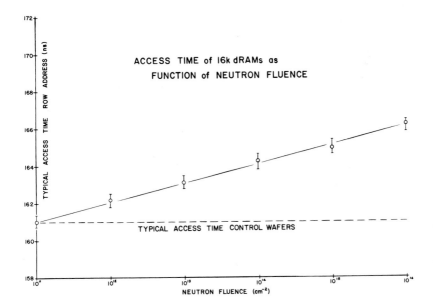

Fig. 4. Increase in access time vs. neutron fluence after annealing. Maximum degradation of this parameter is about 3% for a fluence of 1×10^{16} n/cm^2.

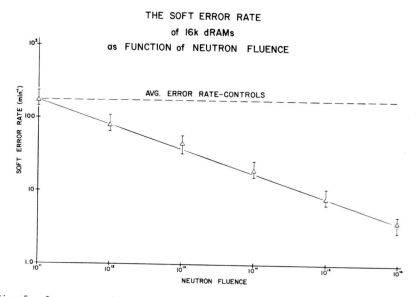

THE SOFT ERROR RATE
of 16k dRAMs
as FUNCTION of NEUTRON FLUENCE

Fig. 5. Improvement in soft error rate vs. fluence after annealing. For a fluence of 1×10^{16} n/cm^2, the mean soft error reduction is about a factor of 45.

spectra is shown in Figs. 6 and 7 for hole and electron traps respectively. The energy levels and capture cross sections for this experiment are given elsewhere in this volume [9].

For the highest d-RAM fluence, annealing to above 400°C was required. The hole trap spectra for this annealing temperature is dominated by the level H_3 (0.35 eV) and to a lesser extent by H_{10} (0.25 eV). The concentration of this level is about 7×10^{14} cm^{-3} for the fast fluence used here. If production of this defect is linear with fluence, then the concentration of H_3 levels should be somewhat below 4×10^{15} cm^{-3} in the d-RAMs irradiated to 1×10^{16} n/cm^2 and annealed to above 400°C. The dominant electron trap, E_1 (0.16 eV) occurs after 400°C annealing in a concentration of 5×10^{13} cm^{-3} for a fluence of 2×10^{14} n/cm^2. For the d-RAM fluence of 1×10^{16} n/cm^2, the trap concentration of this defect should be below 2×10^{15} cm^{-3}. The concentration of 'deeper' electron traps, E_4^I (0.34 eV) and E_3^{III} (0.42 eV), should occur in concentrations of about 2×10^{14} cm^{-3} for the d-RAMs irradiated to 1×10^{16} n/cm^2. The capture cross section for all of these levels is in the low 10^{15} cm^{-3} range. We therefore conclude that several levels in midgap in concentrations of 2×10^{14} cm^{-3} to 2×10^{15} cm^{-3} are acting as the recombination centers for electron-hole pairs in the d-RAMs irradiated to 1×10^{16} n/cm^2. The range of concentrations above agrees well with the defect concentration of 6×10^{14} cm^{-3} estimated from threshold voltage shifts and used to calculate increase in cell capacitance discussed previously.

For the d-RAMs irradiated to lower fluences, lower annealing temperatures were required to restore functionality of the devices. Adams and Sokel have argued that divacancies should dominate the lifetime in the annealing range up to about 300°C [5]. It can be seen from Fig. 7 that the electron trap spectra changes very little in this temperature range. The hole trap spectra is like-

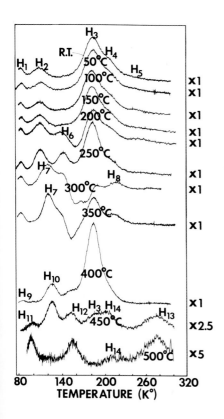

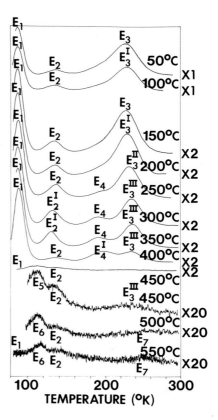

Fig. 6. DLTS spectra of hole traps in neutron irradiated CZ Si as a function of 15 min. isochronal anneals. Fast neutron fluence was about 2×10^{15} n/cm^2. Trap concentration of H$_3$ at room temperature was about 4×10^{14} cm^{-3}.

Fig. 7. DLTS spectra of electron traps as a function of 15 min. isochronal anneals. Fast neutron fluence was about 2×10^{14} n/cm^2 while the concentration of E$_1$ at room temperature was about 4.5×10^{14} cm^{-3}.

wise rather stable below 250°C. Because of the number of levels involved, it is difficult to estimate which levels are the most efficient recombination centers, however, the H$_3$ and E$_3$ levels again appear to dominate the spectra at these lower annealing temperatures. This perhaps explains why the annealing temperatures of Fig. 1 are so different for different fluences while the soft error rate curves of Fig. 6 appear to be independent of the species of defect or annealing temperature used.

Finally, these experiments have been used to model expected soft error improvements for 64K and 256K d-RAMs using the device parameter changes obtained experimentally in this work. This modeling is sufficiently complicated to warrant separate publication and will not be presented, however, the results are shown in Fig. 8. The solid line in this figure represents the data presented here for 16K devices while the dotted lines represent estimates of improvements

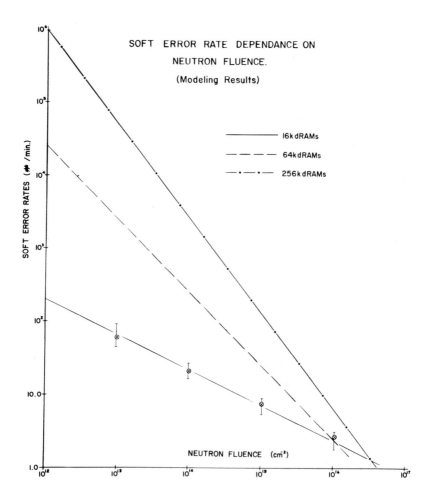

Fig. 8. Results of modeling soft error rate improvement vs. neutron fluence. The solid line fits the 16K experimental results. Estimated improvement for 64K d-RAMs and 256K d-RAMs are factors of 10^4 and 10^6 respectively.

on higher density devices. If experiments follow these models, it appears that neutron damage and "defect engineering" will become tools of the semiconductor memory manufacturer in the future.

SUMMARY AND CONCLUSIONS

The use of fast neutron irradiation to introduce recombination centers has been shown to be a successful technique to reduce α-particle induced soft errors in 16K d-RAMs. Reduction in soft error rate of about a factor of 50 was obtained for a fast fluence of 1×10^{16} n/cm^2 while increasing access time by only

3%. The question of radioactivity has been tested experimentally and it is found to be negligible provided neutron processing is done before Au evaporation and packaging. Irradiation techniques have been developed for wafer processing which can be done both economically and in large quantity. Defect concentrations obtained for DLTS measurements have been shown to agree with those estimates obtained from threshold voltage shifts. Several levels have been suggested as possible recombination sites. Modeling results have been presented which suggest that the above irradiation and annealing procedure will yield additional benefits to still higher density devices.

REFERENCES

1. S. Hershberger and D. Hanson, Electron. News $\underline{24}$, 1 (28 August 1978); S. Hershberger, ibid. $\underline{24}$, 1 (18 December 1978).

2. T. C. May and M. H. Woods in Proceedings of the 1978 International Reliability Physics Symposium (I.E.E.E. 78CH 1294-8PHY, New York, 1978), p. 33; I.E.E.E. Trans. Electron. Devices $\underline{ED-26}$, 1 (1979).

3. D. S. Yaney, J. T. Nelson, L. L. Vanskike, I.E.E.E. Trans. Electron. Devices $\underline{ED-26}$, 10 (1979).

4. J. F. Ziegler and W. L. Lanford, Science $\underline{206}$, 776 (1979).

5. J. R. Adams and R. J. Sokel, I.E.E.E. Trans. Nuc. Sci. $\underline{NS-26}$, 5069 (1979).

6. J. W. Corbett, invited talk at the Annealing Symposium sponsored by the Electrochemical Society, Hollywood, Florida, October 1980.

7. L. G. Heller, D. P. Spampinato and Y. L. Yoo, I.E.E.E. J. Solid State Circuits $\underline{SC-11}$, 596 (1976).

8. T. W. Sigmon and R. Swanson, Solid State Electronics $\underline{16}$,

9. J. M. Meese, this volume.

Published 1981 by North-Holland, Inc.
Narayan, and Tan, eds.
Defects in Semiconductors

ELECTRONIC AND MECHANICAL PROPERTIES OF DISLOCATIONS IN SEMICONDUCTORS

P B HIRSCH
Department of Metallurgy and Science of Materials, University of Oxford,
Parks Road, Oxford, England.

ABSTRACT

It has been known for some time that the velocity of
dislocations in semiconductors depends strongly on the
concentration of electrically active impurities. Various
explanations based on different models of the electronic
structure of the dislocations have been proposed; they
involve the dependence of the formation energy and/or
activation energy of migration of dislocation kinks on the
Fermi level. A review will be presented of these theories
and of recent structural models of dislocation cores and
kinks.

INTRODUCTION

Following the original observations of Patel [1] it has been known for many
years that the velocity of dislocations in Ge and Si, and in some III-V
compounds, depends on the concentration of electrically active impurities [2-9].
Experiments using etch pit techniques and x-ray topography have been
carried out on screws and 60° dislocations, and show that doping affects mainly
the activation energy controlling the dislocation velocity, and to a lesser
extent the preexponential factor. There are at present four theories which
have been proposed to explain the effects, and these will be discussed below.
An understanding of the phenomenon is important for at least three reasons:-
It may illuminate the questions as to 1) whether the cores of dislocations are
reconstructed, or whether they contain dangling bonds, 2) whether dislocations
and/or kinks are associated with deep levels, 3) the effect in Si and Ge is
sufficiently large to expect changes in brittle-ductile transition temperatures
of 50°-100°C, due to quite small additions of electrically active impurities,
and this suggests the possibility of modifying the mechanical properties of
strong (wide band gap) ceramics by this effect. It is plausible that the
hydrolytic weakening effect in quartz, well known to geologists, may in part be
due to a similar mechanism [10].
In this paper we shall review the theories proposed, the calculations which
have been carried out so far on the core structures of dislocations, the models
of kinks and some of the relevant experimental evidence.

THEORIES

The doping effect on dislocation velocity has been interpreted in terms of
the mechanism of motion of dislocations by the formation and migration of kinks,
and its dependence on the position of the Fermi level. The four theories which
have been proposed to account for the phenomenon are:-

Theory of Frisch and Patel [3]
In this theory the authors assume that the dislocation velocity is proportional to the fraction of charged dislocation sites. The fraction of charged
states in a cylinder with radius equal to the Debye screening length about the

dislocation, is calculated from the Fermi Dirac statistics. The Fermi level is determined by assuming electrical neutrality in the cylinder. Two fitting parameters are used:- a) the total volume density of acceptor (or donor) states inside the cylinder, and b) the energy level in the gap associated with the charged dislocation level. For n and p type Ge the results can be explained in terms of an acceptor level E_d at $E_d \sim 0.7\ E_g$ above the valence band, where E_g is the gap energy. For silicon however the results imply the existence of acceptor levels (at $0.6\ E_g$ above the valence band) for n type, and donor levels (at about the same energy) for p type. Schröter, Labusch and Haasen [11] have criticised the model on the grounds, inter alia, that when proceeding to the intrinsic limit from n and p type silicon, oppositely charged dislocations would be predicted. The theory is also phenomenological, i.e. no microscopic mechanism has been proposed for the dependence of velocity on charge.

Theory of Haasen [12, 13]
    In this theory the dislocation is assumed to be charged, to a line charge q. Then the formation energy of a double kink is reduced by the change in electrostatic energy.

$$\Delta E = - \frac{2q^2h}{\varepsilon} \exp\ (- {}^h/2\lambda_D) \tag{1}$$

where h is the height of the kinks, ε the static dielectric constant, λ_D the Debye screening length. The value of q is determined by the relative positions of the Fermi level and the dislocation levels. While the dependence of the observed activation energy for dislocation motion on the Fermi level is qualitatively reproduced, quantitatively the calculated variations are too small by a factor of about six.

Theory of Hirsch [14]
    The basic idea of the theory is that the energy bands associated with screws and 60° dislocations are split, either because of reconstruction or because of a Mott-Hubbard gap, and that kinks are associated with deep lying acceptor and donor levels in the gap between the dislocation level (fig. 1). The dislocation velocity depends on the concentration of kinks and their mobility along the dislocation line. Charged as well as neutral kinks can be formed, and the concentration of charged kinks depends on the relative positions of the Fermi level and the kink acceptor/donor levels. The velocity of a dislocation can be changed by doping a) because the concentration of charged kinks varies, b) because the mobility of charged kinks may be different from that of uncharged kinks.
    The detailed theory follows along the lines described by Hirth and Lothe [15] for uncharged kinks. Hirth and Lothe develop the theory for two limiting cases a) the drift model applicable to high temperatures and low stresses b) the nucleation controlled model, applicable to low temperatures and high stresses. Assuming that the mean free path of the kinks is determined by mutual annihilation, these two treatments lead to the same result, i.e.

$$v_o = \frac{2\sigma_{eff}h^2}{\sqrt{3}kT} D_k \exp\ (- {}^{F_k}/kT) \tag{2}$$

where v_o is the steady state velocity of the dislocation (considered to consist of two partials), h is the kink height, σ_{eff} is an effective stress which takes into account a contribution due to the deviation of the dislocation width from that in static equilibrium, F_k is the formation energy of a kink, D_k is the kink diffusivity, i.e.

$$D_k = b_T{}^2 \nu_D \exp\ (- {}^{W_m}/kT) \tag{3}$$

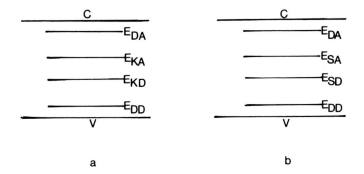

Fig. 1. Electronic energy level scheme assumed for theories of (a) Hirsch, (b) Jones. C, V are conduction and valence bands respectively; E_{DA}, E_{DD} are dislocation acceptor/donor levels; E_{KA}, E_{KD} are kink acceptor/donor levels; E_{SA}, E_{SD} are migration saddle point acceptor/donor levels.

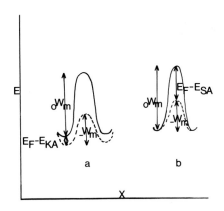

Fig. 2. Energy E as a function of displacement X of kink over the energy barrier for migration for (a) Hirsch (b) Jones models. In (a) the kink formation energy of a neutral kink if effectively lowered if it is charged (dotted line), and the activation energy for migration may also be lowered. In (b) only the activation energy for migration is lowered.

where b_T is the total Burgers vector, ν_D the Debye frequency, W_m = activation energy of migration of an uncharged kink. Equation (2) shows that the velocity is proportional to the equilibrium concentration of uncharged kinks, $_oC_k$ = exp $(- F_k/kT)$. If we now include charged kinks, then, for the drift model, assuming the electron gas to be in thermodynamic equilibrium with the kinks, the concentration of, say, negatively charged kinks $_-C_k$ is

$$\frac{_-C_k}{_oC_k} = \exp\ (E_F - E_{ka} - eV)/kT \tag{4}$$

where E_F is the Fermi level, E_{ka} the kink acceptor level, eV the potential energy of the dislocation due to all charges (including those from kinks). There is a similar expression for positively charged kinks. If, for example, negative kinks predominate, the dislocation velocity will be given by

$$\frac{v_-}{v_o} = \exp\ (E_F - E_{ka} - eV - \Delta W_m)/kT \tag{5}$$

where ΔW_m is the difference between migration activation energy for charged and uncharged kinks. Thus the velocity differs from v_o due to the change in kink concentration, which depends on E_F, and due to a possible change in migration energy. Effectively the kink formation energy is lowered if $E_F > E_{ka}$ and the activation energy for migration changed by a constant amount; the situation is illustrated in fig. 2. In general the dislocation velocity is determined by $_oC_k$, $_-C_k$ and $_+C_k$, and possible changes in migration energy must also be allowed for. Fig. 3 shows a fit of the theory to velocity measurements for 60° dislocations in Si due to George, on the assumption that $\Delta W_m = 0$, that the velocity is therefore proportional to $(_oC_k + _-C_k + _+C_k)$, and that there are four adjustable parameters, i.e. E_{ka}, E_{kd} and their temperature dependences. The fitting was carried out by Professor Schröter, and the results give E_{kd} = 0.28 ± 0.17 ev; $\frac{\partial E_{kd}}{\partial T}$ = - (2.0 ± 2.1) 10^{-4} evK^{-1}; E_{ka} = 0.67 ± 0.04 ev, $\frac{\partial E_{ka}}{\partial T}$ = - (1.6 ± 0.5) x 10^{-4} evK^{-1}. Other tests of the theory are described by Hirsch [14]. It should be noted that contrary to the results following from the theory of Frisch and Patel [3] the energy levels for Ge are found to be acceptors close to the valence band.

The theory has been criticised by the Göttingen group on the grounds that the drift model is not applicable to most of the experimental range of stress and temperature. As has been pointed out above, the theory applicable under conditions of high stress and low temperature, when the velocity is controlled by the nucleation of double kinks, gives essentially the same expression (2), where F_k is now half the formation energy of the double kink, and the weak stress dependence has been neglected [15]. Provided the concentration of charged double kinks can be assumed to be that in thermodynamic equilibrium with the electron gas, the theory under nucleation controlled conditions becomes similar to that under drift conditions. The nucleation rate of double kinks is given according to rate theory by the concentration of double kinks at the saddle point (ignoring the subsidiary saddle points due to migration) times the rate at which the kink moves to the next position (point A in fig. 4). The lifetime of a double kink at the saddle point is therefore ν^{-1} exp W_m/kT. To ensure thermodynamic equilibrium with the electron gas over the whole range of temperatures and dopant concentrations, $W_m \gtrsim 0.75$ ev. However this ignores the possibility that thermodynamic equilibrium may be achieved more rapidly by transfer of charge along the dislocation line.

Theory of Jones [17]

Jones assumes that the kinks are reconstructed and that the doping effect is due to the change in activation energy for migration. He assumes that the

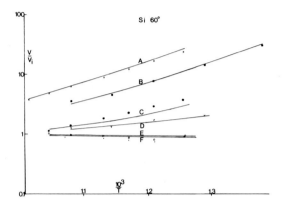

Fig. 3. Ratio of velocities v of 60° dislocations in Si to that of intrinsic material, v_i, as a function of (temperature)$^{-1}$ for various doping concentrations, donors:- A = 14 x 10^{18} cm^{-3}, B = 6 x 10^{18} cm^{-3}, C = 1.2 x 10^{18} cm^{-3}; acceptors:- D = 80 x 10^{17} cm^{-3}, E = 6.8 x 10^{17} cm^{-3}, F = 2.7 x 10^{17} cm^{-3}. Measurements of George [16]; fitting carried out by Schröter.

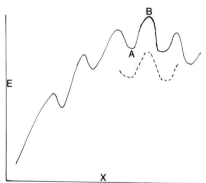

Fig. 4. Energy as a function of distance between two kinks constituting a nucleated double kink. A partially formed double kink may wait in the subsidiary valleys (e.g. A) and equilibrate with the electron gas, reducing effectively the nucleation energy (dotted line). In Jones' theory equilibrium has to be established at B.

electron gas equilibrates with the kink at the subsidiary saddle point configuration for migration (see point B in fig. 4). This treatment leads to a dependence of dislocation velocity on Fermi level similar to that in equation (5), except that E_{ka} has to be replaced by E_{sa}, an electronic level associated with the saddle point configuration (and $\Delta W_m = 0$). Two points require clarification in this theory:- Firstly, whether it is possible to establish thermodynamic equilibrium with the electron gas during the lifetime of the saddle point configuration ($\sim 10^{-12}$ secs), secondly, what the fate of the electron (or hole) is after the saddle point configuration is passed; the kink remains charged until the charge is removed by recombination or by some other mechanism. It should be emphasised that both the Hirsch and Jones treatments involve changes in activation energy for migration; in Hirsch's theory a change ΔW_m in migration energy is postulated if the kink is charged; in Jones' treatment the migration energy is changed through the charged saddle point state, even though the kink ground state is neutral. If in fact a charged saddle point configuration can be attained only if the kink is already charged in the ground state, and if $\Delta W_m = E_{ka} - E_{sa}$, the two theories become identical.

The following points should be noted:- 1) The energy levels deduced from velocity data are not necessarily directly comparable with those deduced from electrical or optical measurements, since the former may involve contributions from the changes in migration energy. 2) In both theories the effect of doping arises through a change in electron energy associated with a charged configuration, the strain energy of the defect being assumed to be independent of charge. However, charged and uncharged kinks could correspond to different configurations of kinks, with different strain energies. This introduces further complexities and uncertainties into the theories, and makes unambiguous interpretation of the energy levels deduced difficult. 3) The expressionsderived for dislocation velocity apply only if the mean free path of the kinks is determined by annihilation by others moving in the opposite direction, and kink formation is uncorrelated on the two partials. Under some conditions these assumptions are not justified and this leads to changes in the expressions (for review see [13]).

Structural Models of Dislocations and Kinks

Dislocations. It is well known that slip dislocations in tetrahedrally coordinated semiconductors can lie either on the narrowly spaced (glide) or widely spaced (shuffle) (111) planes. Weak beam electron microscopy has shown clearly that dislocations introduced by deformation are generally dissociated into partials [18-24]. This means that the dislocations lie either on the glide planes, or alternatively take up the so called "associated shuffle" configuration introduced by Hornstra [25]. The basic idea underlying this type of configuration is that the undissociated shuffle dislocation, which was thought to have a higher mobility because fewer bonds have to be broken during motion, might associate itself on coming to rest with a pair of partials of opposite sign on the neighbouring glide plane, thus reducing the strain energy of the dislocation by dissociation on the neighbouring glide plane. The reaction is

$$\tfrac{1}{2}[1\bar{1}0] + \tfrac{1}{6}[\bar{1}2\bar{1}] + \tfrac{1}{6}[1\bar{2}1] = \tfrac{1}{6}[2\bar{1}\bar{1}] + \tfrac{1}{6}[1\bar{2}1] \qquad (6)$$

The following points should be noted:-

1) There is good evidence that dislocations glide in the dissociated configuration. This is clear a) from some weak beam studies in Ge and Si in which dissociated dislocations were observed to move in the dissociated configuration [26], although the experimental conditions allowed only very small movements to be observed, and b) from experiments due to Wessel and Alexander [27] on Ge and Si in which specimens are quenched under load after deformation. Cockayne, Hons and Spence [28] have recently reported some beautiful electron microscope observations of the motion of dislocations in CdS in the dissociated form.

These results leave no doubt that the dislocations are intrinsically dissociated, whether stationary or moving, and since one of the partials is an ordinary glide partial, and the other effectively a climbed glide partial with a complex core structure possibly with dangling bonds, the advantage of the Hornstra configuration is not clear.

2) Hornstra considered only the 60° associated shuffle dislocation, in which the 30° partial has the complex (climbed glide partial) structure and discussion in the literature is confined to this configuration. Glide dislocation loops of the type formed by low temperature high stress deformation [27] consist of hexagons whose sides are along [110] directions; the Burgers vectors of the partial dislocations are shown in fig. 5; the character of the partials change at the corners. Suppose that the 30° partials making up one of the partials of the 60° dislocation were always to have the Hornstra configuration, the others the glide configuration, then there must be a jog at the corner and expansion of the loop implies climb of the jog. This should be visible as a cusp, but the corners of the loops are generally smooth (see fig. 6). We conclude that either one of the partials (leading or trailing) is a glide and the other an "associated shuffle" partial over the whole of the loop, or both partials are of the glide type. According to Wessel and Alexander [27] the mobilities of the partials differ according to whether they are 30° or 90°, and leading or trailing; the Cologne group [27, 29] suggest that the trailing 30° partial may be associated with point defects, forming over small parts of its length the Hornstra associated shuffle configuration. This interpretation is consistent with EPR measurements [30] which indicate that less than 1% of the possible sites along the dislocation cores are associated with unpaired electrons, and an angular dependence of the EPR signal which suggests that it is associated with a point defect at the core of the 30° partial. This interpretation, and the small magnitude of the signal, suggest that over most of the length of the dislocation, it has simple glide character. It should also be pointed out that the low temperature high stress hexagonal loop structures are formed after predeformation at a higher temperature (say 700°C for Si). The structure consists of tangles and dipoles. On subsequent deformation at low temperatures (say 420°C for Si), apparently no increase in the EPR signal is found (private communication from Prof. Alexander and Dr Weber), suggesting that no measurable increase in point defects occurs either in the bulk or at the dislocations. These results are compatible with slip at low temperatures occurring predominantly on the glide planes.

3) At the present time then the simplest interpretation of the dissociated dislocations, namely that they are basically both glide partials, is also the most plausible. In this connection it should also be noted that recent lattice resolution experiments by Olsen and Spence [31] on a particular 30° partial in Si suggest that it lies on the glide plane. While the simple Hornstra model must be abandoned, it is possible that the partials may climb locally by absorption [or less likely emission] of point defects from glide into the shuffle plane. Addition of a second row of point defects causes the glide partial to climb into the next glide plane. This structure can transform into an extrinsically faulted dislocation by a mechanism described in [32], to account for the formation under certain conditions of extrinsically dissociated screws in Si [21]. Under conditions of high supersaturation climb occurs by a mechanism of nucleation of prismatic loops similar to that observed in Cu-Al alloys [33]. Results on climb experiments will be reported elsewhere [34].

4) There are two reasons why the glide configuration should be favoured. Firstly, because the strain energy is reduced by dissociation, secondly because bond reconstruction is possible along the core of the partials. Fig. 7 shows the diamond cubic lattice projected normal to (1$\bar{1}$0). A 60° shuffle dislocation is made by removing the extra half plane 1234; the top part of the crystal is joined up, leaving a row of dangling bonds normal to the

264

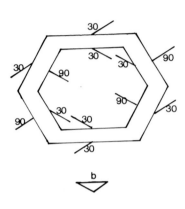

220 Wk-bm

100nm

Fig. 5. Burgers vectors of partials (30° and 90°) for a dissociated hexagonal glide loop, total Burgers vector along b.

Fig. 6. Part of a dissociated hexagonal glide loop in Si, on (111) plane, Burgers vector b, produced by low temperature high stress deformation; 2̄20 weak beam image taken by Ourmazd; specimen produced by Professor Alexander's group.

[1̄1̄2]

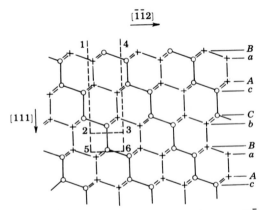

Fig. 7. Diamond cubic lattice projected on to (1̄10) plane. o represents atoms in plane of the paper and + atoms in the plane below (from Hirth and Lothe 1968).

glide plane first proposed by Shockley [35]. A 60° dislocation on the glide plane is made by removing the extra half plane 1564, and joining up the top part of the crystal. In this case the dangling bonds are seen to be inclined at shallow angles to the glide plane, and this makes it possible for bond reconstruction to occur across and along the core. Fig. 8 shows the unreconstructed core structures of a 30° partial (on the left) and a 90° partial (on the right); only atoms above and below the slip plane are shown [36]. Both partials also have unreconstructed kinks. For the 30° partials small displacements of the atoms along the core leads to the reconstructed configuration shown in fig. 9. These structures also show three alternative kink configurations, one with and two without dangling bonds. Calculations due to Marklund [37] and Jones [38] suggest that the cores of 30° partials are in fact reconstructed, the increase in strain energy being small compared to the reduction in the dangling bond energy. Fig. 10 shows reconstruction in the core of the 90° partial. The white bonds indicate the reconstructed bonds across the core; both structures contain a double kink, (a) with and (b) without dangling bonds at kinks. In this case the calculations at the present time are too uncertain to reach definite conclusions about reconstruction [39].

No unambiguous direct evidence is available on whether glide partials are reconstructed along the core. However, according to Wagner and Haasen [40] screw dislocations in deformed Ge, and therefore 30° partials, give rise to a Hall effect which can be interpreted in terms of a full donor band close to the valence band, and an empty acceptor band close to the conduction band. This is as expected from a reconstructed partial [32, 37], and compatible with the level scheme in fig. 1. It is also known that the activation energies for velocities of 60° and screw dislocations in Si are similar (for discussion see [14]). This suggests that the energy levels for the 90° partial are also split, possibly because of reconstruction, with the kink levels between these levels, again as in fig. 1.

Kinks. Models of elementary kinks have been constructed for 30° and 90° partials, and these are shown in figs. 9 and 10. In both cases reconstructed structures are possible as well as structures with dangling bonds. For the reconstructed 30° partial a dangling bond occurs where the one-dimensional superlattice along the partial is shifted by half the superlattic period. For the reconstructed 90° partial similarly a dangling bond occurs where the reconstructed bonds change from + 45° to - 45° to the line direction. These dangling bond defects have been termed "antiphase defects" (APD). The unreconstructed kinks can be considered as reconstructed kinks plus an APD [36].

The energy of a reconstructed kink is expected to be low, and a recent calculation due to Jones [17] for a kink in a 90° partial shows this to be so. On the other hand motion of the reconstructed kink involves effectively the breaking of two bonds, and this is likely to result in a large activation energy for migration. Jones' calculations again show this to be the case. He finds for a reconstructed kink in a 90° partial in Si that $F_k \sim 0.75$ ev and $W_m \sim 1.3$ ev. The total activation energy for dislocation velocity, $F_k + W_m = 2.05$ ev, is of the same order as the experimental values for 60° dislocations in intrinsic Si [for table see [16]]. On the other hand if the kink has the unreconstructed configuration its formation energy is likely to be considerably larger, but now only oneadditional bond has to be broken to move the kink over the saddle point, i.e. we may expect the migration activation energy to be lower. Furthermore the unreconstructed kink may be stabilised by an electron or hole.

It seems therefore that there is the possibility that charged and uncharged kinks have different configurations, energies (F_k) and mobilities (W_m), leading to the complications mentioned in connection with the earlier discussion on the "kink" theories of dislocation velocity.

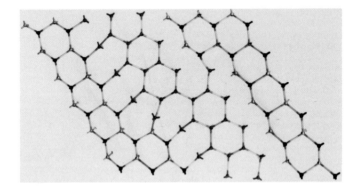

Fig. 8. Dissociated 60° glide dislocation. Core structure of 30°
partial on the left and of 90° partial on the right, without
reconstruction; viewed normal to the slip plane; only one layer
of atoms above and one below slip plane are shown. Kinks are
shown on both partials.

Kink velocities
        It is clear that progress towards a proper understanding of the doping
effect on dislocation velocity will depend on obtaining independent measurements
of the kink velocity (i.e. activation energy for migration) and kink formation
energy. The former could be obtained from internal friction experiments or from
relaxation studies in which it can be ensured that migration of existing kinks
is controlling. There are a few internal friction data available at the present
time [e.g. 41, 42], and the relaxation times of the peaks observed at low
temperatures in Ge in [42] have been interpreted as being due to kink migration
with an activation energy of $\sim$ 0.1 ev. However the energy loss observed at low
temperatures is very small, and the precise interpretation of the data must be
in doubt [see 13].

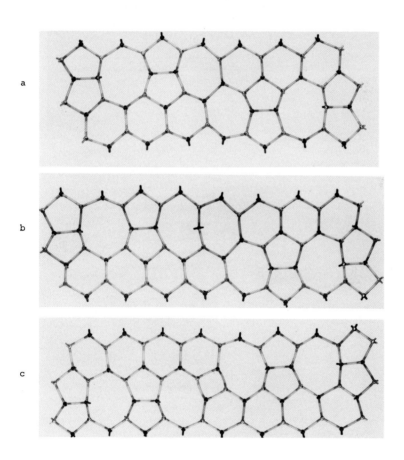

Fig. 9. Reconstructed 30° partials with kinks.
a), c) reconstructed kinks; b) kink with dangling bond.

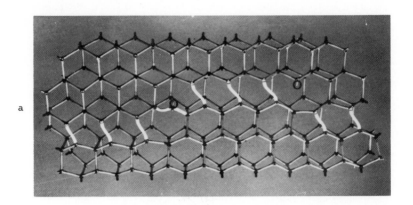

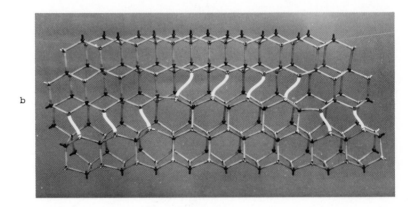

Fig. 10. Reconstructed 90° partials with kinks.
a) kinks with dangling bonds marked by circles,
b) reconstructed kink.

Relaxation experiments on dislocation configurations containing kinks should in principle give unambiguous information. For example, many glide loops produced by low temperature and high stress deformation, according to the recipe of Wessel and Alexander [27] contain rounded corners (see fig. 6). Because of the different mobilities of the partials the stacking fault widths are often much larger than the equilibrium width. On unloading and annealing at the appropriate temperature the corners of the outer partial should become angular by kinks migrating away from the corners, thereby reducing the width of the fault. This process occurs without nucleation of new kinks. The driving force of this mechanism is the line tension and the fault energy. The velocity of the kinks under these conditions is given by

$$v_k = v_D \, a^2 \, (\frac{Sh}{\rho} + \gamma h - F_R)/kT \, \exp \, (- \frac{W_m}{kT}) \tag{7}$$

where v_D is the Debye frequency, a jump distance of the kink, W_m activation energy for migration, S the line tension, h the height of the kink, ρ the local radius of curvature, γ the stacking fault energy and F_R the repulsion from the other partial. If kinks moved with activation energies of 0.1 ev, rounded corners would certainly not be stable at room temperature, when there is a driving force for reducing the width of the fault. Substituting reasonable values of the observed radius of curvature, and for the driving force for reduction of the anomalous fault width in (7) we find that for Si $W_m \gtrsim 0.75$ ev for the curved corners to be stable. Experiments are being carried out at the present time to try to establish W_m by this method. Preliminary in-situ annealing experiments in the high voltage microscope suggest that the configurations are stable to rather high temperatures, but such experiments are subject to contamination and radiation damage, and further experiments are necessary to determine the temperature at which such relaxation occurs without these potential pinning effects.

Möller [43] studied the relaxation in Ge of glide loops intersecting the surface. In this case however the process is controlled by the energy to nucleate a single kink at the surface. It should be emphasised that nucleation of kinks would also need to occur in the experiments on shrinkage of loops lying wholly within a specimen, if the corners were observed to move inwards.

CONCLUSIONS

1.   At the present time complete understanding of the doping effect on dislocation velocity is lacking.
2.   The dependence on Fermi level could be explained within a theoretical framework in which screws and 60° dislocations are associated with split energy levels in the band gap, and kink acceptor and donor levels occur in the gap of the dislocation levels.
3.   The two theories [14, 17] developed along these lines associate the effect either with a chargedkink configuration plus a change in the saddle point energy for migration, or solely with a change in the saddle point energy; both theories predict a similar dependence on Fermi level.
4.   The most plausible interpretation of the observed dissociated glide dislocations is that both partials are on "glide" planes, but that association with point defects could occur at some sites along the dislocation.
5.   The 30° partial is likely to be reconstructed, but for 90° partials this is still uncertain.

6. Kinks in partials can be reconstructed or have dangling bonds. The energies and mobilities of such different structures are expected to be different, leading to complications in the interpretation of the doping effect. Acceptor and donor kink levels could be responsible for the trapping of charge carriers at dislocations.
7. The stability of rounded corners of glide loops with anomalously wide faults at room temperature suggests that the activation energy for kink migration in Si is substantial, probably $\gtrsim$ 0.75 ev. Further experiments are needed to determine the kink mobility directly.

REFERENCES

1. J. R. Patel and A. R. Chaudhuri, Phys. Rev. 143, 601 (1966).

2. J. R. Patel and P. E. Freeland, Phys. Rev. Lett. 18, 833 (1967).

3. H. L. Frisch and J. R. Patel, Phys. Rev. Lett. 18, 784 (1967).

4. J. R. Patel and H. L. Frisch, Appl. Phys. Lett. 13, 32 (1968).

5. V. N. Erofeev and V. I. Nikitenko, Sov. Phys. Solid State 13, 116 (1971).

6. J. R. Patel, L. R. Testardi and P. E. Freeland, Phys. Rev. B13, 3548 (1976).

7. S. B. Kulkarni and W. S. Williams, J. Appl. Phys. 47, 4318 (1976).

8. J. R. Patel and L. R. Testardi, Appl. Phys. Lett. 30, 3 (1977).

9. A. George, Journ. de Physique, Colloque C 6, Supplem. 6, 40, C6-133 (1979).

10. D. T. Griggs and J. D. Blacic, Science, N.Y. 147, 292 (1965).

11. W. Schröter, R Labusch and P Haasen, Phys. Rev. B15, 4121 (1977).

12. P. Haasen, Phys. Stat. Solidi (a) 28, 145 (1975).

13. P. Haasen, Journ. de Physique, Colloque C6, Supplem. 6, 40, C6-111 (1979).

14. P. B. Hirsch, Journ. de Physique, Colloque C6, Supplem. 6, 40, C6-117 (1979).

15. J. P. Hirth and J. Lothe, Theory of Dislocations (New York: McGraw-Hill) 1968.

16. A. George and G. Champier, Phys. Stat. Solidi (a) 53, 529 (1979).

17. R. Jones, Phil. Mag. (1980) in press.

18. I. L. F. Ray and D. J. H. Cockayne, Proc. Roy. Soc. A325, 534 (1971).

19. I. L. F. Ray and D. J. H. Cockayne, J. Microsc. 98, 170 (1973).

20. F. Häussermann and H. Schaumburg, Phil. Mag. 27, 745 (1973).

21. A. Gómez, D. J. H. Cockayne, P. B. Hirsch and V. Vitek, Phil. Mag. 31, 105 (1975).

22. P. L. Gai and A. Howie, Phil. Mag. 30, 939 (1974).

23. H. Gottschalk, G. Patzer and H. Alexander, Phys. Stat. Solidi (a) 45, 207 (1978).

24. A. Gómez and P. B. Hirsch, Phil. Mag. 38, 733 (1978).

25. J. Hornstra, J. Phys. Chem. Solids 5, 129 (1958).

26. A. Gómez and P. B. Hirsch, Phil. Mag. 36, 169 (1977).

27. K. Wessel and H. Alexander, Phil. Mag. 35, 1523 (1977).

28. D. J. H. Cockayne, A. Hons and J. C. H. Spence, Micron Supplem. 1, 11, 32 (1980).

29. H. Alexander, H. Eppenstein, H. Gottschalk and S. Wendler, Journ. Microscopy 118, 13 (1980).

30. E. Weber and H. Alexander, Journ. de Physique, Colloque 6, Supplem. 6, 40, C6-101 (1979).

31. A. Olsen and J. C. H. Spence, Phil. Mag. in press.

32. P. B. Hirsch, Journ. de Physique, Colloque 6, Supplem. 6, 40, C6-27 (1979).

33. D. Cherns, P. B. Hirsch and H. Saka, Proc. Roy. Soc. A371, 213 (1980).

34. D. Cherns and A. Ourmazd, to be published.

35. W. Shockley, Phys. Rev. 91, 228 (1953).

36. P. B. Hirsch, Journ. Microscopy, 118, 3 (1980).

37. D. Marklund, Phys. Stat. Solidi (b) 92, 83 (1979).

38. R. Jones, Journ. de Physique, Colloque 6, Supplem. 6, 40, C6-33 (1979).

39. D. Marklund, Phys. Stat. Solidi, in press.

40. R. Wagner and P. Haasen, Lattice Defects in Semiconductors, Inst. Phys. Conf. Ser. No 23, 387 (1975).

41. H. J. Möller and J. Buchholz, Phys. Stat. Solidi (a) 20, 545 (1973).

42. K. Ohori and K. Sumino, Phys. Stat. Solidi (a) 14, 489 (1972).

43. H. J. Möller, Phil. Mag. 37, 41 (1978).

DISLOCATION DEFECT STATES IN DEFORMED SILICON

J. R. PATEL AND L. C. KIMERLING
Bell Laboratories, Murray Hill, N. J. 07974, USA

ABSTRACT

Transient junction capacitance measurements on deformed silicon reveal a variety of states after deformation. Upon annealing or following more homogeneous deformation the defect spectrum shows a single broad feature. A state at E(0.38) has been tentatively assigned to defect sites along the dislocation. A broad band of states at H(0.35) is postulated to represent states due to the reconstructed dislocation core.

INTRODUCTION

Since the very early work on the electrical behavior of dislocations in semiconductors [1,2,3], there has been a continuous effort to understand their detailed electrical characteristics. Perhaps the most sustained effort in this field has been the work of Haasen, Schröter, Labusch and co-workers [4,5] who have shown that dislocations in deformed p type silicon introduce a band of levels at $E_V + 0.34$ eV for 60° dislocations. Grazhulis et al [6] investigating both n and p type silicon find a band of levels at $E_V + 0.42eV$ and $E_C - 0.67eV$ after deformation. In this paper we consider direct measurements of dislocation levels in deformed silicon using transient junction capacitance measurements performed by Kimerling and Patel [9,10].

EXPERIMENTAL

Details of the experiments have been given in [9,10]. Two types of floating zone crystals were used: (a) dislocation-free crystals both n and p type and high and low resistivity (b) crystals containing $10^5/cm^2$ of dislocations. Samples were deformed either in bending or compression in the range 650° - 750°C.

Specimens for the capacitance measurements were cut parallel to the primary slip plane and polished and cleaned prior to evaporating a Schottky barrier at room temperature: Au—Pd for n type material and Ti for p type silicon.

The principles of the transient junction capacitance technique have been reviewed by Miller et al [11]. In what follows the activation energies to the relevant band edges are denoted as E() for the conduction band and H() for the valence band.

RESULTS

The defect state spectrum of n type silicon is shown in Fig. 1. A large number of defect states which act as majority carrier traps are revealed. The complex spectra suggest that a variety of defect states besides dislocations are generated by plastic deformation. In particular a dominant state E(.63) to E(.68), depending on the doping concentration, is always found after inhomogeneous deformation in both n and p type silicon.

274

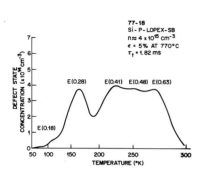

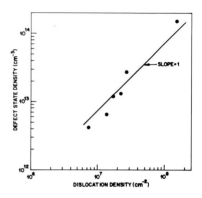

Fig. 1.   Defect state spectrum
of low resistivity n-type
silicon.

Fig. 2.   Defect state concentration as a function of local
dislocation density under
Schottky barrier.

Following an anneal of 1 hr. at 900°C the complex defect spectrum Fig. 1 simplifies and we observe only one broad peak at E(.38) which is stable upon further annealing and can be related to the dislocations which survive the annealing treatment. The argument that most of the peaks observed in Fig. 1 are due to a variety of defects other than dislocations is consistent with EPR observations in similarly deformed crystals in which the EPR signal disappears in the same temperature range [12].

The dislocation morphology does not influence the spectra observed. Two very distinct dislocation morphologies, (1) produced directly after deformation at 750°C showing tangled dislocation networks, and (2) observed after an initial deformation at 750°C followed by a high stress $\tau = 11$ kg/mm^2 at 450°C and resulting in a very regular geometrical pattern of dislocations, showed identical defect spectra similar to those observed in Fig. 1. Thus we postulate that except for the feature at E(0.38) which is stable upon annealing the other peaks in Fig. 1 are related to point defects and defect clusters produced during plastic deformation by dislocation point defect interactions such as non conservative motion of jogs and intersections of moving dislocations.

The observed electron capture kinetics [10] show that the capture process is proportional to the logarithm of time which is characteristic of large defects with a macroscopic space charge region produced by capture of many charges [13,14]. These observations are consistent with the postulate above that plastic deformation in our specimens produces a variety of defects and defect clusters.

In homogeneously deformed crystals with slight deformation, defects other than dislocations generated by plastic deformation should be greatly reduced. This hypothesis was tested in the following experiment. Specimens with an initial density of 10^5/cm^2 were deformed very slightly in bending. Slip is observed to be quite uniform due to the high density of sources available; the dislocation density was in the range $10^7 - 10^8$/cm^2. The spectrum observed after such a deformation showed a single broad peak similar to the peak at E(0.38) which survived a 900°C heat treatment.

Thus this deformation procedure has directly suppressed the production of the complex spectra observed in Fig. 1.

The defect state concentration in crystals deformed homogeneously by bending was measured as a function of dislocation density. The electrically active dislocation line density was measured in SEM charge collection micrographs under the same Schottky barrier used in the transient junction capacitance measurements. Figure 2 shows the defect state density versus the dislocation density. The relation is linear and corresponds to approximately 1% occupation of possible dislocation sites in the dissociated model of a 60° dislocation, Hirsch [15].

To introduce dislocation densities in the range above $10^8/cm^2$, crystals with an initial density of $10^5/cm^2$ were deformed in compression. In Fig. 3 we show the effect of increasing deformation on the defect spectrum. For a deformation of 0.3% the defect spectrum is a little broader but similar to that observed in bending. The dislocation density in this crystal

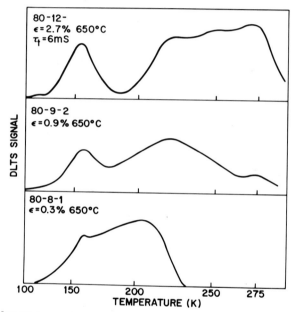

Fig. 3. Defect State spectrum of low resistivity n-type silicon, with initial dislocation density $10^5 cm^{-2}$, deformed in compression.

as observed by etch pits is $5 \times 10^7/cm^2$ and is within the range of the bending experiments Fig. 2. When the deformation is increased to 0.9% the spectra begin to resemble those in Fig. 1. Finally after a deformation of 2.7% the spectrum is very similar to that measured on initially dislocation-free crystals. We postulate that at low deformation in the range 0.3% the deformation is homogeneous with little formation of dislocation debris (point defects, defect clusters etc.). With increasing deformation inhomogeneous slip in local regions generate defects and defect clusters leading to the spectra shown in the lower half of Fig. 3. We conclude that as long as deformation is limited, the capacitance spectrum shows a predominant feature related directly to dislocations. For higher deformations local inhomogeneities in the slip process leads to the generation of point defects and defect clusters resulting in a complex capacitance spectrum.

CONCLUSIONS
The dominant state E(.68) is always present in inhomogenously deformed crystals and completely absent in homogeneously deformed crystals. any direct association of this state with dislocations can therefore be eliminated. The behavior of this state follows closely the the EPR centers observed in similarly deformed samples Erdman & Alexander [17]. It's position agrees well with spin dependent photoconductivity results [8].

The stable defect states observed after annealing are E(.38) and H(.35). We have not examined the H(.35) state in any detail. Its position corresponds well with the half filled band of states Schröter [18]. Such a band of states along dislocation segments could also account for the microwave conductivity along dislocations observed by Grazhulis et al [6].

The E(.38) state density is directly proportional to the dislocation density. The energy position of this state agrees well with the defect state controlling dislocation velocity [19]. Evidence for a similar state at E(.42) has also been obtained from Hall effect studies on deformed n-type silicon. We tentatively assign these states to a defect along a dislocation such as a kink site. Direct tests of this hypothesis are difficult. The doping dependence of dislocation velocity and hence kink site density is not in the range of the capacitance measurements. Specimens in which curved dislocation segments predominate might be fruitful for elucidating the role of kinks.

ACKNOWLEDGEMENTS

We thank P. E. Freeland and J. L. Benton for their able and dedicated assistance in all phases of this work.

REFERENCES

1. W. T. Read, Philos. Mag. *45*, 775 (1954); *45*, 1119 (1954); 46, 111 (1955).

2. W. Schockley, Phys. Rev *91*, 228 (1953).

3. R. A. Logan, G. L. Pearson and D. Kleinman, J. Appl. Phys. *30*, 885 (1959).

4. P. Haasen and W. Schröter, "Fundamentals of Dislocation Theory" J. A. Simmons et al Eds. (Nat. Bur. Stand. (US) Spec. Publ. 317, II 1970) p. 1231.

5. R. Labusch and W. Schröter, "Lattice Defects in Semiconductors" Conference Series No. 23, (Inst. of Physics, London) 56, (1975).

6. V. A. Grazhulis, V. V. Kveder, Yu V. Mukhina, Phys. Stat. Sol. (a) *43*, 407 (1977); *44*, 107 (1977).

7. T. Wosinski and T. Figielski, Phys, Stat. Sol. (b) *83*, K57 (1976) see also review by T. Figielski, J. de Physique, Colloque C6, *40*, 95 (1979).

8. D. Lepine, V. A. Grazhulis, D. Kaplan, Physics of Semicond., Int. Conf. Rome 1081 (1976).

9. L. C. Kimerling and J. R. Patel, Appl. Phys. Lett. *38*, 73 (1979).
10. J. R. Patel and L. C. Kimerling, J. de Physique, Colloque C6, *40*, 67 (1979).
11. G. L. Miller, D. Lang and L. C. Kimerling, Annu. Rev. Mat. Sci. *7*, 377 (1977).
12. F. D. Wohler, H. Alexander and W. Sander, J. Phys. Chem. Solids, *31*, 1381 (1969).
13. T. Figielski, Sol. St. Electr. *21*, 1403 (1978).
14. W. Skielko, O. Breitenstein and R. Pickenheim, Symposium on Defect Induced Phenomena in Semiconductors Krynica, Poland (1980).
15. P. B. Hirsch, J. de Physique Colloque C6 *40*, 27, (1979).
16. E. Weber and H. Alexander, J. de Physique Colloque C6 *40*, 101 (1979).
17. R. Erdmann and H. Alexander to be published Phys. Stat. Sol. (1980).
18. W. Schröter, J. de Physique, Colloque C6 *40*, 51 (1979).
19. J. R. Patel, L. R. Testardi and P. E. Freeland, Phys. Rev. B *13* 3548 (1977).

Published 1981 by North-Holland, Inc.
Narayan, and Tan, eds.
Defects in Semiconductors.

CHARACTERIZATION OF DISLOCATIONS AND INTERFACES IN SEMICONDUCTORS BY HIGH
RESOLUTION ELECTRON MICROSCOPY

J.C.H. SPENCE and A. OLSEN
Dept. of Physics, Arizona State University, Tempe, AZ   85281

ABSTRACT

It is not presently possible to resolve the individual
atoms in any semiconductor by high resolution electron microscopy (HREM). However symmetry arguments may be used to
allow near-atomic resolution lattice images to be interpreted
in rare favorable cases. This method is applied to the
problem of distinguishing shuffle and glide set partial
dislocations in silicon. It is also proposed that two
dimensional characteristic loss energy selected diffraction
patterns be used to reveal the local symmetry about selected
substitutional species implanted in semiconductor lattices.

INTRODUCTION

While structure imaging by high resolution transmission electron microscopy
has given crystal chemists new insight into many defect structures and phase
transformation mechanisms in recent years, there have been few, if any, really
useful results from this technique in the field of semiconductor physics. This
is chiefly because it is not presently possible to obtain a structure image at
atomic resolution of any common semiconductor. Here a structure image is
defined as a lattice image which faithfully reveals a crystal's structure to
some limited resolution and which was recorded under instrumental conditions
which are independant of the structure. "Tuning" the electron microscope to
the structure results, in general, in an uninterpretable lattice image in small
unit cell crystals. Thus the major limitations of the high resolution TEM
technique for semiconductors are (i) the need to extract structural information
from images beyond the current 2.7Å Scherzer point resolution limit, (ii) the
insensitivity of these images to atomic displacements in the electron beam or
projection direction, (iii) the difficulty of extracting chemical or atomic
number information from the images. Approaches to each of these problems are
discussed in the following.

THREE-DIMENSIONAL INFORMATION

Fault vector components in the beam direction may, in favorable cases, be
found from convergent beam patterns recorded from the same area from which a
lattice image has been obtained. The instrumental compatability of the CBED
and structure imaging techniques has been described elsewhere [1]. Frequently,
the destruction by strain of the three dimensional interference condition
needed to produce fine HOLZ and Kikuchi lines in CBED patterns may be used to
obtain several "g.b = 0" conditions from a single CBED pattern, thus allowing
an out-of-ZOLZ g to be found from a single microdiffraction pattern [2]. This
three-dimensional information is usually impossible to extract from lattice
images. Further, if HOLZ lines are used, this technique does not require the
sample to be tilted from the zone axis lattice imaging orientation - an important consideration in view of the very limited tilt available (10°) on modern
HREM machines. There is thus an urgent need for CBED facilities to be fitted

to modern HREM machines so that CBED and near-atomic resolution images can be collected from the same area without moving the specimen.

SHUFFLE AND GLIDE, AND KINKS

The presence of detail beyond the instrumental point resolution is vividly indicated in figure 4. In this image, recorded at 100kV accelerating voltage, each black blob represents a pair of atomic columns about 60Å thick seen along the [110] projection, which is the dislocation line direction. The 30° and 90° partial dislocations are connected by a strip of stacking fault, seen end-on. The extra plane of atoms can be seen by viewing the page at a low angle. By interpreting this [110] axial lattice image of silicon it was hoped to determine whether the end-on 30° partial dislocation seen in the image is an example of the glide or shuffle set. These structures differ by the presence or absence of a single column of atoms (as shown in figure 3) and are expected to show quite different mechanical and electronic properties [3]. However, whereas the image is synthesized chiefly from the crossed 3.1Å (111) lattice fringes, the Scherzer resolution of the microscope used was 3.8Å. The aim of this work was to find some simple image characteristic or "fingerprint" which could be shown by computer image simulation to be insensitive to variations in the parameters of focus Δf, spherical aberration C_s, thickness t and orientation on which lattice images sensitively depend. The difficulties involved in the interpretation of these images are indicated in figure 1, which shows lines of constant phase for the transfer function of a modern electron microscope. All many-beam axial [110] diamond structure lattice images are identical along these lines, and they are periodic in both Δf and C_s. The stationary-phase focus for the (111) reflection Δf_s, the depth of field Δz and the (111) scattering phase ε are all shown. We were able to resolve the black-white ambiguity in the image (to tell whether atoms or tunnels appear black or white) by noting that both two-fold axes in the stacking fault shown in figure 2 lie on a row of atoms (not tunnels), and that, in the ZOLZ approximation, multiple coherent scattering and all instrumental effects preserve point group operations. Thus the two fold axes normal to the page in figure 4 in the stacking fault fall on atomic columns, which are therefore seen to be black. By computing many shuffle and glide core images such as those shown in figure 3 around the estimated values of the image parameters it was found that in all black atom images the shuffle case showed three black dots around the core, while the glide case showed four. We therefore interpret our experimental image shown in figure 4 as giving support for the glide model of the 30° partial dislocation in silicon [4].

It has been suggested that kinks in semiconductors may exert a controlling influence on their mechanical and electronic properties [3]. We have therefore investigated the possibility of imaging kinks directly by HREM. Figure 5 shows a model of a silicon lattice containing two kinks on 30° partial dislocations with opposite fault vectors so arranged as to produce both a center of symmetry and a monoclinic cell which may be periodically continued without severe bond distortion. The dimensions of the artificial cell used for periodic continuation are 22.8Å by 13.9Å, with $\gamma = 106.5°$. A projection of the cell down [111] is shown in figure 6, (plane group p2) which contains two centers of symmetry in projection and a dangling bond at each kink, plus "reconstructed" bonds along the core of the partials [5]. From computed electron microscope images such as that shown in figure 7 for the JEOL 200CX instrument at Scherzer focus we conclude that there is little hope at present of correctly interpreting a HREM image of a kink in silicon. The [111] stacking sequence for the image shown in figure 7 is ABCABCABC.KINK.ABCABCABC. The thickness is 65Å.

ELECTRONIC STRUCTURE OF KINKS AND DISLOCATIONS

In collaboration with M. Cohen, J. Chelikowski and J. Northrup, the electronic structure of both the kink and the 30° glide partial are being computed using the method of Louie [6]. Preliminary results for the unreconstructed partial, based on the atomic co-ordinates used to match the experimental image shown in figure 4, are in good agreement with those of other workers [5] using different algorithms. The glide partial artificial cell contains 48 atoms (5 "orbitals" per atom), while the kink cell contains 78 atoms unrelated by a center of symmetry. In this way it is hoped eventually to understand the influence of doping and temperature on kink motion [3]. Our calculations for the 30° glide partial show P_z-like localized states at the core extending well into the bulk conduction band midway to the B.Z.B. and falling below the Fermi level at the zone boundary.

CHEMICAL INFORMATION IN HREM AND LOCAL SYMMETRY IN INELASTIC SCATTERING.

The sensitivity of structure images to atomic number has been investigated previously [7]. In the [110] projection the Ga/Aℓ and As atomic columns are separated by about 1.3Å in the $Ga_{1-x}Aℓ_xAs$ system. Once instruments capable of resolving these columns become available, microanalysis on an atomic scale may become feasible, provided that thickness and orientation effects can be accurately accounted for. Since research HREM instruments now exist with a point resolution of 1.6Å [8], this goal may well be reached within the next few years. Calculations and experimental images at near atomic resolution [9] suggest that a 1% composition change could be detected at atomic resolution in III-V interface images. This calculation is based on a comparison of the intensity at the Aℓ/Ga column in an HREM image with the intensity at the As column, each of which is assumed to be resolved. The exposure needed is three seconds for an incident current density of 10 Amp cm^{-2}. In view of the sensitivity of coherent elastic Bragg scattering to both composition (using difference reflections) and strain, it is not clear that energy loss spectroscopy (which contains no nuclear contribution) in STEM will provide more useful compositional information than structure imaging. In particular it should be noted that while an "elastics only" image can readily be obtained, an "inelastics only" cannot, since elastic scattering is always present in the inelastic channel [10].

The possibility also exists that valuable chemical information may be obtained from the symmetry of two dimensional characteristic loss energy selected electron diffraction patterns. If the local symmetry about a particular atom projected in the beam direction is given, then it can be shown [11] that the point plane group symmetry of the energy selected electron diffraction pattern formed from fast electrons which have excited a characteristic core loss in that atom is equal to this given local symmetry, with all the crystal point group operations applied to it and the results summed. Here local symmetry refers to the projected symmetry of the co-ordination polyhedra about the selected atom when removed from the crystal. In general it is not a crystal point group operation but may well be the symmetry of the asymmetric unit, taken in isolation. Thus for localized losses from atom A of figure 8 each asymmetric unit possessing 3-fold symmetry produces an energy selected pattern with 3-fold symmetry (Friedel's law does not apply to the dynamical elastic scattering of the inelastically scattered electrons). The result of adding incoherently to this the two-fold rotated pattern from the second asymmetric unit then gives an energy selected pattern with six-fold symmetry. Since the presence of the crystal two-fold axis can be detected by conventional CBED, the q-fold local symmetry about atom A can, in general, be deduced by dividing the n-fold symmetry of the energy filtered pattern by the p-fold projected

282

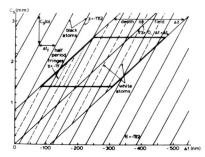

Fig. 1. C_S and Δf values giving black (or white) atom images in silicon.

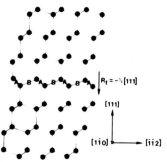

Fig. 2. Intrinsic stacking fault in silicon showing 2-fold axes A and B.

Fig. 3. Computed glide (a) and shuffle (b) images.

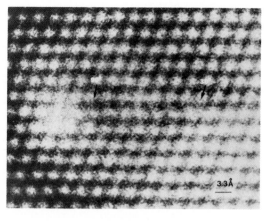

Fig. 4. Experimental 100kV lattice image of dissociated 60° dislocation in silicon, seen in [110] projection.

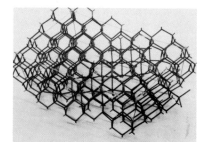

Fig. 5. Model of silicon lattice periodic cell containing two kinks on 30° partials and a center of symmetry.

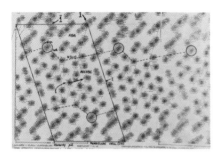

Fig. 6. Projected potential of kink cell along [111]. Stacking ABA and ABCABC.

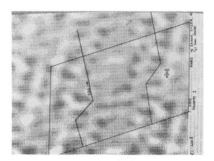

Fig. 7. Computed image at 65Å thickness and 1.7Å resolution of kinks shown in figures 5 and 6.

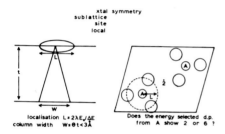

Fig. 8. Inelastic scattering from a center A with three-fold local symmetry.

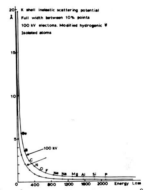

Fig. 9. Localization L in Å as a function of energy loss in eV.

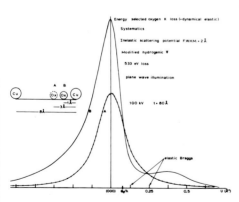

Fig. 10. Oxygen energy filtered K loss intensity for two atom positions A and B.

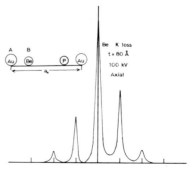

Fig. 11. Similar for Be filtered intensity with larger L.

symmetry of the entire crystal, as obtained by CBED. Here $q = 3$, $n = 6$, $p = 2$, for the example given in figure 8. Similar results hold for mirrors; however there are some intractable cases (e.g. a mirror in the asymmetric unit not parallel to one in the crystal). The range L over which local symmetry is expressed is determined both by column width W and inelastic localization L [10] shown in figure 8. Figure 9 shows calculated values of L for light elements based on the modified hydrogenic wave wave functions and first Born approximation for isolated atoms given by Egerton [12], however the Lorenzian form of the scattering makes a useful value of L difficult to define. Figure 10 shows the calculated inelastic (plus dynamical elastic) intensity along the systematics line for the hypothetical one dimensional CuO crystal shown. This is the intensity which would be recorded in an energy filtered diffraction pattern tuned to the K-shell excitation from oxygen. The thickness is 80Å. Note how the energy selected pattern from the oxygen atom gains a center of symmetry (not possessed by the crystal as a whole) as this atom is moved away from the copper atom from position B to position A. This allows the column width W in figure 8 to be estimated as about 3Å in crystals less than 100Å thick for 100kV electrons. Figure 10 shows similar results for a more delocalized loss at smaller energy loss in which the sharp Bragg reflections have become apparent.

This technique, which requires the use of either an Omega filter or Castaing-Henry filter, may be useful in problems such as ion-implantation, laser annealing and doping generally, where it is required to determine the local environment of a particular atomic species in the lattice. It gives information both on local symmetry and on the orientation of the asymmetric unit and is not subject to the data analysis difficulties of the EXELFS technique.

ACKNOWLEDGMENTS

This work was supported by NSF grant #DMR8002108, ARO grant #DAAG-29-80-C-0080, and the Arizona State University NSF Regional Facility for HREM.

REFERENCES

1. P. Goodman et al., Proc. 38th EMSA Meeting (Claitors, Baton Rouge, 1980)174.

2. R. Carpenter and J.C.H. Spence, Acta Cryst. (1980) in press.

3. For a review, see J. de Phys. C-6 (1979).

4. A. Olsen and J.C.H. Spence, Phil. Mag. (1980) in press.

5. S. Marklund, Phys. Stat. Sol. (b) 92, p. 83 (1979).

6. S. G. Louie, Phys. Rev. B22, p. 1933 (1980).

7. J. Skarnulis, S. Iijima and J. M. Cowley, Acta Cryst. A32, p. 7-9 (1976).

8. N. Uyeda et al., Chemica Scripta 14, p. 47 (1979).

9. A. Olsen et al., Proc. 38th EMSA Meeting (Claitor's, Baton Rouge, 1980) 318.

10. A. J. Craven et al., Phil. Mag. A38, p. 519 (1978).

11. J.C.H. Spence, Optik (1980) in press.

12. R. F. Egerton, Ultramicroscopy 4, p. 169 (1979).

SILICON-SAPPHIRE INTERFACE:
A HIGH RESOLUTION ELECTRON MICROSCOPY STUDY

FERNANDO A. PONCE
Hewlett-Packard Laboratories, Palo Alto, California, USA

ABSTRACT

The structure of the silicon-sapphire interface of CVD
silicon on a $(1\bar{1}02)$ sapphire substrate has been studied in
cross section by high resolution transmission electron
microscopy. Multibeam images of the interface region have
been obtained where both the silicon and sapphire lattices
are directly resolved. The interface is observed to be
planar and abrupt to the instrument resolution limit of
3 Å. No interfacial phase is evident. Defects are
inhomogeneously distributed at the interface: relatively
defect-free regions are observed in the silicon layer in
addition to regions with high concentration of defects.

INTRODUCTION

The silicon on sapphire technology has several applications in microelec-
tronics owing to the good insulation of the substrate with the consequent
reduction in overall parasitic capacitances. High speed field effect devices
and complementary circuits can be fabricated with the added advantage of low
power consumption. Some inherent difficulties exist, however, which manifest
themselves as low MOS channel mobilities and large leakage currents. It is
thought that these difficulties stem from the highly defected nature of the
silicon layer. It is therefore desirable to understand the origin and nature
of these structural defects and the possible effects on the electrical behavior
of the material.

The nature of the silicon-sapphire interface has been the subject of several
studies. Heiman [1] suggested the existence of a thin glassy layer at the
interface in order to explain the appearance of donor surface states. Kühl [2]
and Schlötterer [3] interpreted optical interference experiments by postulating
the existence of an intermediate layer with a composition similar to aluminum
silicates and a thickness between 200 and 400 Å at the interface. SIMS measurements were used in support of the transition layer model showing a continuous
variation in the aluminum composition in the vicinity of the interface
($\sim$ 500 Å) [3-6]. Cullen [5] and Trilhe [6] associated this transition layer
with the thickness necessary to achieve complete coverage of the substrate
surface.

Observations of the defect structure by transmission electron microscopy
(TEM) have been reported [7-9] which indicate the presence of large densities
of stacking faults and twins in the silicon layer. Abrahams and Buiocchi [7]
reported no evidence for the presence of an aluminum-bearing phase in the proximity of the interface but without reference to the resolution of their technique.

In this paper, results from a high resolution TEM study of the silicon-
sapphire interface are presented. Lattice images have been obtained where the
lattice structure of both silicon and sapphire regions are resolved in cross-
section to a best resolution of about 3.0 Å.

EXPERIMENTAL ASPECTS

The material under study is heteroepitaxial [001] silicon grown on a (1$\bar{1}$02) sapphire substrate from the pyrolysis of silane in H_2 at 960°C and at a rate of 1.4 μm per minute to a total thickness of 0.65 μm, followed by a gate oxidation process at 1000°C. The specimens were prepared in cross-section. The silicon-on-sapphire wafer was carefully cut to obtain bars with edges oriented along a <110> direction. Two bars were glued together with the silicon layers facing each other, and then they were embedded in plastic. After mechanical dimpling and polishing, the specimens were ion-milled with Ar^+ at 5 kV at a low incidence angle, followed by further ion milling at 2 kV to remove any surface damage. In order to reduce image drift due to charging and heating effects inside the microscope, the thick regions of the specimens were coated with a gold layer with a thickness of about 1000 Å.

The micrographs were obtained using a Philips EM400 transmission electron microscope equipped with a high resolution stage (spherical aberration coefficient of 1.1 mm) and a LaB_6 filament at an operating voltage of 120 kV. The images were recorded under axial illumination with a beam divergence of 1.3 mrad at Scherzer focus (-729 Å). Under these conditions the expected resolution limit of the instrument is 2.9 Å [10]. The specimen stage in the microscope had no tilting capabilities, thus the experiment relied on the production of specimens which were closely aligned in the appropriate orientation.

THE SILICON AND SAPPHIRE CRYSTAL STRUCTURES

Silicon is a tetrahedrally coordinated material with the diamond cubic structure and a lattice parameter of 5.43 Å. In the [110] projection the atoms are stacked up in columns. For a given atom, two of its bonds to neighboring atoms occur along a plane perpendicular to the [110] direction, while the other two bonds make an angle of 35.4° with [110]. These two oblique bonds correspond to links between {400} planes and give origin to the "dumbell" arrangement observed in figure 1(a). The dumbells are seldom resolved in lattice images and usually appear as bright spots with a {111} interplanar separation of 3.14 Å, as shown in figure 1(b).

The sapphire structure has been clearly described by Kronberg [11]. Figure 2(a) shows the projection of the sapphire lattice along the edge of the morphological rhombohedron [$\bar{2}$021]. Two sets of rhombohedral "r" planes are seen edge-on making an angle of 85.5°, and with interplanar spacing of 3.47 Å. Al_2O_3 coordination polyhedra occur at the intersection of r-planes and correspond to high electronic charge densities that is reflected in the lattice image shown in figure 2(b).

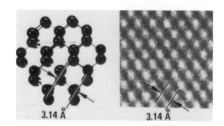

(a)   (b)

Fig. 1. The silicon structure in the [110] projection: (a) model, (b) TEM lattice image.

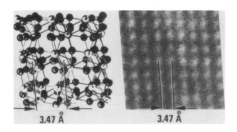

(a)   (b)

Fig. 2. The sapphire structure in the [2̄021] projection: (a) model, (b) TEM lattice image.

    The image characteristics can be understood by considering the diffraction pattern shown in figure 3(a). The pattern corresponds to the superposition of the [110] silicon and the [2̄021] sapphire patterns as shown in figure 3(b). Due to the resolution of the instrument, only five beams each of the silicon and sapphire (shown inside the dashed circumference) contribute significantly in the image formation process. In the sapphire region the image is formed primarily by the central transmitted beam and four {11̄02} reflections corresponding to r-planes. In the silicon region the main contributions are from the transmitted beam and the four {111} reflections present in the [110] projection. The result is a 3.14 Å cross fringe pattern. Defects such as stacking faults and twins on the two sets of {111} planes perpendicular to the foil should be seen edge-on.

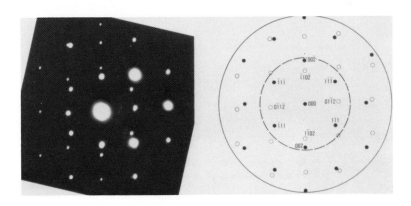

(a)   (b)

Fig. 3. Diffraction pattern at the silicon-sapphire interface. (a) Actual. (b) Indexed pattern: silicon (solid dots) and sapphire (open dots) reflections are shown. Outer circumference is the objective aperture, inner circumference (dashed) represents the effective resolution aperture.

THE STRUCTURE OF THE SILICON-SAPPHIRE INTERFACE

Two specimens belonging to different regions of the same wafer were examined in cross section. The observed defect density in the thick crystal regions was similar to that reported by other workers [7, 12], and both specimens exhibited the same general features. In the thinnest regions of the specimen (estimated to be between 50 and 100 Å thick, from the observation of the first extinction contour) it was possible to resolve the lattices and to obtain sharp images of the defect structure.

Within the resolution of the instrument, the interface is observed to be abrupt and planar, following an r-plane on the sapphire. Structure corresponding to interfacial compounds or phases of more than a monolayer is not seen.

A striking feature is the inhomogeneous defect distribution along the interface. Regions that are virtually defect free such as the one shown in figure 4 were common and extended in segments as large as 500 to 1500 Å along the interface. In these regions no planar nor linear defects are evident in the silicon structure. The interface appears incoherent [13]; there is no continuity of planes across the interface, and a direct correspondence is not evident in the regions of best interfacial resolution. This could be possible if the sapphire surface bonds were relatively free to bend and stretch in length, as well as being free to rotate about the surface normal, thus placing no severe spatial restrictions on the first layer of silicon atoms. The bonding strength of silicon may force the sapphire surface to reconstruct and produce the unitary strength oxygen bonds necessary to bond to the first layer of silicon atoms. Such oxygen bonds at the surface would have the characteristics described above. Further studies of the sapphire surface characteristics may give a better idea of whether this is the case.

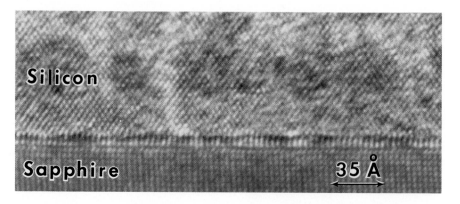

Fig. 4. Lattice structure image of the interface exhibiting a virtually defect free silicon layer.

Other regions exhibit microtwins which may be caused by slip due to thermal and other physically induced lateral stresses subsequent to growth. An illustration of a microtwin at the interface is shown in figure 5. In addition to the two coherent twin boundaries observed edge on, the microtwin is also bound depthwise. This termination in depth is called a "lateral" boundary; and is a second-order, incoherent, twin. In fact, the lower portion of the twin in the photograph extends through most of the film, whereas in the middle portion of the micrograph the image corresponds to a twin superimposed on an untwinned region. The corresponding image has a periodicity of 3/2 the spacing of the

($\bar{1}$11) planes. In the top section of the micrograph, the microtwin emerges out of the specimen and the image region is fully untwinned. Lateral twin boundaries entail a greater deviation from undistorted bonding than those in which twin plane and composition plane are coincident, and are expected to have a somewhat greater effect upon structure-sensitive electrical properties.[13].

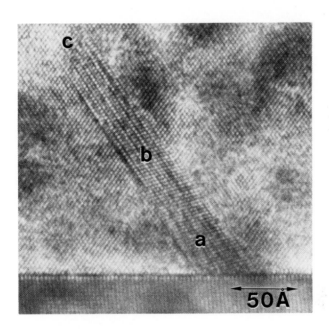

Fig. 5. Microtwin at the interface. In region (a) the microtwin extends through the film. In (b) the specimen is only partially twinned, whereas in (c) the twin is absent.

Regions with high density of defects have also been observed but with less frequency. An example is shown in figure 6, where a number of stacking faults, microtwins and incoherent grain boundaries are present. In particular, a grain with a [1$\bar{2}\bar{1}$] direction close to the perpendicular to the substrate is seen. These regions may result from the coalescence of islands during the growth process, and, due to their incoherence, they ought to have far more deleterious effects on the electrical properties of the interface than would regions showing either no defects or low-order twin boundaries.

In summary, the lattice images presented here indicate that the silicon-sapphire interface is abrupt and planar and that no interfacial phases are present. The defect structure along the interface is inhomogeneous; regions extending up to 1500 Å across the interface were observed to be virtually defect-free. Stacking faults are clearly visible and may correspond to glide on {111} planes in order to relieve lateral stress introduced due to thermal and other physical processes. Other regions exhibit high defect densities, with incoherent grain boundaries which are possibly due to coalescence of islands during growth.

290

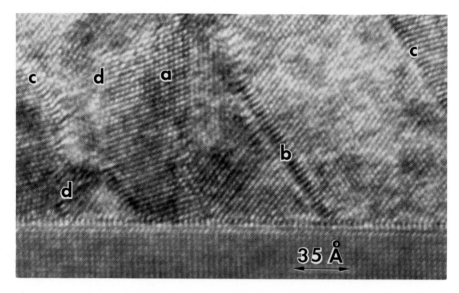

Fig. 6. Highly defected region at the interface. Main features include (a) grain oriented with [12̄1] perpendicular to the interface, (b) stacking fault, (c) twin boundaries and (d) incoherent grain boundaries.

ACKNOWLEDGMENTS

The author would like to thank Tom R. Cass for advice and help in the specimen preparation, Kent Carey, Julio Aranovich and Robert A. Burmeister for their encouragement and valuable discussions, and Prof. Robert Sinclair for the use of the microscope and other facilities at Stanford University.

REFERENCES

1. F. P. Heiman, Appl. Phys. Lett. 11, 132 (1967).
2. Ch. Kühl, H. Schlötterer, and F. Schwidefsky, J. Electrochem. Soc. 123, 97 (1976).
3. H. Schlötterer, J. Vac. Sci. Technol. 13, 29 (1976).
4. J. Mercier, J. Electrochem. Soc. 118, 962 (1971).
5. G. W. Cullen, J.F. Corboy and R.T. Smith, J. Crystal Growth 31, 274 (1975).
6. J. Trilhe, J. Borel and J. P. Duchemin, J. Crystal Growth 45, 439 (1978).
7. M. S. Abrahams and C. J. Buiocchi, Appl. Phys. Lett. 27, 325 (1975).
8. T. Hayashi and S. Kurosawa, J. Crystal Growth 45, 426 (1979).
9. W. E. Ham, M. S. Abrahams, C. J. Buiocchi and J. Blanc, J. Electrochem. Soc. 124, 634 (1977).
10. F. A. Ponce, T. Yamashita, R. Sinclair, Proc. 38th Annual Meeting EMSA (San Francisco, 1980), 320.
11. M. L. Kronberg, Acta Met. 5, 507 (1957).
12. R. Lihl, H. Oppolzer, R. Pongratz, P. Skalicky and W. Sranda, J. Microscopy 118, 89 (1980).
13. G. B. Olson and M. Cohen, Acta Met. 27, 1907 (1979).
14. J. A. Kohn, The American Mineralogist 48, 263 (1958).

DEFECT STRUCTURE OF THE EPITAXIAL Pd$_2$Si-SILICON INTERFACE

D. Cherns[a] and D.A. Smith
IBM T.J. Watson Research Center, Yorktown Heights, N.Y. 10598

ABSTRACT

The crystallography of the epitaxial Pd$_2$Si-silicon interface has been analyzed with special reference to the character of interfacial dislocations. It is proposed that the Pd$_2$Si which forms at the interface does so by the diffusion stimulated glide of transformation dislocations.

INTRODUCTION

Silicon-metal reactions are of great practical importance since they affect the electrical properties of contacts to solid state devices. In addition these reactions have fundamentally interesting kinetic and crystallographic features. In this paper the crystallography and structure of the interface between hexagonal Pd$_2$Si and {111} oriented diamond cubic silicon are analyzed in detail. Previous work has established that the interface is epitaxial with (111)Si $//$ (0001)Pd$_2$Si[1] and [11$\bar{2}$]$_{Si}$//[2$\bar{1}$10] Pd$_2$Si. Transmission electron microscopy has shown that the interface may be stepped locally [2]. The Pd$_2$Si is not a single crystal but is composed of domains [3], some but not all of which are thought to be crystallographically equivalent variants of the Pd$_2$Si structure.

CRYSTAL STRUCTURE

The diamond cubic structure of silicon is well known but it is useful to show the atom arrangement projected on (111), fig. 1a, for comparison with the Pd$_2$Si structure which is shown in figs. 1b and 1c. The stacking of the {111} planes in the diamond cubic structure may be described as ...AaBbCc..... The stacking of the silicide may be described as ....Aα....; the sites designated A,a and α all lie along [111] and are occupied by silicon atoms except that in the A and α layers of the silicide 1/3 and 2/3 respectively of the silicon sites are vacant, figs. 1b and 1c. In the A layer of the silicide each vacant Si site is surrounded by an equilateral triangular cluster of Pd atoms at coordinates of the type 0.26,0,0. In the α layer the Pd atoms again form equilateral triangular clusters around the Si vacancies and have coordinates of the type 0.61, 0, 1/2.[4] The lattice parameters are: silicon a = 5.43Å and for Pd$_2$Si a=6.53Å and c=3.44Å. The misfit parallel to the [11$\bar{2}$]and[111] directions of the silicon is 0.018 and -0.089 respectively, using silicon as the reference lattice.

INTERFACE STRUCTURE AND TRANSFORMATION OF Si TO Pd$_2$Si

Formally transformation of Si to Pd$_2$Si requires the in-diffusion of 6 Pd atoms and the out diffusion of 3 Si atoms respectively per unit cell of the silicide and the transformation of the cubic ....AaBbCc.... stacking to hexagonal ...Aα... stacking. A possible transformation scheme is shown in fig. 2. Transformation dislocations with Burgers vectors of the form $\frac{1}{6}<11\bar{2}> + \frac{\delta}{6}[111]$ i.e. $\frac{1}{6}<1 + \delta, 1 + \delta, \bar{2} + \delta>$, where $\frac{\delta\sqrt{3}}{6}$ is the change in spacing of the (111) planes indicated in fig. 2, are postulated to glide in response to the chemical stress resulting from the fluxes of Pd and Si.

[a] Permanent Address: Oxford University, Dept. of Metallurgy and Science of Materials, Parks Rd., Oxford, England.

The passage of such dislocations between Si planes of the kind Ab, Bc and Ca, figs. 2a-c, changes the stacking sequence to ...Aα.... The interface is stepped at the core of each transformation dislocation and the interface advances an amount c for the passage of each dislocation. There are close parallels with the formation of twins,[5] the fcc-hcp transformation in Co,[6] the growth of hexagonal Ag_2Al precipitates in supersaturated Al-Ag[7] and recent theories of grain boundary migration.[8]

The crystallographically equivalent variants of the silicide can now be discussed. It is not known if factors such as coordination determine whether the first layer of the silicide is A or α. In principle domains stacked Aα Aα may be formed with the c-axes parallel to domains stacked αAαA; the domains are separated by a translational anti-phase domain boundary characterized by a translation c/2. In addition the silicide may nucleate in a twin orientation relative to either of the above variants i.e. rotational domain boundaries characterized by a rotation of 180° about c are formed.

We also note that the origin of the Pd_2Si unit cell can be translated by a/2 <110> referred to the underlying Si. This corresponds to removing different silicon atoms to create the first layer of silicide so that there are three equivalent positions for the origin of the A or α layers of the silicide designated A_1,A_2,A_3 and $α_1$, $α_2$, $α_3$ respectively. $A_iα_j$ stacking i ≠ j does not create the Pd_2Si structure. In total twelve variants of the silicide structure may be formed on a planar (111)-Si surface which implies that 66 different domain boundaries may form in principle from an originally planar Si substrate.

The variants of the Pd_2Si fall into two classes (a) related by rotation (b) related by translation. The former have no long range elastic fields whilst the latter require appropriate dislocations at the Si-Pd_2Si interface. In addition since $c_{Pd_2Si} ≠ a_{Si}/\sqrt{3}$ steps of height $nd_{111_{Si}}$, n = integer, have an elastic field normal to the interface plane. Finally we point out that the edge transformation dislocation is also a misfit dislocation when the extra planes are in the silicide.

EXPERIMENTAL RESULTS

150-400Å films of Pd on (111)Si were prepared by vapor depositing Pd at a rate of 3-4Å sec^{-1} on to cleaned high-purity (111)Si wafers at room temperature in a UHV evaporator operating at background pressures of 10^{-7} torr or better. The silicon wafers were cleaned in sequence in detergent, distilled water and alcohol, dipped for 20-30 secs in buffered HF and finally rinsed in distilled water prior to pumpdown. Films of Pd-(111)Si in the as-prepared state or after various annealing treatments were examined by TEM following backthinning to the Pd surface in HF/HNO_3 using standard techniques.

As-prepared Pd-(111)Si films showed the presence of some epitaxial Pd_2Si grains and polycrystalline Pd. Films annealed at 250°C in a He atmosphere for 2 hrs showed a predominantly epitaxial Pd_2Si layer and no unreacted Pd. Fig. 3 illustrates a 250Å Pd-(111)Si film annealed in this manner, in an area where both Pd_2Si and Si layers are present. Fig. 3(a) is a bright-field micrograph taken under approximately two-beam conditions with g_{Si} = 220 and g_{Pd_2Si}= 600 near the exact Bragg condition. Moiré fringes are visible due to interference between the undeviated beam and a beam doubly diffracted in the silicide and silicon layers. The moiré spacing is given by

$$d = \frac{1}{g_{Si} - g_{Pd_2Si}};$$

using an experimental value of d ~ 110Å where moiré fringes run perpendicular to g in fig. 3(a), we find an approximate misfit of 0.017 between deposit and substrate compared with the natural mismatch of 0.018. Fig. 3(b) is a dark-field micrograph of the area in fig. 3(a) taken in the g-vectors shown and under weak-beam diffracting conditions. In regions where moiré fringes are visible in fig. 3(a), sharp lines of high contrast typical of weak-beam dislocation images appear in fig. 3(b). Close inspection reveals two sets of dislocation lines which cross each other, each crossing point being close to a region of enhanced image contrast. These observations can be explained if we suppose that there are three sets of interface dislocations with Burgers vectors 1/6 <112> with respect to the silicon lattice; on our model these dislocations lie at different (111) levels such that minimal interaction between different dislocations is expected. To account for the observed contrast we note that one set of dislocations has g.b = 0 but |g.b| = 1 for the other two sets which are visible. Where dislocations of the |g.b| = 1 type cross, displacements on the same side of both dislocations are additive in the direction of $\mathbf{g}$ (regions 1)) and displacements in regions adjoining opposite sides of the dislocations (regions 2) tend to cancel in the direction of $\mathbf{g}$. Remembering that images will lie on that side of both dislocations where lattice plane rotations are towards the Bragg position, enhanced image contrast is expected on one side of the dislocations in region 1) and diminished image contrast is expected in regions 2). This is in good agreement with the contrast in the area arrowed. Making the assumption that the dislocations visible in fig. 3(b) are near edge dislocations and act to relieve misfit strains a careful inspection shows that the misfit achieved is consistent with that calculated above.

We also note in fig. 3(a) that whilst some regions such as A exhibit moiré or dislocation contrast this is not always the case. A more extended analysis suggested that most of these other regions contained Pd_2Si grains which were substantially misoriented from the silicon substrate. In this case strong diffraction in the Pd_2Si layers would be unlikely thus reducing the intensity of the double diffracted beam required for the moire contrast (fig. 3(a)). The interpretation of the weak-beam image is less straightforward. We note that the image will only be sensitive to displacements in the silicon and that high angle boundaries are almost certainly involved. However, whether the absence of significant weak-beam contrast signifies too small displacements due to the boundary or whether the Pd_2Si and Si layers are not in intimate contact remains to be resolved.

Fig. 4 shows a weak-beam image of an area adjacent to that in fig. 3. The arrowed feature is the boundary between two regions which were found to have interfacial dislocation arrays. It is clear therefore that the boundary is one between two crystallographically equivalent variants of the Pd_2Si structure as previously discussed. It is tempting to interpret the domain boundary contrast solely in terms of a dislocation at the Pd_2Si-Si interface. However, examination of regions of crystal containing only Pd_2Si (no evidence of silicon or Pd_2Si-Si interface dislocations) showed similar weak-beam contrast from domain boundaries. The effect of boundary structure on such images thus requires further investigation.

DISCUSSION

It has been shown by analyzing the crystallography of the Pd_2Si-Si(111) interface that interfacial steps with height $a_{Si}/\sqrt{3}$ in the [111] direction are associated with dislocations. These dislocations may have a dual role acting on the one hand, as transformation dislocations which convert the cubic silicon structure to a skeleton hexagonal Pd_2Si structure, and on the other as misfit dislocations. Experimental observations confirm the presence of misfit dislocations of the proposed type. We note that steps of height $a_{Si}\sqrt{3}$ may also exist in the interface, the presence of which is associated with a dislocation with a Burgers vector wholly out of the interface plane since $a_{Si}\sqrt{3} \neq 3c_{Pd_2Si}$ the presence of this step would be difficult to detect by the experimental technique here.

If we assume that silicide growth occurs at the Pd_2Si-Si interface, and that growth occurs by movement of transformation dislocations, the growth rate of the silicide v_i is given by

$$v_i = \rho v c \qquad (2)$$

where ρ and v are the density and velocity respectively of the transformation dislocations and c is the step height. The only unknown quantity in equation (2) is the dislocation velocity. We note that the measured thickness of Pd_2Si films on (111) silicon is proportional to the square root of time suggesting that dislocation motion is limited by diffusion (i.e. the arrival of Pd and the removal of Si) rather than an energy barrier for dislocation glide. As stated earlier silicide growth by our mechanism requires in-diffusion of Pd and out-diffusion of silicon in the ratio 2:1.

Experiments to observe silicide growth by observing dislocation movement have been attempted. As-prepared Pd-(111)Si films have been observed under both weak-and strong-beam imaging conditions on a video type display during in-situ heating experiments in a Philips 301 electron microscope. Although some changes in interface structure have been observed, conclusive evidence for the proposed growth mechanism has not been obtained to date.

294

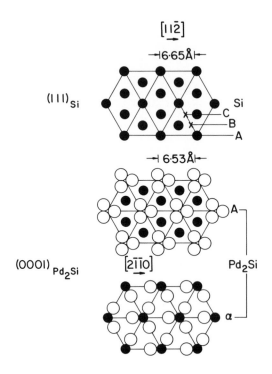

Fig. 1. (a) The diamond cubic structure of Si in a projection on (111) (b) the A-layer of hexagonal Pd₂Si and (c) the α-layer of Pd₂Si.

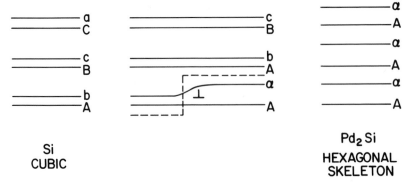

Fig. 2. A possible mechanism for transformation of cubic silicon to a hexagonal silicon skeleton by glide of transformation dislocations; after diffusion of Pd, Pd₂Si is produced.

295

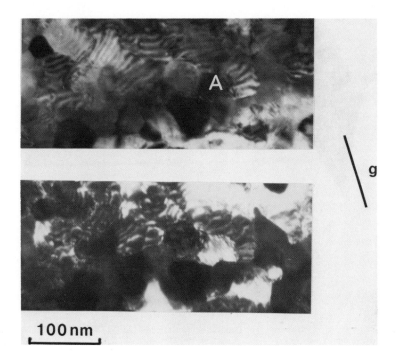

Fig. 3 a & b.   Electron micrographs showning predominantly epitaxial Pd$_2$Si on silicon substrate.

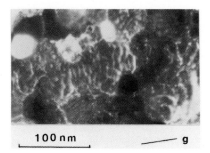

Fig. 4.   A weak beam image of an area of interface adjacent to that shown in fig. 3.

296

ACKNOWLEDGMENTS

We are grateful to Drs. T.Y. Tan and K.N. Tu for useful discussions and to K. Asai for technical assistance. We both wish to thank Dr. K.N. Tu for his work in making possible our collaboration. David Cherns was supported through the IBM Visiting Scientist scheme.

REFERENCES

1. T.S. Kuan and K.N. Tu, Ninth Internation Congress on Electron Microscopy, Toronto, *1*, 304 1978.

2. H. Föll, P.S. Ho and K.N. Tu, "Cross-sectional TEM of Silicon-Silicide Interfaces", IBM report RC8285.

3. T.S. Kuan and J.L. Freeouf, " The Formation of Pd Silicide on Reconstructed <111> Si Surface", (1979).

4. A. Nylund, "Some Notes on the Palladium-Silicon System", Acta Chem. Scand. *20*, 2381, (1966).

5. A.H. Cottrell and B.A. Bilby, "Mechanism for Growth of Deformation Twins in Crystal", Phil. Mag. *42*, 573 (1951).

6. J.W. Christian, "The Theory of Transformation in Metals and Alloys", Pergamon, London, p. 809, (1965).

7. C. Laird and H.I. Aaronson, "The Dislocation Structures of the Broad Faces of Widmanstatten γ Plates in an Al- 15.% Ag Alloy", Acta Met, *15*, 73, (1967).

8. C.M.F. Rae and D.A. Smith, "On the Mechanism of Grain Boundary Migration", Phil. Mag. A, *41*, 477.

Published 1981 by North-Holland, Inc.
Narayan, and Tan, eds.
Defects in Semiconductors

CHARACTERIZATION OF DEFECTS IN SILICON RIBBONS BY COMBINED
EBIC AND HVEM

HORST STRUNK,[†] BRIAN CUNNINGHAM AND DIETER AST
Materials Science and Engineering, Cornell University, Bard Hall, Ithaca,
New York 14853

ABSTRACT

   The electrical properties and crystallographic nature of
linear and planar defects in EFG silicon ribbons were studied.
A direct correlation between electrical and structural properties was obtained by imaging the same areas first with EBIC
and then with HVEM. Coherent twin boundaries were found to be
electrically inactive, but higher order twins and other grain
boundaries generally enhanced minority carrier recombination.
Partial dislocations confined to coherent twin boundaries
were usually electrically active, but in certain instances
partial dislocations were observed which had no apparent
EBIC contrast.

INTRODUCTION

   Rapid progress has recently been made in the production process of edge defined film-fed grown (EFG) silicon ribbon, a description of which can be found
in ref. [1]. EFG ribbon is a promising material for the production of inex-
pensive solar cells and therefore an understanding of the relationship between
the crystallographic nature of the defects and their electrical activity is
desirable. This information can then hopefully be used to modify the ribbon
growth process to reduce the density of electrically active defects which
detract from solar cell efficiency.
   EFG ribbons typically contain a large number of linear boundaries, the
majority of which is parallel to the growth direction. The electrical activity
observed at these boundaries can vary greatly. Linear boundaries can be electrically active along their entire length, or show no detectable activity. A
large number of boundaries show enhanced carrier recombination only along certain sections, resulting in a "dotted" appearance in EBIC. The differing
electrical activity of these linear boundaries has previously been ascribed to
the dislocation content of the boundaries [2] or to the amount of impurity
decoration at boundary dislocations [3]. In the present study two cases of
dotted EBIC contrast are considered. Although the EBIC appearance of both is
similar it will be shown that the defects producing the effect are markedly
different.

EXPERIMENTAL

   The electrical properties of EFG silicon ribbons were investigated in the
scanning electron microscope (SEM) operated in the electron-beam induced current (EBIC) mode. Schottky diodes were produced by evaporating a thin film
(∿500 Å) of Al onto the surface. A high voltage electron microscope (HVEM)

[†]On leave from Max-Planck-Institut für Metallforschung, Institut für Physik,
7000 Stuttgart 80, West Germany.

operating at an accelerating voltage of 1 MV was used to study the crystallographic nature of the defects in the ribbons.

To correlate the electrical and structural properties of defects, areas were selected and mapped out in EBIC. Specimens 3 mm in diameter were then cut out from the ribbon and ion-milled from the back side until the EBIC mapped areas were contained in electron transparent regions. These specimens were subsequently examined by HVEM.

In the present study HVEM has the following advantages over conventional TEM.

(i)   Specimens with thicknesses of up to several μm can be examined, providing large electron transparent regions.

(ii) Extended defects such as twin or other grain boundaries can easily be analyzed.

(iii) The volumes investigated by HVEM and EBIC are comparable.

RESULTS

Case I

Figure 1(c) is part of an EBIC micrograph showing dotted contrast. The same area imaged by HVEM, Figure 1(a), reveals that the EBIC active areas, or "dots," mark the trace of a microtwin with a thickness of $\sim$200 nm. The twin boundaries in Figure 1(a) are visible through fringe contrast. Figure 1(b) shows the same microtwin with the twin boundaries out of contrast. Partial dislocations with different crystallographic character are contained in both twin boundaries, and a comparison with Figure 1(c) shows a one to one correspondence between the position of dots and the position of dislocations. Except for the central spot, the EBIC dots have similar contrast and correspond to individual dislocations. This observation suggests that the dislocations have comparable electrical activities despite having different crystallographic character. HVEM shows that the central dot corresponds to a group of dislocations which are too closely spaced to be individually resolved by EBIC ($\sim$1-2 μm). The EBIC signal in this case is therefore the sum of the signals from the individual dislocations.

Case II

The second example of dotted EBIC contrast is shown in Figure 2(a). An optical micrograph of the etched surface, Figure 2(b), shows that the EBIC contrast results from the interaction of linear boundaries. HVEM micrographs of part of the dotted row are shown in Figure 3. Figure 4 is a sketch showing the geometrical relationship between the different areas. Analysis shows that T_1 is a microtwin $\sim$100 nm thick. T_2 is also twinned with respect to the matrix M but with a different {111} twinning plane. The heavy lines in Figure 4 therefore represent traces of boundaries separating grains with a misorientation resulting from two non-parallel twinning operations. Boundaries of this type have been termed "second order twin joins" by Kohn [4]. In the present case a {111} plane of T_1 matches a {115} plane of T_2. To the authors knowledge this {111}//{115} second order twin has not previously been observed: a coincidence lattice model for this boundary was proposed by Kohn [5]. The dislocation model proposed by Hornstra [6] for twin boundaries resulting from tilts about <110> can also be extended to describe the present case.

Twin boundary dislocations were revealed by tilting the specimen, Figure 3(b). Using a reflection common to grains T_2 and M, the boundaries T_2/M are rendered invisible, whereas the boundaries of the microtwin T_1 are visible in fringe contrast. The dislocations in the T_2/M twin boundary were determined to be Schockley partials lying along a <110> direction and with either 30° or 90° character. The character of the dislocations in the {111}//{115} twin boundaries has not been completely determined so far since these boundaries are steeply inclined in the foil.

Correlation between EBIC and HVEM shows that the areas of high recombination in Figure 2(a) correspond to the {111}//{115} twin boundary joins. The most interesting features however are the partial dislocations in the T_2/M boundaries which are apparently nonactive.

DISCUSSION

The first point to note is that similar features observed in EBIC can result from different crystallographic structures. Care must therefore be exercised when interpreting EBIC micrographs. Direct correlations between EBIC and TEM are required until statistically significant results can be obtained.

Partial dislocations have been observed with apparently two different levels of electrical activity: dislocations giving rise to a dotted EBIC contrast, Figure 1(c), and dislocations with no, or at least considerably lower electrical activity, Figure 3(b). Although a detailed discussion relating the activity of dislocations to their core structures requires a large number of observations at crystallographically identical defects, it is interesting to speculate about the possible significance of the present results. Dislocations introduced by plastic processes are regarded as glide dislocations and are therefore assigned to the "glide set" [7]. In addition the "shuffle set" dislocations [8] must also be considered. The core structure of a dislocation will depend on whether it belongs to the glide or shuffle and on the way the dangling bonds reconstruct [9]. It is conceivable that the core structure of a dislocation will depend on how the dislocation is generated. The present observations, that the electrically active dislocations do not follow <110> directions, and are in fact sometimes curved, and that the non-active dislocations tend to be aligned along a <110> Peierls valley tends to support this view. EFG ribbons contain both ingrown dislocations formed during solidification and dislocations generated by thermal stresses during cooling. It is possible, although not proven, that these two types of dislocations correspond to the non-active and active dislocations observed by EBIC.

Since the present experiments investigate single partial dislocations, measurements are not complicated by dissociation and constriction which occur at single perfect dislocations. If dissociated the two partials cannot be resolved by EBIC, thereby impeding the correlation with HVEM.

The effects of impurity atoms has not been considered explicitly since the present results can be discussed in geometrical terms alone. Both carbon and oxygen are present in varying quantities in EFG ribbons, however decoration of defects by precipitates was not observed in the EFG ribbons examined. Since the carbon content in ribbons is high but cannot be found as precipitates it is possible that the carbon is incorporated into the higher order twins and general grain boundaries. This model is consistent with the observation that an increase in carbon content leads to a decrease in grain size and an increase in the number of electrically active boundaries. A detailed investigation of the role of carbon would require doping of the ribbon during growth with C^{13} to allow the use of detection methods such as SIMS without interference from the C^{12} background.

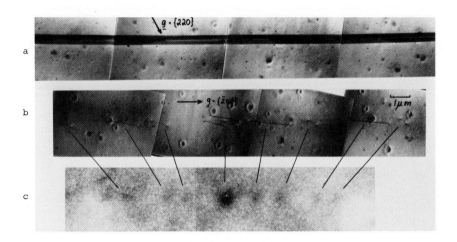

Fig. 1. a) Bright field HVEM image of a microtwin containing dislocations, b) Same area with twin out of contrast, c) EBIC image of the same area.

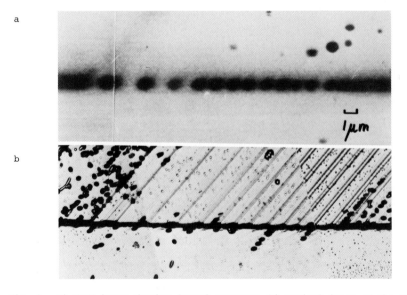

Fig. 2. a) EBIC image showing dotted contrast, b) Optical image of the same area showing interaction of microtwins.

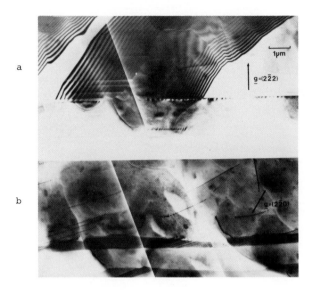

Fig. 3. a) Bright field HVEM image of twin boundaries shown in Fig. 2, b) same area with one set of twin boundaries out of contrast.

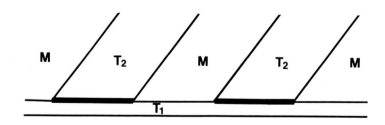

Fig. 4. Schematic sketch of the twin boundaries in Fig. 3 showing the geometrical relationship between the different areas.

302

ACKNOWLEDGMENTS

This work was funded by DOE under contract No. DE-AC02-76ER02899 and by JPL under contract No. 954852. The authors would like to thank Argonne National Laboratories for the use of their HVEM facilities.

REFERENCES

1. J. P. Kalejs, B. H. Mackintosh and T. Surek, J. Cryst. Growth, $\underline{50}$, 175 (1980).

2. K. Yang, G. H. Schwuttke and T. F. Ciszek, J. Cryst. Growth, $\underline{50}$, 301 (1980).

3. J. I. Hanoka, Photovoltaic Material and Device Measurement Workshop, "Focus on Polycrystalline Thin Film Cells," Arlington, Virginia (1979).

4. J. A. Kohn, Am. Mineral., $\underline{41}$, 778 (1956).

5. J. A. Kohn, Am. Mineral., $\underline{43}$, 263 (1958).

6. J. Hornstra, Physica, $\underline{25}$, 409 (1959).

7. H. Alexander, J. de Physique C6, $\underline{40}$, 1 (1979).

8. A. Bourret and J. Dessaux, J. Physique Colloq. C6, $\underline{40}$, 7 (1979).

9. P. B. Hirsch, J. Microscopy, $\underline{118}$, 3 (1980).

Published 1981 by North-Holland, Inc.
Narayan, and Tan, eds.
Defects in Semiconductors

EBIC INVESTIGATION OF HYDROGEN PASSIVATED STRUCTURAL DEFECTS
IN EFG SILICON RIBBON

T. D. SULLIVAN AND D. G. AST
Materials Science and Engineering, Cornell University, Bard Hall, Ithaca,
New York 14853

ABSTRACT

   EBIC contrast of structural defects in as-received and
hydrogen passivated polysilicon ribbon is studied using aluminum Schottky barrier diodes. Enhanced charge collection
after passivation is demonstrated by comparing the EBIC
signal from two sections of a split ribbon, one half of
which was passivated. Electrical activity of specific linear
defects before and after passivation is examined by recording
charge collection profiles across these defects under stan-
dardized conditions of beam voltage and current. Such measurements show that passivation does reduce the electrical
activity of selected defects. Possible reasons for this
behavior will be discussed.

INTRODUCTION

   Hydrogen passivation of electrically active defects in silicon has been
studied by a variety of techniques [1-4] including electron-beam-induced-
current (EBIC) microscopy [5,6]. The present authors [5] observed that
hydrogenation passivated some but not all defects. Seager et al. [6] report
that practically all grain boundaries were passivated. Whether this difference
is due to different hydrogenation procedures or a result of the specimens used
(EFG, SOC) is not known.
   In ref. 5 hydrogenation reduced the electrical activity of certain defect
structures relative to surrounding features. These EBIC images had low magnification (X100) and several relevant parameters (incident beam current, signal
amplification, DC level, etc.) which influence the image were not quantified.
Hence the question remained whether the reduced contrast was due to hydrogenation or merely reflected the difficulty to detect the defect against a brighter
background.
   The present work was undertaken to
   (i)   develop quantitative techniques for assessment of the electrical
         activity of defects observed in EBIC,
   (ii) compare high-resolution-EBIC micrographs with charge collection pro-
         files before and after the hydrogenation of selected defects,
   (iii) image split specimen in order to compare contiguous defects before
         and after hydrogenation and to evaluate matrix activity.

EXPERIMENTAL PROCEDURES AND RESULTS

   300 Å thick aluminum (Al) Schottky barrier diodes were evaporated onto as-
grown undoped (205 Ωcm) EFG ribbon which had been mechanically ground flat,
syton polished and HF etched. Suitable areas were selected for EBIC studies
and the specimen was divided into three pieces (Fig. 1): piece (1) remained
untreated except for Schottky barrier formation and removal, piece (2) was
vacuum annealed, and piece (3) was hydrogenated in a plasma. Diodes 1a and 2a

were positioned such that adjacent areas of pieces (1) and (2) could be imaged simultaneously; similarly diodes 2b and 3b covered both pieces (2) and (3). An additional diode 3a was placed on piece (3) to study specific defect structures. Diodes were always removed prior to vacuum or plasma annealing and redeposited afterwards.

Both pieces (2) and (3) were first vacuum annealed at 200 °C for 8 hours at 8×10^{-5} torr. An EBIC comparison of pieces (1) and (2) (Fig. 2) clearly shows enhanced charge collection in the annealed specimen. The charge collection profile indicates that the collected current is about 75% higher in the annealed specimen. No significant difference was observed in the EBIC image of piece (2) and (3) (not shown). The observed increase in diode efficiency most probably reflects an increase in carrier lifetime, possibly due to the formation of oxygen donor complexes [7].

Piece (3) was then annealed at 200 °C in hydrogen plasma (2×10^{-3} torr for 8 hr. then 0.7 torr for 15 min). This treatment increased the charge collection by about 30% relative to piece (2) (Fig. 3). The effect is clearly seen in the charge collection profile, but is less obvious in the micrograph. Note that piece (3) was held at 200 °C for a total of 16 hr as compared to 8 hr for piece (2). Whether the observed increase is due to hydrogenation alone is therefore not known, since the length of heat treatment also affects carrier lifetime.

Next, piece (3) was annealed at 350 °C for 16 hr. in a 1.5 torr hydrogen plasma. This treatment produced an additional 60% increase in collection, relative to the 200 °C plasma anneal (Fig. 4). As above, thermal effects and hydrogenation cannot be separated.

In an attempt to drive off the hydrogen, piece (3) was annealed at 500 °C in an 8×10^{-5} torr vacuum for 12 hr. Charge collection increased by an additional 70%, to a total of 160% relative to piece (2) after the 200 °C vacuum anneal (Fig. 5). The large increase in collection efficiency could be in part due to the increase in carrier lifetime which reaches a maximum near 500 °C [7]. In addition, Fig. 5 shows some EBIC contrast outside the diode area conceivably due to surface conversion of the lightly ρ-doped matrix by thermal donors.

One particular defect structure in the ribbon is shown at higher magnification (1000X) in Figs. 6, which compare corresponding sections in pieces (2) and (3). The parallel lines are probably composed of closely spaced twin boundaries the majority of which are coherent, and contain partial dislocations. The EBIC contrast of these defects varies considerably with annealing and the influence of hydrogenation will now be examined in more detail.

We first compare the defect contrast in piece (2) after the 200 °C vacuum anneal and piece (3) after the 350 °C hydrogenation (Fig. 6a and b). The profile depth of the darker right-hand boundary is similar in both pieces, even though the matrix level in piece (3) is higher. The contrast of the dislocation-like defects is less than that of the boundaries in piece (2), but is comparable to that of the boundaries in piece (3) after hydrogenation. Hence hydrogenation appears to reduce the electrical activity of the boundaries, but not of the dislocation-like defects (although it does tend to localize them more). The 500 °C vacuum anneal of piece (3) (Fig. 6c) greatly increased the contrast of both boundaries, but again did not significantly alter that of the other defects. Evolution of hydrogen from the boundaries seems the most plausible explanation.

Finally, profile minima corresponding to the dislocation-like defects in all three figures 6a-6c are of similar depth while the DC response (the absolute value of the current collected) varies nearly 300%. A possible interpretation is that these defects have a maximum recombination rate which is saturated at relatively low levels of DC response. Hence as the DC response increases, the relative depth of the minima decreases, rendering it progressively less detectable. Interestingly, no defects initially observed disappeared in toto after any later treatment; indeed defects not initially observed later became visible.

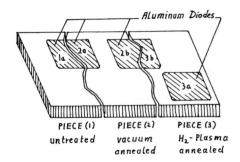

Fig. 1. EFG ribbon, polished and split, with evaporated aluminum diodes..

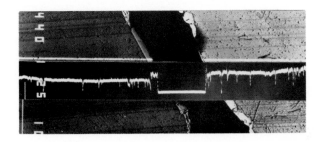

Fig. 2. EBIC image with charge collection profile; piece (1) left, no treatment; piece (2) right, 200°C vacuum anneal.

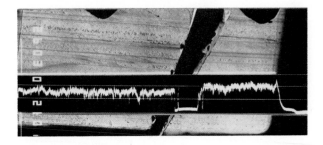

Fig. 3. EBIC image with charge collection profile; piece (2) left,
piece (3) right, 200°C hydrogenation.

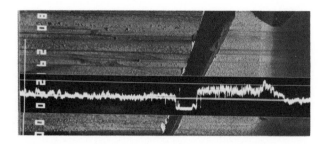

Fig. 4.   EBIC image with charge collection profile; piece (2) left, 200°C anneal in vacuum; piece (3) right, 350°C hydrogenation.

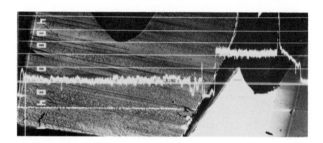

Fig. 5.   EBIC image with charge collection profile; piece (2) left, 200°C vacuum anneal; piece (3) right, 500°C vacuum anneal.

(a)

(b)

(c)

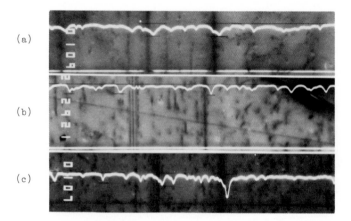

Fig. 6.   EBIC images with charge collection profiles. (a) piece (2), 200°C vacuum anneal; (b) piece (3), 350°C hydrogenation; (c) piece (3), 500°C anneal in vacuum.

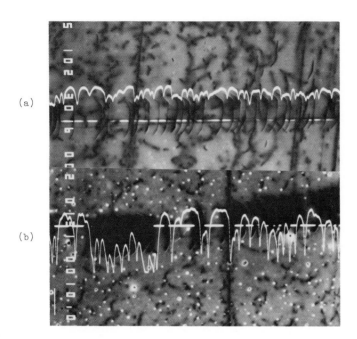

Fig. 7. EBIC image with charge collection profile of scratched region. (a) after 350°C hydrogenation; (b) after 500°C vacuum anneal. Note relative contrast of deformation induced and grown in dislocations.

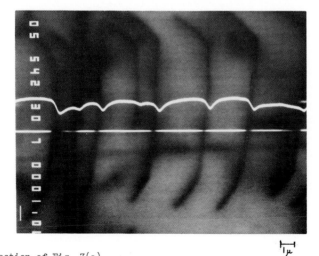

Fig. 8. Section of Fig. 7(a).

On diode 3a (Fig. 1), a scratch was observed to alter appearance after the 350 °C hydrogenation. Examination revealed a family of structures oriented perpendicular to the scratch which appeared to be deformation induced dislocations (Fig. 7a). Prior to the last hydrogenation, these structures, while photographed, were not examined because of their unresolved appearance. High magnification micrographs (5400X) (Fig. 8) at 5 KV exhibit a nominal width of 0.2 µ and a collection profile depth similar to that of other dislocation-like defects. The width and shape of corresponding change collection profiles are consistent with those theoretically predicted for dislocations [8]. After the 500 °C vacuum anneal, associated profile minima increased their depth by a factor of 4, appearing once again rather diffuse (Fig. 7b), presumably due to removal of hydrogen.

The apparent increase in resolution upon hydrogenation is an artifact of the narrow range of contrast which can be displayed in a micrograph. We could image the unpassivated dislocations, at the expense of back ground details, by deviating from our standard imaging conditions.

SUMMARY

A quantitative study of the EBIC contrast of linear defects in annealed, unannealed, and hydrogenated EFG silicon ribbons was carried out using split speci- mens. The major results are :
(i)   heat treatment, per se, increases collection efficiency.
(ii)   plasma hydrogenation reduces the contrast of some defects. However, none of the observed defects showed a total loss of EBIC contrast.
(iii) Hydrogenation is useful to resolve tightly spaced arrangements of strongly active dislocations without loss of background details.

REFERENCES

1. J. I. Pankove, M. A. Lampert and M. L. Tarng, "Hydrogenation and dehydrogenation of amorphous and crystalline silicon," Appl. Phys. Lett. 32 (7), 1 Apr. 1978, p. 439.

2. C. M. Seager and D. S. Ginley, "Passivation of grain boundaries in polycrystalline silicon," Appl. Phys. Lett. 34 (5), 1 March 1979, p 337.

3. T. Makino and H. Nakamura, " Influence of plasma annealing on electrical properties of polycrystalline Si," Appl. Phys. Lett. 35 (7), 1 Oct. 1979 p. 551.

4. L. C. Kimmerling, "Hydrogen passivation of point defects in Si," Appl. Phys Lett. 36 (8), 15 Apr. 1980, p. 670.

5. D. Ast and T. Sullivan, "Preparation of Schottky Diodes for EBIC Investigation of Grain Boundary Passivation in Si Ribbons," DOE/JPL, Report #3, Contract No. 954852, Oct. 1979

6. C. M. Seager, D. S. Guiley and T. D. Zook, "Improvement of polycrystalline silicon solar cells with grain boundary hydrogenation techniques," Appl. Phys. Lett. 36 (1o), 15 May 198), p. 831.

7. H. J. Hovel, Solar Cells, Semiconductors and Semimetals, Vol. II, Academic Press, N.Y. 1975, pp. 10-12.

8. C. Donolato, "Contrast and resolution of SEM charge collection images of dislocations," Appl. Phys. Lett. 34 (1), 1 Jan. 1979, p. 80.

SWIRL DEFECTS IN AS-GROWN SILICON CRYSTALS

A.J.R. DE KOCK
Philips Research Laboratories, Eindhoven, The Netherlands

ABSTRACT

During melt-growth of macroscopically dislocation-
free bulk silicon crystals (floating-zone and
Czochralski technique) microdefects can form due to
the condensation of thermal point defects (self-interstitials, vacancies). The formation of these imperfections,
generally referred to as "swirl defects", is strongly
affected by the growth conditions (e.g. the crystal
pulling rate) and crystal purity. The various reported
defect formation models will be discussed. Special
attention will be paid to the effect of doping on
swirl defect formation.

INTRODUCTION

During growth from the melt of macroscopically dislocation-free silicon
crystals microdefects can form, which have a detrimental influence on device
performance. Due to their typical spiral-type distribution in cross-sectional
crystal slices, these microdefects are generally referred to as "swirl defects".
They have been extensively studied in floating-zone (FZ) and Czochralski-grown
(CZ) silicon crystals. The results of these investigations have been reviewed
in a number of recent articles (1-4).
Several models for the formation of swirl defects have been developed
(2,4-11). In the majority of these models the defect formation is attributed
to the condensation of thermal point defects (self-interstitials and vacancies).
In the present paper the most significant properties of swirl defects are
summarized with some emphasis on recently obtained data related to doping.
The various nucleation models are then briefly described. It is shown that none
of the existing models allows a complete interpretation of all experimental
data. In principle a more consistent model would be obtained by combining
several different elements taken from existing models.

DETECTION OF SWIRL DEFECTS

Swirl defects in silicon crystals have been studied with the aid of a large
number of simple as well as sophisticated diagnostic techniques such as preferential etching (12-17), X-ray transmission topography or infrared transmission microscopy combined with copper or lithium decoration (9,13,14,15,17,
18), double-crystal topgrahic techniques (19,20), section topography (21),
electron beam induced current mode scanning electron microscopy (22) and high
voltage electron microscopy (HVEM) (5,7,23-26). While all these methods give
information about the concentration and distribution of swirl defects, only
HVEM enables the structural characterization of these microdefects.
The presence of swirl defects can also be established more indirectly by
means of device processing steps. They can give rise to the formation of extrinsic stacking faults during oxidation (13,14,27,28) and sometimes during
epitaxial layer growth (29,30). Because swirl defects often adversely affect
device performance, they can be located via e.g. the analysis of leakage current
patterns in diode arrays (13,31,32).

SOME GROWTH-RELATED PROPERTIES OF SWIRL DEFECTS

Numerous investigations have shown that the formation of swirl defects is closely related to the crystal growth conditions (9,14,17,32,33). The most significant properties are summarized below.

Types of swirl defects

In CZ silicon crystals three types of microdefects have been observed, labelled A,B and C (17,26). In FZ material the defect types A and B are also formed (6), but C swirl defects have been detected only occasionally (34). The A swirl defects consist of perfect dislocation loops or loop clusters of interstitial type. The single loops are predominantly located in {111} or {110} planes with Burgers vectors $\frac{1}{2}a < 110 >$ (5,17,25). Their size depends on the crystal pulling rate (21) and crystal purity (5) and varies from about 0.5 μm to about 50 μm. Only in CZ crystals do B swirl defects exhibit a sufficiently large strain field to be analysed by HVEM. This strain was found to be of vacancy type (26), while a small second phase particle ($\leqslant 500$ Å) was present in the centre of the defect. The structure of the still smaller C swirl defects is unknown.

Defect concentration

The concentration of the various types of swirl defects strongly depends on the growth technique : CZ or FZ, the pulling rate V_0 and the concentration of residual carbon in the crystal.

FZ growth. At moderate pulling rates (1-3 mm/min) the concentration of A swirl defects is 10^6-10^7 cm^{-3} (14). In crystals with low carbon content ($< 10^{16}$ cm^{-3}) B swirl defects are not found (5,14); with carbon concentrations between 10^{16} and 10^{17} cm^{-3}, however, the B swirl defect density varies between 10^7 and 10^{10} cm^{-3} (5,14).

At $V_0 \geqslant 4$ mm/min dislocation loop formation is suppressed, and at $V_0 \geqslant 5$ mm/min B swirl defects also disappear (9). However, Roksnoer et al. (35) recently observed that at $V_0 > 6$ mm/min microdefects are formed again (labelled D defects). These defects are distributed homogeneously throughout the crystal with a concentration of about 10^9 cm^{-3}.

When V_0 was lowered to a value of 0.2 mm/min, swirl-free crystals of small diameter (23 mm) were obtained (33).

CZ growth. At the frequently applied pulling rate of about 1 mm/min, A and B swirl defects are formed in concentrations comparable with those present in FZ crystals (17,26). The concentrations of both B and C swirl defects strongly depend on the carbon concentration and are almost zero at carbon concentrations below 10^{16} cm^{-3}.

The critical pulling rate for A swirl defect elimination is about half the value required for FZ growth, i.e. 2 mm/min (17). This low value decreases with increasing crystal diameter and with decreasing axial temperature gradient in the crystal.

Effect of doping

Recently the effect of doping on swirl defect formation in CZ silicon has been investigated (17,26). When doping is done with donor impurities (Sb,P,As) in concentrations above 10^{17} cm^{-3} a complete suppression of A swirl defects results, together with an increase in density of the other two defect types. Doping with acceptors (B,Ga) above 10^{17} cm^{-3} causes the opposite effect i.e. no B or C swirl defects are formed. Dislocation loop formation remains unaffected by the addition of acceptor impurities. Doping with the iso-electronic impurity Sn caused no observable change in the defect generation process.

MODELS FOR SWIRL DEFECT FORMATION

Several models for the nucleation and growth of swirl defects have been proposed. Models I, II and III by Föll et al. (5,6), Petroff et al. (7) and Chikawa et al. (4,10) respectively, attempt to incorporate most of the experimental data summarized in the previous sections. The two remaining models only deal with some aspects of swirl formation. Model IV by Hu (2) considers exclusively the type of thermal point defects participating, while model V by Van Vechten (11) tries to describe the formation and growth of A swirl defects.

Model I
This model (5,6) is based on the assumption that the self-interstitial is the dominant thermal point defect present in silicon at high temperatures $(T \geq 900^{\circ}C)$. The concentration of vacancies is considered to be negligibly small. The self-interstitials are thought to be extended in nature, e.g. 11 atoms occupying 10 lattice sites and having a melt-like structure (36).

The swirl defect nucleation and growth sequence is described to be as follows. During solidification self-interstitials are incorporated into the crystal in a concentration equal to the thermal equilibrium value at the melting point. Supersaturation during crystal cooling causes a rapid agglomeration of self-interstitials to occur resulting in the formation of small droplets. The nuclei necessary for this condensation process consist of a number of self-interstitials and carbon atoms and are therefore heterogeneous in nature. The growth kinetics and stability of the droplets, which are the B swirl defects, are governed by diffusion of additional carbon atoms towards them. Above a certain critical size, which depends on the carbon concentration, a B swirl defect collapses into a faulted dislocation loop which, after a defaulting reaction, finally transforms into an interstitial-type perfect dislocation loop (A swirl defect). For further details of this model the reader is referred to (5,6).

Model II
In this model (7) the formation of B and A swirl defects is again described as a two-stage condensation process of supersaturated self-interstitials similar to model I. However, in contrast to model I, it is assumed that at the solid-liquid interface non-equilibrium concentrations of thermal point defects are incorporated into the crystal. Possible mechanisms for such a trapping of point defects at the growth interface have been treated theoretically by Webb (37) and Fletcher (38). Webb considered trapping due to a deviation of point defect concentration in the outermost layer of the growth interface compared with the equilibrium bulk value, while Fletcher analysed the effect of structural (density) differences between crystal and melt.

Because the temperature distribution in real growth systems is generally not completely cylindrically symmetric, crystal rotation introduces temperature variations at the growth interface, leading to alternative growth and remelt cycles. In model II additional trapping due to such remelt cycles has been proposed. Upon melting the solid-liquid interface might become highly imperfect, as described for instance in the theories of melting by Lennard-Jones et al. (39), Gorécki (40) and Kuhlmann-Wilsdorf (41). If these imperfections are not completely annealed out during the growth cycles following remelt they become trapped in the lattice. Additional mechanisms for trapping related to surface reconstruction or to the capture of liquid inclusions at the growth interface have been proposed by Van Vechten (11,42).

In model II no conclusions are drawn about which point defect is dominant under thermal equilibrium conditions.

Model III

On the basis of in-situ observation of melting and solidification of silicon ribbons on a hot-stage X-ray topographic camera, Chikawa et al. (4,10) developed a model for swirl defect formation. In this model it is assumed that during remelt, liquid droplets are formed in a superheated region just behind the melting solid-liquid interface. The interfacial tension and the volume decrease on melting causes tensile stresses in the lattice surrounding the droplets. This initiates a flow of self-interstitials and impurities into the droplet and a flow of vacancies away from it. The latter point defect is considered to be the dominant species present in equilibrium. When microscopic growth is resumed the droplets solidify. Due to the excess amount of interstitials moved into the droplet an interstitial type of strain field will be built up. Segregation towards the centre of the solidifying droplet of impurities having a covalent radius smaller than that of silicon will tend to reduce this strain field, while impurities larger than silicon will increase the strain. Those solidified droplets whose related strain exceeds a certain value will generated interstitial-type dislocation loops (A swirl defects), while the smaller ones are identical with B swirl defects.

Model IV

Hu (2) assumes that both self-interstitials and vacancies are present in equilibrium at high temperatures. Consequently, the crystal becomes supersaturated with both point defects on cooling. Mutual annihilation of vacancies and self-interstitials, however, is considered to be a relatively slow process with an energy barrier equal to the difference between the displacement threshold and the sum of the formation energies of the point defects, possibly a few to several eV. Therefore, point defect supersaturation is eliminated mainly via two separate condensation processes of self-interstitials and vacancies, resulting in A and B swirl defects respectively. The B swirl defects might be present in various configurations, such as pores (43) or sponge structures (44,14).

Model V

In this model vacancies are considered to be the dominant point defects present in the growing silicon crystal. In order to explain the formation of interstitial-type dislocation loops by vacancy condensation a number of assumptions had to be made (11). The excess vacancies are thought to condense into vacancy discs located on {111} planes. Above a certain diameter (6-8 nm) a (7 x 7) reconstruction of the two opposite internal surfaces of such a planar cavity takes place. This is believed to cause liquid-like waves of expelled atoms to move towards the cavity periphery. On solidification of this "liquid" the related volume increase causes an interstitial type of lattice curvature. After a shearing process, perfect "interstitial-type" dislocation loops with $\frac{1}{2}a \langle 110 \rangle$ Burgers vector would result.

Several objections can be raised against this model and which have recently been summarized by Föll et al. (45). One of the major difficulties with the proposed pseudodislocation loop is that it does not allow for a reorientation of the loop plane by glide, as has been observed by HVEM (5).

In the next section we will confront the four most realistic models (I to IV) with some of the recently obtained growth-related properties of swirl defects.

COMPARISON OF THEORY AND EXPERIMENT

The effect of remelt phenomena

Formation of A swirl defects in FZ silicon crystals is suppressed when pulling rates $\geqslant 4$ mm/min are applied (9). In CZ growth, this already occurs at

a critical pulling rate of 2 mm/min (17). In model I the elimination of A
swirl defects has been related to a certain critical cooling rate which leaves
insufficient time for the B swirl defects to collapse into dislocation loops.
However, the axial temperature gradient in CZ crystals during growth is about
half that measured for FZ crystals of comparable diameter (17, 46). This means
that A swirl-free growth in the CZ case is achieved at a four times lower
cooling rate than required for FZ growth. This cannot be understood on the
basis of model I.

Models II and III relate swirl elimination not only to the cooling rate but
also to remelt suppression (4,7,10). Recently Van Run (47) carried out numerical calculations on the non-steady motion of silicon crystal-melt interfaces
due to temperature fluctuations, and his results provided strong support for
remelt-related swirl formation. He showed that previously published formulae
for the microscopic growth rate variations (48,49) were erroneous because
the purely geometrical approach used, did not take into account the transport
of latent heat. When this was properly done it was established that rotational
remelt is absent if the macroscopic growth rate (pulling rate) exceeds a
critical value which is proportional to the axial temperature gradient in the
crystal at the crystal-melt interface. Using the experimentally determined
axial temperature gradients (17,46) Van Run (47) calculated that for FZ and
CZ silicon rotational remelt cannot occur at pulling rates exceeding 5.3 and
2.7 mm/min respectively.

The observation that additional microdefects (35) are formed when the
pulling rate is still further increased to values above 6 mm/min suggests
that trapping processes as discussed by Webb (37) and Fletcher (38) become
dominant. Both theories predict that the amount of trapped point defects
increases linearly with microscopic growth rate. Because the silicon melt is
about 10% denser than the solid, Fletcher's theory predicts a continuous
increase of the concentration of trapped self-interstitials, while in Webb's
theory vacancies or self-interstitials can be trapped, depending on the perfection of the growth interface on an atomic scale.

The effect of doping
    The complete absence of B swirl defects in p-type CZ crystals (doping le-
vel $> 10^{17}$ cm^{-3}), while A swirl defect formation is not affected (17), is
difficult to explain with the models I and II in which B swirl defects are
considered to be a pre-stage of dislocation loop formation. In model I the
B swirl defects are considered to have a droplet-like structure. Incorporation
into the droplets of a certain fraction of the carbon present in the crystal
will reduce the free enthalpy because the segregation coefficient of carbon
is smaller than unity (k = 0.07). Consequently, high carbon concentrations
result in the formation of large stable B swirl defects, while at low concentrations the droplets readily collapse into dislocation loops. The remaining droplets would be too small to be detected. In this way the absence
of B swirl defects in low-carbon crystals was explained (6). Following
Föll's reasoning, one would expect that, due to the higher solubility of
boron and gallium in liquid silicon than in solid silicon, doping with these
impurities would expand the defect stability region and therefore, in contrast with experimental data, B swirl defects should be very stable. However,
even if we assume that for some unknown reason p-type doping causes a
decrease of the defect stability region to such an extent that the B swirl
defects already collapse at a size at which they are still undetectable
even by copper decoration, one would not expect the formation of A swirl
defects to remain unaffected (17).

    It is more likely, therefore, that A and B swirl defects are not formed by
condensation of only one thermal point defect (the self-interstitial) in a
two-step process but that both vacancies and self-interstitials are present

314

which are removed from superaaturated solid solution via two separate condensation processes as suggested by Hu (model IV (2)). Because the formation of
B and C swirl defects is affected in the same way by doping, it is reasonable
to assume that they both form as a result of vacancy condensation. As already
mentioned, vacancy clusters might indeed be present in various configurations
such as pores (43) and sponge structures (44).

The observed changes in swirl formation on doping are also difficult to
explain with model III. During solidification of a droplet, impurities with
a distribution coefficient k smaller than unity will segregate towards
the centre of the droplet (4,10). Such a segregation of impurities with
covalent radii larger than the radius of silicon (r_c(Si) = 1.17 Å) will
increase the interstitial type of strain field, while smaller atoms will reduce the strain, resulting in larger or smaller swirl defects respectively.
However, this is contrary to what has been observed for bulk growth conditions (17,26). Whereas doping,for instance,with the small atom boron
(r_c(B) = 0.88 Å, k = 0.8) prevents the formation of the B and C swirl defects,
the A swirl formation is completely suppressed by doping with the large atom
antimony (r_c(Sb) = 1.36 Å, k = 0.023). In addition, the strain field of the
B swirl defects in Sb-doped crystals is vacancy-type and identical with those
observed in undoped crystals. Furthermore heavy doping with tin (r_c(Sn) =
1.40 Å, k = 0.016) has no effect. Consequently it can be stated that impurity
segregation inside swirl defects does not occur and droplet formation seems
unlikely.

The experimentally established dependence of B and C swirl formation on
carbon content (5,14,17) points to a heterogeneous nucleation process. This
phenomenon has been incorporated in models I and II. It also explains the
striated defect distribution. For A swirl defect formation a similar dependence
on carbon content is not observed. Whether the striated distribution of the
latter defect is due to a heterogeneous nucleation process or due to striation-
related strain (2) is not known.

CONCLUSION

The preceding discussion shows that none of the models for swirl defect
formation known today allows a complete interpretation of all experimental
data. It seems likely, however, that a more consistent model could be
developed by combining several elements of the existing models. The three
major concepts to be considered are :
(i)   A and B, (C) swirl defects are formed via separate condensation pro-
       cesses of self-interstitials and vacancies respectively (model IV).
(ii)   Non-equilibrium concentrations of thermal point defects can be in-
       corporated into the crystal during growth as a result of remelt and
       phenomena discussed by e.g. Webb (37) and Fletcher (38) (model II).
(iii) Carbon plays an important role in the nucleation of B, (C) swirl
       defects (models I and II).

REFERENCES

1. A.J.R. de Kock, in "1976 Crystal Growth and Materials" eds. E. Kaldis
   and H.J. Scheel (North-Holland Publishing Company, 1977) p. 662.

2. S.M. Hu, J. Vac. Sci. Technol. 14, 17 (1977).

3. A.J.R. de Kock, in Handbook on Semiconductors, vol. 3 ed. S.P. Keller
   (North-Holland Publishing Company, 1980).

4. J. Chikawa and S. Shirai, Jap. J. Appl. Phys. 18, Suppl. 18-1, 153 (1979).

5. H. Föll and B.O. Kolbesen, Appl. Phys. 8, 319 (1975).

6. H. Föll, U. Gösele and B.O. Kolbesen, J. Cryst. Growth 40, 90 (1977).

7. P.M. Petroff and A.J.R. de Kock, J. Crystal Growth 35, 4 (1976).

8. A.J.R. de Kock, P.J. Roksnoer and P.G.T. Boonen, J. Crystal Growth 22, 311 (1974).

9. A.J.R. de Kock, P.J. Roksnoer and P.G.T. Boonen, in Semiconductor Silicon 1973. eds. H.R. Huff and R.R. Burgess (The Electrochemical Society, Inc., Princeton) p. 83.

10. J. Chikawa and S. Shirai, J. Crystal Growth 39, 328 (1977).

11. J.A. Van Vechten, Phys. Rev. B 17, 3197 (1978).

12. P.J. Roksnoer and M.M.B. van den Boom, J. Appl. Phys. 51, 2274 (1980).

13. A.J.R. de Kock, Appl. Phys. Letters 16, 100 (1970).

14. A.J.R. de Kock, Philips Res. Rept. Suppl. No. 1 (1973).

15. T. Abe and S. Maruyama, Denki Kagaku 35, 149 (1976).

16. L.I. Bernewitz and K.R. Mayer, Phys. Status Solidi (a) 16, 579 (1973).

17. A.J.R. de Kock and W.M. van de Wijgert, J. Crystal Growth 49, 718 (1980).

18. T.S. Plaskett, Trans. AIME 233, 809 (1965).

19. J. Chikawa, Y. Asaeda and I. Fujimoto, J. Appl. Phys. 41, 1922 (1970).

20. M. Renninger, J. Appl. Cryst. 9, 178 (1976).

21. A.J.R. de Kock, P.J. Roksnoer and P.G.T. Boonen, J. Crystal Growth 30, 279 (1975).

22. A.J.R. de Kock, S.D. Ferris, L.C. Kimerling and H.J. Leamy, Appl. Phys. Letters 27, 313 (1975).

23. L.I. Bernewitz, B.O. Kolbesen, K.R. Mayer and G.E. Schuh, Appl. Phys. Letters 23, 277 (1974).

24. H.S. Grienauer, B.O. Kolbesen and K.R. Mayer, in Lattice Defects in Semiconductors 1974, Institute of Physics Conference Series No. 23 (The Institute of Physics, London) p. 531.

25 P.M. Petroff and A.J.R. de Kock, J. Crystal Growth 30, 117 (1975).

26. A.J.R. de Kock, W.T. Stacy and W.M. van de Wijgert, Appl. Phys. Letters 34, 611 (1979).

27. K.V. Ravi and C.J. Varker, J. Appl. Phys. 45, 263 (1974).

28. J. Matsui and T. Kawamura, Jap. J. Appl. Phys. 11, 197 (1972).

29. P. Rai-Choudhury, J. Electrochem. Soc. 118, 1183 (1971).

30. K.V. Ravi, Thin Solid Films 31, 171 (1976).

31. C.J. Varker and K.V. Ravi, in ref. 9, p. 670.

32. A.J.R. de Kock, J. Electrochem. Soc. 118, 1851 (1971).

33. P.J. Roksnoer, W.J. Bartels and C.W. T. Bulle, J. Crystal Growth 35, 245 (1976).

34. P.J. Roksnoer, private communication.

35. P.J. Roksnoer and M.M.B. van den Boom, to be published.

36. A. Seeger and K.P. Chik, Phys. Status. Solidi 29, 455 (1968).

37. W.W. Webb, J. Appl. Phys. 3, 1962 (1961).

38. N.H. Fletcher, J. Crystal Growth 35, 39 (1976).

39. J.E. Lennard-Jones and A.F. Devonshire, Proc. R. Soc. A, 169, 317 (1939).

40. T. Gorécki, Z. Metallk. 65, 426 (1974).

41. D. Kuhlmann-Wilsdorf, Phys. Rev. A 140, 1599 (1965).

42. J.A. Van Vechten, Appl. Phys. Lett. 26, 593 (1975).

43. V.V. Voronkov, Sov. Phys. Crystallog. 19, 137 (1974).

44. J. Hornstra, Philips Res. Lab. Report No. 3497 (1959) unpublished.

45. H. Föll, U. Gösele and B.O. Kolbsen, Proceedings of the Sixth International Conference on Crystal Growth, Moscow, September, 1980, to be published.

46. H.K. Kuiken and P.J. Roksnoer, J. Crystal Growth 47, 29 (1979).

47. A.M.J.G. van Run, to be published.

48. K. Morizane, A.F. Witt and H.C. Gatos, J. Electrochem. Soc. 114, 738 (1967).

49. J. Barthel and M. Jurisch, Kristall und Technik 8, 199 (1973).

Published 1981 by North-Holland, Inc.
Narayan, and Tan, eds.
Defects in Semiconductors

REAL-TIME X-RAY TOPOGRAPHIC OBSERVATION OF MELTING AND GROWTH
OF SILICON CRYSTALS

JUN-ICHI CHIKAWA AND FUMIO SATO
NHK Broadcasting Science Research Laboratories, 1-10-11, Kinuta, Setagaya-ku,
Tokyo, 157, Japan

ABSTRACT

A technique for the direct viewing of topographic images
was developed based on a television system using an x-ray
sensing PbO-vidicon camera tube (resolution 25 µm). Melting
and growth processes of plate-shaped crystals were observed
in an argon flow by this technique. The observation showed
that dislocations have very high mobilities by superheating
and the equilibrium dislocation density of zero is achieved
just before melting. In melting of dislocation-free crystals,
liquid drops (locally molten regions) were observed inside
the crystals. While, crystallites appeared in the supercooled
melt near the growth interface. It was found that both the
drops and crystallites are not formed in 5% hydrogen/95% argon
environment. Melting behavior of crystals covered with oxide
films indicates that the solid-liquid interfacial free energy is
greatly lowered by oxygen impurity. This oxygen effect may
be responsible for both the drop and crystallite formation.
Origin of swirl defects in dislocation-free crystals is
discussed in the basis of these observations. Both composite
and common $a/2<110>$ dislocations were observed near the
growth interface: The former are composed of three $a/2<110>$
dislocations and are present at the interface so that, as
the crystal grows, they propagate into the new crystal.
They influence the interface morphology, when their Burgers
vectors have the component perpendicular to the interface.
The common dislocations cannot exist stably at the interface
and are driven into the newly grown dislocation-free region
by thermal stresses.

1. INTRODUCTION

In order to grow perfect crystals, we need information on behavior of
defects during growth. For this purpose, in situ observation by x-ray or
electron diffraction techniques is useful. Electron microscopy provides
very high resolution and until recently has been the only technique used
to observe defect structures during the growth process. For electron microscopy,
however, the following disadvantages have been pointed out [1].

(1) The extreme thinness of specimens used may result in an unusual
situation different from any common macroscopic growth process.

(2) Its application is limited to slow growth rates (nearly equilibrium
states) because of the large magnification, and to the specimens with sufficiently low vapour pressure.

If in situ X-ray observation is possible, the growth processes fairly
similar to common macroscopic ones can be investigated owing to the high

penetrating power of X-rays and low-magnification observation. However, intensities of X-rays diffracted from crystals are very weak, and instantaneous observation of topographic images has been impossible. Most topographic observations have been made ex post facto to the actual process of structure changes by a photographic method which requires a long exposure time[2,3]. In 1968, however, the present author and his co-workers[4] made direct viewing of topographic images possible by developing a high power X-ray generator and a video display technique. Since then, live topography was initiated, and various attempts have been made [5]. The authors' technique has been sophisticated [6], and its development is still continuing. The purpose of this paper is to show its capability by in situ observation of melting and growth processes of silicon crystals.

2. VIDEO TOPOGRAPHY

The topographic system developed consits of a high power x-ray generator, goniometer, and television system. A block diagram of the recent system is shown in Fig. 1.

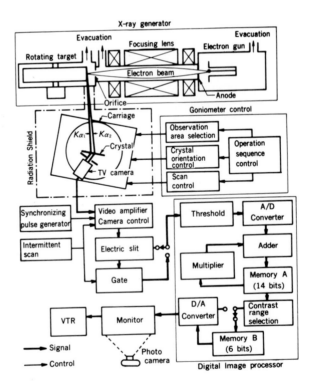

Fig. 1. Block diagram of the video x-ray topography.

2.1. X-ray generator

In order to observe individual dislocations, a resolution better than 30 µm is required. In instantaneous or rapid observation with such a high resolution, photon noise due to the statistical fluctuation of number of photons per picture element has an important effect on image quality [6]; For instance, a typical number of photons received per picture element (30 × 30 µm) per frame (1/30 sec) was only 10 to 15 for perfect regions of a silicon specimen crystal, when a Mo-target x-ray tube was operated at 60 kV and 0.5 A in the geometry shown in Fig. 2. Especially by improving the resolution (smaller picture element), the signal-to-noise ratio of images is lowered, and we need more powerful x-ray generators. As shown in Fig. 1, an x-ray generator was developed in which the electron gun and rotating target are separated by placing a focussing electron lens between them thus making the operation at a high power stable and the focus size adjustable. At present, its maximum power is 100 kW (70 kV, 1.4 A) for a focus size of 1 × 10 mm.

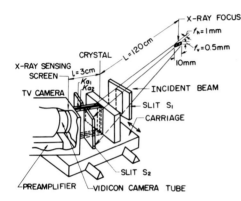

Fig. 2. Geometry for imaging transmission topographs.

2.2. Goniometer

The goniometer used is of a type of Lang camera [7], but large; the scan carriage is 50 × 50 cm. The remote control can be made as shown in fig. 1. The geometry of observation is shown in Fig. 2.

2.3 Television system

The TV system consists of a TV camera unit with an X-ray sensing vidicon-type camera tube, video amplifier, camera control unit, and picture monitor, as operating a usual closed-circuit TV system [Fig. 1]. The TV camera unit and a specimen crystal are placed on the scan carriage of the goniometer. By orienting the crystal so as to satisfy the Bragg condition for the slightly divergent X-ray incident beam $(K\alpha)$, two images due to the diffracted $K\alpha_1$ and $K\alpha_2$ beams with a width of 1 mm are received by the camera tube, as shown schematically in Fig. 2; two band-shaped regions of the crystal are imaged instantaneously at about 30 times enlargement on the picture monitor. Such images are called "direct-view images" for convenience. To display a wide area, the video signals due to the $K\alpha_1$ image is selected electrically and stored in the digital image processor, while the carriage is moved for 3 to 10 sec. At the end of the carriage motion, a Lang topograph with the imaging area of the camera tube (9 × 13 mm) is displayed on the monitor. Hereafter, such images are referred to as "synthesized images".

Improvement of image quality. Digital image processing with a real-time analogue-digital conversion [8] has become popular, and techniques such as noise reducers [9, 10] are very useful for video topography. A digital image processor was designed for the following three basic modes to improve quality of images. (1) Sliding summation of successive frames with the weighting of new to old data adjustable from the control panel. (2) Image integration for a time interval adjustable from the control panel. Moving objects are seen as images changing discretely like animation pictures. (3) Synthesization of Lang topographs from the $K\alpha_1$ images through the electric slit. Acquisition of extremely low intensity images, dramatic improvements in signal-to-noise ratios via frame integration and isolation and enhancement of selected-contrast ranges are possible by digital image processing and intermittent scanning. (For detail, see Ref. [6].) For the image-quality improvement of moving defects, the digital image processor should be used with the optimum integration time. In general, rates of image changes cannot be estimated beforehand. Therefore, the original images should be tape-recorded and afterwards the reproduced images should be improved by the processor.

Resolution. The x-ray sensing PbO camera tube developed by Nishida and Okamoto [11] was designed for $MoK\alpha_1$ radiation, and a resolution of 25 μm was obtained [6]. Recently, a camera tube having an amorphous alloy Se-As with a thickness of 20 μm was developed, and a high resolution of 10 μm was achieved, though the sensitivity is a half of that of the PbO camera tube. Further development is in progress.

3. EXPERIMENTAL PROCEDURE

The results reported here were obtained by the PbO camera tube and Mo- or Nb-rotating target x-ray generator operated at 60 kV and 0.5 A with a focus size of 0.5 × 10 mm.

To observe growth processes of Si crystals, a furnace was mounted on the goniometer, and a crystal was placed between two carbon-plate heaters, as shown schematically in Fig. 3. The middle part of the crystal was melted in an argon gas flow, and then cooled to solidify on the frozen part. Topographic images from the crystal were observed through the carbon heater by the TV camera. Also, growth by a floating zone method was observed by moving both the crystal and TV camera upward at a speed of 1mm/min. Images on the picture monitor were recorded by a 16-mm cinecamera.

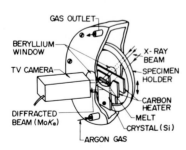

Fig. 3. Schematic illustration of a furnace for the x-ray observation of melting and growth of Si crystals.

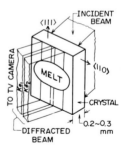

Fig. 4. Orientation of the specimens.

The specimens were plate-shaped crystals with thicknesses of 0.3 to 0.5 mm which were cut from dislocation-free and swirl-free non-doped crystals grown by the conventional floating-zone method. The orientation of the specimens was illustrated in Fig. 4. From some stages of the melting and growth processes, the specimens were cooled down to room temperature with monitoring by the TV camera. They were also examined by conventional topography.

4. OBSERVATION OF MELTING PROCESSES

4.1. Dislocated crystals

Many dislocations are often generated during temperature increase to the melting point (1412°C). A typical melting process of such dislocated crystals is shown with a series of direct-view images in Fig. 5. Time intervals between the images are indicated between them. The black parts in Fig. 5(a) are due to the dislocations. Figures 5(a) and (b) were obtained before melting. The region labeled "M" has the highest temperature and contains a low dislocation density. The continuous observation showed that the dislocations move rapidly toward the lower-temperature regions. The upper region has a small temperature gradient of about 5°C/mm and becomes dislocation-free, while dislocations remain in the lower region where the temperature gradient is about 20°C/mm. In Figs. 5(c) and (d), the surface regions are melted [Fig. 5(g)], and melting through causes a white region on the image in Fig. 5(e) [Fig. 5(h)]. (All photographs are shown in negative in this paper.) Some round fringes appear in Figs. 5(c)-(e). They are "Pendellösung" fringes (equal-thickness fringes)[12] due to thinning of the crystal as shown schematically in Figs. 5(g) and (h). This observation shows that crystals become highly perfect just before melting. The observed sequence shows that melting takes place uniformly from the surfaces in dislocated crystals.

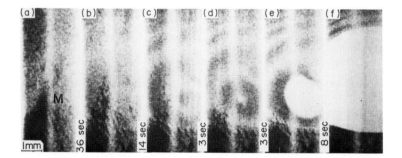

Fig. 5. Melting sequence of a dislocated crystal. (a)-(f) Direct-view images with time intervals indicated between the photographs. $2\bar{2}0$ reflection. Each image consists of two bands due to the MoKα_1 and Kα_2; their intensity difference was adjusted in printing from the movie film, in order to make the Kα_2 image useful. (g)-(i) Schematic illustration of the cross-section of the melting crystal. Melting proceeds from the surfaces, and round Pendellösung fringes are seen. The black dots around the white hole in (e) are due to drops inside the crystal.

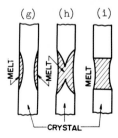

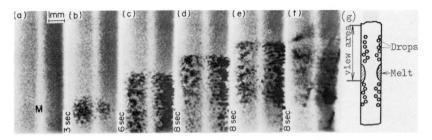

Fig. 6. Rapid melting of a dislocation-free crystal. (a)-(f) Direct-view images by the 2\overline{2}0 reflection. Black spots are due to drops inside the crystal. (g) Schematic illustration of corss-section of the crystal at the stage of (e).

4.2. Dislocation-free crystals

By keeping specimen crystals dislocation-free, however, different melting behaviour was observed, as seen from Fig. 6: Figure 6(a) shows a dislocation-free state just before surface melting begins. In Fig. 6(b), black spots appear in the region where the temperature is highest. (The region with the spots expanded radially out of the view area.) Only the upward expansion can be seen from Figs. 6(c)-(e). Many dislocations were generated suddenly in the interval between Figs. 6(e) and (f) and moved away from the region. Simultaneously, almost all the spots disappeared, as seen from Fig. 6(f). The black spots were found to be due to drops (locally molten regions) inside the crystal by the following observations: (1) Melting proceeds from the black spots (See Fig. 7). (2) They disappear by decreasing the temperature slightly; they appear only in melting processes. (3) Their density is increased by heating rapidly. (4) Their density is decreased markedly by increasing the macroscopic temperature gradient perpendicular to the crystal surfaces. No spots appear for temperature gradients larger than 300°C/cm, unless very rapid melting is made. The spot images are due to the radial strains around drops which are produced by both the interfacial tension and volume difference between the liquid and solid. The sizes of drops were estimated to be 100 to 200 µm. Since these drops were formed near the crystal surfaces, the black spots disappear in the lower region of Fig. 6(e) where the surface regions are melted, as shown schematically in Fig. 6(g).

In slower melting, however, drops are formed even in the inner region. An example is shown by a series of synthesized images in Fig. 7. Figure 7(a) shows the state just before melting begins from the surface. The ellipse-shaped region in Fig. 7(b) is melting from the surfaces (as shown in Fig. 5(g)) and is expanding in Figs. 7(b) to (c). By decreasing the heating power slightly, the ellipse is shrinking in Figs. 7(d) to (f), i.e., growth takes place at the periphery of the ellipse, and the dark contrast along the periphery will be considered in 5.1. However, melting is proceeding at the central region of the ellipse due to thermal inertia, and there are many black spots due to drops. In Fig. 7(e), the crystal is very thin at the center of the ellipse, and many white holes appear due to melting through by some of the drops. In Fig. 7(f), a large hole is seen due to perforation by melting. Many dislocations were generated from the region and propagated downwards, as seen in Figs. 7(e) and (f). Although the dislocations were generated inside the ellipse-shaped region, their images are not seen there, and they stay just out of the region [Fig. 7(e)]. This observation indicates that dislocations cannot exist in melting regions. As seen from comparison between Figs. 7(d) and (e), the black spots along the pass of the dislocations disappear in Fig. 7(e). The melting sequence in Fig. 7 shows that dislocation-free crystals melt inhomogeneously owing to the drop formation, in contrast with the case of dislocated crystals.

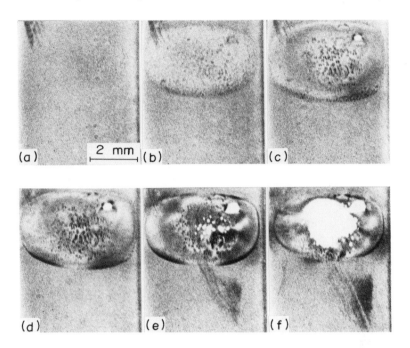

Fig. 7. Slow melting of a dislocation-free crystal. (a)-(f) Synthesized images with an interval of 10 sec. 2$\bar{2}$0 reflection. The ellipse-shaped region is melting from the surfaces, and some Pendellösung fringes are seen due to the thickness variation. Black spots are due to drops inside the crystal. Although dislocations are generated in the ellipse of (e), all of them move downwards, and the ellipse becomes dislocation-free. Observe covering the dislocation images outside of the ellipse, in order to avoid visual illusion.

4.3. Hydrogen effect on drop formation

The same specimens were melted in 5% hydrogen/95% argon environment. The result is shown by synthesized images with an interval of 10 sec in Fig. 8. Figure 8(a) shows the dislocation-free state just before melting. Several Pendellösung fringes are seen vertically. In Fig. 8(b), surface melting proceeds rapidly, as seen from the change in shape of the Pendellösung fringes. Even for such rapid melting, no spot images are seen. This observation shows that hydrogen supresses the drop formation.

4.4. Oxygen effect on melting behavior

In order to examine oxygen effects, the specimen crystals were oxidized and their surfaces were covered with oxide films about 2-μm thick. The series of direct-view images in Fig. 9 shows the melting process. Many dislocations were generated during the temperature increase by stresses due to the oxide films so that individual dislocations are not seen in Fig. 9. Although the crystal was

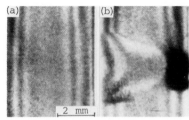

Fig. 8. Hydrogen effect on the drop formation. The specimen crystal was melted in H_2 5%/Ar 95% environment. (a) Just before melting. (b) Melting surface regions. Synthesized images with an interval of 10 sec. $2\bar{2}0$ reflection. Note that the dislocations generated from the right edge do not enter the melting region.

dislocated, inhomogeneous melting occurred: Locally molten regions are seen in Fig. 9(a). Even for such small melting regions, {111} facets appear, and melting proceeds with {111} facetted interfaces. Melting should have taken place spherically, especially for small molten volumes, if the interfacial free energy were appreciable. Therefore, the observed shape of melting interfaces means that interfacial free energy is lowered drastically by oxygen impurity, especially for {111} interfaces.

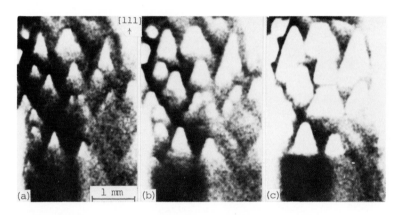

Fig. 9. Oxygen effect on melting. (a)-(c) Direct-view images with an interval of 6 sec. $2\bar{2}0$ reflection. The specimen crystal was coated with oxide films. The melting interfaces are facetted.

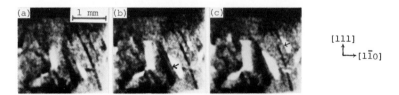

Fig. 10. Dislocation effect on melting observed for an oxygen-doped crystal. (a)-(c) Direct-view images with an interval of 10 sec. $2\bar{2}0$ reflection. Dislocations indicated by the arrows in (b) and (c) are of composite types in the <311> direction.

After the middle part of a specimen crystal is melted, we can grow low-dislocation-density regions by decreasing temperature slightly.* Then, by melting this region again, we can observed effects of individual dislocations on melting under conditions that are nearly free from interfacial energy. A typical example is shown in Fig. 10. It is seen that dislocations have a resistance against melting. Melting begins from dislocation-free regions, and the molten regions stop expanding at the positions of the dislocations indicated by the arrows in Figs. 10(b) and (c).

5. OBSERVATION OF GROWTH PROCESSES

5.1. Crystallites on the melt-side of the interface

As has been seen in Fig. 7, the ellipse-shaped region where surface melting takes place expands from Fig. 7(b) to (c). By decreasing the heating power slightly, the ellipse shrinks from Fig. 7(c) to (f), i.e., growth occurs at the periphery of the ellipse. In Fig. 7(c), the region along the periphery of the ellipse is darker than the maximum of pendellösung fringes seen inside the ellipse. It can be seen that this darkness is due to many black dots. In Figs. 7(d), (e), and (f), the region on the inside of the periphery is seen black.** The contrast is weaker for the upper periphery, where the temperature gradient was higher and the growth rate (shrinking rate) is much smaller. It was observed between Figs. 7(c) and (d) that the black dots are moving by shrinking of the periphery, and the contrast is formed by so many dots that individual dots cannot be seen. The parts on the periphery intersecting with the vertical diameter of the ellipse shows no contrast. The reflecting plane (220 reflection) used was parallel to the vertical direction and perpendicular to the specimen surface. By taking into account the rule that topographic images show no contrast for the reflecting plane to which the displacement of atoms is parallel, one can conclude that crystallites are formed in the supercooled melt along the periphery and are oriented in accord with the crystal lattice. When the growth rate is high or when concentration of doped impurity such as Sb is high, the contrast becomes stronger. This indicates that the crystallite formation becomes dominant by the higher supercooling or lowering the melting point.

In the hydrogen-added argon environment, however, such images were not observed.

5.2. Dislocations near growth interfaces

After the middle part of the plate-shaped crystal was melted completely, as shown schematically in Fig. 5(i), the growth process was observed by decreasing temperature slightly. A typical example of dislocation propagation at the growth interface is shown with a series of direct-view images in Fig.11. The time intervals between them are indicated in the figure. The lower edge of each image is the interface, and the crystal is growing downwards. Figure 11(a) is the state of the seed crystal just before the growth starts. The

---

* In this case, the oxide films were found to be broken, but the oxygen concentration may be still high.

** If the contrast along the periphery is due to interfacial tension, any differences in contrast between the melting and growth cannot be expected. Because the contour of the periphery is very sharp, the contrast cannot be attributed to strains by a local increase of the temperature gradient in the growth process.

crystal is seen black due to many dislocations generated during the temperature increase to the melting point. In Fig. 11(b), a newly grown region is seen where many hairpin-shaped dislocations propagate from the seed crystal, as shown schematically in Fig. 11(g). These dislocations follow the interface motion, touching the interface with their points [Figs. 11(b) and (c)]. In Fig. 11(e), however, the majority of the dislocations are left behind the interface. They were identified to be the common type with Burgers vectors a/2[101] and a/2[011] by examining their image contrast for various reflections after the growth. Fast moving dislocations are seen faintly as broad images, as indicated by the arrows. They are considered to have a/2[1$\bar{1}$0] Burgers vectors perpendicular to the reflecting plane ($\bar{2}$20), because they move so fast that their observation is only possible for such a reflection that gives the maximum contrast. In Figs. 11(e) and (f), two straight dislocations intersect nearly perpendicular to the interface. These were found to be of composite type, each of which consists of three dislocations on different slip planes and is stable enough to intersect the interface [13].

In necking regions of float-zoned crystals, both irregularly curved dislocations and straight composite dislocations were often observed [13]. It was found from their contrast variations with different reflections that the latter have large Burgers vectors such as a[111] = a/2[110] + a/2[10$\bar{1}$] + a/2[01$\bar{1}$] and a[100] = a/2[1$\bar{1}$0] + a/2[101] + a/2[01$\bar{1}$]. The curved dislocations are of the common type with a/2<110> Burgers vectors.

The above observations indicate that common dislocations cannot exist stably at growth interfaces, i.e., a dislocation-free region first grows, and then the dislocations move in this region by thermal stresses.

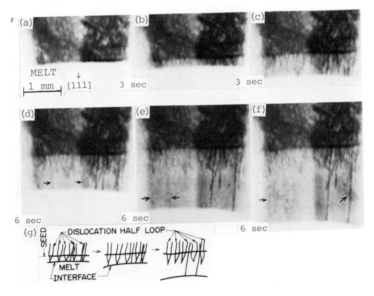

Fig. 11. Dislocations in a growth process. (a)-(f) Direct-view images with intervals indicated between the photographs. 2$\bar{2}$0 reflection. The interface is concave toward the melt. (g) Schematic illustration of the dislocation configurations. Fast moving dislocations have broad images with a faint contrast, and some of them are indicated by the arrows in (d), (e), and (f).

Dislocation effect on facet growth. The dislocation effect has been observed for Czochralski-grown crystals [14, 15]: Interface shapes are known from observation of growth bandings (striations), and facets in dislocation-free growth were found to shrink by dislocation generation [14, 15]. The growth on facets of a dislocation-free and dislocated crystals is explained by two-dimensional nucleation and spiral growth due to dislocations, respectively. In a steady state growth with a certain growth rate, the supercooling ΔT_{max} at the center of the facet is larger for the two-dimensional nucleation than that for the spiral growth. Figure 12 shows direct-view images of a growth process with interfaces convex toward the melt. In Figs. 12(a)-(d) a facet is seen at the central part of the interface. In this case, the temperature gradient near the interface was somewhat higher (20 to 30°C/mm), and many dislocations move in newly grown part from the seed as seen in Fig. 12(a). The interface is convex toward the melt, and a (111) facet is seen at the central part of the interface. The dislocations are moving from the seed to the interface, and some of them look as if they reached the facet in Figs. 12(b) (d). However, the facet does not shrink and is stable. It is concluded that common dislocations cannot intersect with interfaces. When a composite dislocation intersects with the facet, however, it disappears, as seen in Fig. 12(e); a round interface appears, and a composite dislocation is seen at the center of the interface with a high contrast. The dislocation propagates obliquely, and a facet appears again in Fig. 12(f). In this case, the nucleation for growth takes place at the edge of the facet where the dislocation is present [Fig. 12(g)]. Such composite dislocations that influence the interface morphology often propagate in the <311> direction, as seen in Fig. 12(f).

Two types of composite dislocations were observed [13]: One intersects perpendicular to the local interface, and the other propagates obliquely. The former has Burgers vectors nearly parallel to the interface. The latter influences interface morphology, and its Burgers vectors have the component perpendicular to the interface.

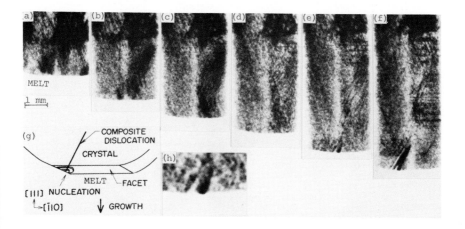

Fig. 12. Effect of composite dislocations on facet growth. (a)-(f) Direct-view images (interval: about 30 sec) of a growth process by the floating zone method. Zone travel rate=1 mm/min. 2̄20 reflection. (g) Schematic illustration of nucleation at the intersection of the dislocation at the interface. (h) Enlargement of the central part of the interface in (e).

Local effect of dislocations on the growth. An enlargement of the
interface in Fig. 12(e) is shown in Fig. 12(h). It is seen that the interface
humps at the intersection with the composite dislocation. We may estimate the
temperature difference between the top of the hump and round interface from the
temperature gradient (20 to 30°C/mm) and the height of the hump (0.1 mm):
Growth on the round interface occurs, at least, with a supercooling of 2°C at
a growth rate of 1 mm/min. Such humps were observed for interfaces concave
toward the melt [13]. This implies that motion of growth steps at the
interface needs a considerably high supercooling, unlike the general view of
step motion in melt growth.

6. DISCUSSION

6.1. Drop and crystallite formation
    Drops and crystallites are formed by opposite processes. However, the
problems of their nucleation are considered in the same way: According to
the simple nucleation theory [16] for a spherical drop (crystallite) having a
radius r , the excess free energy ΔG by their formation is expressed by

$$\Delta G = - (4/3) \pi r^3 L \Delta T / T_m + 4 \pi r^2 \sigma \tag{1}$$

where σ is the interfacial free energy between liquid and solid, L latent
heat of fusion per unit volume, T_m the melting point of the crystal, and ΔT
superheating(supercooling). Using the critical radius

$$r^* = 2 \sigma T_m / (L \Delta T), \tag{2}$$

we obtain the maximum free energy

$$\Delta G^* = (16/3) \pi (T_m / L)^2 \sigma^3 / \Delta T^2 \tag{3}$$

The equilibrium density n of drops is expressed by

$$n = N \exp (- \Delta G^* / k T_m) \tag{4}$$

where N is the number of atoms in unit volume. If we take $n = 10^3 \sim 10^4 \mathrm{cm}^{-3}$,
we have a superheating(supercooling) of ΔT =150, 5, and 0.15°C for σ =200,
20, and 2 erg/cm^2, respectively. For pure silicon, the interfacial free energy
is considered to be around 200 erg/cm [16], and drop (crystallite) formation
is impossible.* However, the observation in Sec. 4.4 shows that σ is greatly
lowered by oxygen impurity: The molten regions surrounded by {111} facets
in Fig. 9 and the irregularly molten regions in Fig. 10 indicate that $\sigma \approx 0$
at the solubility limit of oxygen (2 $\times 10^{18} \mathrm{cm}^{-3}$). This reduction of interfacial
free energy may be related to the fact that the segregation coefficient of
oxygen is larger than unity, k_0=1.25. The observation in Fig. 8 shows that
drops are not formed in hydrogen 5%/argon 95% environment. Therefore, one may
conclude that the drop formation is caused by oxygen impurity. However, drops
were observed in melting of the specimens in argon atmosphere which had been
prepared from crystals float-zoned in hydrogen 5%/argon 95% environment [17].
(Hydrogen in Si crystals diffuses out easily.) It follows from these
observations that trace amounts of oxygen in high-purity argon atmosphere is
enough for drop formation at a reasonable superheating, i.e., the drop
formation is due to homogeneous nucleation with a low interfacial free energy.

---

* Even if drop formation is initiated by some nucleation centers, these
  centers dissolve in small drops which they form and do not act any longer
  to grow the drops.

6.2. Origin of swirl defects

It is well known that microdefects (swirl defects) are formed in striated distribution in dislocation-free floating-zone crystals [18-20]. One of the present authors and co-worker [17, 19] proposed a formation model of swirl defects based on in situ observation of drops. In the conventional floating zone method, crystals are grown in argon environment so that their oxygen concentrations are enough for the drop formation as mentioned above. (Usually, they contains oxygen atoms in order of 10^{16} cm^{-3}.) The temperature fluctuation during growth occurs due to the crystal rotation; the crystals grow by repeated growth and pausing (or remelting). In the pausing periods, the interface region is superheated highly enough to form drops [19]. Since rates of crystal rotation are low ($\lesssim 10$ rpm) for keeping the melt on the crystal solidification of the drops are slow enough for impurity segregation: Most of impurity atoms in Si have segregation coefficients less than unity, and, after solidification of the drops, they segregate at the central region of the drop which is solidified finally. Oxygen atoms, however, are located in the outer regions because of k_0=1.25. Yatsurugi et al.[21] found by both the infrared and radioactive analyes that the segregation coefficient of interstitial oxygen atoms is unity. The excess oxygen atoms were considered to be incorporated as amorphous SiO which is formed at the interface. Namely, 25% of the oxygen atoms in the melt are incorporated into the crystal as SiO, which has a 4% less density than Si. Then, a volume expansion occurs in the outer regions. This produces excess atoms at the central region of the drop which are accommodated in a form of microdefects (dislocation-loops) by the cooling process. In a floating zone in vacuum, crystals have low oxygen concentrations. Therefore, the interfacial free energy become relatively large, and a low density of drops results. However, the well-isolated drops grow largely without influence from neighboring drops. Total amount of impurity and oxygen atoms in the drop volume is larger and results in large defects.[22].

Recent Czochralski technique provides swirl-free crystals in the as-grown state. In post-growth high temperature annealing, however, oxygen precipitates [23] are often formed in striated distribution. Temperature fluctuation during growth is due mainly to convection of the melt and are somewhat smaller compared with the floating-zone method. However, the oxygen concentration is an order-of-magnitude higher. This is very favorable for the drop formation. The drops may form nucleation centers by impurity segregation during their solidification [19]. Since impurity distribution in the recent crystals is very uniform, the striated distribution of oxygen precipitates strongly support the drop model.

6.3. Dislocations in melting crystals

The observations in Secs. 4.1 and 4.2 shows that melting behavior is different between dislocated and dislocation-free crystals, as shown schematically in Fig. 13. When a dislocated crystal is heated from its surface, melting proceeds by forming a dislocation-free layer [Fig. 13(a)] [*], and homogeneous melting occurs from the surface. Whereas, for entirely dislocation-free crystals, inhomogeneous melting occurs by forming drops, Figure 7(e) shows that dislocations cannot be present in melting (superheated) regions. In other words, the dislocated region in Fig. 13(a) is not so superheated, i.e., dislocations prevent superheating. This is supported by the experimental results: (1) Drops disappear suddenly when dislocations pass

---

* Small drops are formed in the dislocation-free layer of a dislocated crystal melting: In Fig. 5(e), some small black dots are seen around the white hole. These are images of small drops.

nearby them (Fig. 6). (a) A dislocation makes a resistance against melting
(Fig. 10).* It follows from the above argument that thermal energy in
superheating is dissipated by dislocation motion (vibrations). It is concluded
that very high mobilities of dislocations are achieved by superheating, and the
free energy of the crystal is lowered by materializing the equilibrium
dislocation-density of zero.

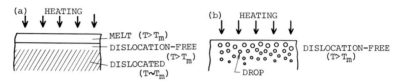

Fig. 13. Difference in melting behavior between dislocated (a) and dislocation-
free crystals (b).

6.4. Dislocations in growth processes

Although dislocations are more stable at growth interfaces, dislocation
mobilities are still quite high, and the equilibrium dislocation density
of zero is achieved easily. Unless there is some resistance against
their motion such as interaction between three dislocations forming a composite
type, dislocations move very fast: For instance, dislocations with $a/2[1\bar{1}0]$
Burgers vectors parallel to the (111) interface move very fast, as seen in
Fig. 11 (the dislocations indicated by the arrows). For dislocations having
Burgers vectors with components perpendicular to the interface, if spiral
growth by the dislocations is assumed to occur in melt growth, the formation
of spiral steps increases free energy of the interface and prevents the
dislocations from intersecting with the interface. This may be responsible
for the formation of hairpin-shaped dislocation half-loops (Fig. 11). If
dislocations intersect with the interface, spiral steps (humps) are formed,
and dislocations are forced to move so as to shorten the step length and to
reduce excess free energy due to the step formation. When such a moving
dislocation meets another dislocation having a Burgers vector with the
opposite sign, they link and stay under the interface by forming a half-loop.
When dislocation density is high, however, spiral steps from neighbouring
dislocations combine before they become long enough to contribute to free
energy of the interface. Consequently, dislocations can intersect with the
interface, when their density exceeds a critical value which increases with the
supercooling at the interface [independent of σ (or step energy per unit
length)], as has been discussed previously [13]. Generally, impurity addition
increases the supercooling which is required for keeping the growth rate at a
certain value and, consequently, may prevent dislocation existence at the
interface. Recently, dislocation-free compound semiconductor crystals such as
GaAs and InP were grown without necking by impurity addition [24]. In such
crystals, dislocations are charged (α and β dislocations), and the
probability for formation of composite dislocations is extremely low.
Therefore, the impurity effect may play an important role in the dislocation-
free growth.

---

* At the melting point, it is unlikely that dislocations are decorated by
  impurity atoms. If we assume a higher concentration of oxygen along a
  dislocation line, we should have a local melting along it, because oxygen
  lowers the interfacial energy.

6.5. Formation of the humps

The observation in Fig. 12 showed that the interface is, at least, at a supercooling of 2°C, i.e., the motion of steps is by a relatively strong driving force due to the supercooling. In order to explain this, it is proposed that the interface is covered with melt layer which contains clusters of atoms like small crystallites. They are hard to incorporate into the crystal lattice, and the step motion is due mainly to incorporation of liquid atoms among the clusters. The higher supercooling is required to compensate the smaller cross-section of the flow of atoms from liquid to crystal. Existence of such an intermediary layer at the interface may be supported by characteristics of impurity incorporation [25]. The hump formation can be understood by assuming that the tip of spiral growth extrudes out of the intermediary layer.

7. CONCLUSION

Our conclusion is summarized as follows:

Melting process. The thermal energy in superheating is dissipated by dislocation motion (vibration), and very high dislocation mobilities are achieved to result in the dislocation density of zero: Melting proceeds by forming a dislocation-free layer. In dislocation-free crystals, locally molten regions (drops) are formed inside them. The drop formation is explained by the observation that the solid-liquid interfacial free energy σ is greatly decreased by oxygen impurity. It is proposed that drops are formed by thermal fluctuation during the conventional melt growth and become an origin of swirl defects.

Growth process. Crystallites are formed in the supercooled region on the melt-side of the interface. Their formation is also considered to be due to oxygen impurity. Both composite and common $a/2<110>$ dislocations exist near the growth interface: The former are composed of three $a/2<110>$ dislocations and are present at the interface so that, as the crystal grows, they propagate into the new crystal. They influence the interface morphology, when their Burgers vectors have a component perpendicular to the interface. The common dislocations cannot exist stably at the interface and are driven into the newly grown dislocation-free region by thermal stresses.

The present status of the video topography may be understood from the in situ study on the crystal growth described above. For more rapid observation with a higher resolution (10 μm), further improvement is in progress.

ACKNOWLEDGMENT

The authors would like to express their sincere thanks to Dr. K. Hayashi and Dr. Tatuoka for their encouragement throughtout this work.

REFERENCES

1. M.E. Glicksman in: Solidification, T.J. Hughel G.F. Bolling Eds. (American Society for Metals, Metal Park 1970) pp. 155-200.

2. G. Champier in: The Characterization of Crystal Growth Defects by X-Ray Methods (Plenum, New York 1980) to be published

3. H. Klapper in Ref. [2]

4. J. Chikawa and I. Fujimoto, Appl. Phys. Lett. 13, 387 (1968).

5. W. Hartmann in: X-Ray Optics, H. -J. Queisser ed. (Springer-Verlag, New York, 1977) pp. 191-217.

6. J. Chikawa in Ref. [2]

7. A. R. Lang in: Modern Diffraction and Imaging Techniques in Material Science, S. Amelinckx, R. Gevers, G Remaut, J. Van Landuyt eds. (North-Holland, Amsterdam 1970) pp. 407-479.

8. For example, see G.D. Heynes, SMPTE Journal 86, 6 (1977).

9. R.H. McMann et al., SMPTE Journal 87, 129 (1978)

10. J. Rossi, SMPTE Journal 87, 134 (1978).

11. R. Nishida and S. Okamoto, Rept. Res. Inst. Electron. Shizuoka Univ. 1, 21, 185 (1966).

12. N. Kato, Acta Cryst. 14, 526, 627 (1961).

13. J. Chikawa and F. Sato in: Proceedings of 11th International Conference on Defects and Radiation Effects in Semiconductors, To be published.

14. T. Abe, J. Cryst. Growth 24/25, 463 (1974)

15. V.V. Voronkov, Kristallografiya 17, 909 (1972), Soviet Physics-Crystallo Crystallography 17, 807 (1973).

16. For example, see B. Chalmers, Principles of Solidification (John Wiley & Sons, New York 1964) pp. 65-76.

17. J. Chikawa and S. Shirai, J. Cryst. Growth 39, 328 (1977).

18. A.J.R. de Kock, in : 1976 Crystal Growth and Materials eds. E. Kaldis and H.J. Scheel (North-Holland) (1977) pp. 662-703

19. J. Chikawa and S. Shirai, Jpn. J. Appl. Phys. 18, Suppl. 18-1, 153 (1979).

20. H. Föll et al., J. Cryst. Growth 40, 90 (1977).

21. Y. Yatsurugi et al., J. Electrochem. Soc. 120, 975 (1973).

22. H. Föll and B.O. Kolbesen: Appl. Phys. 8, 319 (1975).

23. For example, see J.R. Patel, in: Semiconductor Silicon 1977, H.R. Huff, E. Sirtl eds. (Electrochemical Society, Princeton, N.J. 1977) pp. 521-545.

24. Y. Seki et al., J. Appl. Phys. 49, 822 (1978).

25. J. Chikawa and F. Sato, Jpn. J. Appl. Phys. 19, L577 (1980).

OXYGEN, OXIDATION STACKING FAULTS, AND RELATED PHENOMENA IN SILICON

S. M. HU
IBM General Technology Division, East Fishkill, Hopewell Junction, NY 12533,
USA

ABSTRACT

Some salient features of the phenomena of oxidation
stacking faults (OSF) and oxidation enhanced diffusion
(OED) are summarized, and some theories of OSF growth are
critiqued. Then a new theory is developed which shows
the growth to be reaction controlled, and the retrogrowth
to be naturally a regime in the entire OSF growth process.
The theory has provided an estimated bound on the activation energy of self-diffusion in silicon. It is shown
that the observed power law growth kinetics can be explained quite naturally by a bimolecular annihilation
process of the excess self-interstitials. A possible
physical model of such an annihilation process is discussed.
A number of phenomena related to the nucleation of oxygen
precipitates in silicon are reported. Then a model of
nucleation is presented which is shown to be consistent
with all observed phenomena. In this model, the homogeneous nucleation proceeds not merely via the agglomeration
of oxygen atoms, but in combination with small vacancy
clusters. Evidence from the OSF and OED phenomena, and
the nucleation of oxygen precipitates thus necessitates
the dualism of vacancies and interstitials as thermal
defects in silicon at high temperatures.

INTRODUCTION

The study of oxidation stacking faults (OSF) in silicon has been largely
motivated by the recognition of the role of such faults in the defect-limited
yield of silicon microelectronic circuits. Just as importantly, the study
has contributed to the understanding of basic aspects of point defects and
diffusion in silicon. This follows from the recognition that the phenomenon
of oxidation enhanced diffusion (OED) and that of OSF are different manifestations of nonequilibrium point defects in consequence of thermal oxidation
[1, 2]. Due to extensive investigations over the past several years, a
considerable body of information has now been accumulated on the phenomena of
OSF and OED; this will be briefly summarized. A number of theories have been
proposed for the growth of OSF [2-5]; these will be briefly discussed. We
will then develop a new theory of OSF growth, in a very general and natural
way, so that the observed power-law kinetics [6-8] has a very reasonable
explanation, and the retrogrowth phenomenon [6-8] simply emerges as one
regime of the entire OSF growth process.

The topic of oxygen precipitation in silicon has been studied particularly
intensively in the last several years. Nonetheless, the aspect of nucleation
of oxygen precipitates has remained little understood. We will report some
phenomena of the nucleation of oxygen precipitates, and then propose a model
of nucleation that is consistent with all these and previously observed

phenomena. The acceptance both of the model of the nucleation of oxygen
precipitates, and the model of OSF and OED, leads us to a dualism of the
vacancy and the interstitial as the thermal point defects in silicon at high
temperatures.

SOME FEATURES OF OSF AND OED

To provide a perspective of the phenomena of OSF and OED, and to lay a
ground for our development of a theory of OSF growth, we will first summarize
in a list of salient features the experimental results accumulated from
extensive investigations in the past several years:

(1) The growth of OSF varies with time as t^n, where n is between 0.77 and 0.84
[6-9]. The shrinkage of OSF is linear with time [10-13].

(2) Both the growth rate and the shrinkage rate of OSF are independent of
the fault size.

(3) The growth rate and the srinkage rate are unaffected by the proximity to
other OSF.

(4) There exists a retrogrowth regime [6-8], in which the OSF size decreases
with time at a given oxidation temperature, and with temperature for a given
oxidation time.

(5) The oxidation enhancement of impurity (e.g. boron) diffusivity varies
approximately $t^{-1/4}$ [14, 15].

(6) The activation energy for growth is 2.3 eV [6, 9], and the activation
energy for shrinkage is 5.2 ev [10, 13]. There is no significant dependence
of these activation energies on oxidation ambients, or on substrate surface
orientations.

(7) Both OSF and OED are larger on {100} than {111} surfaces, but only by a
factor of $\sim$ 2.

(8) The retrogrowth regime expands considerably into lower temperature
domain for certain vicinal planes of, say, {100} surface [6].

(9) Chloro-ambients (containing e.g., HCl, trichloroethylene, trichloroethane)
decrease OSF growth rate and exapand the retrogrowth regime into shorter time
and lower temperature domain [7, 16-18]. Such ambients also reduce the
effect of OED.

(10) Backside gettering (using, e.g., mechanical lapping, poly-crystalline
silicon films, etc.) does not decrease OSF growth rate, nor expand the retrogrowth regime. It also does not affect OED. [13]

(11) However, both backside gettering and chloro-ambients often greatly
reduce the density of OSF. [13, 16, 19, 20]

(12) The rate of OSF growth increases with oxygen partial pressure [8, 21].
But it also increases with oxidation rate, such as in steam ambients [6].

(13) The growth rate is increased by intermittent removal of the SiO_2 films,
for a fixed length of total oxidation time. [13]

(14) The OSF shrinkage behavior can be greatly affected by the precipitation of oxygen [22].

(15) The OED is effected through excess native point defects, rather than the direct displacement of the impurity atoms themselves into the interstitial position, as demonstrated by the OED of buried boron layer under a 10-μm epi.

(16) The OED of a buried layer at both diffusion fronts are directed down their respective concentration gradients, thus ruling out a defect-impurity counter-flux mechanism, whose effect is unidirectional.

(17) The OED has been reported in one case to have a range of 25 μm from the substrate surface [23].

(18) N-type doping $\stackrel{\sim}{>} 10^{18}$ cm^{-3} has been reported to suppress OSF formation [24].

(19) At sufficiently high temperatures, the OED of phosphorous switches over to an oxidation retarded diffusion [25].

(20) The structure of OSF has been shown [26] to be of the extrinsic type depicted by Hornstra [27].

KINETICS OF OSF GROWTH

From the concept of oxidation generation of excess self-interstitials to a formal theory of stacking fault growth, one needs detailed knowledge of (a) the relationship between thermal oxidation and the kinetics of the generation of excess self-interstitials; (b) the mechanism by which these excess self-interstitials are annihilated in the meanwhile; (c) the controlling mechanism and the kinetics of the incorporation of these excess self-interstitials into the SF.

Abundant experimental evidence has shown that the surface OSF themselves, while certainly capable of capturing excess silicon interstitials to nurture their own growth, are not strong sinks and play an insignificant role in the overall annihilation process of the excess self-interstitials. This conclusion is inevitable since both the OSF growth rate and the OED are not affected by the density of OSF present. In fact, the observation that the rate of OSF growth is essentially unaffected by the proximity of neighboring OSF separated by distances considerably smaller than the size of OSF themselves implies also that diffusion of the excess self-interstitials to the OSF cannot even be the rate-controlling step of the OSF growth, as will be shown later, and that there must be some significant reaction barrier to the construction of an OSF. One can also rule out the role of bulk OSF as the dominant sink since the OSF growth has been observed to be quite independent of the variation in the substrate properties which control the formation of bulk OSF. Then, the prevailing annihilation process must involve any of such processes as clustering of the self-interstitals, recombination with vacancies, and surface regrowth. The first two are bulk processes, and will lead to a decay of the concentration of excess self-interstitials depthwise from surface. A decay length of about 25 μm has been reported [23] from experiments with OED. This could explain my observation [13] that the conventional gettering technique of lapping of substrate back surfaces did not reduce OED, while it had eliminated OSF; the excess self-interstitials could not have reached the back surfaces in any case. The elimination of OSF by the gettering must then be the consequence of the removal of certain fast diffusing species that help

nucleate the OSF. This is in contrast to the effect of oxidation in chloro-
ambients (containing HCl, trichloroethylene, trichloroethane, etc.), which
reduces not only the density of OSF, but also the growth rate of OSF as well
as the magnitude of OED [7, 16-18]. It must then be concluded that oxidation in
chloro-ambients results in a reduced concentration of excess self-interstitials.

The fact that the stacking faults themselves play an insignificant role
in the annihilation of excess self-interstitials allows us to analyze the
transient behavior of the excess interstitials independently of the growth of
the stacking faults. My original OSF growth model has the form [2]

$$dr/dt = \pi a_0^2 D_I (C_I - C_I^*),$$ (1)

where r is the radius of the stacking fault, a_0 the caputre radius of interstitials by the bounding dislocation of the fault, D_I the diffusivity of
self-interstitials, C_I and C_I^* are the concentrations of the self-interstitials
in transient and at thermal equilibrium respectively. As pointed out by Lin
et al. [14], this equation predicts the growth rate to be independent of the
size of OSF, in agreement with Shiraki's experiments [16]. In contrast,
Murarka's model, which adapted Kahlweit's theory of growth of spherical
nuclei [28], predicts a growth rate that is inversely proportional to its
sizes, in contradiction with Shiraki's experimental results. Many points in
Murarka's model have already been criticized by Leroy [3] and by Lin et al.
[4]. However, three serious faults should be commented on here. The first
is the assumption of a time-independent concentration of the excess interstitials throughout the course of OSF growth. Kahlweit intended his precipitate growth theory for the beginning period of nucleation in an infinite
reservoir, which justifies the assumption of a time-independent supersaturation
and leads to a volume-time relationship of

$$V(t) = (4\pi/3) [2v_m D_I (C_I - C_I^*)]^{3/2} t^{3/2},$$ (2)

where v_m is the average atomic volume of the silicon lattice. The same
assumption of a time-independent supersaturation is, however, inappropriate
in the case of OSF growth. The phenomenon of OSF retrogrowth at large time
[6-8] and the decrease of OED with time [14, 15] can only be explained if
the supersaturation decreases with time. Our rationale would also call for
the generation rate of excess self-interstitials to vary with the oxidation
rate, which decreases with time.

The second fault of the model relates to a crucial mistake made in
grafting Kahlweit's model of growth of a spherical precipitate to that of a
disc precipitate. Murarka simply replaced the volume of a sphere in Eq. 2 by
an equivalent volume of a disc $\pi r^2 b$ (b is Burger's vector), without regard to
the fact that a change in the geometry of the precipitate also changes the
boundary condition that governs the diffusion-controlled growth of the precipitate. The boundary of a disc moves much faster than does that of a
sphere. But this V(t) relationship came from the integration of

$$d(4\pi r^3/3)/dt = 4\pi r^2 v_m D_I (grad \, C_I)_r = 4\pi r v_m D_I (C_I - C_I^*),$$ (3)

where the rightside term is derived from the boundary condition governing the
diffusion to a sphere. It is easy to see that had the volume $(4\pi r^3/3)$ in
Eq. 3, instead of that in Eq. 2, been replaced by an equivalent volume of a

disc πbr^2, one would have obtained an r(t) relation which is different from that given by Murarka. The falsehood of the "equivalent volume" concept becomes obscured when replacement is made in the integrated expression (Eq. 2), in which the boundary condition is lost by integration. It may be pointed out that, in order to obtain the growth kinetics of a disc precipitate, it is unnecessary to go the way of juggling the growth kinetics of a spherical precipitate through the device of the unjustified "equivalent volume concept". Analytical expressions for the growth of a disc precipitate are available in the literature, and will be encountered later (Eq. 17). The third serious fault of the model is the assumption of an "equilibrium supersaturation" of self-interstitials in mass action with the ambient oxygen pressure. This concept is thermodynamically unsound: any "supersaturation" (over the intrinsic equilibrium concentration of self-interstitials) induced by the boundary condition is held in electrochemical balance with the outside phase at the boundary, and is not available for OSF growth. This situation should be viewed as analogous to the case of ambient-induced surface charge or space charge, which is not available for recombination. The intended purpose of this concept was clearly for the explanation of the effect of oxygen partial pressure, which however, can be more consistently explained as a consequence of oxidation rate. Only a model in which the concentration of excess interstitials increases with oxidation rate can be consistent with all phenomena, including item (13) in the list of features of OSF growth.

A modification of Hu's derivation of OSF growth has recently been given by Leroy [3], who has also combined it with the shrinkage phenomenon based on the experimental data of Hashimoto et al. [10]. We will comment on Leroy's model subsequently, both in connection with his concept of an "equilibrium excess interstitial concentration", and in connection with his model of diffusion-controlled growth.

Concentration of Excess Self Interstitials

More rigorously, $(C_I - C_I{}^*)$ in Eq. 1 should be replaced by $(C_I - C')$, where C' is the concentration of excess self-interstitials in equilibrium with the stacking fault. C' will be determined in the next section, where it will be shown that, when the elastic energy associated with the stacking fault is small compared to the faulting energy (valid for $r > 100$ nm), C' will be independent of r and hence of t. One expects $C' > C^*$ because of the extra energy associated with the stacking fault. With this replacement, the growth of OSF can be obtained from the time integral of Eq. 1 as

$$r(t) = r(0) + K\int_0^t [(C_I - C_I{}^*) - (C' - C_I{}^*)]dt, \qquad (4)$$

where we have also replaced the factor $\pi a^2 D_I$ by a lumped parameter K, which will be determined in the next section more rigorously than in my previous work [2]. Clearly, the term $(C_I - C_I{}^*)$ inside the brackets in Eq. 4 vanishes in an inert ambient (or when $t \to \infty$). Since $(C' - C_I{}^*)$ is independent of t, the shrinkage in an inert ambient (or for very large t) is, from Eq. 4

$$r(t) = r(0) - K(C' - C_I{}^*)t. \qquad (5)$$

The shrinkage is thus linear, and independent of the size of OSF.

Our main concern in this section is the determination of C_I as a function of t, such that it will be consistent with the observed kinetics of OSF growth. In my original theory [2], I assumed that the rate of generation of excess interstitials is directly proportional to the rate of thermal oxidation, which decreases with time as $t^{-1/2}$. This, as readily seen from Eq. 4 led to a parabolic law of OSF growth, which was to be contradicted later by my own experimental results [6] of a 0.8-th power law kinetics. Subsequent investigators have generally reported a power law kinetics with the exponent n ranging from 0.75 to 0.84 [6-9]. To arrive at such a power law kinetics of OSF growth, one sees that $[C_I(t) - C_I{}^*]$ must vary as $t^{n-1} = t^{-m}$, where m is between 0.16 to 0.25. A simple way to obtain the desired value of m is as done by Antoniadis et al. [14] and by Lin et al [4], who assumed the concentration of the excess interstitials to vary as some power of oxidation rate

$$C_I - C_I{}^* \propto t^{-m} \propto (dX/dt)^{2m}. \tag{6}$$

This expression involves a concept of segregation of the excess interstitials between the SiO_2 and the Si phases, and that the excess interstitials are annihilated only in the SiO_2 phase. On the other hand, Leroy [3] assumed an equilibrium between C_I and the oxidant at the interface through a "law of mass action" that involves raising the oxidant concentration to an empirical, noninteger power. The conventional mass action law as we know it is based on probabilistic reasoning, and is meaningful only for integer power of concentrations. Furthermore, the concept of an "equilibrium supersaturation" of self-interstitials is somewhat similar to that of Murarka, which we have already criticized.

A more reasonable way to arrive at the correct power law is to simply assume that G, the rate of generation of self-interstitials, is nonlinearly related to the oxidation rate as

$$G \propto (dX/dt)^{2m} = At^{-m}, \tag{7}$$

where A is an empirical constant. In contrast to the mass action law, there is no rationale against m being non-integer in this expression, and the question of why the relationship should be nonlinear is merely as relevant as why it should be linear. We extend the reasoning given in [2] and find $C_I(t)$ from the solution of

$$dC_I/dt = At^{-m} - k_1(C_I - C_I{}^*), \tag{8}$$

where k_1 is a rate constant for the annihilation of excess interstitials. With the initial condition of $C_I(0) = 0$, Eq. 8 admits the solution

$$C_I(t) - C_I{}^* = Ak_1{}^{-1}t^{-m}\exp(-k_1t) \sum_{n=0}^{\infty} \frac{1}{n+1-m} \frac{(k_1t)^{n+1}}{n!}, \tag{9}$$

$$\simeq Ak_1{}^{-1}t^{-m} \quad \text{for large t.} \tag{10}$$

Substituting Eq. 9 into Eq. 4 and integrating, one obtains

$$r(t) = r(0) + Kk_1{}^{-1}[(1 - m)^{-1}At^{1-m} - (C_I - C_I{}^*)] - K(C' - C_I{}^*)t \tag{11}$$

$$\simeq r(0) + KA[(1 - m)k_1]^{-1}t^{1-m} - K(C' - C_I{}^*)t \quad \text{for large t.} \tag{12}$$

We now show that a model of bimolecular annihilation of the excess interstitials will also lead to the correct power law kinetics, while allowing us to retain the intuitively appealing assumption of G being directly proportional to the oxidation rate. Assuming C_I to be much larger than C_I^*, we write

$$dC_I/dt = At^{-1/2} - k_2 C_I^2. \tag{13}$$

This is an equation of Riccati type and can be solved analytically. However, the transient behavior of $C_I(t)$ is not readily apparent from the complex analytical expression. We give instead the asymptotic solution at large time,

$$C_I(t) \simeq (Ak_2^{-1})^{1/2} t^{-1/4}, \qquad \text{for} \quad t > (4Ak_2)^{-2/3}. \tag{14}$$

The assumption of a bimolecular kinetics of interstitial annihilation is not as artificial as it might seem. Although we do not have sufficient basis to be specific about any particular process actually operating, two illustrative examples can be given: (a) The first is a bimolecular clustering of the excess interstitials in the bulk, which at the same time would lead to a decay of the excess concentration with distance from the surface, in agreement with the experimental observation of Taniguchi et al. [23]. (b) The surface regrowth process postulated earlier [2], which proceeds via the propagation of surface steps or kinks, may conceivably involve the simultaneous capture of two interstitials. Fig. 1 depicts a likely structure of steps on a silicon

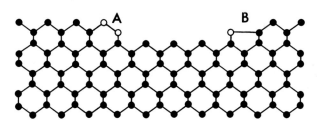

Fig. 1 Schematic rendition of step structures A and B for diatomic and monatomic mechanisms of step propagation, respectively.

{100} surface, generated by thermodynamical fluctuations. The "epitaxial growth" then proceeds via the attachment of extra silicon atoms on such steps, causing the latter to move in a direction that expands the area of surface terraces. It can readily be seen from Fig. 1 that the simultaneous capture of two silicon atoms (open circles attached to the left step in the figure) is required in order not to alter the step structure. The capture of a single atom would result in a new step structure (the open circle attached to the right step in the figure). Since the equilibrium step structure, A, that prevails on the original surface is the minimum energy configuration, the new, nonequilibrium step structure B is necessarily a higher energy configuration. Therefore, the bimolecular mechanism of step propagation will dominate provided

$$C_I/C_s > \exp[(H_A - H_B)/kT], \tag{15}$$

where C_s is the concentration of lattice sites ($C_s = 1/v_m$), and H_A and H_B are the enthalpies of formation of surface steps of structures A and B respectively. This condition may likely be met in the initial period of oxidation. As C_I decreases with time (because of decreasing oxidation rate), the probability of two excess silicon atoms neighboring each other eventually becomes so small that the bimolecular process of step propagation becomes negligible. The monomolecular process of step propagation, eventually becoming more important, consists of two consecutive processes in cycle: the capture of a single silicon atom to form structure B, and the capture later of another silicon atom to recover the equilibrium structure A. Clearly, the formation of the higher-energy structure B is by far the slower of the two consecutive processes, and therefore controls the overall rate of monomolecular annihilation. We extend Eq. 13 to include the monatomic annihilation process that may become significant at long oxidation time and/or high oxidation temperature

$$dC_I/dt = At^{-1/2} - k_1(C_I - C_I^*) - k_2(C_I^2 - C_I^{*2}). \tag{16}$$

The growth of OSF resulting from first the solution of Eq. 16 and then the integration of Eq. 4 would be expected to exhibit three regimes as rendered in Fig. 2; but in reality, only two regimes can be unambiguously distinguished: the growth and the retrogrowth regimes. The retrogrowth comes from the second term in the integrand of Eq. 4, and will be analyzed in the next section.

The dependence of the OSF growth on oxygen partial pressure, p, as we have suggested earlier, comes from the dependence of oxidation rate, which varies with the oxygen partial pressure approximately $p^{1/2}$. For either a monatomic annihilation mechanism in which $C_I \propto A \propto (dX/dt)^{1/2}$, or a bimolecular annihilation mechanism in which $C_I \propto A^{1/2} \propto (dX/dt)^{1/2}$, the same pressure dependence of $C_I \propto p^{1/4}$ is obtained.

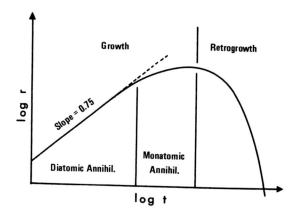

Fig. 2   Three regimes in the growth of oxidation stacking faults.

341

The Growth of OSF

The fact that the growth rate of OSF is quite independent of the distances separating them would suggest a reaction-controlled rather than a diffusion-controlled mechanism of growth. This is the assumption already made in my original theory, in which the concentration of excess interstitials is taken to be uniform up to the dislocation core. This ensures that each stacking fault has the same uniform environment in which to grow, regardless of how closely it is neighboring other SF. Leroy [3] also recognized the independence of the OSF growth on d, the spacing between stacking faults, and have in fact highlighted this aspect of OSF growth; yet his model shows 10/11 of the drop in interstitial concentration to be outside the dislocation core, a feature that can only arise from a diffusion-controlled mechanism. This contradiction of his model is, however, obscured by his adaptation of the steady-state radial diffusion through a hallow cylinder in his diffusion-controlled growth model. Such an adaptation is not new, and is merely an expediency to circumvent the difficulty that a steady-state solution does not exist for diffusion to an infinitely long cylindrical sink in an infinite medium. This adaptaion necessitates the designation of a fictitious "outside radius", r_0, of the imaginary cylinder, whose inside radius is set by the core radius of the dislocation. Devoid of a physical meaning, r_0 is in all known cases given an arbitrary value, with the usual apology that any value of r_0 assigned is of no great consequence since its effect is logarithmic. By assigning a value of $r_0 < d$, and letting $C(r \geq r_0)$ = constant, as Leroy did, one creates a fictitious condition in which the concentration of excess interstitials surrounding each cylinder is uniform in the space between stacking faults. Being fictitious, however, such a condition cannot in reality insure the OSF growth to be independent of proximity. There is just no actual physical means to enforce that condition of an imaginary cylinder. The usefulness of the mathematical expediency of an imaginary hallow cylinder is strictly limited to affording a simple way for an order-of-magnitude estimate of the diffusional flux to a line sink, and may never be for the fictitious spatial distribution of the diffusant.

We now show that the growth of OSF can be independent of d only if the growth is reaction-controlled. The solution for steady-state diffusion to a sink of a closed geometry in an infinite space is always bounded, in contrast to the problem for a sink of an infinite cylinder. The case of a spherical sink is well known. As a sink, the bounding dislocation of an oxidation stacking fault can be represented by a torus. The solution to the electrostatic problem of a torus in an infinite medium is available in the literature [29]. Flynn [30] pointed out that the effective radius of a localized sink of any geometry is equal to the electrostatic capacity of a conducting body having the same dimension as the sink, the electrostatic problem and the steady-state diffusion problem being governed by the same Laplace equation. Consider a torus having a core radius r_c and a loop radius r. Following Flynn [30] (with some minor modifications and a correction), we equate the rate of increase of the volume of the SF to the product of the atomic volume v_m and the total flux across the surface of the torus. The full expression thus obtained consists of an infinite series of Legendre polynomials. The asymptotic approximation to the full expression, valid for $r > 3r_c$ (a condition that prevails in OSF growth), gives

$$d(\pi r^2 b)/dt = 4\pi^2 v_m (r^2 - r_c^2)^{1/2} [\ln(8r/r_c)]^{-1} D_I (C_B - C_c). \qquad (17)$$

A somewhat similar expression has been given by Seidman and Balluffi [31].

In Eq. 17, C_B is the concentration of interstitials at an infinite distance from the dislocation core, and is hence identical to $C_I(t)$ obtained in the preceding section for a uniform substrate devoid of OSF. C_c is the interstitial concentration at the surface of the core. C_c is to be determined by simultaneously satisfying the reaction occurring at the core. We may extend Eq. 17 to the case of n neighboring stacking faults separated by small distances. If $(n - 1)d < r$, (a case that often happens, as in the example given by Leroy [3]) one may approximate the SF assemblage as a single torus with a radius r and an effective core radius nr_c and obtain

$$d(n\pi r^2 b)/dt = 4\pi^2 v_m (r^2 - n^2 r_c^2)^{1/2} [\ln(8r/nr_c)]^{-1} D_I (C_I - C_c). \tag{18}$$

$$\simeq [4\pi^2 v_m r/\ln(8r/nr_c)] D_I (C_I - C_c). \tag{19}$$

Since the total diffusional flux across the surface of the torus must equal the rate at which the interstitial atoms in the core have surmounted the reaction barrier and become part of the stacking fault, we may write

$$d(\pi r^2 b)/dt = v_m (2\pi^2 rr_c^2)(C_c - C')f^*$$
$$= v_m 2\pi^2 r(r_c^2/r_a^2)(C_c - C')D_I \exp(-\Delta H/kT). \tag{20}$$

In Eq. 20, $2\pi^2 rr_c^2 (C_c - C')$ gives the number of excess (over C') interstitials within the torus. r_a is the interatomic distance in the lattice. f^* is the frequency with which an interstitial will surmount the reaction barrier. ΔH denotes the difference between the reaction energy barrier and the activation energy of interstitial migration (see Fig. 3). Combining Eqs. 17 and 20, one has

$$C_c = (C_I + BC')/(1 + B); \tag{21}$$

$$B = (n/2)[\ln(8r/nr_c)](r_c/r_a)^2 \exp(-\Delta H/kT). \tag{22}$$

Eq. 19 then becomes

$$dr/dt = (\pi v_m/b)(r_c/r_a)^2 (1 + B)^{-1} D_I (C_I - C') \exp(-\Delta H/kT). \tag{23}$$

The growth of OSF will be diffusion-controlled if $\Delta H \overset{\sim}{<} 0$, so that $B \gg 1$. Note that r is usually many orders of magnitude larger than r_c. C_c may then approach C' according to Eq. 21. Eq. 19 then reduces to

$$dr/dt \simeq (\pi v_m/b) D_I (C_I - C')/(n/2)\ln(8r/nr_c) \tag{24}$$

Eq. 24 indicates that if the rate of OSF growth is diffusion-controlled, it would be approximately inversely proportional to the number of SF in a dense assemblage--a conclusion that is in contradiction to experimental observations. On the other hand, if $\Delta H \overset{\sim}{>} 4kT$, so that $B \ll 1$, the OSF growth becomes reaction-controlled. C_c then approaches C_I according to Eq. 21. Eq. 20 then reduces to

$$dr/dt = (\pi v_m r_c^2/br_a^2)(C_I - C')D_I \exp(-\Delta H/kT), \tag{25}$$

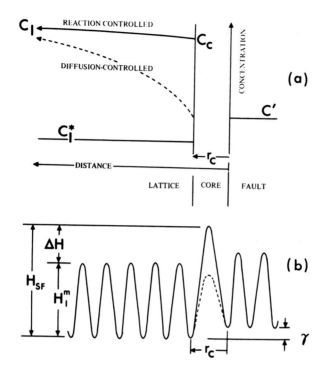

Fig 3 Concentrations (a) and potential energy (b) of silicon self-interstitials as functions of distance from an OSF in a one-dimensional abstraction.
The dashed curves represent a hypothetical situation of diffusion-controlled
growth of OSF.

which gives a rate of OSF growth that is independent of the number of SF in a
dense assemblage, in agreement with experimental observations.

It now remains to determine C'. We may express the total energy of a
stacking fault as

$$E = \pi r^2 \gamma + [\mu b^2 r / 2(1 - \nu)][\ln(8r/b) - 1] - (\pi r^2 b / v_m) kT \ln(C'/C_I^*) - \pi r^2 b \sigma_n. \quad (26)$$

In Eq. 26, the first term is the faulting energy, the second term the elastic
energy due to the dislocation loop [32] bounding the stacking fault, the third
term the chemical potential for the $\pi r^2 b / v_m$ interstitial atoms extracted from
solution into the stacking fault, and the fourth term the strain energy due
to an external stress (e.g., due to a surface film). σ_n is the component of
the external stress normal to the SF. It can be estimated that for the size
of SF of about 100 nm or larger, the elastic energy term becomes unimportant
compared to the faulting energy term. Furthermore for substrate stresses

limited to about 10 MPa (any stress above this will be truncated by plastic flow of the substrate), $|b\sigma_n| \ll \gamma$, and the last term in Eq. 26 can also be neglected. We seek to obtain C' as a function of SF size r by minimizing E in Eq. 26 with respect to r for a given C'. After the omission of the second and the last energy terms because of their smallness, we obtain

$$C' = C_I^* \exp(v_m \gamma / bkT), \tag{27}$$

which turns out to be r-independent. (C' would be r-dependent if the elastic energy term dominated). Now we can substitute Eq. 27 into Eq. 25 and, on assumption of a monatomic annihilation process, express $(C_I - C_I^*)$ in terms of Eq. 10:

$$dr/dt = (\pi v_m r_c^2 / r_a^2 b) e^{-\Delta H/kT} [D_I (C_I - C_I^*) - D_I C_I^* (e^{v_m \gamma / bkT} - 1)]$$

$$= (\pi v_m r_c^2 / r_a^2 b) e^{-\Delta H/kT} [Ak_1^{-1} D_I t^{-m} - D_s C_s (e^{v_m \gamma / bkT} - 1)], \tag{28}$$

where C_s is the concentration of lattice sites $(C_s = 1/v_m)$, and where we have also made the substitution $D_s C_s = D_I C_I^*$, on the assumption of an interstitialcy mechanism of self-diffusion [2, 33-36]. Inside the brackets on the righthand side of Eq. 28, the first term drops out either in an inert ambient (A = 0), or at very large time $(t^{-m} \rightarrow 0)$. Furthermore, because D_s has a much larger activation energy (about 5 eV) than does D_I (about 2 eV), the second term in the brackets may dominate at sufficiently high temperatures. This is, of course, just the retrogrowth phenomenon that I observed some years ago [6]. The rate of shrinkage of OSF can be given its own expression by dropping the first term in the brackets in Eq. 28. Since $\gamma v_m/b < kT$, we may take a linear approximation for the exponential term inside the brackets in Eq. 28. We thus obtain the OSF shrinkage rate as

$$dr/dt = -(\pi v_m \gamma r_c^2 / r_a^2 b^2 kT) D_s \exp(-\Delta H/kT). \tag{29}$$

This expression indicates that the OSF shrinkage rate is independent of size and time, which is in agreement with experimental observations. Eq. 28 may be integrated to give the final expression for OSF growth:

$$r(t) = r(0) + (\pi v_m r_c^2 / r_a^2 b) e^{-\Delta H/kT} [(1 - m)^{-1} Ak_1^{-1} D_I t^{1-m} - (\gamma/bkT) D_s t], \tag{30}$$

where we have again made a linear approximation for the exponential term involving the faulting energy. If the bimolecular mechanism of excess interstitial annihilation, treated in the preceding section, is considered to be the more likely, the reader can readily modify both Eqs. 28 and 30 by replacing the first term inside the brackets in both equations with $C_I(t)$ from Eq. 14, and with its time integral, respectively.

The following parameters are assumed:

$$v_m = 1.97 \times 10^{-23} \text{ cm}^3; \qquad \gamma = 70 \text{ ergs/cm}^2;$$

$$r_c \simeq b = 0.32 \text{ nm}; \qquad r_a = 0.23 \text{ nm}.$$

The unknown parameters are D_I, ΔH, A, and k_1 (or k_2 if the annihilation mechanism is bimolecular). While these can be analyzed in conjunction with the retrogrowth and the enhanced diffusion, we will not attempt to discuss this here. The activation energy of silicon self-diffusion reported in the literature ranges from 4.7 to 5.13 ev [37-41]. The author's estimate of the activation energy of OSF shrinkage is 5.2 eV [10, 13], out of which at least 0.4 eV must be allocated to ΔH, as discussed earlier. To best accommodate this requirement and the range of activation energy of self-diffusion available, we assign a value of 0.4 eV for ΔH, thus yielding an activation energy of self-diffusion of 4.8 eV from the shrinkage data. This is close to the values given by Peart [37], Ghoshtagore [39] (our criticism of Ghoshtagore's data, given in [34], remains unchanged nonetheless), and Kalinowski and Sequin [41]. If we now substitute into Eq. 29 the values of parameters given above, and the silicon self-diffusivity of Peart, one obtains

$$dL/dt \simeq 2dr/dt = - 2.2 \times 10^{14}T^{-1}\exp(-5.2 \ eV/kT), \quad cm/s, \quad (31)$$

where L is the chord of a stacking fault intersecting the substrate surface, and is assumed to be approximately given by the diameter of a circular stacking fault. For illustration, Eq. 31 predicts a shrinkage rate of 60.4 nm s^{-1} at 1150°C, whereas the experimental results in Fig. 4 of [22] gives a value of 98 nm s^{-1}. The growth part of Eq. 30 (or Eq. 28) on the other hand, can only be determined empirically because of the factor Ak_1^{-1}.

NUCLEATION OF OXYGEN PRECIPITATES

Both heterogeneous [42, 43] and homogeneous [44, 45] nucleation of oxygen precipitates in silicon have been reported in the literature. These have been based on limited observations, carried out under limited experimental conditions. Consequently, the models from such reports are not consistent with all the phenomena observed. We will not comment on these reports in the limited space of this article. Rather, I will first present a number of phenomena related to the nucleation of oxygen in silicon, from my own observations made in the last five years, and will then propose a model of nucleation that is consistent with all such observations.

Ambient Effect

It has been observed that [46] the precipitation of oxygen in silicon is retarded in oxidizing ambients. More recent investigations have shown the extent of the retardation to vary considerably, depending on such variables as the carbon content of the silicon crystal, how the crystal was grown, the position of the specimen in the crystal, and the oxygen content. The model, suggested earlier [46], that small vacancy clusters take part in the formation of nuclei is consistent with this phenomenon. The thermal oxidation of silicon generates excess self-interstitials, which in turn can be expected to annihilate the vacany clusters, at least to some extent. It would seem reasonable that the existence and the distribution of such vacancy clusters are dependent on the crystal variables just mentioned.

Banded Distribution of Oxygen Precipitates

Frequently, oxygen precipitates out very inhomogeneously in silicon substrates, exhibiting banded structures. This occurs despite that the initial concentration of dissolved oxygen was quite uniform. Quantitative study of such inhomogeneous precipitation can be made by one-dimensional profiling across the bands, using either the technique of scanning infrared laser tuned at the oxygen absorption peak (9 μm) [47], or stepping spreading resistance measurements of thermal donors converted from those oxygen atoms remaining in solution after thermal processings. A common procedure for the conversion of dissolved oxygen into thermal donors is to heat-treat the sample at 450°C for 100 hours. The formation of oxygen thermal donors has been extensively discussed in the literature [48-50]. Two-dimensional mappings, on the other hand, would provide a better perspective of the structural features of inhomogeneity of oxygen precipitation. This can be accomplished in two ways. After each thermal annealing (for precipitation), the sample is subjected to thermal conversion treatment. Since the thermal conversion is carried out at such a low temperature, no further precipitation will occur. The specimen is then suitably beveled (5° to 10°) and polished. In the first way, the sample is then copper-plated. Regions where oxygen has precipitated out would retain the P-type conductivity of the original substrate, and are therefore not copper-plated. Other regions where substantial amount of oxygen had remained in solution exhibit an N-type conductivity due to the thermal donors, and are copper-plated. Fig. 4a is a montage of photomicrographs showing a strip of beveled

Fig. 4 Zonal oxygen precipitation revealed on a 5° bevel by (a) thermal conversion and copper staining, and (b) modified Sirtl etching. The light regions in (a) were copper-plated, where much oxygen had remained in solution. Oxygen precipitates were revealed as etch mounds in (b). The sample had been annealed at 1000°C in vacuo for 24 hrs. Reference bar = 2 mm.

sample. The banded regions which are copper-plated (light regions) are where most of the original oxygen had not precipitated, whereas regions which are not plated (dark regions) are where the oxygen had been substantially exhausted by precipitation. In the second way, the sample of Fig. 4a, say, can be reused by first removing the plated copper in nitric acid, and then preferentially

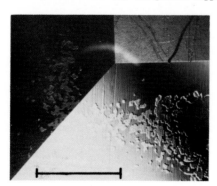

Fig. 5 Thermal conversion/copper plating on two penpendicular 5°-bevels, showing a perspective view of the precipitation inhomogeneity. The horizontal bevel corresponds to region R in Fig. 4a. Reference bar = 0.4 mm.

etched in a modified Sirtl etch, by which oxygen precipitates are revealed as etch mounds (Fig. 4b). Fig. 5 and Fig. 6 show magnified views of selected regions corresponding to Fig. 4a and Fig. 4b, respectively. Since the dissolved oxygen was initially uniform, it is clear that some other factor must

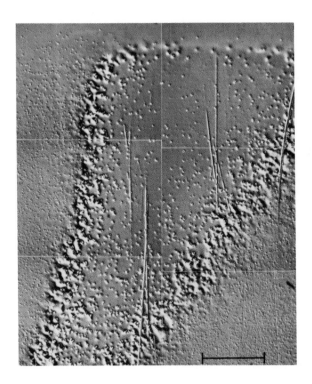

Fig. 6 Magnified view of region S in Fig. 4b, showing a zone of low precipitate density amid zones of high precipitate density. The precipitates were revealed as etch mounds. Reference bar = 0.2 mm.

have effected the oxygen nucleation, and that this factor must be spatially inhomogeneous. Since the spatial distribution of alternating zones of slow and fast precipitation is reminiscent of the distribution of swirl defects [51 - 55], one may suggest that the banded distribution of precipitates could correspond to zones alternatingly richer and leaner in small vacancy clusters, or oxygen-vacancy clusters, and that such clusters are an essential ingredient in the formation of precipitate nuclei.

Density-Time-Temperature Relation

By collecting data on the density of precipitates in the high density zones from samples annealed at temperatures ranging from 800 to 1150°C, for various annealing times up to 96 h, we obtain a density-time-temperature relationship as shown in Fig. 7. Such a relationship, in which the density of precipitates increases with time at a given temperature, and increases with

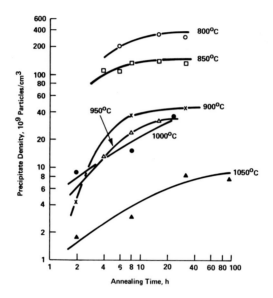

Fig. 7 Density of precipitates as a function of annealing time and temperature (in vacuo). All wafers had initial oxygen concentrations of 1.3 - 1.4 x 10^{18} atoms/cm^3.

the decrease of temperature at a sufficiently large time, can only indicate a homogeneous nucleation mechanism, in which nuclei are continuously assembled with the passage of time, and the critical nucleus size decreases with the decrease of temperature. Note that the highest density of precipitates observed in the temperature and time range covered in this study is a few orders of magnitude higher than the highest density of B-clusters of swirl defects so far reported [51 - 53].

Sequential Heat Treatments

Sequential heat treatments in steps of different temperatures can help provide insight into the mechanism of nucleation. The overall rate of precipitation, defined as the total amount of oxygen that has come out of solution, is seen in Fig. 8 to be greatly enhanced by a prior low-temperature annealing. This is easily understood in terms of the conventional theory of homogeneous thermal nucleation. The first-step low-temperature heat treatment produces a high density of nuclei, according to Fig. 7. The second step of a higher-temperature heat treatment then promotes the diffusion-controlled growth of precipitates on those nuclei already created in the first step. One perhaps would, according to conventional wisdom, predict the results of a high-low two-step heat treatment to be essentially the same as a single-step heat treatment, with the high temperature first step omitted. The reasoning runs

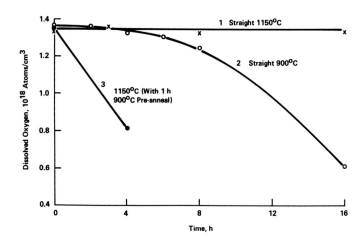

Fig. 8 The effect of low temperature (900°C) prior heat treatment on the accelerated precipitation in subsequent higher temperature heat treatments (1150°C).

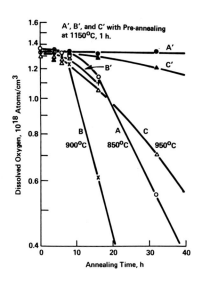

Fig. 9 Effect of high temperature (1150°C) heat treatments on retard-ation of precipitation in subsequent lower temperature (850, 900 and 950°C) heat treatments. Curves A', B' and C' are for samples with 1-h prior heat treatment at 1150°C in vacuo, and then heat-treated together with samples A, B and C respectively.

350

this way: Since the density-time-temperature relationship (Fig. 7) of one-step heat treatments shows that the nuclei are formed thermally, and their density increases with time, a simple backward extrapolation suggests that they were not present in the as-grown crystals. Then, in the high-temperature heat treatment step, there are few nuclei for oxygen to precipitate on. Therefore, little or nothing should have occurred in the first step. The actual results, shown in Fig. 9, contradict this prediction. The high-tempera-ture step appeared to have irreversibly retarded the precipitation occurring in the low-temperature step that followed.

Nucleation Model

     We now suggest a model of nucleation that is consistent with all these four phenomena. While the density-time-temperature relationship suggests a homogeneous (in the fast precipitation zones) thermal nucleation, the nuclea-tion involves not just the agglomeration of oxygen atoms themselves, but also involves some forms of very small vacancy clusters. Such small clusters can be thought of as the relatives of the more sizable B-clusters of the swirl defects, the B-clusters probably being vacancy clusters [53, 55, 56]. The spatial distribution of such small vacancy clusters explains the occurrence of the fast and the slow precipitation zones. Such vacancy clusters must be so small that they would not act as infinite sinks for the oxygen, nor be able to accommodate the number of oxygen atoms needed in a nucleus of a critical size. For otherwise, the nucleation would have been a heterogeneous one, and the precipitate density would have been a constant independent of time and tempera-ture. Oxygen dissolves in silicon interstitially, and it diffuses in silicon also interstitially. This means that there is no counter flux of lattic silicon atoms when the interstitial oxygen atoms are amassing into a nucleus, thus creating an enormous local excess of matter. If the precipitate phase is SiO_2, as recently identified [57], there will be a local volume expansion of

about 127%. This situation changes the rule of the game in the play of energy of nucleation, from one of surface energy vs. chemical potential to one of primarily strain energy vs. chemical potential, with the possibility that the strain energy may dominate all the way up to an unreasonably large nucleus size. What this means is that a homogeneous nucleation may not be feasible, unless it is accompanied by a simultaneous ejection of lattice silicon atoms. The latter process means an additional amount of energy for the formation of SiO_2 precipitates of subcritical size in a silicon matrix, against that for SiO_2 precipitates of super-critical size, the latter energy being determinable

from solubility. It is clear that much of the strain energy could be allevi-ated if a vacancy cluster participates in the formation of the nucleus, and thereby accommodates a fraction of the excess material. It should also be clear that if the vacancy clusters are so large as to able to accommodate more than the number of oxygen atoms in a nucleus of critical size, the nucleation would be a heterogeneous one. Inasmuch as an "initial state" is almost impos-sible to define in a given crystal, some of the small vacancy clusters in the starting substrate could already have some oxygen atoms in them. While the numbers of oxygen atoms in these vacancy clusters may be small, they could easily become supercritical at sufficiently low temperatures. Should one then say that there are nuclei already existing in the as-grown crystal? This is merely a matter of definition. The dividing line disappears at sufficiently low temperatures, at which the supersaturation is so enormous that the critical size of a nucleus is that of a single oxygen atom!

     In a two-stage high-low heat treatment, the preexisting vacancy clusters are broken up in the first-stage high-temperature heat treatment, into monovacancies which then readily escape to the surface of the thin substrate ($\sim$ 400 μm). This process is irreversible. On the other hand, a breakup of clusters that consist only of oxygen atoms is clearly reversible, since most of the oxygen atoms would remain in the substrate.

In summary, only a homogeneous nucleation model which involves the participation of very small vacancy clusters can explain all the observed phenomena consistently. The model is also very rational, and could even be deduced from the properties of oxygen and SiO_2 in silicon--admittedly an after-thought. But it should be added that the vacancy cluster may not have a monopoly on the oxygen homogeneous nucleation process. Conceivably, other agencies (e.g., carbon [58]) could also have the requisite property of reducing the energy of nucleation. Their dominance could lead to a different set of phenomena.

CONCLUSION

A refined theory of the growth of OSF has been developed, in which the retrogrowth is a natural regime in the growth process. On the other hand, the phenomenon of retarded phosphorous diffusion [25] does not have the same natural explanation. Our derivation has shown that, to be free of a proximity effect, the OSF growth must be reaction-controlled, rather than diffusion-controlled as assumed in other published models. The theory leads to a reaction barrier of at least 0.4 eV higher than the activation energy of migration of self-interstitials, and also to an activation energy of self-diffusion of 4.8 eV as an upper limit. It is shown that the power-law kinetics of growth can be very reasonably explained via a bimolecular annihilation of self-interstitials.

The nucleation of oxygen precipitates in silicon is shown to be a special kind of homogeneous thermal nucleation and involves small vacancy clusters as an ingredient. The homogeneous nucleation is deduced from, among others, the density-time-temperature relation of the precipitates. The involvement of vacancy clusters is essential to the explanation of such other phenomena as the retardation effect of oxidizing ambients, and the irreversible inhibition effect of high-low temperature sequential heat-treatments; but it can also explain the banded distribution of oxygen precipitates.

Taking at once both the model of nucleation of oxygen precipitates, and the model of OSF/OED, as enunciated here, one comes to recognize the dualism of the vacany and the interstitial as the thermal point defects in silicon at high temperatures. This reinforces the conclusion reached previously [2, 53, 55].

REFERENCES

1.   I. R. Sanders and P. S. Dobson, Philos. Mag. **20**, 881 (1969).

2.   S. M. Hu, J. Appl. Phys. **45**, 1567 (1974).

3.   B. Leroy, J. Appl. Phys. **50**, 7996 (1979).

4.   A. M. Lin, R. W. Dutton, D. A. Antoniadis, and W. A. Tiller, J. Electrochem. Soc. (to be published).

5.   S. P. Murarka, Phys. Rev. **B16**, 2849 (1977).

6.   S. M. Hu, Appl. Phys. Lett. **27**, 165 (1975).

7. C. L. Claeys, E. L. Laes, G. J. Declerck, and R. J. van Overstraeten, in Semiconductor Silicon 1977, H. R. Huff and E. Sirtl, eds. (The Electrochemical Society, Princeton, 1977), pp. 773-784.

8. S. P. Murarka, J. Appl. Phys. 49, 2513 (1978).

9. K. H. Yang, Recent News, Electrochem. Soc. Meeting, Toronto, May 11-16, 1975.

10. H. Hashimoto, H. Shibayama, H. Masaki, and H. Ishikawa, J. Electrochem. Soc. 123, 1899 (1976).

11. Y. Sugita, H. Shimizu, A. Yoshinaka, and T. Aoshima, J. Vac. Sci. Technol. 14, 44 (1977).

12. C. L. Claeys, G. J. Declerck, and R. J. van Overstraeten, Appl. Phys. Lett. 35, 797 (1979).

13. S. M. Hu, Mater. Res. Soc. Meeting, Boston, November 15-17, 1976.

14. D. A. Antoniadis, A. G. Gonzalez, and R. W. Dutton, J. Electrochem. Soc. 125, 813 (1978).

15. D. A. Antoniadis, A. M. Lin, and R. W. Dutton, Appl. Phys. Lett. 33, 1030 (1978).

16. H. Shriraki, Jpn. J. Appl. Phys. 15, 1 (1976).

17. T. Hattori, J. Electrochem. Soc., 123, 945 (1976); Denki Kagaku, 46, 122 (1979).

18. Y. Nabeta, T. Uno, S. Kubo, and H. Tsukamoto, J. Electrochem. Soc., 123, 1417 (1976).

19. T. Hattori and T. Suzuki, Appl. Phys. Lett. 33, 347 (1978).

20. Y. Hokari, Jpn. J. Appl. Phys. 18, 873 (1979); 16 1899 (1977).

21. S. P. Murarka, J. Appl. Phys. 48, 5020 (1977).

22. S. M. Hu, J. Appl. Phys. 51, 3666 (1980).

23. K. Taniguchi, K. Kurosawa, and M. Kashiyagi, J. Electrochem. Soc., 127, 2243 (1980).

24. T. Kato, Y. Sugita, and A. Yoshinaka, Jpn. J. Appl. Phys. 11, 1066 (1972).

25. R. Francis and P. S. Dobson, J. Appl. Phys. 50, 280 (1979).

26. O. L. Krivanek and D. M. Maher, Appl. Phys. Lett. 32, 451 (1978).

27. J. Hornstra, J. Phys. Chem. Solids, 5, 129 (1958).

28. M. Kahlweit, in Progress in Solid State Chemistry, H. Reiss, ed., (Pergamon, New York, 1965), Vol. 2, P. 134.

29. H. Buchholz, Elektrische und Magnetische Potentialfelder (Springer Verlag,

Berlin, 1957), p. 233.

30. C. P. Flynn, Phys. Rev. A587, 133 (1964); Point Defects and Diffusion (Clarendon Press, Oxford, 1972), p. 486.

31. D. N. Seidman and R. W. Balluffi, Philos. Mag. 13, 649 (1966).

32. F. Kroupa, Czech. J. Phys. B10, 284 (1960).

33. A. Seeger and K. P. Chik, Phys. Stat. Solidi, 29, 455 (1968).

34. S. M. Hu, in Atomic Diffusion in Semiconductors, D. Shaw, ed., (Plenum, New York, 1973), Chap. 5.

35. A. Seeger and W. Frank, these proceedings.

36. U. Goesele, these proceedings.

37. R. F. Peart, Phys. Stat. Solidi, 15 K119 (1966).

38. B. J. Masters and J. M. Fairfield, Appl. Phys. Lett. 8, 280 (1966).

39. R. N. Ghoshtagore, Phys. Rev. Lett. 16, 890 (1966).

40. H. J. Mayer, H. Mehrer, and K. Maier, Intl. Conf. on Radiation Effects in Semiconductors, Dubrovnick, Yugoslavia, 1976.

41. L. Kalinowski and R. Sequin, Appl. Phys. Lett. 35, 211 (1979).

42. K. V. Ravi, J. Electrochem. Soc. 121, 1090 (1974).

43. F. Shimura, T. Tsuya, and T. Kawamura, App. Phys. Lett. 37, 483 (1980).

44. P. E. Freeland, K. A. Jackson, C. W. Lowe, and J. R. Patel., Appl. Phys. Lett. 30, 31 (1977).

45. J. Osaka, N. Inoue, and K. Wada, Appl. Phys. Lett. 36, 288 (1980).

46. S. M. Hu, Appl. Phys. Lett. 36, 561 (1980).

47. A. Ohsawa, K. Honda, S. Ohkawa, and R. Ueda, Appl. Phys. Lett. 36, 147 (1980).

48. C. S. Fuller and R. A. Logan, J. Appl. Phys. 28, 1427 (1957).

49. W. Kaiser, H. L. Frisch, and H. Reiss, Phys. Rev. 112, 1546 (1958).

50. V. Cazcarra and P. Zunino, J. Appl. Phys. 51, 4206 (1980).

51. A. J. R. deKock, Philips Res. Rept. Suppl. No. 1 (1973).

52. A. J. R. deKock, in Semiconductor Silicon 1977, H. R. Huff and E. Sirtl., ed. (Electrochem. Soc., Princeton, 1977), p. 508.

53. A. J. R. deKock and W. M. van der Wijgert, J. Crystal Growth, 49, 718 (1980).

354

54. J. Chikawa and S. Shirai, J. Crystal Growth $\underline{39}$, 328 (1977).

55. S. M. Hu, J. Vac. Sci. Technol. $\underline{14}$, 17 (1977).

56. S. Yasuami, J. Harada, and K. Wakamatsu, J. Appl. Phys. $\underline{50}$, 6860 (1979).

57. S. M. Hu, J. Appl. Phys. $\underline{51}$, 5945 (1980).

58. For example, the transfer of one carbon atom from a substitutional site in the silicon matrix to a growing silicon carbide precipitate will release one vacancy, and the precipitation of two interstitial oxygen atoms requires the absorption of only a single vacancy to produce an unstrained particle of silica. See R. Bullough and R. C. Newman, Rep. Prog. Phys. $\underline{33}$, 101 (1970); especially p. 144.

Published 1981 by North-Holland, Inc.
Narayan, and Tan, eds.
Defects in Semiconductors

ENHANCED DIFFUSION AND FORMATION OF DEFECTS DURING THERMAL OXIDATION*

J. NARAYAN, J. FLETCHER, B. R. APPLETON AND W. H. CHRISTIE[1]
Solid State Division, Oak Ridge National Laboratory, Oak Ridge, TN   37830

ABSTRACT

Enhanced diffusion of dopants and the formation of
defects during thermal oxidation of silicon has been
investigated using electron microscopy, Rutherford
backscattering, and secondary ion mass spectrometry
techniques. Enhanced diffusion of boron was clearly
demonstrated in laser annealed specimens in which secondary
defects were not present. In the presence of secondary
defects, such as precipitates, enhanced diffusion of boron
was not observed. The absence of enhanced diffusion during
thermal oxidation was also observed for arsenic in silicon.
The mechanisms associated with thermal-oxidation enhanced
diffusion are discussed briefly.

INTRODUCTION

Thermal oxidation to form a passivating oxide layer constitutes a very
important step in device manufacturing processes [1]. During thermal oxidation
stacking faults are generated, which are often connected to the oxide layer
[1,2]. Oxidation enhanced diffusion (OED) has been reported for boron but there
is controversy with respect to OED of arsenic in silicon [1]. The purpose of
this paper is to show that secondary defects often play an important role in
determining the enhancement of diffusion. While OED is observed for boron in
the absence of secondary defects, it is not observed for arsenic in silicon.
These results and observed changes in microstructures due to oxidation are critically evaluated in view of the following mechanisms: (1) undersaturation of
vacancies created by oxidation, (2) the generation of self interstitials, and
(3) generation of dopant interstitials which may create self interstitials.

EXPERIMENTAL

Silicon (2-6 Ω-cm) specimens with $\langle 100 \rangle$ and $\langle 111 \rangle$ orientations were
implanted with $^{11}B^+$ (35 keV, doses 2.0 x 10^{15} and 1.0 x 10^{16} cm^{-2}), $^{75}As^+$
(100 keV, 1.0 x 10^{14} and 1.0 x 10^{16} cm^{-2}), and $^{28}Si^+$ (80 keV, dose 1.0 x 10^{15}
cm^{-2}) ions. Some of these specimens were laser annealed with single ruby laser
pulses (λ = 0.694 μm, Energy density E = 1.5-2.0 J cm^{-2} and pulse duration
τ = 15-20 x 10^{-9} s) with a laser beam diameter of about 2 cm. In order to
determine the role of defects in thermal oxidation, the defect-free laser
annealed specimens, and the ion implanted specimens containing dislocation
loops, were thermally oxidized under wet steam at 900°C. Companion specimens
were treated in dry nitrogen at 900°C and these results were compared with those
obtained under steam oxidation. Steam oxidation was accomplished by evaporating
extremely pure water and passing it through the furnace tube. Defect
microstructures were investigated using plan-view and x-section electron
microscopy. The dopant distributions and the lattice location of dopants were
monitored using Rutherford ion backscattering and channeling, and dopant

*Research sponsored by the Division of Materials Sciences, U.S. Department of
Energy under contract W-7405-eng-26 with Union Carbide Corporation.
[1]Analytical Chemistry Division.

356

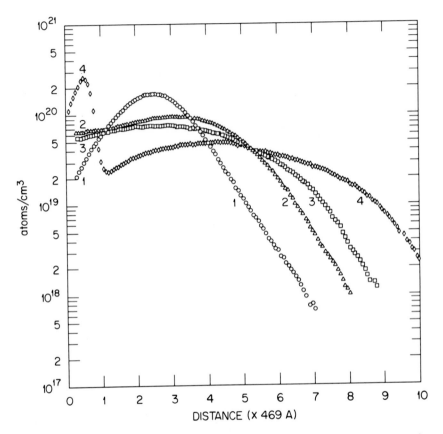

Fig. 1. Boron concentration profiles after various treatments: Curve 1 is as-implanted (dose = 2.0 x 10^15 cm^-2) profile; Curve 2 is after laser annealing (LA) at E = 1.5 J cm^-2; Curve 3 is LA + thermal annealing in dry N_2 (13.5 min); and Curve 4 is LA + steam oxidation.

distribution profiles were compared with those obtained using secondary ion mass spectrometry (SIMS) techniques.

RESULTS AND DISCUSSION

Figure 1 shows boron distribution profiles in a (100) silicon specimen, implanted to a dose of 2.0 x 10^15 cm^-2 after various treatments. The as-implanted profile (labelled as 1) is Gaussian, as expected. After laser annealing (profile labelled 2) the boron distribution is broadened both toward the surface and deep into the crystal. The laser annealed profile has been shown to be consistent with a melting model [3]. Thermal annealing of laser annealed specimens in dry nitrogen at 900°C for 13.5 min leads to further broadening of the profile (see curve 3), this being consistent with high tem-perature diffusion [4]. When laser annealed specimens were thermally oxidized

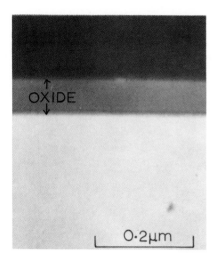

Fig. 2. X-section electron micrograph
from the specimen corresponding to curve
4 in Fig. 1, showing the oxide layer but
no defects.

at 900°C for 13.5 min, oxidation enhanced diffusion (the difference between
curves 3 and 4) was observed in the deeper regions (beyond 2350 Å). From TEM x-
section micrographs (Fig. 2) the thickness of the oxide layer was determined to
be 850 Å. The segregation of boron, up to a concentration of 2.5×10^{20} cm^{-3},
in the oxide layer was observed. In the oxidized specimens, neither defect
clusters nor dislocations were present below the oxide layer. Similarly no
defects were observed in dry N_2 annealed specimens.

Boron concentration profiles in specimens implanted to a dose of 1.0×10^{16}
cm^{-2} are shown in Fig. 3. After laser annealing it is evident from profile 2 in
Fig. 3 that in certain regions the boron concentration has exceeded the
equilibrium solubility limit [4] (3×10^{20} cm^{-3}) at 900°C. Therefore, dry N_2
annealing at 900°C leads to the precipitation of excess boron as shown in the x-
section micrograph in Fig. 4. The concentration profiles after dry N_2 annealing
and steam oxidation are denoted as 3 and 4, respectively in Fig. 3. A comparison between these two profiles shows that the enhancement of diffusion does
not occur in the 1.0×10^{16} cm^{-2} case. The x-section micrograph (Fig. 5a) of
the laser annealed specimens after thermal oxidation also shows the presence of
precipitates. Detailed characterization of defects in plan-view micrographs
(Fig. 5b) indicated the presence of only precipitates and no dislocation loops.

Figure 6 shows the arsenic distribution profile (labelled 1) in a (111)
specimen implanted with a dose of 1.0×10^{16} cm^{-2}. After laser annealing dopant
redistribution occurs (profile 2) similar to that observed for boron implanted
specimens. Thermal annealing in dry N_2 for 13.5 min leads to the distribution
of dopants (profile 3) in the deeper regions, however, similar treatment in
steam leads to piling-up of arsenic (profile 4) underneath the oxide layer, but
no difference between the N_2 and steam profiles is observed in the deeper
regions. TEM x-section micrographs showed the presence of a high density of
stacking faults below the oxide layer (as shown in Fig. 7). The number density
of these faults was found to be much higher in <111> than in <100> specimens.
The stacking faults were found to be connected to the Si-SiO$_2$ interface during
the growth.

Rutherford backscattering and channeling measurements were made on thermally

358

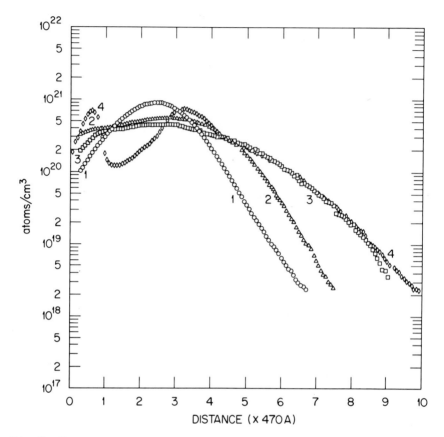

Fig. 3. Boron concentration profiles after various treatments: (1) as implanted (dose = $1.0 \times 10^{16} cm^{-2}$), (2) after laser annealing at E = 1.5 J cm^{-2}, (3) LA + thermal annealing in dry N_2, and (4) LA + steam oxidation.

oxidized specimens to determine the relative concentration of arsenic in substitutional and nonsubstitutional sites. After oxidation at 900°C for 7 min, ~70% of the arsenic in the 200 Å region below the oxide layer is nonsubstitutional, (see Fig. 8). Even though considerable overlap exists in the total and substitutional fractions, a slight nonsubstitutionality seems to persist into the deeper regions (~ 1000 Å) of the oxidized specimens. However, after 13.5 min of thermal oxidation most of the arsenic is in substitutional sites (Fig. 9), some nonsubstitutionality seems to exist very close (~ 100 Å) to the surface. It should be pointed out that the spreading resistance technique [5] commonly used for OED would not be able to detect 200 Å layers containing nonsubstitutional arsenic, since the sensitivity of this technique is > 500 Å.

In order to study the growth of dislocation loops during oxidation, some of the arsenic implanted specimens (implanted to a dose 1.0×10^{14} cm^{-2}) were thermally annealed at 850°C for 20 min in dry nitrogen to obtain an average loop size of 300 Å with a number density of 1.0×10^{10} cm^{-2}. A cross-section

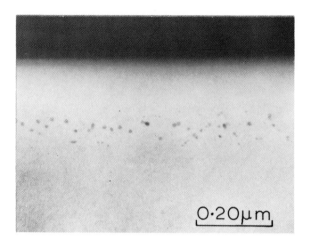

Fig. 4. X-section micrograph showing boron precipitates after LA + annealing in dry N_2.

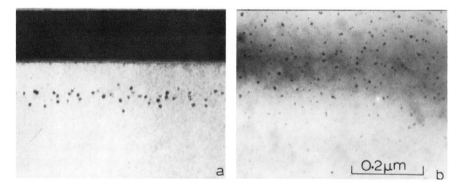

Fig. 5. (a) X-section micrograph after LA + steam oxidation (b) Plan-view after LA + steam oxidation.

micrograph showed that these loops are distributed in a region 600 ± 200 Å from the surface. One batch of these samples was annealed in dry N_2 for 6 min. at 900°C and the other batch was annealed in steam for 6 min. at 900°C. A comparison of microstructures is shown in Fig. 10, where growth of the loops during oxidation is observed. A similar batch of samples was oxidized for 20 min. at 900°C indicating further growth of the loops.

In order to study the effect of dopants on the growth of dislocation loops, some of the Si^+ implanted (dose = 1.0 x 10^{15} cm^{-2}) specimens were annealed at 900°C for 20 min. in dry N_2 and then one batch was annealed at 900°C for 13.5 min in dry N_2, and a similar batch was oxidized at 900°C for 13.5 min. The

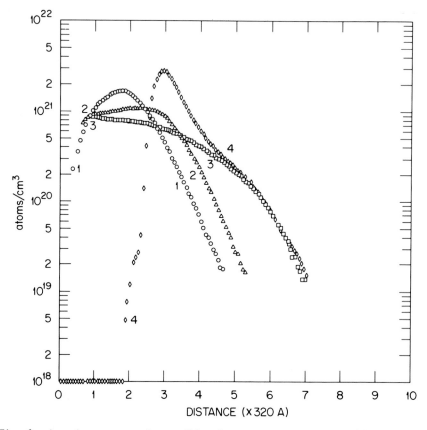

Fig. 6. Arsenic concentration profile after various treatments: (1) as implanted (dose = 1.0 x $10^{16} cm^{-2}$), (2) after laser annealing at E = 1.5 J cm^{-2}, (3) LA + thermal annealing in dry N_2, and (4) LA + steam oxidation.

Fig. 7. X-section micrograph after LA + steam oxidation showing the oxide layer and the stacking faults underneath the oxide.

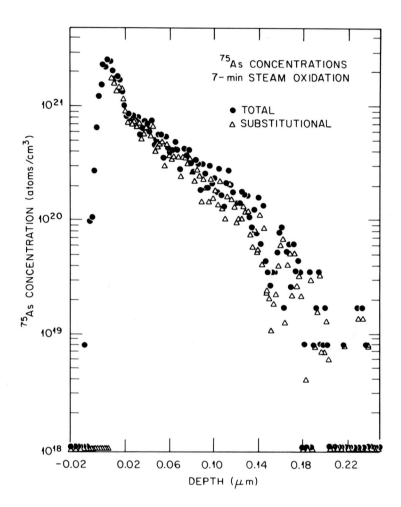

Fig. 8. Arsenic concentration profiles showing substitutional arsenic after LA + 7 min. oxidation. Note the logarthmic scale for the concentration in which the differences between the total and substitutional arsenic concentrations in the near surfaces regions, although large, get less emphasized.

362

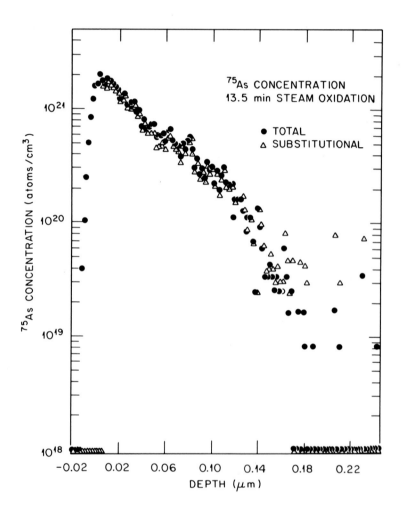

Fig. 9. Arsenic concentration profiles showing substitutional and non-substitutional arsenic after LA + 13.5 min oxidation.

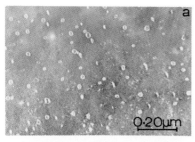

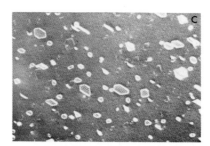

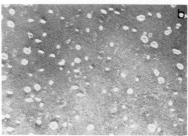

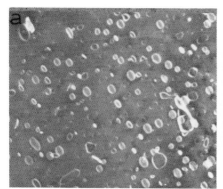

Fig. 10. Dislocation loops in arsenic doped (1.0×10^{14} cm^{-2} Si + annealing at 850°C for 20 min in dry N_2) specimens after the following annealing treatments: (a) at 900°C for 6 min in dry N_2, (b) another sample at 900°C for 6 min in steam, (c) another sample at 900°C for 20 min in steam.

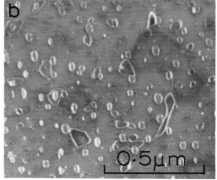

Fig. 11. Dislocation loops in Si^+ implanted (1.0×10^{15} cm^{-2} Si^+ + 900°C for 20 min. in dry N_2) and annealed specimens after the following annealing treatments: (a) at 900°C for 13.5 min in dry N_2, (b) at 900°C for 13.5 min in steam.

364

results are shown in Fig. 11 where no preferential growth of loops was observed in the oxidized specimens.

There are presently two models available to explain the phenomenon of OED and the formation of stacking faults during thermal oxidation. The first one assumes that thermal oxidation consumes vacancies in the silicon lattice under the oxide layer [6,7]. In response to this vacancy undersaturation, Frenkel pairs are generated, with the vacancies trying to make up for the deficiency and thus creating a supersaturation of self-interstitials. It is assumed that these interstitials enhance the thermal diffusion of dopants, and they could also take part in the nucleation and growth of interstitial type clusters such as extrinsic stacking faults. One related aspect of this mechanism is that interstitial type defects such as extrinsic stacking faults are generated due to an undersaturation of vacancies and they grow via vacancy emission thus replenishing the vacancy deficiency. In this process, interstitials are not generated in the lattice, and should not result in OED. The major objection to this model is the assumption of generation of Frenkel-pairs. Since the energy of formation of a Frenkel-pair is of the order of 6-7 eV in Si, the rate of Frenkel-pair generation should be vanishingly small.

The second model assumes the generation of excess interstitials because of incomplete oxidation in a thin layer underneath the oxide layer [1,2]. These excess interstitials take part in the OED as well as in the nucleation and growth of stacking faults. It should be emphasized that the process by which a stacking fault grows by absorbing self-interstitials, and the process by which the fault grows by vacancy emission, are not equivalent.

The present results have shown that oxidation enhanced diffusion occurs for boron in Si in the absence of secondary defects such as precipitates. In the presence of precipitates, no clustering of interstitials in the form of dislocation loops, or around the precipitates, is observed. This tends to suggest that oxidation forces boron into interstitial sites, which then diffuses faster than it would via a vacancy mechanism and OED is observed. The most important conclusion from these results is that in the presence of secondary defects OED is modified or suppressed completely. The lattice location studies in the case of arsenic suggest that arsenic may be going into nonsubstitutional sites during oxidation in the beginning. However the observations of loop growth suggests that self interstitials may be generated by a "kick-out mechanism" [8] where dopant atoms occupy substitutional sites by kicking out self interstitials which then cluster, resulting in the observed loop growth. A comparison of the diffusion of arsenic with that of boron also suggests that, unlike boron, once arsenic is in an interstitial site it cannot diffuse rapidly and hence OED does not occur. The absence of loop growth during the oxidation of silicon implanted and thermally annealed specimens, also lends support to our suggestion that self interstitials are not directly generated during thermal oxidation. Finally, it should be remarked that the role of oxidation-induced stacking faults in the OED is unknown at this point. More work is needed to establish the role of these defects in the OED phenomenon.

REFERENCES

1. S. M. Hu, J. Vac. Sci. Technol. 14, 17 (1977).

2. S. M. Hu, J. Appl. Phys. 45, 1567 (1978).

3. C. W. White, J. Narayan and R. T. Young, Science 204, 461 (1979).

4. A. S. Grove, Physics and Technology of Semiconductor Devices (Wiley, New York, 1967).

5. D. A. Antoniadis, A. M. Lin and R. W. Dutton, Appl. Phys. Lett. <u>33</u>, 1030 (1978).

6. R. J. Jaccodine and C. M. Drum, Appl. Phys. Lett. <u>8</u>, 29 (1966).

7. I. R. Sanders and P. S. Dobson, Phil. Mag. <u>20</u>, 881 (1969).

8. J. Narayan, these proceedings.

Published 1981 by North-Holland, Inc.
Narayan, and Tan, eds.
Defects in Semiconductors

PRECIPITATION OF OXYGEN AND INTRINSIC GETTERING IN SILICON

W. K. TICE

IBM General Technology Division, Essex Junction, VT   05452

T. Y. TAN

IBM Research Division, Yorktown Heights, NY   10598

ABSTRACT

In this review, dislocations introduced by prismatic punching at SiO_2 precipitate sites in Czochralski silicon are shown to act as metal adsorption centers. Conditions necessary to localize SiO_2 precipitate and dislocation complexes in wafer regions remote from semiconductor devices are discussed. This localization achieves an intrinsic gettering effect. The application of the intrinsic gettering mechanism to bipolar and MOS device technologies is shown to improve device performance and leakage limited yields.

INTRODUCTION

Recently, it has been shown that the amount and distribution of oxygen in single crystal silicon wafers can have a profound influence on semiconductor device behavior and yields. This review deals with the mechanism of oxygen precipitation and consequent dislocation generation in silicon. The effects of localizing such SiO_2 precipitate-dislocation complexes (SiO_2 PDC) in wafer regions remote from active semiconductor devices is discussed in terms of bipolar device leakage limited yields (LLY) and MOS storage time. It will be shown that such regions perform a gettering role, thereby enhancing device performance and yields.

The subject will be treated in the following sequence:

1. Introduction of oxygen during Czochralski (CZ) crystal growth.

2. Evidence for heterogeneous nucleation of SiO_2 precipitates in silicon.

3. Growth of SiO_2 precipitates.

4. Prismatic punching of dislocations by a compressive misfit stress field introduced into the silicon matrix by differential contraction between precipitates and matrix during wafer cooling.

5. Adsorption of metallic impurities by SiO_2 PDC.

6. Localization of SiO_2 PDC in wafer regions remote from active devices to achieve an intrinsic gettering effect.

7. Experimental results showing beneficial results of intrinsic gettering on semiconductor device yields/performance.

INTRODUCTION OF OXYGEN INTO SILICON

During silicon single crystal growth using the CZ method, the quartz crucible and silicon undergo the reaction:

$$Si + SiO_2 \rightarrow 2SiO \qquad\qquad [1-4]$$

At typical crystal pulling temperatures ($\sim$1412 $^{\circ}$C), SiO is volatile and passes out of the liquid-solid interface. This results in a nonhomogeneous distribution of oxygen at the crystal-liquid interface, giving a nonuniform distribution across the boule diameter. Depletion of SiO during crystal pulling can also lead to nonuniform oxygen distribution along the length of the silicon boule. Typical oxygen concentrations for a single crystal are 2-20 x 10^{17} at./cm^3. The oxygen is present initially in Si-O-Si bound interstitial form, which gives an infrared absorption band at 9 μm [5,6].

NUCLEATION OF OXIDE PRECIPITATES

The possibility of SiO_2 precipitation due to decomposition of silicon-oxygen solid solution had been discussed by Patel in 1964 [7]. Later, Batavin [8] published curves of oxygen concentrations versus heat treatment time at 1000 $^{\circ}$C, which showed that oxide precipitation took place after an incubation period of about 10 hr. The incubation time is reduced with an increase in annealing temperature. Heat treatment of similar oxygen-rich material in the temperature range of 300 $^{\circ}$C - 500 $^{\circ}$C produces donors, believed to be SiO_4 complexes [9, 10]. Higher-temperature annealing removes these electrically active complexes. At termination of the incubation period, during which oxide nucleation sites of critical size are formed, interstitial oxygen level decreases to the equilibrium value at that particular temperature.

According to Batavin [8], precipitation can be either heterogeneous or homogeneous. Recently, Shimura et al [11, 12] proved by differential infrared absorption that oxide precipitation occurs by heterogeneous nucleation. Here it was shown that the cristobalite SiO_2 spectrum increases inversely with the spectrum of interstitial oxygen, according to annealing temperature in the 600 $^{\circ}$C - 1100 $^{\circ}$C range. Cristobalite precipitates tend to form distinct bodies with increases in time/temperature. During subsequent high-temperature annealing (i.e., 1230 $^{\circ}$C/2 hr), oxide precipitates redissolve, and the oxygen returns to interstitial positions in the silicon lattice. Conversely, $SiO2$ precipitation in samples annealed 2 hr/1230 $^{\circ}$C does not occur during subsequent low-temperature annealing despite the presence of the same or higher, supersaturated ratio of oxygen. It follows that oxygen precipitation always takes place by heterogeneous nucleation.

The same authors also showed that oxide precipitate nuclei consist of small cristobalite bodies formed during initial growth/annealing processes. Primary defect nuclei are believed to be impurity clusters or condensed point defects. These results are to be expected since precipitate nucleation in solid solutions is accompanied by lattice distortions that increase activation energy for homogeneous nucleation. On the other hand, lattice strains associated with crystal imperfections can be lessened by heterogeneous nucleation of a new phase, thereby lowering the activation energy for precipitation.

FORMATION OF SiO_2/DISLOCATION COMPLEXES

The possibility that dislocations are generated at SiO_2 precipitate sites was first proposed by Patel [7]. Later, studies of SiO_2 precipitate morphology and nature of dislocations generated at the precipitate-matrix interface were carried out by Tan and Tice [13] and Maher et al [14]. To generate SiO_2 PDC, silicon wafers containing 2 x 10^{18} at./cc of oxygen were annealed in N_2 at 1050 °C for 60 hr. Silicon material from near the center of wafer cross sections was then observed using transmission electron microscopy (TEM). IR spectroscopy, X-ray topography, and TEM techniques were also employed to show overall microstructural differences between annealed high (i.e., 2 x 10^{18} at./cc) and low (i.e., 3 x 10^{17} at./cc) oxygen-content wafers.

For annealed high-oxygen samples, approximately 1000 Å - 1 µm precipitates with a density of 10^4 to 10^5 ppt/cc were observed under all diffraction contrast TEM conditions. Precipitate size was uniform in any one wafer but was quite variable from sample to sample. Figure 1 depicts initial interstitial oxygen content, as measured by IR spectroscopy, and X-ray topographs of the same wafers after annealing. Note the formation of large imperfection clusters near the center of the oxygen-rich material. By using the analytical procedures of Kaiser and Keck [15], it was found that such defect clusters contain precipitates of the chemical composition SiO_2.

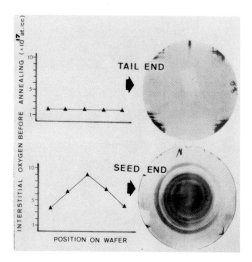

Fig. 1. Typical concentration and distribution of interstitial oxygen measurement by IR in unprocessed CZ Si wafers from tail and seed ends of boule. Transmission X-ray topographs of same wafers after 1050°C/60 hr annealing are shown by arrows.

By careful TEM observation of precipitates in various stages of development, a growth model was proposed. For the case of a $(001)_{Si}$ precipitate habit plane, nucleation of a cross-shaped ribbon on the (001) plane first takes place. This ribbon is composed of two branches of equal length and situated parallel to [100] and [010]. Widening of ribbons proceeds by dendritic growth until a square-shaped precipitate with sides parallel to <110> directions is formed. This sequence is illustrated in Fig. 2.

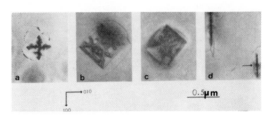

Fig. 2. TEM micrographs showing morphology and growth process of SiO_2 precipitates. (a) Formation of cross-shaped ribbon, (b) dendritic growth of ribbon, (c) completed thin-square precipitate, and (d) edge-on observation of precipitate, confirming plate-like morphology. (Reprinted from Philosophical Magazine, Vol. 34, No. 4, Taylor and Francis, Ltd. with permission.)

Standard TEM diffraction contrast techniques were used to detect dislocations, determine dislocation geometry, and nature (i.e., interstitial or vacancy). For isolated precipitates, these dislocations consist of rows of prismatic loops situated along any of the six possible <110> glide directions (Fig. 3).

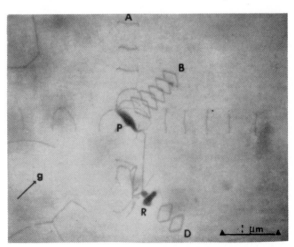

Fig. 3. Dislocation loops generated in wafer center region by prismatic punching. SiO_2 precipitates are designated P and R. Loop series PB, PA, etc., represent glide along four of the six possible <110> directions. Si habit planes of P and R are {100} type. (Reprinted from Philosophical Magazine, Vol. 34, No. 4, Taylor and Francis, Ltd. with permission.)

Such dislocation arrays were predicted by Seitz for the mechanism of prismatic punching [16]. When particles are compressed within a matrix, to relieve the resulting misfit strain, interstitial loops are propagated along glide directions. This mechanism was subsequently confirmed for the case of spherical particles [17] in metals.

For non-spherical SiO_2 plates in silicon, dislocation loops were the same size and shape as a projection of the parent precipitate onto the (110) plane. Unlike spherical precipitates in an isotropic matrix, square SiO_2 plates in silicon cause anisotrophy in the misfit stress field. Consequently, loops are generated along only a few of the possible glide directions. For example, when a precipitate had the (001) habit plane, loops were seldom observed along [110] but are generated along the [011] axis. This situation is shown schematically in Fig. 4. Note that the square precipitate morphology gives rise to interstitial loops along only four of the six possible glide directions not situated in the precipitate habit plane. In general, however, variation of geometry always occurs owing to dislocation interactions. A complex tangle of dislocations appears in Fig. 5.

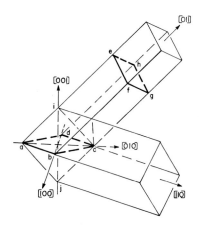

Fig. 4. The precipitate-dislocation loop geometry. The square plate-like precipitate morphology limits the loops generated to along only four of the six possible glide directions that do not lie in the precipitate habit plane. Thus for the case shown, dislocation loops such as efgh are generated along [011] axis but none along [110] since the precipitate abcd has a (001) habit plane. (Reprinted from Philosophical Magazine, Vol. 34, No. 4, Taylor and Francis, Ltd. with permission.)

In Fig. 3, TEM analysis showed that these loops are pure edge dislocations with Burgers vectors of $\frac{1}{2}<110>$. Furthermore, such crystal imperfections are all interstitial in nature. These findings are consistent with the fact that SiO_2 precipitates compress the silicon matrix.

As a final part of dislocation characterization, the loop nucleation mechanism was determined as follows. During plastic deformation of fcc crystals, prismatic loops are created directly by cross-slip of preexisting dislocations upon intersecting a particle. On the other hand, Orowan shear loops can be cre-

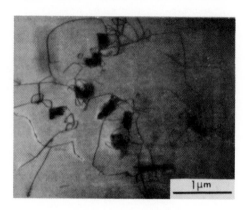

Fig. 5. Dislocation network resulting from loop interactions. (Reprinted from Philosophical Magazine, Vol. 34, No. 4, Taylor and Francis, Ltd. with permission.)

ated around a particle, subsequently assuming prismatic shape by cross-slip [19, 20, 21]. In both of these instances where stress is externally applied to form preexisting dislocations, prismatic loops can form without nucleation of generator loops. When stress fields are entirely internal as in the case with SiO_2 precipitates in silicon, generator loops must be nucleated prior to formation of prismatic loops. In fact, to accomplish significant reduction of misfit strain, generator loops must be created a number of times. Weatherby [22] has indicated that such generator loops should be shear loops for fcc crystals. According to the proposed shear mechanism, shear loops nucleate at the particle-matrix interface and extend outward on a slip plane to assume semicircular shape. These loops subsequently become rhombus-shaped by repeated cross-slip.

In previously mentioned work [13], semicircular generator loops were observed (Fig. 6). These dislocations are all situated on {111} glide planes, as expected for diamond-centered cubic material, and have Burgers vectors lying in the loop plane. These observations confirm Weatherby's [22] shear model for generator loop nucleation.

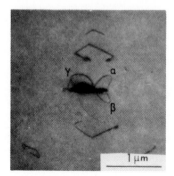

Fig. 6 Generator loops γ, α, β situated on {111} type habit planes. (Reprinted from Philosophical Magazine, Vol. 34, No. 4, Taylor and Francis, Ltd. with permission.)

For silicon wafers annealed at 1050 $^\circ$C in N_2, stacking faults are not generated at SiO_2 PDC [13]. However, annealing in H_2 at 1200 $^\circ$C produces both stacking faults and prismatic dislocations around the precipitates [14]. Detailed mechanisms that would differentiate generation of dislocations alone to those which would generate dislocations plus stacking faults are presently not clearly understood.

METAL GETTERING BY SiO_2 PDC

Infrared microscopy and X-ray topography have demonstrated that fast-diffusing contaminant impurities, especially copper, tend to precipitate predominantly in silicon wafer regions of high-oxygen concentration [23-26]. Fiermans and Vennik [24, 25] proposed that this phenomenon might be explained by heterogeneous nucleation of copper on dislocations generated at the silicon-oxide precipitate interface. A study of the precise nature of copper precipitation in silicon containing a large amount of precipitated SiO_2 was subsequently carried out by the present authors [27].

In this experiment, interstitial oxygen concentration of p-type [001] wafers was first measured by the infrared spectroscopy method. Samples with 2×10^{17} - 2×10^{18} at./cc of oxygen were subsequently annealed in N_2 at 1050 $^\circ$C for 60 hr to induce oxygen precipitation. The presence or absence of SiO_2 PDC was determined by TEM and chemical-etching techniques. The same samples were next copper-coated from a saturated $Cu(NO_3)_2$-HF solution. Copper was then diffused into the sample by annealing in air at approximately 1100 $^\circ$C for 15 sec. Growth of copper precipitate colonies was enhanced by rapidly cooling the wafer to room temperature [28].

A TEM study of copper-diffused samples that were known to have SiO_2 PDC confirmed the heterogeneous nucleation of secondary precipitate colonies at SiO_2 PDC (Fig. 7). Virtually no small secondary precipitates were observed in the matrix silicon remote from SiO_2 PDC.

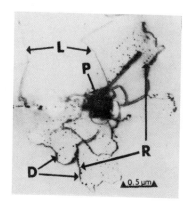

Fig. 7. SiO_2 PDC after deliberate diffusion with Cu. SiO_2 precipitate is designated P, and primary prismatic dislocation loops as L. Secondary CuSi precipitate colonies designated R with surrounding predominately edge dislocation D are situated in regions previously occupied by prismatic loops whose original shape/size has been altered by climb. (Reprinted by permission of Applied Physics Letters.)

374

When subjected to strong g = 220_{Si} two-beam diffraction TEM conditions, 40 Å moiré fringes were exhibited by the small secondary precipitates (Fig. 8). This spacing is close to that reported by Das [29] for CuSi precipitates in silicon. Finally, the presence of copper within the small precipitates was confirmed by energy dispersive X-ray analysis (EDA), using a TEM operating in the STEM/EDA mode. Colonies of such CuSi precipitates are surrounded by CuSi-decorated dislocations of predominantly edge nature. The geometry of original prismatic loops within SiO_2 PDC is mainly destroyed after copper diffusion. This is because vacancies are given off by loops to help accommodate the extra volume of copper being deposited at secondary precipitate sites. Rows of precipitates within CuSi colonies indicate sequential positions of the climbing dislocations.

These observations prove that prismatic loops within SiO_2 PDC act as sites for the heterogeneous nucleation of CuSi precipitates. Similar gettering behavior by SiO_2 PDC has been observed for iron and nickel [30]. Adsorption of metal impurities takes place at SiO_2 PDC because the free energy of a precipitate-decorated dislocation is less than that of dispersed metal atoms plus undecorated dislocation [31, 32].

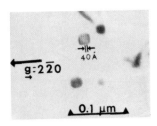

Fig. 8. Moiré fringe contrast obtained from single CuSi secondary precipitate using strong g = 2$\overline{2}$0 Si two-beam diffraction conditions. (Reprinted by permission of Applied Physics Letters.)

INTRINSIC GETTERING BY SiO_2 PDC

The term "gettering," as used in semiconductor processing, denotes removal of metallic precipitates from active device regions by a variety of techniques [33, 34]. Gettering is important to device behavior/yields since metals can form electrically active generation-recombination sites within device active regions, thereby increasing leakage currents. In bipolar devices, metal precipitation can also cause emitter-to-collector near shorts termed "pipes." In the preceding sections, it was demonstrated that SiO_2 PDC can act as centers for metal precipitation.

Let us now examine how this situation can be used to achieve gettering. This mechanism has been termed "intrinsic gettering" to distinguish it from externally imposed, or "extrinsic gettering," techniques [38]. In passing, it should be noted that oxygen in CZ silicon can induce a materials hardening effect. This effect also can increase semiconductor device leakage limited yields (LLY) by reducing wafer plastic deformation [36, 37].

The mechanism of intrinsic gettering by SiO_2 PDC is now explored in detail. First, Batavin et al [18] have shown that device leakage currents are increased when active device regions are formed in SiO_2 PDC-rich silicon wafer surfaces. This condition was achieved by annealing silicon single crystal boules to form SiO_2 PDC prior to wafer preparation. However, this situation can be avoided in

actual processing conditions. For example, Kaiser and Keck [14] noted oxygen-
denuded layers adjacent to the surface of silicon boules after crystal growth.
These layers were attributed to evaporation of SiO during cyrstal pulling.   It
has also been observed that annealed silicon wafers tend to have SiO_2 PDC confined mainly to wafer regions well below the surface [13].   Figure 9 shows the
in-depth distribution of the SiO_2 PDC in a wafer annealed at 1050 °C for 60 hr.
This sample was cleaved along a <110> direction and Dash-etched for 10 min.
The SiO_2 PDC region lies about 30 µm below both front and back wafer surfaces.
Yue and Ruiz [38] have demonstrated that the depth at which SiO_2 PDC begins to
form in the subsurface region increases with annealing temperature.   It follows
that, if sufficient high temperature and times are used early in a semiconductor device processing sequence, the active device regions will be entirely free
of SiO_2 PDC.   By proper tailoring of annealing treatments, SiO_2 PDC can be segregated in the wafer volume.   In this location, SiO_2 PDC act as beneficial
gettering sites to reduce metal precipitate-induced device leakage in the near
surface silicon.
     The most important physical phenomenon occurring here is the creation of
wafer surface regions free from SiO_2 PDC during normal device processing by
out-diffusion of oxygen.   Such out-diffusion occurs in wafer surface regions
during annealing in any ambient.   Thus, the oxygen content in these areas can
become sufficiently low so that SiO_2 precipitates need not form.   In inert ambi-

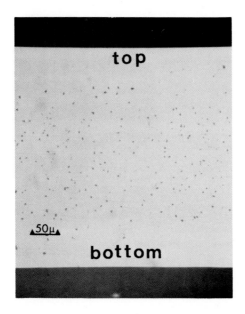

Fig. 9.   Photomicrograph showing in-depth distribution of SiO_2PDC
in Dash-etched wafer cross section.

ents, the wafer surface oxygen region content tends to reach zero, while in O_2 ambient oxygen concentration tends to reach C_e, the thermal equilibrium value [39]. In the latter case, oxygen out-diffuses to the SiO_2-Si interface where it helps to grow the SiO_2 layer. A schematic for the out-diffusion of oxygen in both cases is shown in Fig. 10.

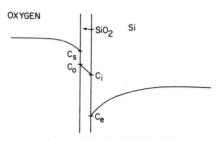

A. IN THE OXIDIZING AMBIENT

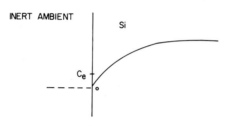

B. IN THE INERT AMBIENT

Fig. 10. Oxygen content near Si wafer surface during heat treatment. (a) In oxidizing ambient. C_s, C_o equilibrium value of oxygen at the gas – SiO_2 interface in the gas and SiO_2 (respectively); C_i, C_e: equilibrium value of oxygen at the SiO_2 – Si interface in SiO_2 and Si (respectively). (b) in inert ambient.

INFLUENCE OF INTRINSIC GETTERING ON YIELDS

To confirm the effectiveness of intrinsic gettering, bipolar devices were fabricated on two groups of wafers [35]. The first group initially contained interstitial oxygen in excess of 1×10^{18} at./cm^3 while the other wafers had an oxygen content of 3×10^{17} at./cm^3. The LLY, based on a 5 A/mil^2 emitter-base leakage criterion, was significantly different for the two groups. Wafers from the high-oxygen group all gave LLY greater than 50%, while those of the low-oxygen group gave less than 10%. The presence of SiO_2 PDC in the volume of the high-yield wafers was confirmed by Dash-etching cross sections. Conversely, almost no SiO_2 PDC were visible in the low-oxygen/low-LLY group. The number of slip dislocations in the two groups was approximately equal, which indicates that the LLY differential was due to intrinsic gettering.

Yue and Ruiz [38] subsequently investigated the influence of intrinsic gettering on storage time in MOS capacitors. These devices were fabricated both with and without back side phosphorous diffusion (i.e., extrinsic) gettering. For both experimental portions, maximum storage time, measured by the pulsed C-V method, was found when a well-defined SiO_2 PDC region was present at least 25 μm below the surface. These results confirm that intrinsic gettering is also an important factor in influencing non-bipolar device performance/yields.

It is known that dislocations can become saturated with impurities. It follows that SiO_2 PDC can lose their ability to adsorb metals during a long series of process heat treatments in contaminated ambients. However, an annealing experiment has shown that dislocations around SiO_2 PDC are generated each time wafers are heat-cycled [30]. The SiO_2 PDC zone thus retains its gettering ability due to continuous generation of new dislocations during hot processing.

Having established the mechanism of intrinsic gettering and its effectiveness, it should be pointed out that there are technology issues which dictate optimum oxygen levels. For example, low-temperature processing would require a relatively lower level of oxygen than a high-temperature technology to ensure low SiO_2 PDC density near the wafer surface. Optimum interstitial oxygen content of starting silicon wafers is thus not clearcut; each application of the intrinsic gettering technique must be judged with the unique process technology in mind.

Also, the expansion of prismatic loops normal to their slip plane can take place by climb induced by a nonequilibrium concentration of silicon self-interstitials generated at secondary precipitate sites (Fig. 11). Thus, when secondary precipitation is excessive, dislocations from SiO_2 PDC climb into active device regions with catastrophic results to semiconductor device yields.

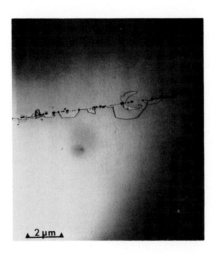

Fig. 11.   Expansion of SiO_2 precipitate-induced dislocation by climb associated with secondary precipitation.   Multibeam, bright-field TEM condition.

378

Generation of SiO_2 PDC is naturally occurring and quite unavoidable in complicated processing schemes such as bipolar integrated circuit fabrication. Because of the "screening" or "masking" effects of SiO_2 PDC in such cases, routine gettering techniques utilizing damage on wafer back sides may be rendered ineffective.

SUMMARY

In this paper, we have reviewed recent developments concerning SiO_2 precipitation and intrinsic gettering in silicon. Oxygen from crucibles is initially introduced into interstitial positions in the silicon lattice during crystal growth. During hot processing, to fabricate semiconductor devices, such wafers tend to be in a supersaturated condition with respect to oxygen. This results in heterogeneous nucleation of cristobalite (SiO_2). Primary defect nuclei are most likely condensed point defects or impurity clusters in the silicon lattice. After significant growth in $(001)_{Si}$ wafers, precipitate morphology tends to be that of a thin square plate with $\{100\}_{Si}$ habit planes and $<110>_{Si}$ sides. Since the silicon matrix is compressed by SiO_2 precipitates, prismatic loops of interstitial character are generated to relieve misfit strain. In the initial nucleation stage, shear-type generator loops are situated on $\{111\}_{Si}$ planes. Final prismatic loops are created by cross-slips of these initial generator loops.

SiO_2 PDC act as centers of metal adsorption during hot processing. By localizing these defects in the center of wafer cross sections, an intrinsic gettering action results.

Intrinsic gettering has been applied to both bipolar and MOS process technologies. Use of this technique increased bipolar LLY, as defined earlier, by a factor of approximately five. MOS storage time was also maximized by intrinsic gettering. Thus, the sequential events, oxygen introduction into silicon$\rightarrow$$SiO_2$ nucleation/growth$\rightarrow$prismatic punching of dislocations$\rightarrow$gettering of metallic impurities, are critical in relating oxygen to semiconductor device yields in CZ silicon.

REFERENCES

1. Y. Yatsurugi, N. Akiyama, Y. Endo, T. Nozaki, J. Electrochem. Soc. 120, 976 (1973).

2. T. Nozaki, Y. Yatsurugi, N. Akiyama, J. Electrochem. Soc. 117, 1567 (1970).

3. A. J. R. DeKock, Philips Res. Rept. Suppl. 1, 7 (1973).

4. M. Oliver, thesis, Grenoble U. (1975).

5. H. J. Hrotowski and R. H. Kaiser, J. Phys. Chem. Solids 9, 214 (1959).

6. P. Capper, A. W. Jones, E. J. Wallhouse, J. G. Wilkes, J. Appl. Phys. 48, 1646 (1977).

7. J. R. Patel, Discuss. Faraday Soc. 38, 201 (1964).

8. V. V. Batavin, Soviet Physics - Crystallography 15, 100 (1970).

379

9. W. Kaiser, Phys. Rev. 105, 1751 (1957).

10. W. Kaiser, H. L. Frisch, H. Reiss, Phys. Rev. 112, 1546 (1958).

11. F. Shimura, H. Tsuya, T. Kawamura, Appl. Phys. Lett. 37, 483 (1980).

12. F. Shimura, H. Tsuya, T. Kawamura, J. Appl. Phys. 51, 269 (1980).

13. T. Y. Tan and W. K. Tice, Phil. Mag. 34, 615 (1976).

14. D. M. Maher, A. Staudinger and J. R. Patel, J. Appl. Phys. 47, 3813 (1976).

15. W. Kaiser and P. H. Keck, J. Appl. Phys. 28, 882 (1957).

16. F. Seitz, Phys. Rev. 79, 723 (1950).

17. D. A. Jones and E. W. J. Mitchell, Phil. Mag. 3, 1 (1958).

18. V. V. Batavin, Soviet Phys. Solid St. 8, 2478 (1967).

19. P. B. Hirsch and F. J. Humphreys, Proc. R. Soc. A, 318, 45, 73 (1970).

20. L. M. Brown and W. M. Stobbs, Phil. Mag. 23, 1201 (1971).

21. F. J. Humphreys and A. T. Stewart, Surf. Sci. 31, 389 (1972).

22. G. C. Weatherby, Phil. Mag. 17, 791 (1968).

23. G. Schwuttke, J. Electrochem. Soc. 108, 163 (1961).

24. L. Fiermans and J. Vennik, Phys. Status Solids 21, 627 (1967).

25. L. Fiermans and J. Vennik, Phys. Status Solids 22, 463 (1967).

26. S. M. Hu and M. P. Poponick, J. Appl. Phys. 43, 2067 (1972).

27. W. K. Tice and T. Y. Tan, Appl. Phys. Lett. 28, 564 (1976).

28. E. Nes and J. Washburn, J. Appl. Phys. 42, 3652 (1971); 43, 2005 (1972).

29. G. Das, J. Appl. Phys. 44, 4459 (1973).

30. W. K. Tice, unpublished data (1980).

31. W. C. Dash, J. Appl. Phys. 27, 1193 (1956).

32. A. H. Cottrell, Strength of Solids (Physical Soc., London, 1948), p. 32.

33. J. E. Lawrence (for example), Tran. Metall. Soc. AIME 242, 484 (1968).

34. W. K. Tice and H. J. Geipel, Electrochem. Soc. Abs. 79-2, 1257 (1979); IBM J. Res. and Dev. 24, 310 (1980).

35. T. Y. Tan, E. E. Gardner, W. K. Tice, Appl. Phys. Lett. 30, 175 (1977).

36. W. D. Sylwetrowicz, Phil. Mag. 7, 1825 (1962).

37. S. M. Hu and W. J. Patrick, J. Appl. Phys. 46, 1869 (1975).

38. J. T. Yue and H. J. Ruiz, Electrochem. Soc. Ex. Abs. 77-1, 535 (1977).

39. T. Y. Tan and W. K. Tice, Abs. Fourth Am. Conf. on Crystal Growth, p. 79 (NBS-Gaithersburg, Md., July 16-19, 1978).

Narayan, and Tan, eds.
Defects in Semiconductors

THE EFFECT OF ANNEALING TEMPERATURE ON THE MORPHOLOGY OF
STACKING FAULTS IN CZOCHRALSKI SILICON

R. F. PINIZZOTTO and H. F. SCHAAKE
Central Research Laboratories, Texas Instruments, Inc., Dallas, TX 75265

ABSTRACT

Nucleation and growth of stacking faults formed in CZ silicon
during oxygen precipitation have been studied using x-ray topography, TEM and FTIR. Samples were annealed in argon for various
times at 550°C and 750°C followed by a 16 hour anneal in dry oxygen at 1000°C. In samples annealed at 550°C, the stacking faults
were several layers thick with colonies of precipitates at their
centers. The faults in samples annealed at 750°C contained only
one particle and were single in nature. It is proposed that the
faults are formed by thin oxygen precipitate platelets and that
the different morphologies are due to different oxygen precipitation rates. The platelets are probably a modified cristobalite,
as determined by micro-diffraction results.

INTRODUCTION

Stacking faults in Czochralski silicon have been studied for many years and
are the subject of numerous scientific papers. One way these stacking faults
are formed is during the precipitation of oxygen in silicon due to various heat
treatment cycles. Some authors report that the stacking faults are generated
and nucleated at one oxygen precipitate particle and then grow by the capture
of the interstitials formed to accomodate the volume change caused by the oxidation progress [1]. Others have said that colonies or groups of particles are
found on the faults [2]. No one has proposed a mechanism that can reconcile
the two types of morphologies. In this paper, we report the results of our
studies of both types of faults. A model explaining their possible relationship is proposed. We also report on micro-diffraction results that imply the
material on the faults is a modified cristobalite phase of SiO_2.

EXPERIMENTAL DETAILS

The material used in these experiments was (100) p-type boron doped,
5-10 Ω -cm silicon grown by the Czochralski process in a commercially built
crystal puller. The slices were all cut from the top of a single boule. A
resistivity stabilization anneal was not performed. Other than the crystal
growth process itself, the wafers were not subjected to any heat treatments not
specifically mentioned in this report. Fourier transform infrared spectroscopy
was used to measure the oxygen and carbon concentrations. The radial distribution of these impurities was measured by stepwise scanning a 3-mm aperture
across the slices. The oxygen concentration was found to vary by a factor of
2 from the edge to the center of the slice with the maximum concentration being
1.0×10^{18} cm^{-3}, using the calibration of Kaiser and Keck [3]. The carbon concentration was below the detectability of our FTIR apparatus and is estimated
to be $< 2 \times 10^{16}$ cm^{-3}. The samples were annealed in slice form in a two step
process. The first step was designed to form nuclei which could survive latter
higher temperature anneals. These were done in an argon atmosphere between
550 and 750°C for times ranging from 15 minutes to 128 hours. The times at

each temperature were chosen to give equivalent oxygen diffusion distances. The second step was a 16 hour anneal at 1000°C in a dry oxygen atmosphere. This step was designed to be a reasonable facimile of a typical MOS process flow. TEM samples were ultrasonically machined from the center region of each slice, which was found by x-ray topography to have the greatest amount of precipitation. The disks were chemically jet-thinned from one side with HF-HNO₃ solutions for a time approximately half that needed for complete perforation. They were then flipped over and thinned to electron transparency. The observed area was thus characteristic of the bulk properties of the sample and not the surface region. TEM was performed with a JEOL 100CX STEM equipped with a field emission gun, x-ray energy dispersive spectrometer and electron energy loss spectrometer.

RESULTS AND DISCUSSION

The morphologies of the stacking faults were found to depend strongly on the temperature of the first nucleation anneal. In samples annealed at 550°C, the faults were centered around colonies of particles (Figure 1). The colonies could contain as many as 30 members, typically 100 - 300Å in diameter. In contrast, the samples annealed at 750°C contained single stacking faults with only one particle at the center of each (Figure 2). In both cases, the faults were all 2.1 to 2.3 μm in diameter. This result implies that the growth of the faults takes place at 1000°C and that differences in nucleation are probably responsible for the differences in the structures. In the sample annealed for 8 hours at 750°C, there were several very large stacking faults between 17 and 23 μm long. These faults did not have any specific nucleation sites that could be observed, and no reasonable explanation for their existence could be found.

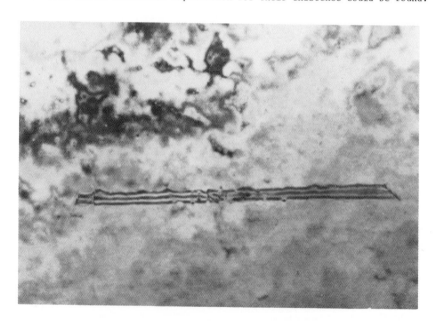

Figure 1. Typical stacking fault found in silicon annealed for 128 hours at 550°C plus 16 hours at 1000°C.

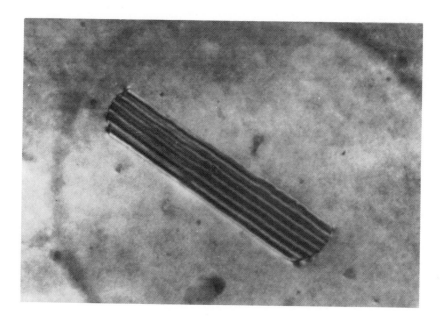

Figure 2. Typical stacking fault found in silicon annealed for 8 hours at 750°C plus 16 hours at 1000°C.

Several techniques were used in attempts to identify the nature of the particles apparently responsible for stacking fault nucleation. Micro-micro diffraction is JEOL terminology for diffraction patterns formed by placing the fully focused electron beam in one area without scanning it across the sample. The electron beam is $\sim$ 8Å in diameter and the analyzed areas are thus $\sim$ 100Å in diameter or less. When a micro-micro diffraction pattern is formed in an area of the stacking fault without particles, only the spots due to the silicon matrix are observed. However, when the beam is placed on one of the particles, extra spots are found in the patterns (Figure 3). The spots are in a rectangular array and the distances correspond to lattice spacings of 2.94 and 2.54Å. The spots do not fit any SiO_2 phase, hcp Si or paraffin. The most reasonable structure is a tetragonal lattice with a = 7.14Å and C = 7.12Å. One of the [100] axes of the silicon is parallel to a [a00] axis of the particle. The simplest interpretation of this data is that the precipitates are a modified cristobalite. The differences in the lattice parameters may be due to the stresses on the particle due to the tremendous volume change associated with changing Si into SiO_2. The chemical composition of these particles could not be determined with either XEDS or EELS in spite of literally hundreds of attempts.

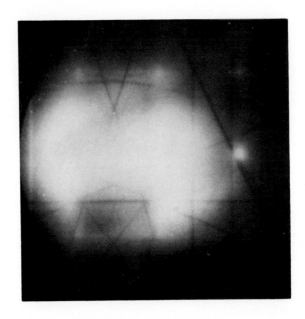

Figure 3. Micro-micro diffraction pattern of particles found on a stacking
fault after 128 hours at 550°C plus 16 hours at 1000°C.

If the particles of the colonies are imaged using weak beam techniques, it
is seen that each particle is surrounded by a bright ring, indicative of either
a dislocation loop or a stress field different from that of the fault itself.
In addition, the fringes of the stacking fault are usually continuous across
the particles. This implies that the lattice strains caused by the particle
are of the same type and sign and close to the same magnitude as that of the
stacking fault. Supportive evidence is the observation that when the small
particles associated with oxygen precipitates are imaged in weak beam with
the same deviation parameters, fringes, with the same spacing as stacking
faults are observed. A model which can reconcile all of these observations
is the replacement of a section of the stacking fault with a very thin, 1 to 2
atomic layer thick, SiO_2 platelet. The platelet is formed as the primary defect
of oxygen precipitation. The accomodation of the volume change causes interstitials to be formed which grow into stacking faults. The stacking fault is
discontinuous at the precipitate interface which gives rise to a strain field
around the particle. This model is similar to that first proposed by Patel,
except for the details of stacking fault nucleation. The existence of the
multiple nature of the defects at low nucleation temperatures can be explained
by the difference in nucleation rates at low and high temperatures. At low
temperatures, the increase in the rate of precipitate particle formation leads

to more numerous, smaller particles than at high temperatures. The increased
supersaturation at low temperatures can cause secondary precipitation of
oxygen on the Frank partial as well. If more than one particle nucleates a
fault on neighboring planes, then multiple faults can be formed. It should
be pointed out that the images of faults separated by one to ten interplanar
distances are essentially the same. At the higher temperatures, one particle
is formed and grows into a more complex shape, due to the lower nucleation
rate. Once this occurs there are no nucleation sites for multiple fault formation.

SUMMARY

The morphology of stacking faults formed during the precipitation of oxygen
in CZ silicon depends on the nucleation temperature. At 550°C, the faults
are multiple and contain precipitate colonies. At 750°C, the faults are single
with one particle. The particles were found to be SiO_2 in a modified cristobalite form. The faults are formed due to silicon interstitials caused by
accomodation of the volume change of Si changing to SiO_2. More particles are
formed at the low temperature either due to the increased nucleation rate or
due to precipitation on the Frank partial. The multiple fault is formed by
neighboring fault planes on different precipitate particles.

ACKNOWLEDGEMENTS

The authors would like to thank Dr. S. C. Baber for the FTIR measurements
and M. Jarvis, S. Johnston, K. Madden, R. Rose and C. Youts for assistance
in the experimental procedures.

REFERENCES

1. J. R. Patel, K. A. Jackson and H. Reiss, J. Appl. Phys., 48, 5279 (1977).

2. W. Patrick, E. Hearn, W. Westdorp and A. Bohg, J. Appl. Phys., 50, 7156
   (1979).

3. W. Kaiser and P. H. Keck, J. Appl. Phus., 28, 882 (1957).

THE EFFECT OF CARBON ON THE PRECIPITATION OF OXYGEN IN
CZOCHRALSKI SILICON

R. F. PINIZZOTTO AND H. F. SCHAAKE
Central Research Laboratories, Texas Instruments Inc., Dallas, TX 75265

ABSTRACT

Researchers in both the U. S. and Japan have reported
that carbon impurities can significantly alter the precipitation of oxygen in CZ silicon. We have employed FTIR,
x-ray topography and TEM to study this phenomenon in sili-
con containing < 2 x 10^{16} C cm^{-3} or 1.3 x 10^{17} C cm^{-3}. A
single step anneal of 16 hours at 1000°C will cause oxygen
precipitation to occur in the high carbon material, while
a 32 hour pre-anneal at 600°C is necessary for precipitation in the low carbon material. If a single anneal of
120 hours at 750°C is used, light precipitation occurs in
both types of material. A pre-anneal of 15 minutes at
1000°C followed by 120 hours at 750°C reduces precipitation in the high carbon material only slightly, but completely eliminates precipitation in the low carbon material.
It can be concluded that carbon causes heterogeneous nucleation of oxygen precipitation in CZ silicon and it is
proposed that it does so by lowering the interfacial energy
of the precipitates.

INTRODUCTION

The precipitation of oxygen in Czochralski silicon has recently become a very
important research topic for materials scientists working in the semiconductor
industry. Oxygen precipitation has been shown to be an effective intrinsic
gettering mechanism which can reduce the deleterious effects of heavy metal
contamination during integrated circuit processing. Several companies have
published process modifications that were specifically designed to promote
oxygen precipitation during semiconductor device fabrication.
    Transmission electron misroscopy (TEM) studies of oxygen precipitation in
CZ silicon have identified over 20 unique microdefect morphologies. These
include many familiar structures such as stacking faults and prismatic dislocation loops, and some highly unusual ones such as {100} platelets and
small precipitate clusters. At the present time there is no comprehensive
model capable of explaining why so many types of defects occur or how material
processing affects the microstructure. There is yet to be agreement on the
nucleation mechanism responsible for precipitate formation or on the growth
kinetics. Several workers have suggested that carbon plays an important role
in the process by promoting heterogeneous nucleation which enhances the rate
of precipitation [1,2,3]. Many of the results are not based on one-to-one
comparisons of material with high and low carbon concentrations and are speculative in nature. Even worse, many of the reports do not supply the details
of the type of material used, whether or not it received a resistivity stabilization anneal after crystal growth, or the anneal ambient used in the experiments.
    The goals of this study are to examine the effect of carbon concentration
on both the morphology of the oxygen precipitates and on their nucleation

and growth. The material used was particularly tightly controlled in an effort to separate the effects due to carbon concentration and heat treatment from the normal variation in oxygen concentration and doping found in most starting material.

EXPERIMENTAL DETAILS

Two types of material were used for these experiments, each from a crystal grown from a different type of polysilicon. In both cases, the crystals were boron doped, p-type, 5-10 Ω -cm (100) silicon grown by the Czochralski process in a commercially built crystal puller. Only the top of each boule was used. Slices were prepared using standard techniques and 3 mm thick slabs were cut from the crystal both above and below the section used for slices. The slabs were used to improve the sensitivity of the carbon concentration measurements. The first type or "low carbon" material had an oxygen concentration of 1.0×10^{18} cm^{-3} in the center of the slice as measured by Fourier Transform Infrared Spectroscopy (FTIR). The calibration constant determined by Kaiser and Keck was used to convert form IR absorption to atomic concentration [4]. The carbon concentration was below the detectability limit of the FTIR technique in the slabs as well as the slices. It is estimated to be $< 2 \times 10^{16}$ cm^{-3}. The second type or "high carbon" material contained 1.15×10^{18} cm^{-3} oxygen and 1.3×10^{17} cm^{-3} carbon in the center of the slice. The radial concentrations were measured in a step-wise manner by scanning a 3 mm aperture across the slice. The carbon concentrations were found to be essentially independent of position, but the oxygen concentrations decreased to half the center value 3 mm from the slice edge. The slices used in these experiments were initally in the as-grown condition, that is no resistivity stabilization anneal or other high temperature heat treatment was performed other than explicitly stated in this report.

Slices of both types of material were annealed for various times and temperatures designed to examine several aspects of oxygen precipitation in silicon. In the first experiment, samples were annealed for 16 hours at 1000°C in dry oxygen to simulate in a simple manner a typical MOS process. In particular, the diffusion distance of oxygen for this heat treatment is the same as the total diffusion distance calculated for a complete 16k dynamic RAM process flow. Other slices were annealed for 32 hours at 600°C in Ar, to promote oxygen precipitation, followed by the 16 hours at 1000°C growth anneal. Since the results from these experiments suggested that carbon was playing an active role in the nucleation process and was substantially altering the nucleation mechanism, the third experiment was an extended nucleation anneal of 120 hours at 750°C in N$_2$. The last experiment was to anneal the slices for 15 minutes at 1000°C in Ar followed by 120 hours at 750°C in N$_2$. This last process was designed to dissolve any grown-in defects or nucleation sites and then demonstrate true nucleation of oxygen precipitates from a homogeneous solution. The anneal matrix is summarized in Table 1.

TABLE 1.
Anneal Matrix

Experiment Number	Anneal	Comments
1	16 hrs, 1000°C	Defect Growth; MOS Process Simulation
2	32 hrs, 600°C + 16 hrs, 1000°C	Nucleation + Growth
3	120 hrs, 750°C	Extended Nucleation
4	15 min, 1000°C + 120 hrs, 750°C	Dissolution + Extended Nucleation

Transmission x-ray topography was used to examine the amount and spatial distribution of the oxygen precipitation. For a given high temperature heat treatment, the diffracted x-ray intensity is proportional to the amount of oxygen precipitated [5]. This assumes that the defect type and size distributions are the same for both areas being compared, either within a slice or from slice to slice. If the defect morphology changes, the diffracted x-ray intensity per defect will change as well. Therefore, it is not possible to directly relate diffracted x-ray intensity from slices annealed differently unless the defect morphology is known.

TEM samples were ultrasonically machined from each slice taking care to note the radial position of each in order to estimate its oxygen concentration. The samples were chemically jet thinned equally from both sides using various HF-HNO$_3$ solutions until they were electron transparent. The observed microstructures are thus characteristic of the bulk properties of the samples. TEM examination was performed using a JEOL 100CX Scanning Transmission Electron Microscope equipped with a field emmission gun, electron energy loss spectrometer and energy dispensive x-ray spectrometer.

RESULTS AND DISCUSSION

No evidence of oxygen precipitation could be found in the low carbon material annealed only for 16 hours in dry oxygen at 1000°C by either x-ray topography or TEM. However, when the same material is annealed for 32 hours at 600°C plus 16 hours at 1000°C, precipitation is observed in the control region of the slice where the oxygen concentration is the highest. There are several types of defects, but the majority of these are square platelets with{100} habit planes. The platelets are less than 10 nm thick and are surrounded by square, perfect dislocation loops with their sides oriented along <110> directions. These results were expected and similar reports have been made by other workers [6,7]. The high carbon material behaved differently, however, and the results were unpredicted. The samples annealed only for 16 hours at 1000°C still displayed oxygen precipitation in the center of the slice. This result indicates that heterogeneous nucleation is an important process in silicon and conclusively demonstrates that carbon does affect oxygen precipitation. It is not possible from our experiments to determine if enhanced nucleation occurs during crystal growth itself or during the subsequent high temperature anneals. The defects in this sample consist mainly of colonies of small precipitate particles intertwined with perfect dislocation loops. When the high carbon material is annealed with the two step process (Experiment 2), two distinct regions with different diffracted x-ray intensities are observed in the transmission topographs (Figure 1). The zone in the center of the slice has a very high diffracted x-ray intensity. TEM found the defect density to be very high as well, > 10^{13} cm^{-3}. The defects are very small and not as fully developed as in the other samples. Most of the precipitates cannot be analyzed because their strain fields are complicated and not amenable to TEM diffraction contrast analysis. The outer concentric ring had a lower x-ray diffraction intensity and the defects were the same type of colonies with dislocations found in the high carbon material after only the 1000°C anneal.

The precipitates in the high carbon material annealed with the two-step process have probably all been formed by a heterogeneous nucleation process. The effects of the carbon addition are two-fold. It can enhance the nucleation rate as evidenced by the larger number of defects in the high carbon material. It also allows precipitation to occur at lower oxygen concentrations, since the radius of precipation is larger in the high carbon slices.

Since the initial experiment indicated that carbon was significantly altering the nucleation process, slices of both types were annealed for 120 hours at 750°C. This heat treatment was designed to be an extended nucleation anneal,

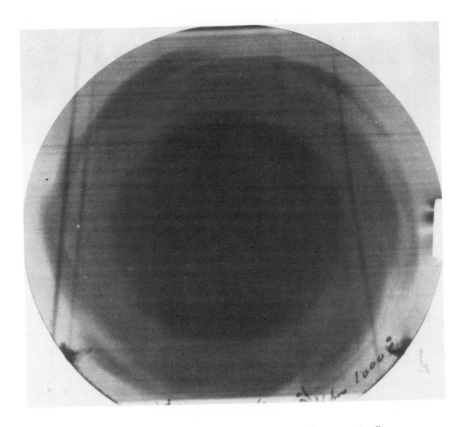

Figure 1. Transmission X-ray Topograph of "high carbon"
material annealed for 32 hrs at 600°C plus 16 hrs at 1000°C.

since 750°C has been found to be an efficient nucleation tempeature. The x-ray
topographs of slices of both types of material are identical. The central half
of each slice shows an increase in diffracted x-ray intensity. TEM of the low
carbon slice showed that the only defects formed were {100} platelets in an
early stage of development. In the high carbon material, the main defects were
long rods aligned along <113> directions. These are very similar to the defects found in Si after irradiation and subsequent annealing in the same temperature range. In addition to the rods, {100} platelets, perfect dislocation
loops, prismatic loops and stacking faults were observed. A very high density
of small, unanalyzable defects were found as well. These observations again
support the theory that carbon is affecting the nucleation stage of precipation.

One of the major questions which remains unanswered by these expreriments is whether the nucleation of oxygen precipitates takes place during the crystal growth process or if it is caused by the anneals between 600 and 750°C. To answer this question, slices of both low and high carbon material were annealed for 15 minutes at 1000°C followed by 120 hours at 750°C. The results of this experiment were astonishing! In the low carbon material, no defects or any evidence of oxygen precipitation could be found by either x-ray topography or TEM. Evidently, the nucleation sites formed during crystal growth were dissolved by the short high temperature anneal. In the high carbon material, on the other hand, x-ray topography showed that precipitation had occurred in the center of the slice to about the same extent as in the 750°C for 120 hours only sample. TEM found these defects to be mainly small <113> rods and small unanalyzable precipitates. The defect density was lower than in the sample annealed for only 120 hours at 750°C. Interestingly, no dislocations were observed.

The results described above suggest that the nucleation of oxygen precipitates does not take place at the intermediate temperature of 750°C, but probably occurs during crystal growth. The 750°C anneal is necessary to allow the nuclei to grow to a large enough size to be stable during the 1000°C heat treatment. Carbon affects nucleation of oxygen precipitates by increasing the number density of the defects and their morphology. The simplest model that can explain both phenomena is the reduction of the interfacial energy of the precipitates either by incorporation of the carbon into the particle or its segregation to the particle surface. The nucleation rate is an extremely sensitive function of the surface energy and can easily change by orders of magnitude with small changes in this parameter. The morphologies of the defects may change either because the relative energy of each crystallographic plane of the precipitate changes or because the point defect supersaturations are modified due to enhanced nucleation. If the true nucleation temperature regime is reduced to below 750°C, the interfacial energy of the precipitates must be quite high, probably > 500 mJ/m^2. This value is of the same order as the value proposed by the Japanese and in a previous paper from Texas Instruments [8,9], but is not in agreement with the value of 100 mJ/m^2 proposed by workers from Bell Laboratories [6]. It should be noted that the 450°C donor level associated with oxygen in silicon is related to the earliest stages of precipitate formation [5]. This supports the theory that the donor level is due to SiO$_4$ formation rather than being due to the presence of substitutional oxygen.

A final point is that we have observed the effects of annealing ambients on defect morphology and formation, as recently reported by researchers at IBM [10]. However, the effects of the ambient were minor compared to the effects of carbon concentration and annealing parameters in our experiments. A cleverly designed experiment with controlled ambients could be used to test for the effect of changes in point defect concentration on defect morphology since the ambient can affect the density of defects at the silicon surface or interface.

SUMMARY

Carbon impurities in Czochralski silicon affect both the kinetics of oxygen precipitation and the morphologies of the defects formed. In high carbon material, the precipitates are more numerous, smaller and more complicated structurally than in low carbon material. So called nucleation anneals done at 750°C probably cause nuclei formed during crystal growth to become large enough to survive subsequent high temperature processing. The simplest explanation of why carbon changes oxygen precipitation is that it reduces the interfacial energy of the particles formed. Elucidation of the mechanism of carbon and oxygen precipitation in CZ silicon could enhance the efficacy of intrinsic gettering currently used in some integrated circuit processing.

ACKNOWLEDGEMENTS

The authors would like to thank Dr. S. C. Baber for the FTIR measurements and M. Jarvis, S. Johnston, K. Madden, R. Rose and C. Youts for assistance in experimental procedures.

REFERENCES

1. R. F. Pinizzotto and H. F. Schaake, Electrochem. Soc. Fall Meeting Extended Abstracts, October 1980, pp. 1127–1129.

2. A. Staudinger, J. Appl. Phys. $\underline{49}$, 3870 (1978).

3. S. Kishino, Y. Matsushita and M. Kanamori, Appl. Phys. Lett., $\underline{35}$, 213 (1979).

4. W. Kaiser and P. H. Keck, J. Appl. Phys., $\underline{28}$, 882 (1957).

5. H. F. Schaake, S. C. Baber and R. F. Pinizzotto, to be published in Semiconductor Silicon/1981, the Electrochem. Soc., Inc., Princeton, N. J.

6. P. E. Freeland, K. A. Jackson, C. W. Lowe and J. R. Patel, Appl. Phys. Lett., $\underline{30}$, 31 (1977).

7. K. Yamamoto, S. Kishino, Y. Matsushita and T. Iizuka, Appl. Phys. Lett., $\underline{36}$, 195 (1980).

8. J. Ōsaka, N. Inoue and K. Wada, Appl. Phys. Lett., $\underline{36}$, 288 (1980).

9. H. F. Schaake, Electrochem. Soc. Fall Meeting Extended Abstracts, October 1980, pp. 1133–1135.

10. S. M. Hu, Appl. Phys. Lett., $\underline{36}$, 561 (1980).

Published 1981 by North-Holland, Inc.
Narayan, and Tan, eds.
Defects in Semiconductors

ELECTRON MICROSCOPE CHARACTERIZATION OF PULSE ANNEALED SEMICONDUCTORS

A. G. CULLIS

Royal Signals and Radar Establishment, St. Andrews Road, Malvern, England

ABSTRACT

The pulse processing techniques that have assumed promin-
ence over the past few years offer various important advantages for device fabrication technology. However, the usefulness of each individual method depends substantially upon
the specific annealing mechanism involved. This article
demonstrates the role of electron microscopy in elucidating
such mechanisms and in analysing annealed semiconductor
structures of importance to both research workers and semiconductor technologists. The range of laser and electron
beam pulse annealing methods is covered and defect structure
transitions observed are related to the solid and liquid
phase processes occurring. Characteristic impurity trapping
and segregation phenomena are described.

INTRODUCTION

The tremendous increase in world-wide interest in the pulse annealing of
semiconductors reflects, in part, the very great potential that exists for novel
applications in the field of device processing. However, the various pulse
annealing techniques also permit the solid-state physicist to probe new regimes
of crystal growth. Therefore, their importance in the area of basic materials
research is also most substantial. Due to these considerations, a wide range of
physical analytical techniques has been used to study the properties of pulse
annealed semiconductor layers. In particular, work based on electron microscopy
has provided a very large amount of structural information and it is the purpose
of the present article to outline many of the general results obtained. Some
emphasis will be placed on studies of Q-switched laser annealed material since
it is here that most detailed analytical work has been carried out. Nevertheless, the effects of scanning CW laser annealing and rapid pulsed and scanning
electron beam annealing also will be described.

Q-SWITCHED LASER ANNEALING

The use [1-4] of Q-switched laser radiation to anneal semiconductor layers
has received in-depth study. Short pulses ($\sim$10-100nsec) of radiation at
typically 694nm or 1060nm from, respectively, ruby and Nd-YAG lasers are often
employed. The laser spots may be small and overlapped by a scanning process [3]
or single, large area pulses can be utilized. In the latter case, to obtain
maximum uniformity in the annealed region, it is desirable to use a laser beam
diffuser which can give a uniform energy density of irradiation [5].
Most studies have concerned the annealing behaviour of ion implanted Si
specimens. The general structural changes that take place in initial ion-
induced amorphous layers have been elucidated by transmission electron microscope (TEM) investigations of plan-view specimens [6]. For a single laser pulse
anneal, as is clear from Fig. 1, increasing the radiation energy density leads
to the production of a range of residual layer defects from polycrystals to dislocated single crystal and, finally, to single crystal free of extended defects.

394

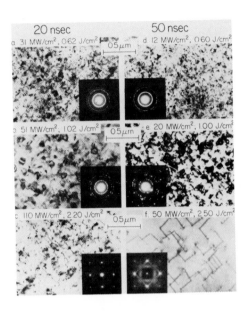

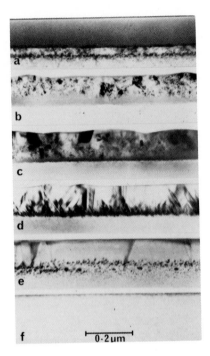

Fig. 1. TEM images and diffraction patterns for Si$^+$ ion implanted (001) Si samples irradiated with 20nsec and 50nsec ruby laser pulses. (After Tseng et al. [6])

Fig. 2. TEM images of cross-sections of (001) Si implanted with 4x10^{15} 150keV-As$^+$/cm^2: a) as-implanted, b) 0.20J/cm^2, c) 0.35J/cm^2, d) 0.85J/cm^2, e) 1.00J/cm^2 and f) 1.20J/cm^2 (After Cullis et al [8])

The basic mechanism for the annealing is transient surface melting, this having been determined by the study of various physical phenomena including surface reflectivity changes [7] and dopant diffusion behaviour [3]. The progressive penetration of the melt front through an ion implanted amorphous Si layer as the incident radiation energy density is increased is, perhaps, best demonstrated by the TEM study of cross-sectional specimens [8]. A typical sequence for ruby laser annealed As$^+$ ion implanted Si is shown in Fig. 2. Here, for the lower radiation densities, we see the progressive penetration of polycrystals through the initial amorphous layer. Then at about 0.8J/cm^2, the laser-induced melt reaches the single crystal interface permitting single crystal material to form during recrystallization. However, it is likely that the melt formed up to this stage is significantly undercooled [9] since, even for annealing at 0.85J/cm^2, the original internal interface is a source of defects in the recrystallized layer. Nevertheless, for a radiation energy density of 1.0J/cm^2 the melt does penetrate partially through this interface to leave a buried point defect cluster band with occasional V-shaped dislocation pairs propagating up to the external surface. Complete elimination of extended defects is achieved by the

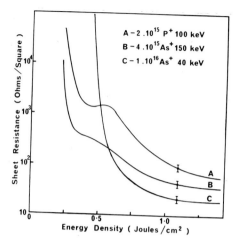

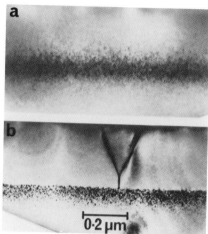

Fig. 3. Sheet resistance of As$^+$ and P$^+$ ion implanted (001) Si (p-type wafers) as a function of ruby laser pulse energy density. (After Cullis et al. [8])

Fig. 4. TEM images of cross-sections of (001) Si implanted with 10^{16} 100keV-B$^+$/cm^2: a) as-implanted, b) 1.5J/cm^2. (After McMahon et al. [10])

increased melt depth due to annealing at 1.2J/cm^2.

It is important to relate the changes in structure, due to laser annealing, of the type shown in Fig. 2 to the variations in electrical properties exhibited by the layer [8]. Curve B in Fig. 3 demonstrates that the progressive change from a polycrystal to a single crystal structure leads to a 2-stage electrical activation characteristic. The higher resistance intermediate state (near 0.5J/cm^2) is produced by the polycrystal layer structure, where carrier scattering effects are very important. Annealing above ∿1J/cm^2 gives very efficient As dopant activation in single crystal material. A similar 2-stage annealing characteristic is given by a P$^+$ ion implanted layer (curve A). Figure 3 also illustrates, in curve C, the importance of the thickness of the initial amorphous Si layer in determining the annealing onset. Decreasing the amorphous Si thickness also decreases the coupling of the radiation to the layer so that higher energy densities are needed for onset.

When an initial ion implanted layer has been only partially amorphized, the penetration of the melt produced by laser annealing can also be observed with great clarity by cross-sectional TEM studies. Figure 4a shows the initial buried band of displacement damage produced by high dose B$^+$ ion implantation in Si. Annealing with Q-switched ruby laser radiation at 1.5J/cm^2 melts down to the middle of the band [10], recrystallization then giving very good Si with few residual defects (Fig. 4b). The principal newly produced defects are occasional dislocation pairs nucleated in the remaining portion of the buried damage band. Laser annealing at higher radiation energy densities can eliminate all residual extended defects.

Annealing with Q-switched laser radiation pulses can introduce conventional implanted dopants onto Si lattice sites with high efficiency [2-4]. However, if the implanted impurity exhibits a low equilibrium solubility in the lattice,

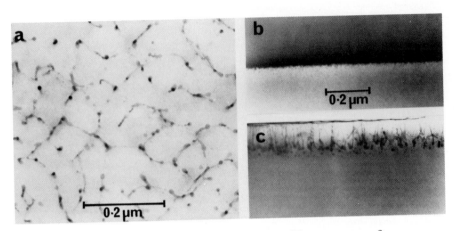

Fig. 5.   TEM images of (001) Si_2implanted with 5×10^{15} 150keV-Fe^+/cm^2:
a) plan view, annealed at $2J/cm^2$, b) and c) cross-sections of as-implanted and
$2J/cm^2$ annealed material, respectively. (After Cullis et al. [13])

excess impurity can precipitate upon annealing. Two characteristic phenomena
are often observed. First, the impurity can be segregated towards the external
surface if it exhibits a very low solid-liquid distribution coefficient [11,12].
This phenomenon is analogous to the process of zone refining. Second, the
impurity can segregate laterally to give channels of a precipitate phase which
form the boundaries of small crystalline Si cells [12]. This latter effect is
observed when instabilities due to constitutional supercooling occur in the re-
solidifying melt [13,14]. Figure 5 illustrates the nature of the cells formed
when Si ion implanted with Fe^+ is ruby laser annealed. Also shown in Fig. 5c is
the manner in which the cell walls can penetrate well into the Si matrix. The
phenomenon of constitutional supercooling is well known in the field of crystal
growth [15] but has not before been studied under conditions giving such high
resolidification velocities. The latter can be varied through at least the
range 1-5 m/sec by varying the background substrate temperature of a Si specimen
during the laser annealing process [14] and it has been found [16] that the cell
size depends critically upon this parameter. This is demonstrated in Fig. 6 for
the cell structure produced by laser annealing In^+ ion implanted Si. In this
example, the cell size changes from ~450Å to ~850Å as the resolidification
velocity is changed from 4 to 2 m/sec. This variation in cell size and, indeed,
the regimes under which cells are generally stable can be well accounted for by
the application [16,17] of the full interface morphological stability theory of
Mullins and Sekerka [18]. This takes into account the effect of a number of
important phenomena including capillarity.
    Also important for the analysis just mentioned is the effect of non-
equilibrium solute trapping [19] which can occur during high velocity recrystallization. This phenomenon effectively increases the impurity distribution
coefficient and can lead to the incorporation of metastable high concentrations
of dopant on substitutional matrix lattice sites [12,14,20]. The effect is
dramatically illustrated by the laser annealing of Pt^+ ion implanted Si.
Although the equilibrium solid solubility of Pt in Si has a maximum value of
about $10^{17}/cm^3$, as is clear from the channelled and random backscattering

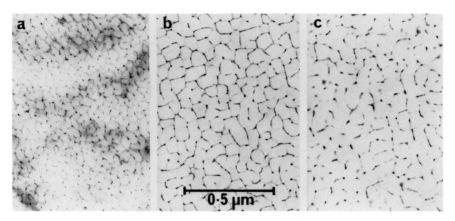

Fig. 6. TEM images of (OOl) Si implanted with 3×10^{15} 170keV-In$^+$/cm^2 and annealed at 1.5J/cm^2. Note dependence of cell·size on substrate background temperature: a) 77°K; b) 300°K and c) 410°K. (After Cullis et al. [16])

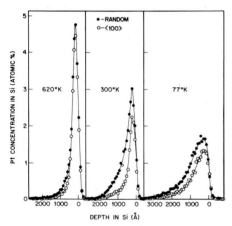

Fig. 7. Random and channelled 2-MeV He$^+$ ion backscattering spectra showing variation in dopant substitutionality for Pt$^+$ ion implanted and laser annealed Si. Background substrate temperatures were 620, 300 and 77°K, respectively. (After Cullis et al. [14])

spectra of Fig. 7 the amount of the impurity on substitutional lattice sites in the laser recrystallized matrix can exceed this by approximately 3 orders of magnitude [14]. The non-equilibrium impurity trapping effect is strongly dependent upon the velocity of the resolidification interface, increasing as the velocity increases. This behaviour is demonstrated in Fig. 7 and has also been observed for In impurity in Si when the recrystallization velocity was changed by varying both the background substrate temperature and the laser pulse length [21].

Not all implanted low solubility impurities exhibit the effects just described. For example, if an impurity is actually insoluble in the initial laser-induced melt precipitation can occur at this stage. This is thought to

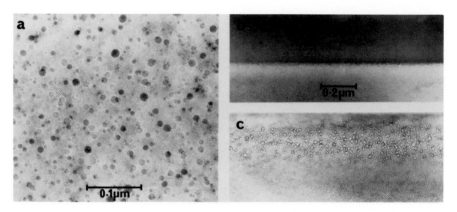

Fig. 8. TEM images of (001) Si implanted with 2×10^{15} 150keV-Ar^+/cm^2: a) plan view, annealed at $2J/cm^2$; b) and c) cross-sections of as-implanted and $2J/cm^2$ annealed material, respectively. (After Cullis et al. [13])

occur when Ar^+ ion implanted Si is subjected to ruby laser annealing [13]. As is shown in Fig. 8a, the resulting surface layer consists of high quality single crystal Si, although this contains a very high density of bubbles containing the precipitated Ar. When such a specimen is examined in cross-section (Fig. 8c) it is found that the bubble distribution has a number density which rises to a maximum at approximately the projected range in the Si of the original Ar^+ ions. Thus, only very limited impurity diffusion has apparently occurred. The final state is consistent with the precipitation of Ar in the transient liquid Si, the relatively immobile bubbles then being trapped near their initial locations upon resolidification [13].

In addition to the work described above, other investigations have demonstrated the way in which deposited layers of disordered Si can be recrystallized by use of Q-switched laser irradiation. Both deposited amorphous Si [22] and deposited polycrystalline Si [23] have been studied. In the former case, free volume initially incorporated into the as-deposited film gives rise to the presence of cavities in the final recrystallized matrix [24]. Of course, certain types of deposited Si film can have initial high perfection as in the case of Si on sapphire which is used for commercial device fabrication. However, once again laser annealing can dramatically change the basic Si defect structure. This is demonstrated in Fig. 9 where cross-sectional TEM images show the structure transitions which occur in As^+ ion implanted Si on sapphire [25]. The initial surface amorphous Si layer recrystallizes at a markedly lower radiation energy density than for homoepitaxial Si, this being due principally to the relatively low thermal conductivity of the sapphire substrate. It is also evident that microtwins initially present in the Si often reform and penetrate completely through the laser recrystallized layer. For a radiation energy density of ~$0.8J/cm^2$ the penetration of the Si melt is more than 3000Å into the epitaxial layer with the formation of an increased density of dislocations in the recrystallized material. The melt penetrates completely through the epitaxial layer for irradiation at $1.1J/cm^2$. This leads to a transformation of the complete layer structure from an initial situation where microtwins dominate to a final condition with a high density of dislocations. The latter often originate from the region of the heteroepitaxial interface undoubtedly due to

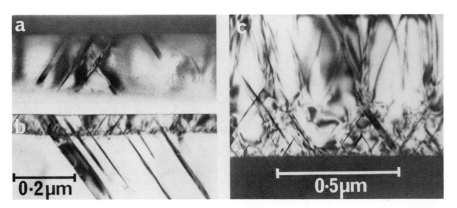

Fig. 9. TEM cross-sectional images of (001) Si on sapphire implanted with 10^{16} 40keV-As$^+$/cm^2: a) as-implanted; b) and c) ruby laser annealed at 0.4J/cm^2 and 0.8J/cm^2, respectively. (After Hill et al. [25])

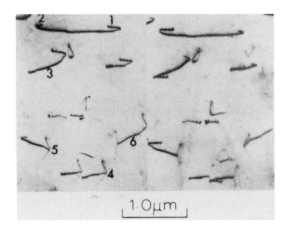

Fig. 10. TEM stereo-pair (g 220) showing dislocation regrowth following partial dissolution by limited laser melting. (After Narayan and Young [27])

the lattice irregularity at this location. Other studies of laser annealed Si on sapphire have, for example, demonstrated [26] the edge rounding that can occur for islands fabricated in the Si layer.

When dislocations are initially present in laser annealed Si, intersection of these by the melt front at its position of deepest penetration can give rise to a characteristic dislocation splitting behaviour during growth [27]. Typical annealed structures are shown in Fig.10. For example, at 4,5 and 6 the a/2 [110] dislocations split at one end during growth from the melt to give two other

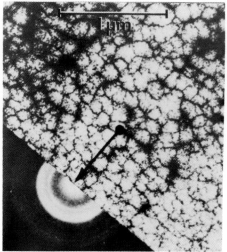

Fig. 11. TEM image and diffraction pattern from a Pt film on Si after Nd-YAG laser annealing at 18MW/cm^2. (After Poate et al. [28])

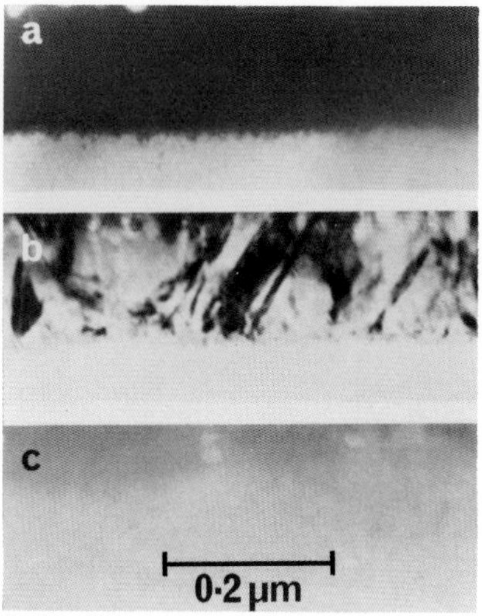

Fig. 12. TEM cross-sectional images of (001) Ge implanted with 5×10^{15} 150keV-As$^+$/cm^2: a) as-implanted; b) and c) ruby laser annealed at 0.2J/cm^2 and 0.6J/cm^2, respectively.

whole dislocations with Burgers vectors along different $\langle 110\rangle$ type directions. Dislocations which did not split (for example, at 1,2 and 3) also yield regrown segments along $\langle 113\rangle$ directions and, since these are not normal to the layer surface, a reduced line tension along [113] presumably could be important.

A further prominent class of structures which have been studied after laser annealing treatment consists of deposited metal layers on Si. This type of processing has clear potential importance for device contact fabrication. An example of the laser alloying of a Pt film into Si [28] is given in Fig. 11. This shows the result of irradiating a 450Å Pt film on Si at a photon power density of 18MW/cm^2 using a Nd-YAG laser. The final structure once again exhibits cell formation due to the effect of melt supercooling. The cell walls within the recrystallized Si in this case [28] contain predominantly PtSi, although amorphous phases can be formed under other conditions. The production of cellular segregated phases in laser-reacted layers of Co and Pd on Si has also been observed [29].

Although Q-switched laser annealing has been applied mainly to the processing of Si layers and alloys, other semiconducting materials have also received investigation. Of the Group IV elements, Ge exhibits many properties that are similar to those of Si and the laser annealing of ion implanted layers gives melting with structure transitions which are analogous to those of Figs. 1 and 2. For example, Fig. 12 provides cross-sectional TEM images which illustrate the ruby laser annealing of initially amorphous, As$^+$ ion implanted Ge. Here, it is clear that for even a radiation energy density of only 0.6J/cm^2, full recrystallization occurs yielding single crystal Ge with no residual extended defects. An intermediate state showing mostly microtwins with some polycrystals is produced at just 0.2J/cm^2. The relatively low radiation energy densities

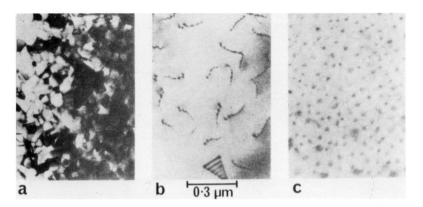

Fig. 13. TEM images of GaAs implanted with 10^{15} 400keV-Te$^+$/cm^2 and ruby laser
annealed at a) 0.6J/cm^2; b) 1.0J/cm^2 and c) 1.4J/cm^2.   (After Campisano
et al. [30])

required to produce these structure changes are a consequence of the increased
radiation absorption coefficient of Ge compared to that of Si.

In addition to the elemental semiconductors, III-V compound materials have
great technological importance. The Q-switched laser annealing of ion implanted
GaAs has received detailed attention and it has been found [30,31] that the
general structural changes which occur are similar once again to those observed
for Si. This is illustrated [30] in Fig. 13 which shows the polycrystal and
dislocated single crystal stages for the ruby laser annealing of Te$^+$ ion
implanted GaAs. However, even the well annealed layer (1.4J/cm^2) is not free
from structural features and, indeed, it has been shown [32] that islands of
metallic Ga can remain on the final annealed surface due to preferential loss
of relatively volatile As. A corresponding sequence of structure transitions is
also seen [33] for the laser annealing of ion implanted InP. These are demonstrated in Fig. 14 where an amorphous S$^+$ ion implanted InP surface transforms
either to a polycrystalline state or single crystal depending upon the energy
density of the incident ruby laser radiation. In this case, the transient
surface melting can lead to loss of volatile P and this is illustrated in
Fig. 14d, which shows the metallic In islands that can remain on an InP surface
after repeated laser pulse anneals at a low radiation energy density.

SCANNING CW LASER ANNEALING

Another method of laser annealing, distinct from that described above,
employs a high power CW laser beam to transiently heat the semiconductor sample
[34,35]. Since the radiation power densities achievable at the sample surface
are lower than for Q-switched laser annealing, the CW beam (often from an Ar$^+$
ion laser) is generally scanned over the surface [34] with a local dwell time of
the order of a few milliseconds. The relatively extended heat pulse duration so
produced ensures that the dominant annealing mechanism is solid phase regrowth
of the irradiated layer [36].

The application of this method to the annealing of ion implanted Si layers
has demonstrated its ability to produce high quality epitaxial regrowth. TEM
studies [37] have shown that amorphous As$^+$ ion implanted Si can be induced to

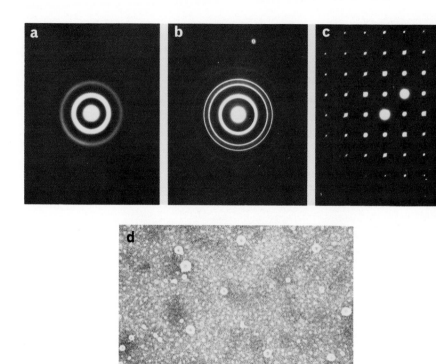

Fig. 14. a), b) and c) Transmission electron diffraction patterns from 180keV
S^+ ($6 \times 10^{14}/cm^2$) ion implanted InP, respectively, as-implanted, and ruby laser
annealed at $0.2J/cm^2$ and $0.5J/cm^2$. Note transition to single crystal structure
with increasing energy density of irradiation. d) SEM image of In islands
formed on InP subjected to multiple pulse annealing. (After Cullis et al. [33])

regrow with the formation of few extended lattice defects in the annealed
region, while the As dopant exhibits high electrical activity. When the ion
implantation does not give an initial complete amorphous layer excellent recrystallization is also possible. For example, B^+ ion implant doses in this
category can yield very low numbers of extended defects upon scanning laser
annealing [35] while conventional furnace annealing gives dislocation loops and
rod defects [38].

A particularly important area of application for scanning laser annealing
lies in the recrystallization of doped polycrystalline Si which is used for
gates and interconnects in Si-gate MOS devices. For example, when B^+ ion
implanted polycrystalline Si on Si_3N_4 is scanned with a laser beam [39] it is
possible to produce substantial grain growth and good activation of the dopant.
This is illustrated in Fig. 15, the annealing in this case being produced by
working at a temperature just above the layer melting threshold.

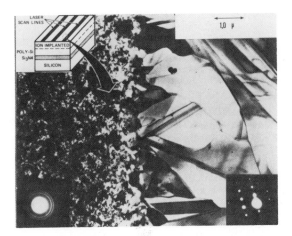

Fig. 15. TEM image of ion implanted and CW laser annealed polycrystalline Si
film at the boundary of the laser scan line: insets show diffraction patterns
characteristic of each region. (After Gat et al. [39])

ELECTRON BEAM ANNEALING

When one considers the importance of electron beam techniques for advanced
device lithography applications it is, perhaps, not surprising that electron
beam annealing techniques are also the subject of detailed study. The first to
be proposed is analogous in many ways to Q-switched laser annealing and relies
upon the deposition of sufficient energy in the sample surface by a single,
large area, short duration (~100nsec) pulse of electrons [40]. The irradiation
takes place in a vacuum chamber and the basic annealing mechanism is transient
surface melting. The structure changes that take place in an ion implanted
layer are well illustrated by cross-sectional TEM work and a series of images
[41] corresponding to a range of electron beam energy densities are shown in
Fig. 16. Melt penetration sufficient to give almost complete epitaxial recrystallization occurs for an electron beam pulse at $0.55J/cm^2$, while the formation of a strongly undercooled melt is consistent with the multi-level defect
structure which occurs at lower irradiation energy density levels. Indeed, the
reduced melting temperature and melting enthalpy of amorphous Si have been
measured [41]. When the rapid pulse electron beam annealing technique is
applied to the fabrication of device structures efficient activation of ion
implanted dopants is generally observed. However, dislocation formation occurs
in the annealed regions under certain conditions [42] and this, together with
anomalous dopant redistribution effects and pulse nonuniformity phenomena [42],
requires further study.

The second important electron beam annealing method involves scanning the
sample surface with a focussed electron probe. This is analogous to the
scanning CW laser technique and, once again, it relies upon the thermal pulse
delivered by the beam to give solid phase recrystallization [43,44]. Excellent
recrystallization of ion implanted layers can be obtained and the direct writing
of ~1μm, discrete crystalline lines in such a layer has been demonstrated [45].
While this shows great promise for device fabrication, there is a second
technique which also has great potential importance. This uses rapid repeated

404

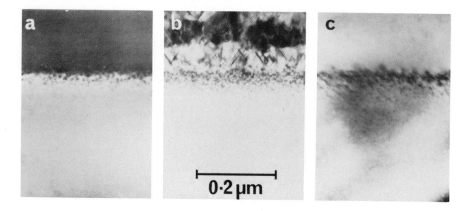

Fig. 16. TEM cross-sectional images of P+ ion implanted (001) Si: a) as-implanted, showing initial amorphous layer; b) and c) pulse electron beam annealed at 0.5J/cm² and 0.55J/cm², respectively. (After Baeri et al. [41])

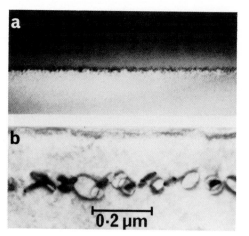

Fig. 17. TEM cross-sectional images of P+ ion implanted (001) Si: a) as-implanted, showing initial amorphous layer and b) multiple scan electron beam annealed at 30W for approximately 5sec. (After McMahon et al. [10])

scans of a device structure to raise its overall temperature to the level at which recrystallization can take place in a few seconds [46]. This multi-scanning electron beam approach effectively provides a 'fast furnace' facility and the defect structures in annealed ion implanted layers can more closely resemble their conventionally annealed counterparts. A typical example is given in Fig. 17 where it is seen that an initial amorphous layer yields a buried array of dislocation loops after the anneal [10]. A high level of P dopant activation is observed in the recrystallized Si demonstrating the potential usefulness of the technique to the device manufacturer.

ACKNOWLEDGEMENTS

The author is indebted to a number of colleagues for their close collaboration during studies in some of the areas described above. Particular thanks are due to Dr. J. M. Poate (Bell Labs.), Prof. G. Foti and Dr. P. Baeri (University of Catania) and Dr. D. T. J. Hurle, Mr. H. C. Webber and Mr. N. G. Chew (RSRE).

REFERENCES

1. I. B. Khaibullin, E. I. Shtyrkov, M. M. Zaripov, M. F. Galyautdinov and G. G. Zakirov, Sov. Phys. Semicond. 11, 190 (1977).

2. G. Foti, S. U. Campisano, E. Rimini and G. Vitali, J. Appl. Phys. 49, 2569 (1978).

3. G. K. Celler, J. M. Poate and L. C. Kimerling, Appl. Phys. Lett. 32, 464 (1978).

4. J. Narayan, R. T. Young and C. W. White, J. Appl. Phys. 49, 3912 (1978).

5. A. G. Cullis, H. C. Webber and P. Bailey, J. Phys. E: Sci. Instrum. 12, 688 (1979).

6. W. F. Tseng, J. W. Mayer, S. U. Campisano, G. Foti and E. Rimini, Appl. Phys. Lett. 32, 824 (1978).

7. D. H. Auston, C. M. Surko, T. N. C. Venkatesan, R. E. Slusher and J. A. Golovchenko, Appl. Phys. Lett. 33, 437 (1978).

8. A. G. Cullis, H. C. Webber and N. G. Chew, Appl. Phys. Lett. 36, 547 (1980).

9. B. G. Bagley and H. S. Chen in: Laser Solid Interactions and Laser Processing - 1978, S. D. Ferris, H. J. Leamy, J. M. Poate eds. (Am. Inst. Phys. New York 1979) pp. 97-102.

10. R. A. McMahon, H. Ahmed and A. G. Cullis, Appl. Phys. Lett. 37 (1980).

11. P. Baeri, S. U. Campisano, G. Foti and E. Rimini, Phys. Rev. Lett. 41, 1246 (1978).

12. A. G. Cullis, J. M. Poate and G. K. Celler, in: Laser Solid Interactions and Laser Processing - 1978, S. D. Ferris, H. J. Leamy, J. M. Poate eds. (Am. Inst. Phys., New York 1979) pp. 311-316.

13. A. G. Cullis, H. C. Webber, J. M. Poate and N. G. Chew, J. Microsc. 118, 41 (1980).

14. A. G. Cullis, H. C. Webber, J. M. Poate and A. L. Simons, Appl. Phys. Lett. 36, 320 (1980).

15. W. Bardsley, J. S. Boulton and D. T. J. Hurle, Solid-State Electron. 5, 395 (1962).

16. A. G. Cullis, D. T. J. Hurle, H. C. Webber, N. G. Chew, J. M. Poate, P. Baeri and G. Foti, Appl. Phys. Lett., in the press.

406

17. D. T. J. Hurle, A. G. Cullis, H. C. Webber, J. M. Poate, P. Baeri and G. Foti, to be published.

18. W. W. Mullins and R. F. Sekerka, J. Appl. Phys. 35, 444 (1964).

19. K. A. Jackson, G. H. Gilmer and H. J. Leamy, in: Laser and Electron Beam Processing of Materials, C. W. White, P. S. Peercy eds. (Academic Press, New York 1980) pp. 104-109.

20. C. W. White, S. R. Wilson, B. R. Appleton and F. W. Young, J. Appl. Phys. 51, 738 (1980).

21. P. Baeri, J. M. Poate, G. Foti, S. U. Campisano, E. Rimini and A. G. Cullis, Appl. Phys. Lett. 37, (1980).

22. J. C. Bean, H. J. Leamy, J. M. Poate, G. A. Rozgonyi, T. T. Sheng, J. S. Williams and G. K. Celler, Appl. Phys. Lett. 33, 227 (1978)

23. M. Tamura, H. Tamura and T. Tokuyama, Jap. J. Appl. Phys. 19, L23 (1980).

24. H. J. Leamy, G.A. Rozgonyi, T. T. Sheng and G. K. Celler, in: Laser and Electron Beam Processing of Electronic Materials, C. L. Anderson, G. K. Celler and G. A. Rozgonyi eds. (Electrochem. Soc., Princeton 1980) pp. 333-343.

25. C. Hill, D. J. Godfrey, A. G. Cullis, H. C. Webber and N. G. Chew, Appl. Phys. Lett. in the press.

26. C. P. Wu and G. L. Schnable, RCA Review 40, 339 (1979).

27. J. Narayan and F. W. Young Jr., Appl. Phys. Lett. 35, 330 (1979).

28. J. M. Poate, H. J. Leamy, T. T. Sheng and G. K. Celler, Appl. Phys. Lett. 33, 918 (1978).

29. G. J. van Gurp, E. J. Eggermont, Y. Tamminga, W. T. Stacy and J. R. M. Gijsbers, Appl. Phys. Lett. 35, 273 (1979).

30. S. U. Campisano, G. Foti, E. Rimini, F. H. Eisen, W. F. Tseng, M-A. Nicolet and J. L. Tandon, J. Appl. Phys. 51, 295 (1980).

31. D. K. Sadana, M. C. Wilson and G. R. Booker, J. Microsc. 118, 51 (1980).

32. S. S. Kular, B. J. Sealy, M. H. Badawi, K. G. Stephens, D. K. Sadana and G. R. Booker, Electron. Lett. 15, 413 (1979).

33. A. G. Cullis, H. C. Webber and D. S. Robertson, in Laser Solid Interactions and Laser Processing - 1978, S. D. Ferris, H. J. Leamy, J. M. Poate eds.. (Am. Inst. Phys., New York 1979) pp. 653-658.

34. A. Gat and J. F. Gibbons, Appl. Phys. Lett. 32, 142 (1978).

35. A. Gat, J. F. Gibbons, T. J. Magee, J. Peng, P. Williams and C. A. Evans, Jr., Appl. Phys. Lett. 33, 389 (1978).

36. J. S. Williams, W. L. Brown, H. J. Leamy, J. M. Poate, J. W. Rodgers, D. Rousseau, G. A. Rozgonyi, J. A. Shelnutt and T. T. Sheng, Appl. Phys. Lett. $\underline{33}$, 542 (1978).

37. A. Gat, J. F. Gibbons, T. J. Magee, J. Peng, V. R. Deline, P. Williams and C. A. Evans, Jr., Appl. Phys. Lett. $\underline{32}$, 276 (1978).

38. R. W. Bicknell and R. M. Allen, Rad. Effects $\underline{6}$, 45 (1970).

39. A. Gat, L. Gerzberg, J. F. Gibbons, T. J. Magee, J. Peng and J. D. Hong, Appl. Phys. Lett. $\underline{33}$, 775 (1978).

40. A. C. Greenwald, A. R. Kirkpatrick, R. G. Little and J. A. Minnucci, J. Appl. Phys. $\underline{50}$, 783 (1979).

41. P. Baeri, G. Foti, J. M. Poate and A. G. Cullis, Phys. Rev. Lett. in the press.

42. J. M. Leas, P. J. Smith, A. Nagarajan and A. Leighton, in Laser and Electron Beam Processing of Electronic Materials, C. L. Anderson, G. K. Celler and G. A. Rozgonyi eds. (Electrochem. Soc., Princeton 1980) pp. 141-151.

43. R. A. McMahon and H. Ahmed, Electron. Lett. $\underline{15}$, 2 (1979).

44. J. L. Regolini, J. F. Gibbons, T. W. Sigmon, R. F. W. Pease, T. J. Magee and J. Peng, Appl. Phys. Lett. $\underline{34}$, 410 (1979).

45. K. N. Ratnakumar, R. F. W. Pease, D. J. Bartelink and N. M. Johnson, J. Vac. Sci. Technol. $\underline{16}$ (1979).

46. R. A. McMahon and H. Ahmed, J. Vac. Sci. Technol. $\underline{16}$, 1840 (1979).

Published 1981 by North-Holland, Inc.
Narayan, and Tan, eds.
Defects in Semiconductors

MELTING AND LASER ANNEALING IN SEMICONDUCTORS USING 0.485 μm AND 0.193 μm
PULSED LASERS*

J. NARAYAN, J. FLETCHER, AND R. E. EBY[1]
Solid State Division, Oak Ridge National Laboratory, Oak Ridge, TN 37830

ABSTRACT

Annealing of displacement damage, the dissolution of
boron precipitates, the formation of constitution supercooling cells, and the broadening of dopant profiles have
been studied in laser annealed silicon. These samples were
irradiated with a dye laser (λ = 0.485 μm, τ = 9 ns, E =
0.7-1.2 J cm^{-2}) and an Excimer laser (λ = 0.193 μm, τ = 9 ns,
E = 0.5-0.7 J cm^{-2}) pulses. These results can be consistently interpreted by invoking melting during pulsed
laser irradiation. Thus these results provide convincing
evidence for the melting phenomenon.

INTRODUCTION

High power ruby and Nd:YAG laser pulses have been used to remove displacement damage, dislocation loops and precipitates [1,2]. These results as well as
those on dopant-profile broadening, and the formation of constitution supercooling cells have been successfully explained on the basis of a melting model
[3,4]. The basic assumption in this model is that the energy from the electronic system can be rapidly ($<$ 10^{-9} s) transferred to the lattice and this
energy is then utilized to heat and melt thin ($<$ 1 μm) surface layers of
materials. Some authors notably Khaibullin et al. [5] and Van Vechten et al.
[6] have speculated on a plasma model. According to the plasma model, the
energy transfer from the electronic system to the lattice is delayed ($\sim$ 200 ns)
through the screening of phonon emission, and the high concentration of
electrons and holes produced during pulsed laser irradiation may cause annealing
via enhanced diffusion of point defects, and glide and climb of dislocations.
At a still higher plasma density ($\sim$ 8 x 10^{21} cm^{-3}) electrons excited into antibonding states are proposed to lead to a weak first order phase transition
resulting in softening of the lattice. The crystal would then be in a fluid-
like state with energy retained primarily in the electronic system. This energy
would then be distributed over a much greater depth as carrier diffusion becomes
important. As the plasma density declines, the material passes back through the
phase transition leading to recrystallization.
Recently Lo and Compaan [7] have measured the lattice temperature of silicon
during pulsed laser irradiation by Raman scattering within 10 ns of a dye laser
pulse (λ = 0.485 μm, τ = 9 ns) in the energy range 0.7 to 1.0 J cm^{-2}. They
reported a temperature rise of $\sim$ 300°C for a 1.0 J cm^{-2} laser pulse, thus
lending support to the plasma annealing model. In this paper we have used dye
laser pulses (supplied by A. Compaan) and Excimer laser pulses, to study
annealing of displacment damage and boron precipitates, dopant profile
broadening, and the formation of constitution supercooling cells. These results
can only be interpreted on the basis of the melting model.

*Research sponsored by the Division of Materials Sciences, U.S. Department of
Energy under contract W-7405-eng-26 with Union Carbide Corporation.
[1]Analytical Chemistry Division.

EXPERIMENTAL

Silicon specimens (2-6 Ω-cm) with <100> and <111> orientations were implanted with $^{75}As^+$ or $^{121}In^+$ ions. The energy of the ions ranged from 100 to 200 keV and the doses from 1.0 to 10.0 x 10^{15} ions cm^{-2}. Some of the silicon specimens were diffused with boron using B_2O_3 at 950°C creating boron precipitates in the top ~ 300 Å layer. These samples were irradiated with the Q-switched output of a dye laser (wavelength λ = 0.485 µm, pulse duration τ = 9 x 10^{-9} s) with pulse energy density 0.7-1.2 J cm^{-2}. These laser pulses were similar to those used by Lo and Compaan for their Raman temperature measurements [7]. The shape of the laser spot was elliptical with major and minor axes being 180 µm and 130 µm respectively. Some of the ion implanted specimens were treated with 100 XR - UV - EXCIMER (ArF) LASER pulses (λ = 0.193 µm, τ = 9 x 10^{-9} s, E ≈ 0.5 J cm^{-2}). The beam size was 1.9 x 0.7 cm. The detailed changes in the microstructure of laser annealed specimens were studied using a Philips (EM-400) analytical microscope. Concomittant variations in dopant concentration profiles were investigated using Secondary Ion Mass Spectrometry, using an ion beam ~ 40 µm diameter.

RESULTS AND DISCUSSION

Figure 1 shows x-section TEM micrographs from an arsenic implanted, and dye laser annealed specimen. As-implanted specimens contained an amorphous layer ~ 1800 Å thick followed by a band (~ 700 Å thick) of dislocations loops. After irradiating with a 0.75 J cm^{-2} laser pulse (Gaussian energy distribution across the spot), the central regions (20-100 µm across) were found to be defect-free as shown in Fig. 1(d). This annealed region was surrounded by a narrow region containing V-shaped dislocations [Fig. 1(c)], which was in turn followed by an annular polycrystalline region [Fig. 1(b)]. The polycrystalline grain size was very small in the outer regions where the melt-front did not penetrate completely through the amorphous layer, as shown in Fig. 1(a). Figure 2 shows plan-view electron micrographs where the annealed regions in Fig. 2(a) [corresponding to Fig. 1(d)], V-shaped dislocations in Fig. 2(b) [corresponding to Fig. 1(c)], and polycrystalline structures in Fig. 2(c) [corresponding to Figs. 1(a) and 1(b)] are clearly shown. From the comparison of these results with those obtained using uniform laser pulses, the following can be concluded. In the central region the pulse energy density was high enough such that the melt-front exceeded the thickness of the damaged layer, and it grew defect-free with the same perfection as the substrate. In the region immediately surrounding the central region the melt-front reached the dislocation band, and the loops intersected by the melt-front grew back as two segments of V-shaped dislocations. The dislocations of a given Burgers vector were found to grow in a specific direction, which was characteristic of crystal growth from the melt [8]. In the outer regions where the melt-front did not reach the crystalline substrate, polycrystals were observed.

The dopant profiles from the central and the unannealed regions for 0.75 and 0.92 J cm^{-2} laser pulses are shown in Fig. 3. The as-implanted profile from the unannealed regions is Gaussian as expected. In the laser annealed region, (profile 1) significant dopant redistribution occurs both toward the surface and into the deeper regions of the crystal. The drastic changes in the dopant profile after laser annealing are consistent with melting. The inset in Fig. 3(b) shows the experimentally deduced melt-front penetration vs time, which was constructed from the estimated velocity of solidification (~ 6.0 ms^{-1}), total melt-lifetime (60 ns), and the maximum depth of melt-front penetration of 3300 Å obtained from dislocation growth experiments [9]. From Fig. 3(b), the average diffusion distance at a depth of 2000 Å was estimated to be 250 Å, and the corresponding melt-lifetime from the inset was 20 ns. Assuming unidirectional

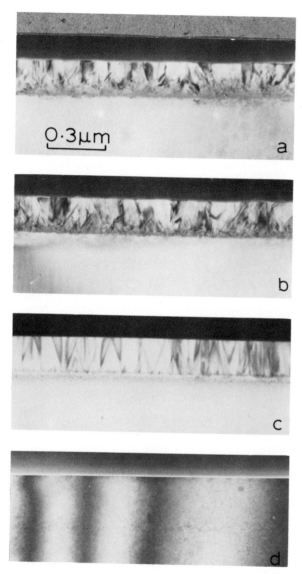

Fig. 1. X-section electron micrographs showing microstructures (for E = 0.75 J cm^{-2} pulse) from outer regions to the middle of the laser spot (as a function of increasing pulse energy density). (a) Polycrystalline layer with underlaying amorphous layer; (b) polycrystalline layer with no amorphous layer; (c) V-shaped dislocations growing from the dislocation band below the amorphous layer; and (d) almost a "complete" annealing.

412

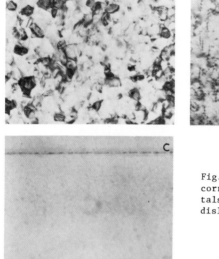

Fig. 2. Plan-view electron micrographs corresponding to Fig. 1. (a) Polycrystals similar to Fig. 1(b), (b) V-shaped dislocations, and (c) "complete" annealing.

0·40μm

diffusion and equating $\sqrt{D\tau} = 250$ Å, we obtained the value of D as 3.0 ($\pm$ 1.0) x 10^{-4} cm^{2}s^{-1}, which was found to be in good agreement with the value of the diffusion coefficient of arsenic in liquid silicon [10,11]. This value of D was also found to be in good agreement with detailed numerical calculations to be reported later [12].

Figures 4 and 5 provide results on the dissolution of boron precipitates in silicon. Boron diffusion at 950°C using B$_2$O$_3$ produced boron precipitates in the top ~ 300 Å layer with average precipitate size 100 Å. An electron micrograph showing the precipitates is given in Fig. 4(a). A dye laser pulse of 1.17 J cm^{-2} leads to dissolution of these precipitates so that no defects are observed as shown in Fig. 4(b). The boron concentration profiles before and after laser annealing are shown in Fig. 5. The redistribution or the transport, of boron which was contained in the precipitates near the surface occurred over a distance of about ~ 1500 Å. It should be remarked that using known diffusion coefficients, the drastic profile changes observed after laser annealing cannot be explained by diffusion in the solid state. For example, the diffusion coefficient of B in the solid state even close to the melting point is about 10^{12} cm^2 s^{-1}. If we take the characteristic time τ of the laser annealing to be approximately ten times the pulse duration (a large overestimate), the diffusion length is still estimated to be about 3 x 10^{-2} Å. The presence of a plasma could enhance (via ionization effects) the migration of already present point defects, and could enhance the glide of dislocations; but there is no realistic plasma model to explain the break-up and dissolution of precipitates [13].

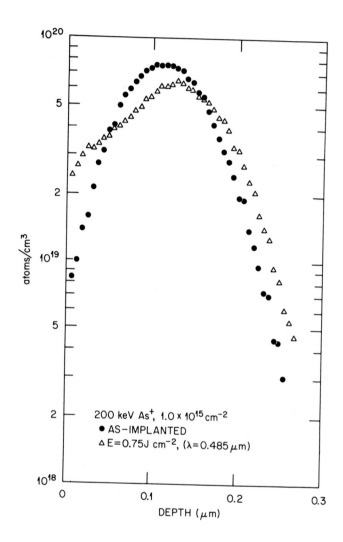

Fig. 3(a) Arsenic concentration profiles before and after the dye laser annealing: E = 0.75 J cm^{-2}.

414

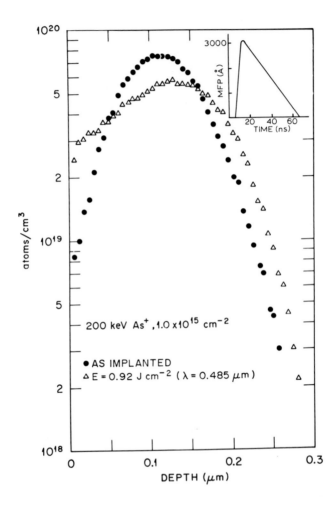

Fig. 3(b) The arsenic concentration profile before and after the dye laser annealing; E = 0.92 J cm^{-2}. The inset shows melt-front penetration as a function of time.

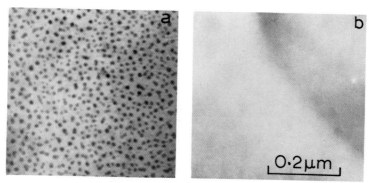

Fig. 4. Dissolution of boron precipitates in laser-annealed silicon: (a) boron precipitates in as-diffused (at 950°C, 30 min) silicon, and (b) precipitates dissolved after laser annealing.

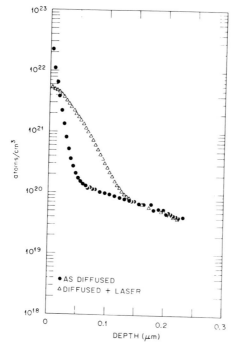

Fig. 5. Boron concentration profile before and after laser annealing (E = 1.17 J cm^{-2}).

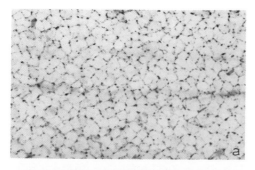

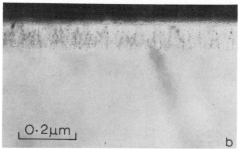

Fig. 6. Cell formation in
indium-implanted silicon:
(a) Plan-view micrograph,
(b) corresponding x-section
micrograph.

Figure 6(a) shows constitutional supercooling cells in an indium implanted
125 keV In$^+$, 1.0×10^{16} cm^{-2}) specimen after dye laser annealing (E = 1.17
J cm^{-2}). The average cell size was determined to be 350 Å. The interior of
each cell was found to be defect-free silicon with pure indium segregating at
the cell walls. From the cross-section micrograph of this specimen [Fig. 6(b)],
it was found that the formation of the cells started at 900 Å depth, with the
cell walls being nearly normal to the growing interface. The formation of cells
occurs because, during the solidification of a binary alloy, there is a limiting
concentration above which a planar solid-liquid interface becomes unstable; the
wavelength of this instability corresponds to the cell size. The primary
driving force for this instability is the constitutional supercooling which is
defined as the local difference between the actual temperature and the temperature of the liquidus curve in the phase diagram. Following the perturbation
theory originally developed by Mullins and Sekerka [14,15], we have solved the
problem of cell formation taking into account the dependence of distribution
coefficient upon the velocity of solidification. The perturbation theory of
instability includes constitutional supercooling, surface tension, the transport
of heat and the evolution of latent heat from the interface. From the observed
cell size of 350 Å, a solidification theory of 6 ms^{-1} was estimated from these
calculations. For this velocity the concentration of indium above which a
planar solid-liquid interface would become unstable was calculated to be 3 x
10^{20} cm^{-3}, which is in good agreement with the measured indium concentration
profiles in the laser annealed regions [9].

LASER ANNEALING USING UV LASERS

In this section, we present results on the annealing of displacement damage,
and dopant-profile broadening using UV lasers. We had proposed that the higher

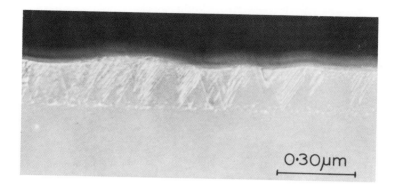

Fig. 7. X-section electron micrograph showing the characteristics of annealing, using a UV laser (λ = 0.19 µm, τ = 9 ns), in an arsenic implanted (100 keV, ^{75}As$^+$, dose = 1.0 x 10^{16} cm^{-2}). The melt-front reached the dislocation band present under the amorphous layer.

absorption coefficient associated with the shorter wavelength may lead to shorter melt-lifetimes, and therefore less profile broadening. However, the following results show that absorption coefficients above a certain value ($\sim$ 6.0 x 10^4 cm^{-1}) do not change the amount of broadening significantly.

Figure 7 shows x-section micrographs showing V-shaped dislocations indicating that the melt-front reached the region (below the amorphous layer) containing dislocation loops. The as-implanted specimen (100 keV As$^+$, 1.0 x 10^{16} cm^{-2}) contained a $\sim$ 1600 Å amorphous surface layer with a region containing dislocation loops (about 200 Å thick) below the amorphous layer. Complete annealing of displacement damage was observed upon increasing the pulse energy density so that the melt-front penetrated beyond the region containing dislocation loops. However, considerable irregularities at the surface are observed (Fig. 7) probably due to surface evaporation. This effect is enhanced by increasing the pulse energy density. The dopant profiles before and after laser annealing are shown in Fig. 8. Considerable dopant redistribution occurred, which was found to be consistent with diffusion in liquid silicon. It is noteworthy that the amount of broadening is considerably less for UV laser that for ruby laser annealing [1], but it is comparable to dye laser annealing. This indicates that beyond a certain limit a further increase in the absorption coefficient does not change the amount of dopant-profile broadening observed.

In conclusion, pulsed dye and UV laser annealing can be used to anneal displacement damage in ion implanted silicon. The dopant profile broadenings are much smaller than broadening produced by ruby laser annealing, due to the reduced melt-lifetimes. The dye laser pulses have also been used to remove dopant precipitates in boron diffused silicon. These results, as well as those on the formation of constitutional supercooling cells, are completely consistent with the melting model.

418

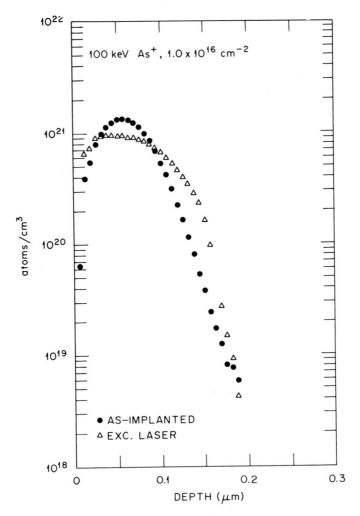

Fig. 8. Arsenic concentration profile before and after laser annealing from a specimen corresponding to Fig. 7.

ACKNOWLEDGMENT

We are grateful to Prof. A. Compaan for sending his laser irradiated specimens and irradiating our samples, and to Tachisto Inc. (Dr. U. K. Sengupta) for the use of his UV laser.

REFERENCES

1. C. W. White, J. Narayan and R. T. Young, Science 204, 461 (1979).

2. J. Narayan, J. of Metals 32, 15 (1980).

3. J. C. Wang, R. F. Wood and P. P. Pronko, Appl. Phys. Lett. 33, 455 (1978).

4. P. Baeri, S. U. Campisano, G. Foti and E. Rimini, J. Appl. Phys. 50, 788 (1979).

5. I. B. Khaibullin, B. I. Shtyrkov, M. M. Zaripov, R. M. Bayazitov and M. F. Galjautdinov, Rad. Eff. 36, 225 (1978).

6. J. A. Van Vechten, R. Tsu, F. W. Saris and D. Hoonhout, Phys. Lett. 74a, 417 (1979); J. A. Van Vecten, R. Tsu and F. W. Saris, Phys. Lett. 74a, 422 (1979).

7. H. W. Lo and A. Compaan, Phys. Rev. Lett. 44, 1604 (1980).

8. J. Narayan and F. W. Young, Jr., Appl. Phys. Lett. 35, 330 (1979).

9. J. Narayan, (unpublished).

10. H. Kodera, Jap. J. Appl. Phys. 2, 212 (1965).

11. Yu. M. Shashkov and V. M. Gurevich, Russ. J. Phys. Chem. 42, 1082 (1968).

12. J. Narayan and J. C. Wang, to be published.

13. J. Narayan, Appl. Phys. Lett. 34, 312 (1979).

14. W. W. Mullins and R. F. Sekerka, J. Appl. Phys. 35, 444 (1964).

15. J. Narayan, J. Appl. Phys., in press.

Published 1981 by North-Holland, Inc.
Narayan, and Tan, eds.
Defects in Semiconductors

421

ANALYSIS OF DEFECTS IN LASER ANNEALED GaAs*

JOHN FLETCHER, J. NARAYAN and D. H. LOWNDES
Solid State Division, Oak Ridge National Laboratory, Oak Ridge, Tennessee 37830

ABSTRACT

The nature and depth distributions of residual damage in ion implanted and pulsed ruby laser annealed GaAs have been studied using both plan-view and cross-section transmission electron microscopy (TEM) specimens for high dose (1.0 x 10^{15} cm^{-2}) Zn^+, Se^+ and Mg^+ implants. It was found that laser energy densities above 0.36 J/cm^2 were required to remove the implantation damage, this threshold energy density giving good agreement with that indicated by electrical activation measurements. Laser induced surface degradation of the GaAs was present even for energy densities as low as 0.25 J/cm^2, and more severe damage, with the introduction of dislocations near the surface, was present for energy densities above 0.8 J/cm^2. The use of thin SiO$_2$ layers for encapsulation during laser annealing was found to substantially reduce this surface degradation.

INTRODUCTION

Good electrical activation in ion implanted GaAs usually requires annealing temperatures above 800°C. It is then necessary to use some form of encapsulation in order to prevent dissociation of As and degradation of the surface of the wafers. However, for furnace annealing, none of the present used encapsulants give good electrical reproducibility. Short laser pulses have successfully been used to anneal ion implanted layers in silicon [1,2]. Because of the short duration for which the annealed region remains at a high temperature during laser annealing, the possibility of laser annealing GaAs without using an encapsulant exists. Previous studies of laser annealed ion implanted GaAs [e.g. 3,4] have been with electrical, Rutherford backscattering, and Raman scattering measurements. Only limited electron microscope studies of the annealing behavior have been made [5,6]. In the present study cross-section, as well as plan-view, TEM specimens have been used to study pulsed laser annealing of Zn^+, Se^+ and Mg^+ implanted GaAs. The results of electrical activation measurements and SIMS studies on the same samples will be discussed briefly. These have been presented elsewhere in more detail [7].

EXPERIMENTAL

Single crystal (100) semiconducting GaAs wafers were implanted with either 150 keV Zn^+, 160 keV Se^+, or 35 keV Mg^+ ions. The dose in each case was 5 x 10^{15} cm^{-2} and the implants were performed with single pulses from a pulsed ruby laser (λ=0.694 µm, τ=20-25 ns). The studies were made for energy densities in the range 0.3 to 1.0 J/cm^2, the laser annealing being performed in air. In most cases no encapsulation was used. However, in one case the effect of using

*Research sponsored by the Division of Materials Science, U. S. Department of Energy under contract W-7405-eng-26 with the Union Carbide Corporation.

a thin layer of SiO_2 as an encapsulant for laser annealing was investigated. The SiO_2 was evaporated onto the GaAs wafer in high vacuum. For all of the laser annealing, the laser beam was first homogenized by passing the laser light through a bent fused silica light pipe [8], the resultant homogenized beam being a spot size ~1 cm^2. The TEM studies were performed on a Philips EM400 electron microscope operating at 100 kV. In order to study in detail the annealing effects of the laser pulse on the ion implantation displacement damage, both plan-view and cross-section TEM specimens [9] were used in order to obtain the full three-dimensional distribution of the residual defects. The plan-view specimens were prepared by chemical jet polishing from the reverse side with a solution of 2% bromine in methanol. The corresponding cross-section specimens were taken from adjacent regions in the center of each laser annealed region.

ANNEALING OF IMPLANTATION DAMAGE

Although for an implant dose as high as 5×10^{15} cm^{-2} an amorphous layer would normally be formed in GaAs, in the present study this was found not to be so. This was most probably the result of beam heating effects during implantation, which are likely to be significant for such high dose shallow implants. The plan-view micrograph and electron diffraction pattern for the as-implanted Se^+ implanted sample are shown in Fig. 1. The heavy damage consisted of polycrystalline regions, twins, dislocation tangles and small loops. The diffraction pattern shows the characteristic ring pattern formed by the polycrystalline GaAs regions. Figure 2 shows the corresponding cross-section

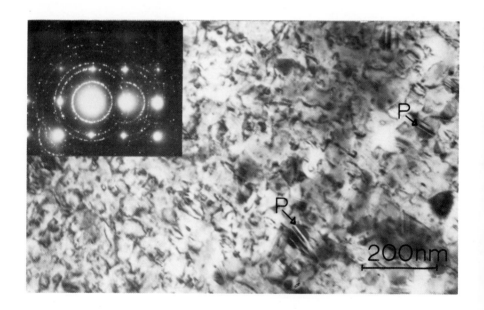

Fig. 1. Se^+ implanted GaAs. Plan-view electron micrograph. Polycrystalline regions (P) are seen in the thinner regions of the specimen. Inset: Electron diffraction pattern showing polycrystalline ring pattern.

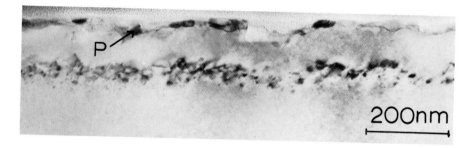

Fig. 2. Cross-section micrograph corresponding to Fig. 1.

TEM micrograph. The polycrystalline region is seen to extend from the surface to a depth of ~40 nm. The region from 40 nm to 100 nm is relatively defect free, containing only some very small loops and defect clusters. The region from 100 nm to ~180 nm contains a high density of loops, dislocation tangles, and small twins (~30 nm across). On the deeper side of this layer the damage consists mainly of small loops. The projected range and standard deviation for 160 keV Se$^+$ ions are 56.4 nm and 26.0 nm, respectively. Hence, in this case the peak region of the implant profile lies in relatively damage free material. This is in contrast to the behavior of ion-implanted silicon, where for high dose implants at elevated temperatures a broad band of heavy damage is formed [10].

Two different aspects of the laser annealing need to be considered. First, the removal of the ion-implantation damage by the pulsed laser, and second, the actual production of damage by the laser pulse due to, for example, loss of arsenic from the surface of the GaAs wafer. Consider first the damage removal. For the lowest energy density considered, 0.25 J/cm^2, a high density of residual defects remained. Figure 3 shows the residual damage for the laser annealed Se$^+$ implanted sample. The laser melt front had penetrated to a depth of just over 100 nm, therefore penetrating into the deeper damage layer. The subsequent epitaxial regrowth onto this damage layer resulted in a high density of dislocations, and some faulted regions, being propogated into the regrown layer. The polycrystalline rings were now absent from the electron diffraction pattern, and the twin spots, orginating from the small twinned regions in the deeper damage layer, were more easily distinguished. When the laser energy density was increased to 0.36 J/cm^2, Fig. 4, the melt front penetrated to a depth of 175 nm, leaving only a narrow band of the small loops that were originally present in the deeper region of the as implanted damage distribution. A low density of dislocation pairs are propagated into the laser annealed layer, originating from where the melt front had penetrated partway through some of the small loops. Sometimes, these took the form of two partial dislocations enclosing a faulted region, this latter case originating from the partial removal and regrowth of small faulted loops by the melt front. Annealing with laser pulse energy densities greater than 0.36 J/cm^2 resulted in complete removal of the original ion-implantation damage.

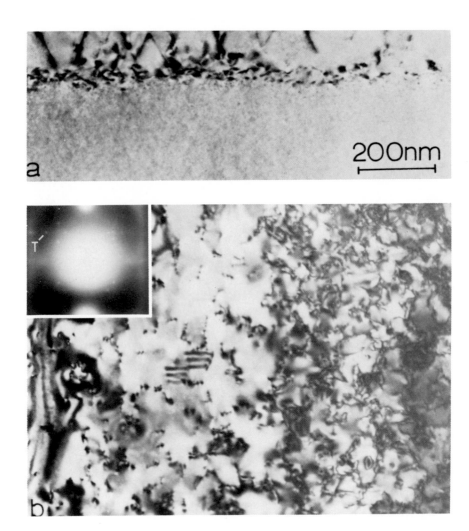

Fig. 3. Laser annealed Se⁺ implanted GaAs. Energy density 0.25 J/cm². (a)
Cross-section electron micrograph; (b) Plan-view electron micrograph. Inset:
Electron diffraction pattern showing twin spots.

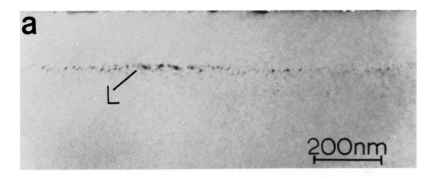

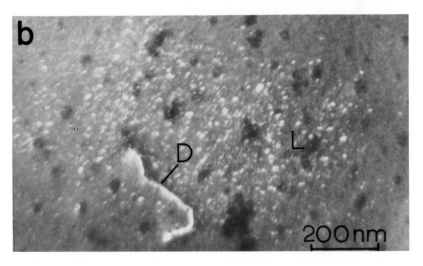

Fig. 4. Laser annealed Se$^+$ implanted GaAs. Energy density 0.36 J/cm^2. (a) Cross-section electron micrograph. (b) Plan-view electron micrograph showing loops (L), and dislocations (D) in the regrown layer.

For the magnesium implanted samples additional defects were present after laser annealing. These consisted of a high density of very small defects, Fig. 5, and were present even after laser annealing at 1.0 J/cm^2. The cross-section micrograph shows that these defects lie in a narrow band at the surface. The results of SIMS measurements on these samples have been presented elsewhere [7]. For the Mg$^+$ implanted samples there was evidence of surface segregation of the Mg after laser annealing, and these small defects may be associated with this.

426

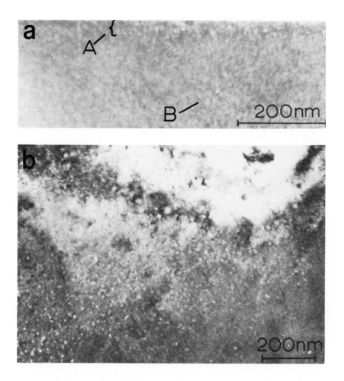

Fig. 5. Laser annealed Mg$^+$ implanted GaAs. Energy density 0.7 J/cm^2. (a) Cross-section electron micrograph showing (A) small defects near the surface and (B) defects remaining from the original implantation damage.

LASER INDUCED DAMAGE

For conventional thermal annealing of GaAs, it is necessary to protect the surface of the specimen with an encapsulant in order to prevent thermal decomposition due to loss of arsenic. With pulsed laser annealing the loss of arsenic ions should be much less because of the short time for which the surface region remains at a high temperature. However, in the present investigation, degradation of the surface region was evident even for the lowest energy density used (0.25 J/cm^2). Optical Nomarski interference micrographs of the surface of the laser annealed specimens, Fig. 6, showed the presence of surface ripples, which became more pronounced with increasing laser energy density. For the higher energy densities there was also a higher density of finer background structure, and occassional large "craters" were present. Figure 7 shows TEM micrographs for the Zn$^+$ implanted specimens showing the presence of dark regions on the specimen surface after laser annealing. X-ray analysis in the TEM indicated that these were gallium-rich regions. For a laser energy density of 0.25 J/cm^2, these regions were small but the density was high. For laser energy densities of 0.36 and 0.49 J/cm^2 it was found that the size of the regions increased, and

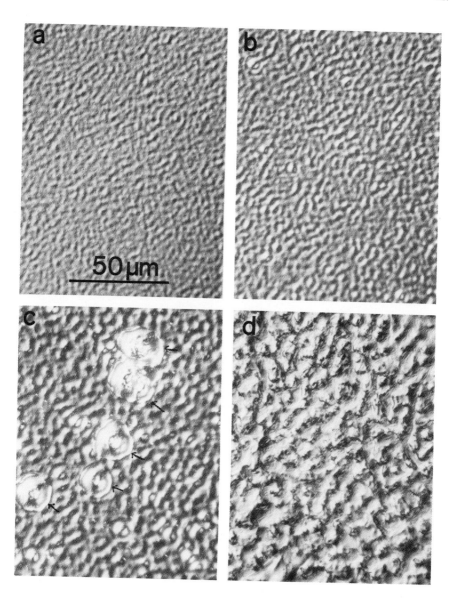

Fig. 6. Optical Nomarski micrographs of the surface of the laser annealed GaAs. (a) 0.49 J/cm^2, (b) 0.62 J/cm^2. (c) 0.81 J/cm^2. Arrow indicates craters on the surface. (d) 0.98 J/cm^2.

428

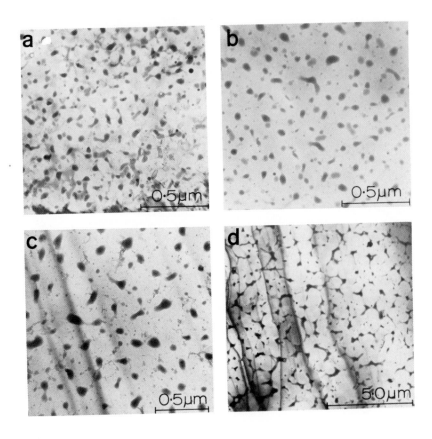

Fig. 7. Plan-view electron micrographs showing the presence of gallium rich regions (darker regions) after laser annealing. (a) 0.25 J/cm^2. (b) 0.36 J/cm^2. (c) 0.49 J/cm^2. (d) 0.81 J/cm^2.

the number density decreased, with increasing energy density. For the 0.81 J/cm^2 laser annealed specimen, the regions started to form a "cell-type" structure with a cell size of 1-2 µm. Some areas of the specimen showed a much finer "network" of gallium rich regions, as shown in Fig. 8. For specimens annealed at 1.0 J/cm^2 this finer structure predominated. Hence, the localized regions of fine structure in the 0.81 J/cm^2 annealed specimens were probably due to localized hot spots. Some of the gallium rich regions in these finer structures had dislocations associated with them, representing a form of laser induced

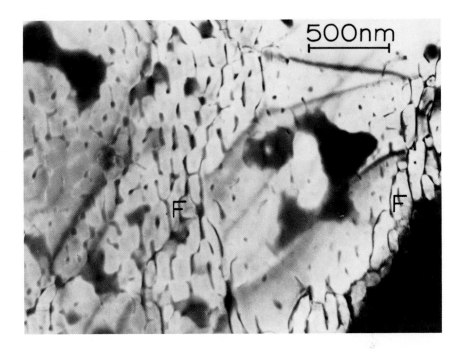

Fig. 8. Finer structure (F) present in some regions after laser annealing GaAs with an energy density of 0.81 J/cm^2.

damage for energy densities above 0.81 J/cm^2. Cross-section TEM specimens prepared from such regions showed that these dislocations were very shallow, being formed in the top 50 nm of the sample.

One of the Mg$^+$ implanted GaAs specimens was coated with a thin SiO$_2$ layer before laser annealing (0.7 J/cm^2). The small defects present in the unencapsulated laser annealed specimens were still present. However, the gallium-rich regions were no longer present on the surface.

CONCLUSIONS

TEM studies on high-dose ion-implanted GaAs showed that if, due to beam heating during implantation, the samples did not become amorphous, then a large proportion of the implant profile lay in relatively defect free material, in contrast to the situation usually found in silicon. Hence, the annealing behavior could be expected to be similar to that of low-dose ion-implanted GaAs. It was found that for energy densities greater than 0.36 J/cm^2 the laser melt front penetrated beyond the ion-implantation damage region, resulting in a defect free regrown layer. However, there was a secondary effect of the laser annealing; the degradation of the surface of the GaAs and the production of gallium-rich regions. This was found to be present even for the lowest energy density used, 0.25 J/cm^2. For energy densities above 0.81 J/cm^2 the surface degradation became more severe with the presence of laser induced damage and dislocations. The use of a SiO$_2$ encapsulant was found to reduce thin surface degradation.

In the case of Mg$^+$ implanted GaAs a high density of small defects was also present in the surface region of the laser annealed GaAs, these possibly being associated with the segregation of Mg towards the surface of the GaAs during the laser annealing. Electrical activation measurements on pulsed ruby laser annealed, ion implanted GaAs [7] showed that there is an energy density range, $0.5 \leq E_\ell \leq 0.8$ J/cm^2, in which high activation of the implanted ions could be achieved. This gives a good correlation with the TEM observations that energy densities above 0.36 J/cm^2 are required to remove the as-implanted damage, and energy densities above 0.81 J/cm^2 produce severe degradation of the surface with resultant laser induced damage.

REFERENCES

1. J. Narayan, C. W. White, and R. T. Young, Rad. Effects 47, 167 (1980).

2. C. W. White, J. Narayan and R. T. Young, Science 204, 461 (1979).

3. Laser-Solid Interactions and Laser Processing-1978 (AIP Conf. Proc. 50) ed. by S. D. Ferris, H. J. Leamy and J. M. Poate, (American Institute of Physics, New York, 1979).

4. Laser and Electron Beam Processing of Materials, ed. by C. W. White and P. S. Peercy, (Academic Press, New York 1980).

5. D. K. Sadana, M. C. Wilson and G. R. Booker, J. of Microscopy 118, 51 (1980)

6. S. U. Campisano, G. Foti, E. Rimini, F. H. Eisen, W. F. Tseng, M-A. Nicolet and J. L. Tandon, J. Appl. Phys. 51, 295 (1980).

7. D. H. Lowndes, J. W. Cleland, W. H. Christie and R. E. Eby, to be published in Laser and Electron Beam Solid Interactions and Materials Processing, Boston, Massachusetts, 1980.

8. A. G. Cullis, H. C. Weber and P. Bailey, J. Phys. E: Sci. Instr. 12, 688 (1979).

9. J. Fletcher, J. M. Titchmarsh and G. R. Booker, EMAG 1979 (Institute of Physics Conf. Series No. 52) p. 153, 1980.

10. J. Fletcher and G. R. Booker, J. Electrochem. Soc. 80-2, 1155 (1980).

Published 1981 by North-Holland, Inc.
Narayan, and Tan, eds.
Defects in Semiconductors

CONVECTION AND CONSTITUTIONAL SUPERCOOLING CELLS IN LASER ANNEALED SILICON*

J. NARAYAN AND J. FLETCHER
Solid State Division, Oak Ridge National Laboratory, Oak Ridge, TN 37830

ABSTRACT

The formation of convection and constitutional super-cooling induced cells has been studied in indium implanted, laser annealed silicon using plan-view and cross-section electron microscopy. The convection cells were associated with the spatial inhomogeneity in the laser pulse, which leads to large temperature gradients in the lateral direction. The average size of constitutional supercooling cells decreases with increasing velocity of solidification, and it also decreases from (111) orientation to (100). The results of a perturbation theory will be discussed, which predicts the cell size and limiting concentration of instability as a function of velocity of solidification.

INTRODUCTION

High-power laser pulses have been successfully used to remove displacement damage in ion implanted semiconductors. Annealing of dislocations, loops, and precipitates in silicon, using laser pulses similar to those used in displacement damage annealing, provides rather convincing evidence for the melting phenomenon [1-4]. In this paper we report results on the formation of cells due to convection, and constitutional supercooling (CS). From the study of cell sizes and limiting concentrations ($C_{s(min)}$) above which instability sets in, as a function of velocity of solidification, we can obtain detailed information on rapid solidification. A theory predicting cell size and $C_{s(min)}$ as a function of solidification velocity will be discussed briefly.

EXPERIMENTAL

Silicon (2-6 Ω-cm, n-type) single crystals having $\langle 100 \rangle$ and $\langle 111 \rangle$ orientations were implanted with indium ($^{115}In^+$, 125 keV), gallium ($^{69}Ga^+$, 100 keV), and bismuth ($^{209}Bi^+$, 250 keV) to doses in the range $10^{15} - 10^{16}$ ions cm^{-2}. The implanted specimens were irradiated with a single pulse of a ruby laser (wavelength, λ = 0.694µm; pulse duration, τ = 15 x 10^{-9} s) operated in the single (TM00) mode. The pulse energy density was varied from 1.2 to 2.2 J cm^{-2}. Some of the samples were treated with spatially inhomogeneous laser pulses in order to obtain large temperature gradients in the lateral directions and to produce convection in the laser-melted layers. The diameter of the laser beam was about 2 cm, which could be varied through the use of lens. The ion implanted, laser annealed specimens were studied using plan-view and cross-section microscopy in a Philips (EM-400) analytical microscope.

RESULTS AND DISCUSSION

Although we have obtained results on cell formation in In$^+$, Ga$^+$, Bi$^+$, and Fe$^+$ implanted silicon as a function of laser parameters and substrate temperature

*Research sponsored by the Division of Materials Sciences, U.S. Department of Energy under contract W-7405-eng-26 with Union Carbide Corporation.

432

we have chosen, in view of space, to present results only on the In-Si System. Fig. 1(a) shows a bright-field (B-F) electron micrograph from a (100) indium (dose = 1.0×10^{16} cm^{-2}) implanted specimen, which was treated with a laser pulse that was spatially inhomogeneous. The spatial inhomogeneity in pulse energy density leads to large temperature gradients in lateral directions, resulting in convection in the laser-melted layer. This drives indium toward the outer regions, and the indium concentration in the middle of the convection-induced cell [5] is depleted. Direct evidence for the formation of large (~1μm) convection cells during inhomogeneous laser treatment is shown in Fig. 1(a), where constitutional supercooling cells (~500Å) are observed near the convection cell walls. In some samples the spatial inhomogeneity was in the form of rings, which led to a convection-induced cell structure as shown in Fig. 1(b).

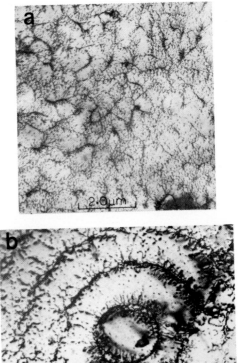

Fig. 1. Formation of convection-induced patterns in In$^+$ implanted (dose = 1.0×10^{16} cm^{-2}), laser annealed silicon.

The results on constitutional supercooling cells obtained using a uniform laser pulse are shown in Fig. 2. During the freezing of a binary alloy there

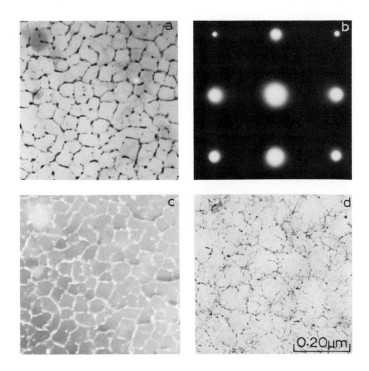

Fig. 2. Plan-view electron micrographs showing constitutional supercooling cells in the indium implanted specimens: a) Bright-field micrograph of cells in an (001) specimen, b) (001) diffraction pattern showing silicon as well as indium spots, c) dark-field micrograph using one of the indium spots, and d) Bright-field micrograph of cells in an (111) specimen.

is a limiting concentration above which a planar solid-liquid interface becomes unstable and the instability results in the formation of cell structures. Fig. 2(a) shows a plan-view micrograph from a (100) specimen irradiated with a 1.60 J cm^{-2} pulse. The average cell size was determined to be 500 Å. The interior of each cell was found to be defect-free silicon. Fig. 2(b) shows a diffraction pattern which contains, in addition to 200 and 220 Silicon spots, 110 indium diffraction spots, the latter appearing near missing 200 silicon spots. The diffraction results also indicate that indium has grown epitaxially with the silicon. A dark-field micrograph obtained using one of the In spots is shown in Fig. 1(c), which indicates that virtually all the indium has segregated at the cell walls. Under similar laser irradiation conditions, the cells for the ⟨111⟩ orientation were observed to have a different morphology (Fig. 1(d)) from that for the ⟨100⟩ orientation. The average cell size was found to increase from 500 Å in the ⟨100⟩ to 700 Å in the ⟨111⟩ orientation. The

434

cross-section micrograph in Fig. 3 from a ⟨100⟩ specimen showed that the cells started at a depth of 950 Å with the cell walls aligned close to the normal. The increase in cell size in the ⟨111⟩ orientation compared to the ⟨100⟩ orientation was similar to the increase in cell size observed for the ⟨100⟩ orientation case when the velocity of solidification was reduced by increasing the pulse energy density or the substrate temperature. These results were found to be consistent with the calculations to be described below.

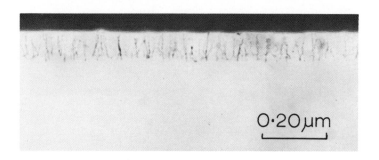

Fig. 3. X-Section electron micrographs showing the depth at which cells started forming.

In the perturbation theory originally developed by Mullins and Sekerka [6], the problem of morphological instability and formation of cells has been solved taking into account constitutional supercooling, surface tension, and transport of heat, including the evolution of latent heat from the interface. Assuming local equilibrium at an isotropic solid-liquid interface, the limiting concentration C_s in the solid, above which a planar solid-liquid interface is unstable, is given by [4]

$$C_s' = \frac{D_L k'}{mV\bar{\alpha}(k'-1)} \left[\frac{k_L G_L}{k_s + k_L} \left(\alpha_L - \frac{V}{K_L}\right) + \frac{k_s G_s}{k_s + k_L} \left(\alpha_s + \frac{V}{K_s}\right) \right.$$

$$\left. + \bar{\alpha} \, T_m \, \Gamma \, \omega^2 \right] \left[1 + \frac{2 k'}{(1 + (2D_L\omega/V)^2)^{1/2}-1} \right],$$

(1)

with

$$\alpha_L = (V/2K_L) + \left[(V/2K_L)^2 + \omega^2\right]^{1/2}$$

$$\alpha_s = -(V/2K_s) + \left[(V/2K_s)^2 + \omega^2\right]^{1/2}$$

$$\bar{\alpha} = (k_s \alpha_s + k_L\alpha_\ell) / (k_s + k_L)$$

where D_L is the impurity diffusion coefficient in liquid, V is the velocity of solidification, m is the liquidus slope in the phase diagram, k_L and k_s are the thermal conductivities, and K_L and K_s are the thermal diffusivities, of the liquid and solid, respectively, G_L and G_s are the thermal gradients in the liquid and solid respectively, T_m is the melting point of the planar interface in the absence of solute, Γ is the ratio of the solid-liquid surface energy to the latent heat of fusion per unit volume, ω is the spatial frequency associated with the instability, and $k' = C_s/C_L$ where C_s and C_L are the interface solubilities in the solid and liquid respectively. k' depends upon the velocity of solidification and the following function was used in the calculations [7]

$$k' = k_o \exp (- RT \ln k_o) (1 - \exp - (V/V_o))/ RT) \qquad (2)$$

where k_o is the equilibrium distribution coefficient, and V_o is a constant which can be determined experimentally.

We have solved $C_s (\omega)$ as a function of ω for given values of V. The minimum value of $C_s (\omega)$ gives the stability-instability demarcation. $\lambda_{min} = 2\pi/\omega_{min}$ then represents the cell size where ω_{min} is the value corresponding to the minimum value of C_s. The following values of the materials constants were used in the calculations

$k_L = 70 \text{ Jm}^{-1}\text{s}^{-1}\text{K}^{-1}, k_s = 22 \text{ Jm}^{-1}\text{s}^{-1}\text{k}^{-1}$

$D_L = 1.0 \times 10^{-8}\text{m}^2\text{s}^{-1}$ (curve a) $5.0 \times 10^{-8}\text{m}^2\text{s}^{-1}$ (curve b),

$\quad k_o = 4.0 \times 10^{-4}, V_o = 2.83 \text{ ms}^{-1}$

$T_m = 1.3 \times 10^{-7} \text{ km}$

$k_L = 3.0 \times 10^{-5} \text{ m}^2\text{s}^{-1}, k_s = 9.4 \times 10^{-6} \text{ m}^2\text{s}^{-1},$

$G_s = 6.57 \times 10^8 \text{ km}^{-1}, G_L = 1.33 \times 10^6 \text{ km}^{-1}$

$m = -400k \text{ (at. fract.)}^{-1}$.

The results of the calculations of cell size as a function of solidification velocity are shown in Fig. 4. The calculated cell size increases with decreasing solidification velocity. The observed cell sizes are plotted in Fig. 4 for $D_L = 1.0 \times 10^{-8}\text{m}^2\text{s}^{-1}$ (curve a) and $5.0 \times 10^{-8}\text{m}^2\text{s}^{-1}$ (curve b). For curves (a) and (b) the distribution coefficient $k' = f(v)$ is given by equation (2), but for the other two curves k' was taken to be equal to 0.15, and $k' = k_o = 4.0 \times 10^{-4}$. The agreement between the observed and calculated cell sizes (in (100) orientation) is considered good in view of the uncertainty in D_L. From the calculation in Fig. 4, the increase in cell size, from 500 Å for (100) to 700 Å for (111) orientation, suggests a decrease in solidification velocity by ~40% in (111) compared to (100) orientation.

436

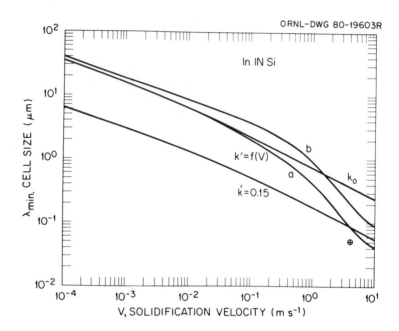

Fig. 4. Cell size as a function of velocity of solidification: Profile (a) $k' = f(v)$, diffusion coefficient in liquid $(D_L) = 1.0 \times 10^{-8} m^2 s^{-1}$; Profile (b) $D_L = 5.0 \times 10^{-8} m^2 s^{-1}$; and profiles for a fixed value of $k' = 0.15$, and for $k' = k_o$. Crossed circle represents experimental cell size obtained from Fig. 1(a).

The velocity of solidification can also be varied by changing the substrate temperature. By cooling the substrate to liquid nitrogen temperature, the velocity is increased by 40%, similarly by heating the substrate to 600°C it decreases by 50%. The cell sizes in indium, gallium and bismuth implanted silicon have also been studied as a function of solidification velocity and good agreement with the theory was again obtained [8].

Recently we have addressed the effect of surface tension Γ on the cell size. Since the atomic density in {111} planes is higher than in {100} planes, the liquid-solid surface tension in the ⟨111⟩ orientation is expected to be correspondingly higher (≤20%) than in the {100} orientation [9]. The calculations show that a change in surface tension of about 20% leads to less than a 5% change in the cell size. Therefore, it is believed that the observed change in cell sizes is primarily due to differences in solidification velocities in the ⟨111⟩ and ⟨100⟩ orientations.

In conclusion, convection and constitutional-supercooling induced cells in indium implanted and laser annealed silicon provide further evidence for the

melting phenomenon. The details of rapid solidification can be obtained by
studying the cell size as a function of solidification velocity. The observation of larger cell sizes in the <111> compared to the <100> orientation leads
to the conclusion that solidification velocity in the <111> orientation is ~40%
less than that in the <100> orientation.

REFERENCES

1. C. W. White, J. Narayan, and R. T. Young, <u>204</u>, 461 (1979).

2. J. Narayan, J. Electrochem. Soc. <u>80-1</u>, 294 (1980).

3. J. Narayan and C. W. White, Phil. Mag. (in press).

4. J. Narayan, J. Appl. Phys. (in press).

5. G. J. Van Garp, G. E. Eggermont, Y. Taminga, W. T. Stacy, and J. R. M.
   Gijsbers, Appl. Phys. Lett. <u>35</u> 273 (1979).

6. W. W. Mullins and R. F. Sekerka, J. Appl. Phys. <u>35</u>, 444 (1964).

7. R. F. Wood, Appl. Phys. Lett. <u>37</u>, 302 (1980).

8. J. Narayan and C. W. White (to be published).

9. K. K. Mon and D. Stroud, Phys. Rev. Lett. <u>45</u>, 817 (1980).

Published 1981 by North-Holland, Inc.
Narayan, and Tan, eds.
Defects in Semiconductors.

TEM INVESTIGATION OF THE MICROSTRUCTURE IN LASER-CRYSTALLIZED Ge FILMS

RONALD P. GALE, JOHN C. C. FAN, RALPH L. CHAPMAN, AND HERBERT J. ZEIGER
Lincoln Laboratory, Massachusetts Institute of Technology
Lexington, Massachusetts 02173

ABSTRACT

The crystallite morphology and defect structure of laser-crystallized Ge films have been characterized by transmission (TEM) and scanning (SEM) electron microscopy. The elongated crystallites present in these films have their long axis in the [001] direction. In regions where these crystallites are well aligned with each other, competitive lateral growth and (110) texture are observed. The defects most frequently found are {111} microtwins. A growth model that accounts for the texture and defects is proposed.

INTRODUCTION

We have previously reported [1] the crystallization of thin amorphous Ge films on fused-silica substrates (without surface relief structures) by scanning with the slit image of a cw Nd:YAG laser. The crystallized films contain well-aligned elongated crystallites with lateral dimensions up to 2-3 x 100 μm arranged in periodic structures with spacings of 40 to about 250 μm along the direction of laser motion. A quantitative theoretical model that predicts the periodic structures has been developed for the dynamics of the crystallization process [2], but the growth mechanism that produces the aligned crystallites is not well understood.

Our model for scanned laser crystallization attributes the periodic structural features to the cyclic motion of the crystallization front. In brief, crystallization begins when the laser heats an amorphous region of the film to a critical temperature, T_c. Due to the resulting release of the latent heat of crystallization, the crystallization front initially moves forward at a velocity much greater than the laser scan rate. The front eventually outdistances the laser so far that its temperature falls below the critical value, and it abruptly comes to rest. When the laser moves far enough for the temperature at the new location of the front to reach the critical value, the cycle is repeated. By fitting this model to the observed periodicity, a value of 700°C has been obtained for T_c. We refer to each cycle as a crystallization event; the structure produced by such an event is unaffected by another laser scan. The cyclic motion of the front has been confirmed experimentally by optical transmission measurements, and the growth velocity has been measured as about 200 cm/sec [3].

In this investigation we have used transmission (TEM) and scanning (SEM) electron microscopy to characterize the crystallite and defect morphologies of laser-crystallized Ge films. On the basis of the results we propose a model for grain growth in these films.

440

CRYSTALLITE AND DEFECT MORPHOLOGY

Figure 1 is a transmission optical micrograph of a section of a crystallized Ge film, showing periodic features with 40 µm spacing. The wide lighter bands consist of elongated crystallites that fan out to produce a chevron-like pattern. Outside the center area the crystallites forming each half of the pattern are aligned roughly parallel to each other, with an orientation of about 55° to the laser scan direction. At the outer edges of the pattern (not shown) this angle increases to over 90°. The narrow darker bands in Fig. 1 are transition regions, formed at the beginning of each crystallization event, in which the film changes from amorphous to fine-grained to coarse-grained.

LASER SCAN DIRECTION ⟶

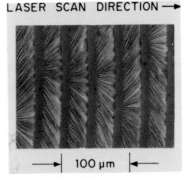

Figure 1 Photomicrograph showing periodic structural features of Ge film (near the center of laser image) after scanned laser crystallization at room temperature.

⟶| 100 µm |⟵

Part of the transition region for one crystallization event is shown in more detail in Fig. 2, a bright-field TEM micrograph. The elongated crystallites seen in the upper right corner were formed during this event, while those seen in the lower left corner were formed at the end of the preceding event. The large grains at the end of the transition region have a maximum diameter of 0.3 µm, the thickness of the Ge film. These grains are randomly oriented both with respect to the substrate surface and with respect to the laser scan direction. The elongated crystallites emerge from this region with an initial width of less than 0.5 µm and a length that is limited by contact with other crystallites in their path. Thus crystallites that grow the most rapidly into an uncrystallized region dominate. This competitive growth is confirmed by SEM observations of surface morphologies where ridges about 2000 Å wide and 500-1000 Å high delineate the crystallites (Fig. 3). It is seen that some crystallites grow wider (as at A), while their neighbors get narrower (as at B) and eventually terminate as a result of competitive lateral growth. The beginning of a ridge is usually seen in the area where a crystallite widens, resulting in the appearance of two crystallites with a common origin. Groups of over 20 crystallites have been traced to a single source. Such groups, rather than individual crystallites, dominate the structure.

Electron diffraction patterns show that the long axis of the elongated crystallites is oriented in the [001] direction, whatever the angle betwen this axis and the laser scan direction. For the well-aligned crystallites, where this angle is ± 55°, the (110) plane is generally parallel to the

Figure 2 Bright-field TEM micrograph showing a transition region in a Ge film that was laser crystallized at a background temperature of 400°C.

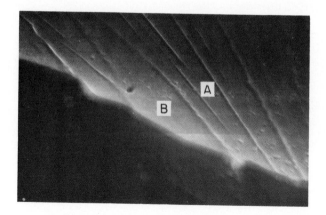

Figure 3 SEM micrograph of the cross-section and surface of the aligned crystallite region of a laser crystallized Ge film on a fused-silica substrate, showing growth ridges. The sample was gold-coated for SEM examination.

442

substrate surface within 10° and the [112] direction is approximately normal to
the laser scan. The aligned crystallites are separated from each other by
low-angle tilt boundaries, with no rotation component observed for the [110]
axis. Large crystallites with other textures are also observed, often
separating groups of the well-aligned crystallites. Independent confirmation
of overall strong (110) texture in the films was obtained by x-ray
diffraction.

The most plentiful defects found in all the crystallites are microtwins
50-200 Å thick with lateral dimensions of 200-2000 Å, as shown in Fig. 4.
These microtwins have their long edges parallel to either the (111) or (1̄11)
planes, which are the twinning planes. In some crystallites the microtwins of
both orientations are distributed uniformly with densities exceeding
10^{10} cm^{-2} (Fig. 4, left) while in others they are concentrated at the grain
boundaries(Fig. 4, right). In the latter structure twins of one orientation
form dense bands along one edge of the crystallite, while twins of the other
orientation form similar bands at the other edge. The origin of these twins
is not certain, although the symmetry of their shape and distribution suggests
that they are formed after crystallization as a result of stresses in the
film.

As shown in Fig. 4, the crystallites contain dislocations that run
roughly parallel to the [001] direction and terminate at the crystallite
boundaries. These boundaries appear as dark lines whose contrast is
independent of diffraction conditions, but in some places these lines split
into two or three finer dislocation lines. Only rarely could an array of
dislocations indicating a classic low-angle tilt boundary be observed.

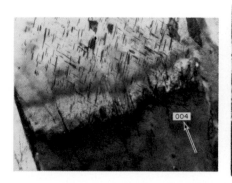
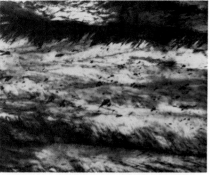

1 µm   1 µm

Figure 4 Bright-field TEM micrographs of two aligned (110) [001] crystallites
in a laser-crystallized Ge film, illustrating (left) the uniform distribution
of twins and (right) bands of twins near each grain boundary. In both
micrographs dislocations can be seen lying parallel to the grain boundaries.

GROWTH MODEL

It is clear from the observed axial orientation of the elongated crystallites that the [001] direction is the fastest growth direction. We attribute the texture in the aligned crystallite region to the competitive lateral growth that occurs in this region. If there is a preferred direction for lateral growth, then the crystallites that propagate axially are those that can grow laterally in this preferred direction. Since these crystallites have their axial orientation and one lateral orientation specified, their orientation in the direction normal to the substrate is also determined. The observed (110) texture would therefore be a consequence of the [001] axial direction and a preferred lateral growth direction as yet unidentified. This explanation also accounts for the absence of texture in the regions of radial growth. In these regions there is no effective preferred lateral direction, since crystallites growing outward radially can expand laterally without impeding the progress of their neighbors.

A non-zero lateral component of the growth velocity results in a convex growth interface for each crystallite. This shape is consistent with the formation of ridges at the grain boundaries. These ridges are produced by a displacement of material from the center of the crystallite to its edges, indicating that the edges crystallized after the center. To account for this transport to the grain boundaries, there must be a buildup of Ge in front of the laterally growing interface. This strongly suggests the presence of a liquid region between the amorphous and crystallized Ge, consistent with theoretical models proposing the formation of a narrow molten zone as an intermediate step in the crystallization process [4,5]. The volume change during solidification of such a zone would account for ridge formation at the crystallite boundaries and would produce large compressive stresses normal to the boundaries. We speculate that the observed microtwins are formed as a result of these stresses.

Observations of the crystallized films reveal the shape of the crystallization front only at the locations reached by the front at the ends of the successive crystallization events. At these locations the front is smooth rather than scalloped (as seen in the lower left corner of Fig. 2). This shape is not inconsistent with our proposal that the individual crystallites have a convex interface, since such an interface is expected to be stable only while the crystallization front is moving rapidly forward.

Figure 5 is a schematic of the growth model, summarizing the orientations and textures observed in the different regions along the crystallization front. The diagrams represent one crystallization event at three successive times: the initial radial growth, where the crystallites have non-parallel [001] axes and (110) texture; the aligned region, with parallel [001] axes and (110) texture; and a peripheral region where again there is radial growth, beyond the end of the laser slit image, with non-parallel [001] and no texture. The crystallization front is represented as scalloped in those regions where it is moving forward.

CONCLUSION

The crystallite morphology and defect structure of laser-crystallized Ge films have been characterized by transmission (TEM) and scanning (SEM) electron microscopy. To explain the observations we have proposed a growth model with the following principal features: axial growth occurs preferentially along the [001] direction; in the aligned crystallite region, competitive lateral growth occurs with an unidentified preferred direction; the elongated crystallites have a convex growth interface; a molten zone is formed at the growth interface; stresses normal to the crystallite boundaries are produced because these boundaries crystallize after the center.

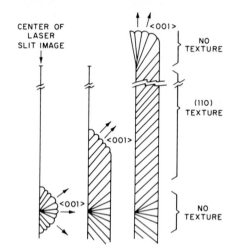

Figure 5 Schematic diagram representing a crystallization event at three successive times.

ACKNOWLEDGEMENTS

We acknowledge helpful discussions with A. J. Strauss, M. W. Geis, B-Y. Tsaur, and J. P. Salerno. The scanning electron microscopy was performed by M. C. Finn and P. M. Nitishin. This work was supported by the Department of the Air Force.

REFERENCES

1. J. C. C. Fan, H. J. Zeiger, R. P. Gale, and R. L. Chapman, Appl. Phys. Lett. 36, 158 (1980).

2. H. J. Zeiger, J. C. C. Fan, B. J. Palm, R. P. Gale, and R. L. Chapman, in Laser and Electron Beam Processing of Materials, eds. C. W. White and P. S. Peercy (Academic Press, New York, 1980), pp. 234-240.

3. R. L. Chapman, J. C. C. Fan, H. J. Zeiger, and R. P. Gale, Appl. Phys. Lett. 37, 292 (1980).

4. G. H. Gilmer and H. J. Leamy in Ref. 2, pp. 227-233.

5. R. B. Gold, J. F. Gibbons, T. J. Magee, J. Peng, R. Ormond, V. R. Deline, and C. A. Evans, Jr., ibid., pp. 221-226.

RAMAN SCATTERING CHARACTERIZATION OF BONDING DEFECTS IN SILICON

RAPHAEL TSU
Energy Conversion Devices, Inc., Troy, Michigan,48084 U.S.A.

ABSTRACT

Raman scattering is used for the characterization of
defects in Si. Damage is produced in single crystal silicon
by ion-implantation of As and Si. The phonon structure of
the damaged layer is that of the typical amorphous Si.
After irradiation by pulsed laser(10ns,532nm) at energy
density of approximately 0.1J/cm^2, a Raman peak appears at
a frequency between 508 cm^{-1} and 517 cm^{-1} depending on
implant dosage. The higher the implant dosage, the lower is
the frequency. We explain this in terms of the residual
bonding defects caused by the presence of extraneous atoms
such as oxygen. On the other hand, irradiation at an energy
density in excess of 0.5 J/cm^2, a Raman peak appears at a
frequency close to that of the single crystal except for
small shifts due to Fano-shift. For implant dosage in excess
of 4×10^{16} As/cm^2, we have found additional peaks at 222 cm^{-1}
and 267 cm^{-1} which are close to the metallic arsenic modes
indicating the presence of arsenic clusters.

INTRODUCTION

It is accepted by many that ns-pulsed-laser annealing of ion-implanted Si is
accompanied by melting and recrystallization. The observed reflectivity rise
which occurs during the laser pulse has been used as an experimental evidence
for the strictly thermal model[1]. On the other hand, there are several papers
arguing for non-thermal models [2],[3].An attempt has been made by Tsu and Jha
[4] to explain the long life-time of excited e-h pairs involved in plasma
oscillation which is crucial in a non-thermal model.Recently, Lo and Compaan[5]
have found that the lattice temperature from Raman scattering during pulsed-
laser heating of Si is only 300°C above the ambient temperature with typical
energy densities used in most annealing. Several years ago, in the course of
investigating the laser recrystallization of ion-implanted Si by a frequency-
doubled Nd:YAG laser, Tsu et al could not explain the appearance of an extra
peak in Raman scattering appearing at 513 cm^{-1}, which is shifted down from
522 cm^{-1} after irradiating the sample at a energy density far below the threshold [6],[7]. In this work, the notion of an intermediate state was introduced. In a subsequent paper, Tsu et al [8] presented several experimental
results to advance the notion of an undercooled liquid state having peculiar
properties, so as to explain the down-shift of the Raman peak caused by strain.
However, further work by Pollak et al [9] has shown that the electroreflectance
peak appears at the usual 3.4 eV, indicating the absence of any appreciable
strain. In this paper, Raman measurements are presented for many samples having
different implant-dosage ranging from 10^{14} As/cm^2 to 10^{17} As/cm^2, as well as
two samples of Si-implanted at 10^{15} and 10^{16} Si/cm^2. The position of the down-
shifted peak varies with implant dosage and energy density of the 10 ns,532nm
pulsed-laser. It is shown here that small particle size-effect cannot explain
the magnitude of the shift. Oxygen atoms may be the cause of the shift. The

446

fully annealed As implanted samples exibit a Fano-shift which allows a direct
measure on the number of substitutional arsenic atoms (10). Furthermore, for
implant-dosage exceeding 4×10^{16} As/cm^2, the R_1 and R_2 peaks for metallic As
show up, indicating the presence of arsenic clusters (11).

RESULTS AND DISCUSSIONS

Raman spectra of three typical samples are shown in Fig.(1). The ordinate

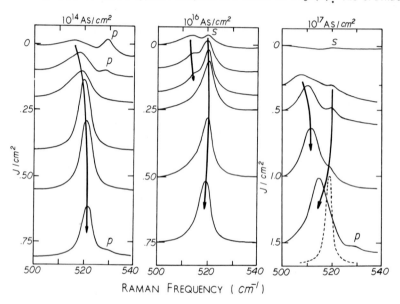

Fig.1. Raman scattering for several ion-implanted pulsed-laser annealed samples.
4880 Å Ar-ion laser is used.

is marked J/cm^2. For example, for the 10^{14} As/cm^2 sample, the top curve starts
from the point of the ordinate marked 0, which means that this spectrum is for
unannealed, and the bottom curve starts at 0.8 indicating that this spectrum
is for irradiation at 0.8 J/cm^2. The arrow shows the peak shift from 517 cm^{-1}
to 521.5 cm^{-1}. The peaks marked "p" indicate the plasma-line of the laser at
4880 Å. The spike filter was left out purposely to show that the high diffuse
scattering of a typical amorphous layer was reduced after transforming into a
good single crystal or crystalline state, and reappeared when the surface morphology exibited the typical damage from the pulsed-laser at high power. He-
backscattering indicated good channeling at 0.4 J/cm^2. For the 10^{16} As/cm^2
implanted sample, (center curves), the intermediate peak moves from 513 cm^{-1} to
515 cm^{-1} and merges into the Fano-shifted peak. The Fano-shifted peak starts
at 520 cm$_0^{-1}$ (marked by "S",for substrate contribution due to penetration of
the 4880 Å light through the damaged layer),and moves to 521.5 cm^{-1} and back

to 519 cm^{-1} due to interaction of the phonon with the continuum, electronic transition between Δ_1 and Δ'_2 of the conduction bands. For the 10^{17} As/cm^2 sample, the intermediate peak first appears at 0.3 J/cm^2 at 508 cm^{-1}. The Fano-shifted peak moves from 521.5 to 516 cm^{-1} corresponding to an electron concentration of 4×10^{21} cm^{-3}. The dotted curve shows the same sample after thermal annealing for five minutes at 950 °C. The peak position is at 519 cm^{-1}, corresponding to n = 2×10^{21} cm^{-3}. Thermal annealing allows the excess As above the solid solubility limit to diffuse deep into the undamaged underlayer, thereby lowering the concentration, leading to a smaller Fano-shift. It may be asked why the intermediate peak shows up in the unannealed cases. Several of our samples after ion-implantation show up a Raman peak around 513 to 517 cm^{-1}. We have purposely chosen these examples to make a point. The usual amorphous peak at 480 cm^{-1} is much weaker whenever the 513 peak shows up, and the 513 peak will more likely show up in lightly implanted Si. This brings up an interesting point. Pulsed-laser annealing can create this intermediate peak if it is not present before annealing, but it can also shift this peak if it is already present in the as-implanted samples. This fact tells us that the origin of this intermediate peak is not restricted to pulsed-laser annealing, ion-damage can also be responsible.

Figure 2 shows the results of all our measurements for more than ten samples of different implant conditions. The dashed line represents the fully annealed.

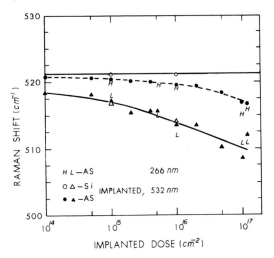

Fig. 2. Raman peak positions for various samples having different implant-dosage. Open points for Si-implanted and close points for As-implanted.

For example, for 10^{14} As/cm^2, it is the 0.5J/cm^2 annealed, and for 10^{17}As/cm^2, the position of the peak is taken from the curve for 1.5 J/cm^2 curve of Fig. 1. The solid curve represents the position of the peak first observed at an energy density increasing from zero. Thus, for 10^{14} As/cm^2, it is the 0.1 J/cm^2 position, and for 10^{17} As/cm^2, it is the position for the 0.3 J/cm^2 curve. The open triangle and circle represent the Si-implanted samples. Since there is no doping in this case, the fully annealed peak appears at 521.5 cm^{-1}. The points "H" and "L" are obtained with a 266 nm UV pulsed-laser. Note that there is hardly any change in the results.

Let us examine if the down-shifted peak can be caused by small particle size-effects. We have plotted in Fig. 3 using the rule $k\ell \sim 2\pi$ and the phonon-dispersion from neutron scattering (11). The shaded region represents the region in question between the TO-phonons along <100> and <111>.

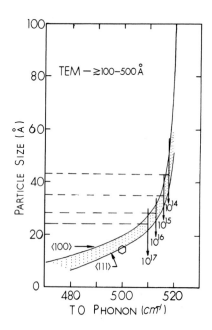

Fig. 3. Particle size vs. TO phonon using $k\ell \sim 2\pi$ and neutron scattering, the open hexagon gives the calculated (13).

The calculated phonon frequency for a cluster of 64 atoms (13) falls fairly close to the region in question shown as an open-hexagon. Taking the measured Raman peak position for four implant-dosage, we found that the corresponding particle size ranges from 24Å to 42Å. The particle size has been measured by Tan (14) and Gibson (15), for a number of these samples. The typical size is 200Å. Recently, a 20Å Si epitaxial layer has been studied by Levine et al. (16), using stimulated Raman scattering. No significant shift has been found. Thus, we conclude here that the shift is not caused by small particle, rather by a change in structure. Graczyk (17) has found that the radial distribution function beyond the second-neighbor is affected by an introduction of 10%

oxygen into a-Ge. We would expect the same result for silicon when oxygen is introduced. In fact, Baglin and Tsu (18) have always found oxygen in the He-channeling studies of our laser annealed samples. The more implant-dosage, the more oxygen was detected. In fact, at times oxygen concentration may be comparable to what was implanted! This explains why in Fig. 2 the intermediate peak position is the same whether As or Si was implanted. It also explains why the shift is larger for higher implant dosage. In the course of studying the dislocation loops, Tan (19) needed the presence of oxygen atoms to explain his TEM micrographs. Up to now, we have not explained why the Raman peak is downshifted. Oxygen atoms may form two bonds with two adjacent Si atoms, increasing their masses, thereby lowering the phonon frequencies. This explains also why there does not appear to have a two-phase behavior in the Raman experiment. Since the Γ-phonon frequencies are lowered, the relatively sharp peak (~ 6 to 10cm^{-1}) is actually the zone-center phonons, and the usual phonon peak at 480cm^{-1} for the amorphous phase disappears.

The last figure gives our results for looking for As-cluster. The R_1 and R_2 lines corresponding to the E_g and A_{1g} modes of the rhombohedral metallic arsenic show up clearly in Fig. 4 for the annealed. Since we found these for an implant-dosage of $\gtrsim 4\times10^{16}$ As/cm^2, and the damage layer is $\sim 1000\text{Å}$, we can estimate that

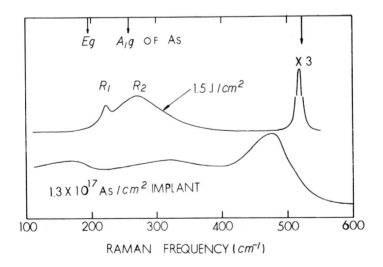

Fig. 4. Raman spectra of a heavily As-implanted Si. The appearance of R_1 and R_2 lines indicate As-clusters.

the solubility limit of As in Si is approximately 4×10^{21} As/cm^3. Therefore, combination of Raman measurements on the R_1 and R_2 lines with Hall measurements should give added understanding on the mechanism of cluster formation for the dopants.

Implicit in our discussion, the notion of simple thermal melting cannot be the mechanism for the gradual structural change, at least for the power densities below the so-called threshold. Nevertheless, it is believed at this stage, an undercooled liquid, still may explain our present results. On the other hand, the notion of gradual reordering through a hot e-h plasma may also be viable. Perhaps, it is semantic whether one calls it an undercooled liquid, or a plasma due to the breaking of a significant percentage of the Si-Si bonds.

Lastly, based on some preliminary laser annealing on MBE grown amorphous Si by Chou (20), no significant shift in Raman peak has been found. The Raman frequency is 520cm^{-1} and a line width of 8cm^{-1} at 0.2 J/cm^2, and 522cm^{-1} with a linewidth of 4.5 cm^{-1} at 0.5 J/cm^2. This further supports our model that the large down-shift observed is due to structural change induced by the presence of oxygen, or even arsenic. The intermediate state is not an intrinsic property, rather, reflects the presence of bonding defects, particularly noticeable when extraneous atoms are present.

Many of the results were obtained at IBM T.J. Watson Research Center before the author joined his present affiliation. The contributions made by J.E. Baglin and J.A. VanVechten, and above all, by T.Y. Tan are gratefully acknowledged.

REFERENCES

1. D.H. Auston, C.M. Surko, T.N.C. Venkatesen, R.E. Slusher, & J.A. Golovchenko, Appl. Phys. Lett. 33, 437 (1978).
2. J.A. VanVechten, R. Tsu, F.W. Saris, & D. Hoonhout, Phys.Lett.74A, 417 (1979).
3. J.A. VanVechten, R. Tsu & F.W. Saris, Phys. Lett. 74A, 422 (1979).
4. R. Tsu & S.S. Jha, Journal de Phys. Collogue C4, Supp. 5, 25 (1980).
5. H.W. Lo & A. Compaan, Phys. Rev. Lett. 44, 1604 (1980).
6. R. Tsu, J.E. Baglin, T.Y. Tan, M.Y. Tsaĩ, K.C. Park & R.T. Hodgson, AIP Conf. Proc. NO. 50. Boston, (1978) 344.
7. The above reference should read:"Amorphous Si has a very broad peak at 475 cm^{-1}. We are not sure of the origin of the peak at 513cm^{-1} for 0.05 J/cm^2."
8. R. Tsu, J.E. Baglin, T.Y. Tan & R.J. vonGutfeld, Proc. 80-1, Laser & Electron Beam Processing on Electronic Materials, C.L. Anderson, G.K. Celler & G.A. Rozgonyi, Electrochemical Soc., Los Angeles, 1979, p. 382.
9. F.H. Pollak, R. Tsu & E. Mendez, Laser & Electron Beam Processing of Materials, C.W. White and P.S. Peercy eds. (Acad. Press, 1980) p. 195.
10. M. Chandrasekhar, J.B. Renucci, & M. Cardona, Phys. Rev.B17, 1623 (1978).
11. R.L. Farrow, R.K. Chang, S. Mroczkowski & F.H. Pollak, Appl. Phys. Lett. 31 768 (1977).
12. G. Dolling, Elastic Scattering of Neutrons, Symp. Chalk River, 2, 37 (1972).
13. R. Alben, D. Weaire, J.E. Smith, M.H. Brodsky, Phys. Rev. B11, 227 (1975).
14. T.Y. Tan, Private communication.
15. J.M. Gibson & R. Tsu, Appl. Phys. Lett. 37, 197 (1980).
16. B.F. Levine, C.G. Bethea, A.R. Tretola & M. Korngor, Appl. Phys. Lett. 37, 595 (1980).
17. J.F. Graczyk, Thin Solid Films, 70, 303 (1980).
18. J.E. Baglin, R. Tsu, Unpublished.
19. T.Y. Tan, Private communication.
20. The MBE amorphous Si sample was grown by N.J. Chou of IBM Research Center.

RESIDUAL DEFECTS AFTER PULSED LASER ANNEALING OF ION IMPLANTED SILICON

A. MESLI, J.C. MULLER, D. SALLES, P. SIFFERT
Centre de Recherches Nucléaires
Groupe de Physique et Applications des Semiconducteurs (PHASE)
67037 STRASBOURG-Cedex,FRANCE

ABSTRACT

Capacitance transient spectroscopy has been used to investigate the electrically active defects subsisting, after a ruby laser pulse annealing, in ion implanted silicon. In contrast to the common view, it is shown that the identified point defects are related to residual implantation related defects buried beyond the dopant distribution and not to the laser effect.

INTRODUCTION

The incorporation of dopant elements into the silicon lattice by a pulsed annealing induced either by lasers or electrons, has been intensively investigated during the last couple of years, with the hope that new device processing techniques can be developed. In principle, these methods offer several advantages related to the local heating and the high thermal gradients. In particular, it has been demonstrated that ion implanted silicon can be pulse annealed to a state of very high electrical activity of the dopants, without any visible crystallographic defects such as dislocations, stacking faults, precipitates, at least within the sensibility of the common techniques like Rutherford backscattering (RBS) or transmission electron microscopy (TEM).

In view of all these advantages, one would expect large application in the device manufacturing technology. However, in practice, no structure has really been prepared having performance well in excess of the conventional structures. This is mainly due to different effects which can degrade the devices characteristics. Essentially, two categories can be considered :

– defects due to the interaction of the laser's energy with the semiconductor : stresses, or quenching [1-3] which limit the carrier lifetime in the regrown layer, or which degrade the Schottky barrier performance when the starting material has been laser illuminated over a certain threshold [4] ;

– defects resulting from the implantation and which subsist after laser treatment, due for example of an epitaxial regrowth on a damaged zone, extending deeper towards the bulk than the dopant distribution.

We have investigated the residual damage subsisting after ion implantation and laser annealing by using the Capacitance Transient Spectroscopy. Strong arguments suggest that the observed damage does not result from the laser bombardment.

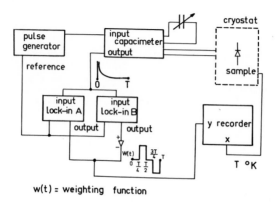

Fig. 1. Schematic of the DLTS experimental set-up.

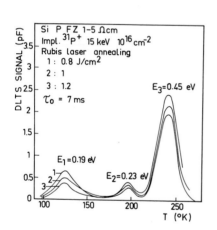

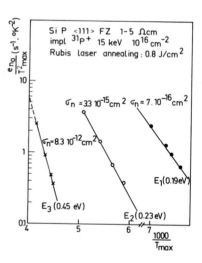

Fig. 2. DLTS signal for a P+ implanted junction (15 keV, 10^{16} cm-2) and annealed with various ruby laser pulse energies.

Fig. 3. Thermal emission rates VS $\dfrac{1000}{T}$ for the three electron traps. The capture cross section are also reported.

EXPERIMENTAL CONDITIONS

Samples preparation
P-type silicon samples, grown by floating zone (FZ) along an <111> axis, having a resistivity in the range 1-5 Ω. cm have been implanted off-axis with 15 keV P^+ ions to a dose of 1.10^{16} cm^{-2}, to produce an N^+P structure.

The damage annealing was performed by a Q-switched ruby laser opera- ting in the monomode regime. The pulse duration is about 20 ns and the power density was in the 0.8-1.2 J/cm^2. The illumination has been done in air.

Measurements
The DLTS measurements have been performed with a double phase lock-in detector with a square weighting function, following the block diagram given in figure 1.

Furthermore, the frequency dependence of the C-V behaviour has been observed [4].

RESULTS

A typical DLTS spectrum is reported on figure 2 ; three electron traps are clearly visible. A plot of the thermal emission rate of these levels versus inverse temperature T, with a correction for the preexponential T^2 dependence of e_n indicates (figure 3) that the levels resulting from the implantation annealed by laser are : $E_c - 0.19$ eV, $E_c - 0.23$ eV and $E_c - 0.45$ eV. All three levels disappear after a thermal annealing at 600°C during 15 min.

It should be noticed on figure 2 that the intensity of the three levels decreases when the laser energy increases. Figure 4 shows that the laser power of 0.8 J/cm^2 is higher than laser energy threshold determined for the minimum defect density [3]. The latter defines the conditions at which the depth of melting equals that of implantation damage. For higher laser powers, up to 1.2 J/cm^2, the defect level remains constant. This is an indication that we have not really introduced laser process related defects.

The depth profile of the three observed defects for an 1 J/cm^2 laser illumination are reported on figure 5. It appear that their extension is much deeper than the expected range of 15 keV P^+ implanted ions.

DISCUSSION

The DLTS measurements have revealed a decrease of the defect level concentration with higher laser energies. This suggests that the laser illumination is not responsible for the observed defects, which may be due to implantation related damage.

The following experiment confirms this hypothesis : an N^+P junction was prepared by first depositing a thin dopant film on top of a P-type wafer and illuminating it with the ruby laser pulse, working in the "melting regime". The diode investigation revealed defect concentration several orders of magnitude below that observed for the implanted device [6].

Taking, further, into account the fact that for the laser energies em- ployed in this work (0.8-1.2 J/cm^2), the molten depth does not exceed 0.2-0.3 μm [6], whereas the defects distribution given in figure 5 exceeds 1 μm, it is highly probable that the measured defects are related to the ion implantation.

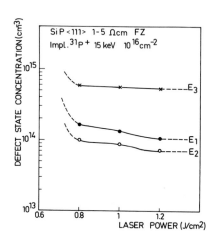

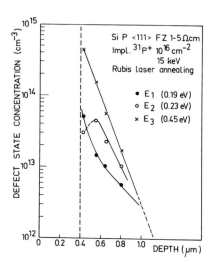

Fig. 4. Total defect state concentration of the three electron traps after various ruby laser pulse energies (τ_o = 7 ms).

Fig. 5. Depth profile of the three defects after 1 J/cm^2 ruby laser annealing. The limit of measurement at zero bias voltage is given by the dashed line (τ_o = 7 ms).

In fact, it is well established, since many years, that the distribution of ion implanted dopants into misaligned silicon, has a deeply penetrating tail extending well below the experimental LSS gaussian shape [7]. Evidence of the tail and tentative explanation of its origin have been discussed by Dearnaley [8]. Migration of substitutional dopant-vacancy pairs [9-10] play a dominant role. For example off axis 40 keV P$^+$ implantation gives a tail extending 1 μm in depth. If the direction is not far off the channelling direction, supertailing can give projectiles penetration up to ten times the expected range [10-11]. Even for 20 keV P$^+$ implantation into <111> oriented silicon, in which the expected range is less than 0.05 μm for random direction, a supertail of 6 μm [10] has been measured.

Due to experimental difficulties, much less is known about the penetration of the damage. However, if the ion penetration is very deep, it may be anticipated that the low speed projectiles, which give rise to intense nuclear collisions, produce severe atom displacements.

Consequently, during the epitaxial regrowth of the implanted zone, the depth reached by the laser is not sufficient and the regrowth occurs from an interface which is not fully free of defects.

These incorporated defects can be annealed by a short time treatment at about 600°C.

CONCLUSION

DLTS measurements have been used to investigate the residual defects existing in silicon after ion implantation and ruby laser annealing. In opposition to the generally accepted view, we have shown that the defects are much more related to the ion implantation process rather than to the laser treatment.

REFERENCES

1. P.M. Mooney, R.T. Young, J. Karins, Y.H. Lee and J.W. Corbett, Phys. Stat. Sol. 48a, K 31 (1978).

2. J.L. Benton, L.C. Kimerling, G.L. Miller, D.A.H. Robinson, G.K. Celler, AIP Proceeding n° 50 on : Laser Solid Interactions and Laser Processing, S.D. Ferris, H.J. Leamy, J.M. Poate eds (American Institute of Physics, New York, 1979) p.p.543-549.

3. L.C. Kimerling and J.L. Benton, AIP Proceeding on Laser and Electron Beam Processing of Materials, C.W. White and P.J. Peercy eds (Academic Press, New York, 1980) p.P.385-396.

4. E. Buttung et al., to be published.

5. E. Fogarassy et al., to be published.

6. R.O. Bell, M. Toulemonde and P. Siffert, Appl. Phys. 19, 313 (1979).

7. P. Blood, G. Dearnaley and M.A. Williams, Rad. Effects 21, 245 (1974).

8. G. Dearnaley, M.A. Wilkins and P.D. Goode, Proc. 2nd Int. Conf. on Ion Implantation in Semiconductors, I. Ruee and J. Graul eds (Springer Verlag, 1971) p. 439-454.

9. M. Hirata, M. Hirata, H. Saito, J. Phys. Soc. Jap. 27, 405 (1969).

10. R. Bader and S. Kalbitzer, Appl. Phys. Lett. 16, 13 (1970).

11. R. Bader and S. Kalbitzer, Rad. Effects 6, 211 (1970).

LUMINESCENCE PROPERTIES OF GaAs EPITAXIAL LAYERS GROWN BY LIQUID PHASE
EPITAXY AND MOLECULAR BEAM EPITAXY

P. M. PETROFF
Bell laboratories, Murray Hill, New Jersey 07974 USA

ABSTRACT

The luminescence properties of undoped GaAs-Ga$_{1-x}$Al$_x$As
layers grown by liquid phase epitaxy (LPE) and molecular
beam epitaxy (MBE) are compared for double heterostructure
and multiquantum well (MQW) superlattices. Low temperature cathodoluminescence and scanning transmission
electron microscopy are used to identify the main luminescence centers and in some instances establish their
origin. The possible effects of the observed luminescence
features on the degradation process in GaAs-Ga$_{1-x}$Al$_x$As
laser devices grown by MBE and LPE are also discussed.

INTRODUCTION

One of the most intriguing problems to unravel in the field of defects in
semiconductor devices has been that of the degradation phenomenon in GaAs-
Ga$_{1-x}$Al$_x$As semiconductor lasers. The process of dislocation climb involving
point defect and dislocation motion has been shown to be responsible for the
rapid degradation process [1,2]. The slow degradation process [3,4] has also
been attributed to motion of point defects resulting in point defect aggregates. Both the slow [3,4] and rapid degradation seem to involve point
defects and their origin has been a subject of controversy. On the one hand
the defects have been considered to be native to the material, e.g., antisite
defects and point defect complexes, resulting from the crystal growth
process [1,2]. While on the other hand their origin was associated with the
minority carrier recombination processes at the dislocation core [5,6]. The
present paper is a first step in addressing this problem. We have used two
defect sensitive techniques: low temperature high resolution cathodoluminescence in conjunction with scanning transmission electron microscopy (STEM)
and transmission electron microscopy (TEM), to analyze double heterostructures
(DH) and multiquantum well (MQW) superlattices [7]. A special importance is
attached to the analysis of interfaces which are thought to be possible sources
of point defects [1] responsible for the rapid degradation. Furthermore, we
have analyzed two types of materials grown by techniques which can yield a
widely different stoichiometry; namely, liquid phase epitaxy (LPE) which gives
a Ga rich growth condition, and molecular beam epitaxy (MBE) which yields
layers grown under As rich conditions. We have examined the possible relationships between the rapid degradation phenomena and the defects related to
crystal growth conditions in lasers made by these two different crystal growth
techniques.

EXPERIMENTAL

Undoped GaAs-Ga$_{1-x}$Al$_x$As DH and MQW superlattice structures [7] have been
grown by MBE. In addition, undoped GaAs-Ga$_{1-x}$Al$_x$As DH structures were also
grown by LPE and compared to the MBE grown ones. The GaAs substrate orientation was (100). Chemical polishing and etching in bromine methanol and
hydrogen peroxide-sulfuric acid and water solution were used prior to the

layer growth. The MBE undoped layers were grown after removal of the GaAs native oxide at 650°C. The As pressure during the layer growth was in the range 10^{-7} to 10^{-6} torr and the growth temperature T = 690°C.

The LPE undoped DH structures were grown on (100) GaAs substrates from a saturated melt at temperatures of 800°C and a cooling rate of 0.2°C/min. The cladding layer composition was $Ga_{1-x}Al_xAs$ with x $\geq$ .30. The substrates for both the LPE and MBE growth contain 10^4 to 10^5 dislocations per cm^2.

The samples were prepared for STEM and CL analysis by selective etching of the substrate with a solution of H_2O_2 and NH_4OH with a pH = 7.05. A uniformly thinned area of ∿2 mm^2 is obtained by this technique. The final sample-electron beam configuration for these experiments which allows for the collection of both the STEM and CL signal was that shown in Fig. 1. This configuration allows to distinguish the CL emission and distribution from the active layer or the inter-faces between the GaAs and $Ga_{1-x}Al_xAs$ cladding layers. We analyse the signal from the A or B interfaces by positioning the sample either in the configuration shown in Fig. 1 (to image the B interface) or inverting it with respect to the e^- beam (to image the A interface). This is made possible by the appropriate choice of the GaAs layer thickness (>4000 A) to absorb photons emitted by the interface or the $Ga_{1-x}Al_xAs$ layer located below the GaAs layer.

The luminescence properties are analyzed on a microscopic scale in a modified scanning transmission electron microscope which allows for a simultaneous collection of the cathodoluminescence (CL) information and STEM or TEM signal of the analyzed area [8]. The CL signal is analyzed with a 50 cm spectrometer coupled to a cooled RCA 31034 photomultiplier tube. A photon counting elec-tronic and computer controlled multichannel analyzer-spectrometer system are used for recording the CL spectra with a fixed electron beam position. The photon energy resolution is 0.5 mev. In addition, CL monochromatic micrographs are obtained by scanning the beam over the sample for a fixed wavelength setting of the spectrometer. The CL analysis is carried out with the samples mounted on a liquid He cooled stage at temperatures 16°K $\leq$ T $\leq$ 400°K. The electron beam energy is 100 KeV for the CL analysis and 200 KeV for the STEM and TEM analysis which is carried out after the CL experiment. The e^- beam spot size can be adjusted from 50Å to 500Å and the useful beam intensity is in the range 5.10^{-11}A to 5.10^{-8}A for the CL analysis.

RESULTS AND DISCUSSION

The CL spectrum of an undoped LPE grown DH structure is shown in Fig. 2. For the e-beam positioned on two different locations on the sample the emission line 4 shifts from 1.581 ev (Fig. 2B) to 1.594 ev (Fig. 2A). This line is found to shift by a large amount (1.530 ev to 1.670 ev) for various beam positions on the sample. This difference in CL emission is related to variations in the Al content in the interface as line A is associated with the $Ga_{1-x}Al_xAs$ CL emission from the interface region [9]. This interpretation is supported by the mono-chromatic images of the A and B interfaces (Fig. 3A and 3B) which relate the spatial variations of the CL intensity at 7500 Å with the Al distribution (at x ∿ 0.11) in the interfacial regions A and B. The compositional grading in the plane of the interface has previously been attributed to the melt down and re-growth processes which are inherent to LPE grown DH structures [9]. In addition to compositionally inhomogeneous interfacial regions the GaAs active layer pro-duces several luminescence lines which have been identified as follows: On the basis of previously published photoluminescence data [10, 11] the luminescence line at 1.490 ev line (1) is identified with the carbon acceptor (e,C°) recom-bination and line 2 at 1.455 ev is tentatively associated with the Ge acceptor center. The (e,C°) luminescence intensity is strongly dependent upon the injection conditions and becomes larger than the bound exciton-donor (D°,x) emission (line (0) at 1.512 ev) with decreasing carrier injection density.

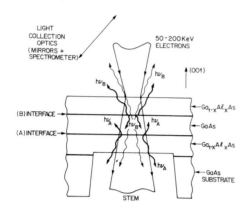

Fig. 1 Schematic of the sample structure
and signal collection configuration.
Heavy lines indicates interfacial
regions.

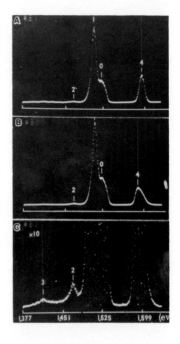

Fig. 2 (A) CL emission spectrum from an LPE
grown DH structure with a .5μm
GaAs active layer. The layers are
undoped and the cladding layer
composition is x = .30. Fig. 2A and
2B correspond to two positions of
the e⁻ beam. Fig. 2C is the same
as Fig. 2B with a gain of 10.

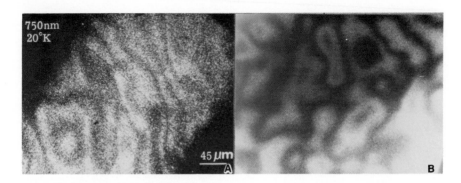

Fig. 3 (A) CL monochromatic micrograph of the A interface for an LPE grown DH structure.

(B) CL monochromatic micrograph of the B interface. The same area as in (a) is imaged. $\lambda = 7500$ Å. The contrast distribution reflects relative variations of the CL intensity with the dark area corresponding to a lower CL signal.

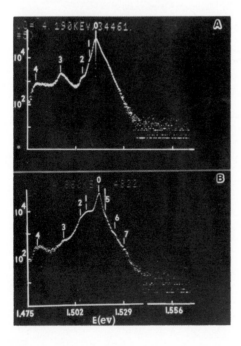

Fig. 4 (A) Logarithmic display of the CL intensity as a function of photon energy for an MBE DH structure.

(B) Logarithmic display of the CL intensity as a function of photon energy for an MBE grown MQW super-lattice structure.

The CL spectrum of an MBE grown DH structure is shown in Fig. 4A. The luminescence of the GaAs active layer is similar to that of the LPE grown DH layers. Line 0 at E = 1.513 ev corresponds to the (D°, x) bound exciton luminescence, while line 3 at 1.493 ev and line 4 at 1.480 ev are associated with the (e, C°) and (e, Si°) centers, respectively. The lines 1 and 2 at 1.510 ev and 1.506 ev are specific to the MBE grown material [12, 13] since they are not observed in material grown by vapor phase epitaxy and metallorganic epitaxy or LPE. The monochromatic CL micrographs indicate that at the main emission energy (Fig. 5A) a number of nonradiative (at 1.513 ev) centers are present in the material. Their density is 10^4 to 10^5 cm^{-2}. Series of CL micrographs at various photon energies below 1.513 ev show that some of these centers behave as radiative centers in the energy range $1.498 \leq h\nu \leq 1.508$ ev (Fig. 5B). The STEM analysis (Fig. 5C) of such centers show them to correspond to dislocations which originate in the substrates. The photon energies at which the dislocations produce radiative centers correspond to those of the centers producing the 1.506 ev and 1.510 ev lines (Fig. 4A lines 1 and 2) in the dislocation free region of the material. We propose that the defects in the two cases are of the same nature and that in regions of material for which their concentration is higher a stronger luminescence is recorded. The temperature dependence of these luminescence centers is consistent with a defect bound exciton luminescence. This luminescence line is tentatively associated with point defects or impurity complexes. Indeed point defects-dislocations associations are likely to occur because of the dislocation gettering action which could yield an increased point defect concentration around the dislocation during the layer growth. The MBE grown interfaces apart from the defects discussed above do not show any evidence of a compositional grading. This finding is consistent with the measurements of the MBE interface abruptness which were previously measured by TEM [14].

The CL spectrum of a representative MQW superlattice sample is shown in Fig. 4B. From published data [7, 15] the free exciton (at 1.516 ev) (line 0) is attributed to the n = 1 electron heavy hole (e-hh) transition. The emission lines 5, 6, and 7, respectively, are attributed to the electron light hole (e-lh) n = 1 transition at 1.519 ev, the 1.524 ev forbidden transition and to the e-hh n = 2 transition at 1.530 ev. The below bandgap luminescence is similar to that of the DH structure in Fig. 4A: the lines 1 and 2 correspond to the 1.510 ev and 1.505 ev emission and are attributed to point defects related to the MBE growth technique. By contrast to the DH structure the monochromatic CL micrographs from the MQW superlattice are very uniform (Fig. 6). There is a complete absence of discrete nonradiative centers in the energy range $1.475 \leq h\nu \leq 1.570$ ev. This result is all the more striking since the TEM analysis and the etch pit study of these MQW superlattices indicate the presence of dislocations in the superlattice. Their density 10^4-10^5 cm^{-2} is identical to that in the substrate.

The effects of the observed luminescence features on the degradation process of laser devices grown by LPE and MBE techniques are now discussed. First, we remark that the nature of the degradation induced dark line defect (DLD) is identical for laser devices produced by LPE or MBE growth techniques. A TEM micrograph of a DLD in a laser grown by MBE is shown in Fig. 7. The climb character of the dislocation network is immediately apparent. The DLD dislocation network just as in LPE grown material [1, 2] has an interstitial character and most of the dislocation dipoles are aligned along the <100> direction although some parts are along <110>. These results could imply that identical defects are responsible for the dislocation climb in LPE or MBE lasers. Since carbon impurities are present in both MBE and LPE lasers, we may speculate that a point defect complex involving C is at the origin of the point defects needed for the dislocation climb during DLD growth. On the other hand, the presence of point defects originating at the interfaces in the case of LPE lasers or point defects intrinsic to the MBE material cannot be set aside. The

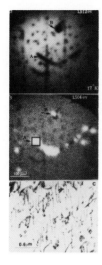

Fig. 5   (A)   CL monochromatic (hν = 1.512 ev) micrograph of the active layer in a DH sample grown by MBE.

            (B)   CL monochromatic micrograph hν = (1.504 ev) of the same area as in A.

            (C)   STEM micrograph of the area framed in B.

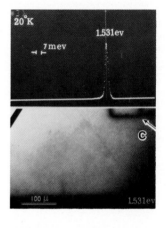

Fig. 6   (A)   CL spectrum of an undoped MQW superlattice grown by MBE. The GaAs layer thickness is 161Å and that of the $Ga_{.66}Al_{.34}As$ is 183Å. The sample thickness is 2.94μm.

            (B)   CL monochromatic (1.531 ev) micrograph of the same MQW superlattice. C indicates recombination centers due to microcracks in the thin film.

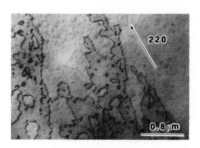

Fig. 7   Bright field TEM micrograph of a part of a DLD in a rapidly degraded GaAs-Ga$_{1-x}$Al$_x$As laser made by MBE.

results indicate that there are several possible sources of point defects in
both the MBE and LPE grown lasers. Finally, we note that the MQW superlattice
material exhibits the interesting effect that dislocations <u>are not</u> behaving as
nonradiative centers. This effect is possibly related to the two dimensional
nature of the carrier confinement. We expect that if indeed carrier recombination at dislocations is reduced and takes place via radiative processes in MQW
lasers, we may have the possibility of establishing the role of dislocations as
a point defect source during the DLD growth.

CONCLUSIONS

We have established several differences between the spectral and spatial
luminescence properties of LPE and MBE grown GaAs-Ga$_{1-x}$Al$_x$As DH structures and
MQW superlattices. The wealth of data provided by the luminescence analysis,
if anything, provides a wider number of choices for the possible origin of the
point defects needed for the rapid degradation process. Impurity centers
common to both LPE and MBE material could provide the necessary point defect
complexes needed for the DLD formation. The compositional grading and inhomogeneity in the interfaces of the LPE (DH) structures and point defect intrinsic
to the MBE (DH) structures are also likely sources of point defects. Finally,
the unusual luminescence quality and the absence of nonradiative carrier
recombination processes at dislocations in MQW superlattices may prove useful
in understanding the role of dislocations during degradation.

ACKNOWLEDGEMENTS

This work was done with the collaboration of R. A. Logan, C. Weisbuch, A. C.
Gossard, W. Wiegmann, A. Y. Cho, W. Tsang, and R. L. Hartman. The technical
help of J. LePore was appreciated throughout this work.

REFERENCES

1. P. M. Petroff and R. L. Hartman, Appl. Phys. Lett., 23, 1969 (1973) and
    J. Appl. Phys., 45, 9, 3899 (1974).

2. P. W. Hutchinson, P. S. Dobson, P. O'Hara and D. H. Newman, Appl. Phys.
    Lett., 26, 250 (1975).

3. P. M. Petroff (unpublished results).

4. I. Hayashi, Proceedings of 15th International Conference on the Physics of
    Semiconductors, Kyoto (1980) (to be published).

5. S. O'Hara, P. W. Hutchinson and P. S. Dobson, Appl. Phys. Lett., 30, 368
    (1977).

6. P. W. Hutchinson and P. S. Dobson, Phil. Mag., 41, 4, 601 (1980).

7. R. Dingle in Festkorperprobleme (Advances in Solid State Physics, Ed.
    H. J. Queisser (Pergamon/Vieweg. Braunschweig), Vol. 15, p. 21 (1975).

8. P. M. Petroff, D. V. Lang, J. L. Strudel, and R. A. Logan, Scanning
    Electron Microscopy (SEM Inc. AMF O'Hare, IL 6066 USA), Vol. 1, p. 325
    (1978).

9. P. M. Petroff and R. A. Logan, J. Vac. Sci. Technol., 17, 5, 1113 (1980).

10. G. B. Stringfellow and R. Linnebach, J. Appl. Phys., $\underline{51}$ (4) 2212 (1980).

11. D. J. Ashen, P. J. Dean, D. T. J. Hurle, J. B. Mullin, and A. M. White, J. Phys. Chem. Solids, $\underline{36}$, 1041 (1975).

12. P. M. Pretroff, C. Weisbuch, R. Dingle, A. C. Gossard, and W. Wiegmann (to be published).

13. H. Kunzle and K. Ploog, Appl. Phys. Lett., $\underline{37}$, 416 (1980).

14. P. M. Petroff, A. C. Gossard, W. Wiegmann, and A. Savage, J. Cryst. Growth, $\underline{44}$, 5 (1978)

15. C. Weisbuch and R. Dingle (to be published).

THE INTERRELATIONSHIP BETWEEN STRUCTURE AND PROPERTIES IN InP AND InGaAsP
MATERIALS

S. MAHAJAN, Bell Laboratories, Murray Hill, NJ 07974

ABSTRACT

Three examples pertaining to the influence of heavy doping
on the microscopic perfection of as-grown InP crystals, thermal decomposition of InP crystals and epi-layers and optically-
induced degradation of InGaAsP epi-layers have been chosen
to delineate some of the correlations observed between
microstructure and properties in these materials. It
is shown that the macroscopic perfection can be significantly improved by highly doping with Zn, S, and Se.
However, in the case of Zn, not all of the dopant atoms
are in solid solution and some of them have clustered to
form precipitates, whereas these features are not seen
in highly S- and Se-doped crystals.
Thermal decomposition of epi-layers introduces areas
which appear dark in a photoluminescence scan. It is
argued that these regions evolve due to the out-
diffusion of P. In addition, when InP epi-layers
are grown on thermally-decomposed substrates, dislocation density in the epi-layers is higher than that
in the substrate. Arguments have been developed to
rationalize these observations.
Non-luminescent regions develop in InGaAsP epi-
layers by optical pumping. These areas are associated with dislocation clusters which appear to
evolve by glide of the existing dislocations.

INTRODUCTION

In recent years, InP, InGaAs and InGaAsP have evolved into a promising group
of semiconducting materials. This stems primarily from their application as
light sources and detectors in the current light-wave-communication system
based on fused silica fibers.
Horiguchi and Osanai [1] and Payne and Gambling [2] have shown that state-
of-the art fibers exhibit low attenuation and minimum material dispersion in
the 1.1 - 1.5 µm region. Consequently, diode lasers and LEDs emitting in this
wavelength regime constitute optimal light sources. Several investigators
[3-6] have recently demonstrated that light sources based on the InP/InGaAsP
system, emitting at these wavelengths, can be fabricated. One of the attractive features of this system is that, over a wide range of compositions, lattice-matched InGaAsP layers can be grown on InP substrates by LPE [6]. It is
therefore feasible to tailor the bandgap and thus the emission wavelength of
opto-electronic devices to be compatible with the spectral properties of the
fiber.
The purpose of this paper is to delineate some of the correlations observed
between structure and properties in InP and InGaAsP materials. Three examples
pertaining to highly doped InP crystals, thermal decomposition of these materials and optically-induced degradation of InGaAsP epi-layers have been chosen
to illustrate these correlations.

HIGHLY DOPED CRYSTALS

Seki and co-workers [7,8] have investigated the influence of different dopants on the perfection of as-grown InP crystals. The quality of these crystals
has been assessed using etch pitting and X-ray topography. The most interesting
observation that emerges from their work concerns the reduction of dislocation
density with heavy doping. They have shown that macroscopically dislocation-
free crystals can be produced by highly doping with Zn, S and Te. We have extended these investigations and have evaluated using etching, X-ray topography,
transmission cathodoluminescence (TCL) [9] and transmission electron microscopy
(TEM) the influence of heavy doping with Zn, S and Se on the microscopic perfection of as-grown crystals.

The structural characterization and hardness evaluation were confined exclusively to crystals grown in Bell Laboratories. For various studies, seeded
single crystals oriented along the <111> direction were grown using the LEC
technique described in detail by Bonner [10]. The crystals were doped with
Zn, S and Se. For different evaluations, slices were cut parallel to the (001)
plane from the grown crystals and were subsequently polished using a chemical-
mechanical technique.

A variety of etchants have been developed for delineating material defects
in InP crystals [11-15]. However, the recent work [16] shows that Huber etch
[14] is the only etchant that reliably reveals dislocations, saucer pits, protrusions and growth striations on the (001) plane. Furthermore, one-to-one
correspondence between dark spots observed on TCL images and etch pits revealed
using Huber etch has previously been established [17].

Figures 1(a) and 1(b) show, respectively, material defect distributions observed on the (001) planes of crystal doped with Zn to 1.2 x 10^{17} and 5.7 x
10^{18} cm^{-3}. Dislocations and saucer pits are delineated by D and S on the micrographs. In addition, growth striations which run nearly horizontally are weakly resolved in the highly doped crystal [Fig. 1(b)], but they are not discer-
nible in Fig. 1(a). It is also apparent that the dislocation density can be
reduced considerably by highly doping with Zn, a result consistent with the
work of Seki and co-workers [7,8].

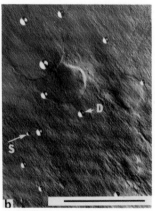

Fig. 1. Huber etched (001) surfaces of as-grown crystals doped with Zn to (a)
1.2 x 10^{17} and (b) 5.7 x 10^{18} cm^{-3}. Dislocations and saucer pits are delineated
and saucer pits are delineated by D and S. The marker represents 0.1 mm.

467

The crystals doped with Zn to different levels have been examined in detail by TCL. At low carrier concentration (~5 x 10^17 cm^-3), growth striations and dislocations are revealed. When the doping level is ~1 x 10^18 cm^-3, a high density of dark spots are also observed and these results are reproduced as Fig. 2. These spots vary in size, do not have comet-like shapes which are characteristic of dislocation images [9] and their density is 1 x 10^4 cm^-2. The spots appear to lie contiguous to dark bands which are dopant-rich regions. With increasing carrier concentration, their distribution becomes more homogeneous. In view of these observations, it is very likely that these features are caused by precipitates. This assessment has been confirmed by TEM [18].

Fig. 2. TCL image of a crystal doped with Zn to 1 x 10^18 cm^-3. The marker represents 50 μm.

The presence of precipitates in Zn-doped crystals is consistent with the results of Galavanov et al. [19], Kundukov et al. [20], Hooper and Tuck [21] and Tuck and Hooper [22]. They have reported that for high Zn-concentrations the number of free carriers does not match the number of Zn atoms introduced by diffusion. Hooper and Tuck [21] and Tuck and Hooper [22] have hypothesized that some of the Zn atoms may form some kind of complex with vacancies in the crystal.

In order to assess the influence of Zn-doping on mechanical strength, hardness was measured on the (001) plane along the <110> direction using a knoop indentor. These results are listed in Table 1. It is clear that the knoop hardness number (KHN) shows minor variations with carrier concentration.

The (001) Huber etched surface of a wafer doped with S to 1 x 10^19 cm^-3 is reproduced as Fig. 3. Again, D and S pits, protrusions (P) and growth striations are quite well delineated. Furthermore, consistent with the work of Seki et al. [7,8], large portions of this slice did not reveal any D and S pits, implying that they are macroscopically dislocation-free. This was confirmed by TCL [16]. In TCL images, only growth striations of varying doping

TABLE I
Carrier concentration versus knoop hardness number (KHN) for Zn-doped crystals

Carrier concentration	KHN
cm^{-3}	
7×10^{17}	438±9
1×10^{18}	438±10
$2\text{-}3 \times 10^{18}$	443±4
1×10^{19}	433±10

concentrations appeared as light and dark bands. No localized dark spots corresponding to dislocations or other defects were apparent. However, a TEM study of this sample reveals the presence of small dislocation clusters as shown in Fig. 4. Fig. 4(a) was taken under the two-beam diffraction condition, whereas Fig. 4(b) was obtained using quasi-kinematical situation. It may be inferred by comparing Figs. 4(a) and 4(b) that dislocations are weakly decorated with impurities or microprecipitates.

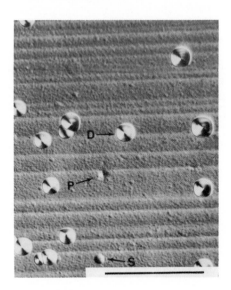

Fig. 3. Optical micrograph showing etch pit pattern observed on the (001) plane of a highly S-doped crystal. The marker represents 0.1 mm.

Shown in Fig. 5 is a TCL image of a wafer doped with Se to 5×10^{18} cm^{-3}. The features apparent in this image are the growth striations and localized dark spots corresponding to dislocations. These spots are approximately of the same size and have comet-like shapes because dislocations intersect the surface at oblique angles. The dislocation density in this sample is ~4×10^{3} cm^{-2}. Again, it appears that the dislocation density can be reduced by highly

doping with Se. However, the TEM examination of these samples showed that dislocation clusters are decorated with microprecipitates whose chemical nature
remains to be ascertained [16].

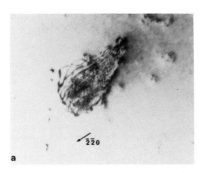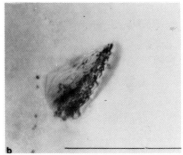

Fig. 4. TEM micrographs obtained from a macroscopically dislocation-free region
of a highly S-doped crystal: (a) two-beam diffraction condition, and (b) quasi-
kinematical situation. The plane of the micrograph is ~ (001) and the marker
represents 1 μm.

Fig. 5. TCL image of a crystal doped with Se to 1×10^{19} cm^{-3}. The marker
represents 100 μm.

The introduction of dislocations during crystal growth could be attributed to three different sources. First, dislocations present in seed crystals could propagate into a growing crystal. Second, dislocations generated at peripheral regions of the growing crystal could, under the influence of thermal gradient-induced stresses, propagate into the crystal interior. Third, supersaturation of point defects could lead to clustering to form dislocation loops during cool-down following growth. Since it is possible to grow from dislocated seeds crystals with large areas which are macroscopically dislocation-free, it appears that the quality of seeds does not have an important influence on the perfection of as-grown crystals [18]. This can be rationalized on the basis of Chikawa's observations pertaining to Si crystals [23]. He has shown that during the growth of crystals from dislocated seeds, dislocation density in the growing crystal is progressively reduced due to the formation of closed dislocation loops.

Seki et al. [7,8] have observed that the order of effectiveness of different dopants in reducing the dislocation density is Zn, S and Te. They have rationalized their results in terms of the differences in bond strengths between the host and dopant atoms. They have argued that the increased bond strength reduces the probability of introducing thermal gradient-induced dislocations. Since S-doped crystals are the hardest [16] and Zn-doped crystals exhibit minor variations in KHN with carrier concentration (Table I), their hypothesis cannot be entirely correct. It appears that in the absence of pronounced thermal gradients, supersaturation and clustering of point defects into potential dislocation sources contributes significantly to impairing the perfection of as-grown crystals. In the present situation, the improved perfection resulting from S and Se doping could result from point defects-dopant interactions. Previously, deKock et al. [24] have invoked this hypothesis to explain the influence of p- and n-type doping on microdefect formation in macroscopically dislocation-free as-grown Czochralski Si crystals. However, in the case of Zn-doped crystals the macroscopic perfection could be due to the presence of precipitate - matrix interfaces and associated dislocation clusters. These features act as sinks for point defects and thus prevent them from clustering to form dislocation sources.

THERMAL DECOMPOSITION AND ASSOCIATED EFFECTS

At temperatures (600° - 680°C) generally used for growing epitaxial layers, P has fairly high vapor pressure and tends to evaporate preferentially. This causes the accumulation of excess In or In-Ga alloy at the bottom of thermally-induced etch pits. For reproducible preparation of high quality InP epitaxial layers, it is essential to remove thermal damage prior to the growth of epitaxial layers. Two techniques are commonly used to prevent or remove such damage: (i) the use of a polished InP cover piece [25] which protects the substrate before growth , and (ii) an in situ In etch [26]. The combination of the two methods has also been reported [27].

Recently, Temkin et al. [28] have investigated thermal decomposition of InP and its influence on the quality of iso-epitaxial layers and InP/InGaAsP device wafers. They have used photoluminescence, photocurrent scanning and etch pitting to evaluate the effectiveness of substrate protection measures described above. Figure 6(a) shows an iso-epitaxial layer grown on the (001) substrate. During the epi growth, the substrate was protected by an InP cover piece. It is clear that the layer exhibits good surface morphology under Nomarski interference contrast. Only in some isolated areas very small dissociation pits and associated In droplets with local density of less than 10^4 cm^{-2} can be seen. These shallow pits probably form during the short interval when the wafer is exposed to high temperature following the termination of growth. Figure 6(b) shows the same area as in Fig. 6(a) after Huber etching for 45 secs

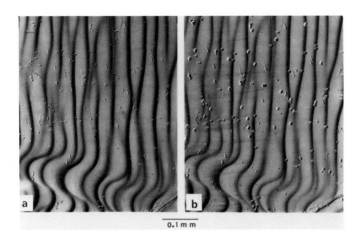

Fig. 6. Optical micrograph of an epi-layer grown with neither the source-piece protection nor the melt-back: (a) as-grown surface, and (b) same area as in (a) except that the epi was Huber etched for 45 secs to reveal dislocations.

to reveal dislocations. The etch pits are either isolated or form pairs, very likely due to the presence of small dislocation half loops. To ascertain whether the dislocations observed in Fig. 6(b) result from the replication of defects present in the bulk of the underlying substrate, the epitaxial layer and the portion of the substrate contiguous to the epi-substrate interface were removed. The subsequent Huber etching of the substrate indicates significantly fewer dislocation density and an absence of very small dislocation half loops.

In addition to the microscopic defects discussed above, the photoluminescence scan of iso-epitaxial layers reveals a number of randomly distributed dark spots. Figure 7 shows an image obtained on a 5 μm thick epitaxial layer. The upper part of the scanned area is uniformly bright, whereas a number of irregularly shaped dark regions can be seen in the remainder of the photograph. The smallest defects that are imaged are approximately 7-10 μm in diameter, their apparent size limited by the extent of the laser beam. The largest dark feature is about 150 μm long. The difference in photoluminescence efficiency between the bright and dark areas is no more than a factor of 2.

The preceding results can be rationalized if it is assumed that a shallow and discontinuous layer of degraded material is formed at the surface of InP. In view of the fact that thermal treatment at temperature higher than 370°C readily degrades the stoichiometry of near-surface InP due to P loss [29], this assumption is reasonable. It is therefore envisaged that the dark regions seen in Fig. 7 could be P-deficient areas of the epi-layer.

Two possibilities exist regarding the origin of some of the dislocations observed in Fig. 6(b). First, the condensation of point defects inherited by an epi-layer during growth may result in the formation of dislocations and dislocation loops. Second, thermal decomposition-induced defects, i.e., P vacancies, in the substrate may aggregate to produce the observed shallow defects which can then be replicated in the epi-layer. The second mechanism is more

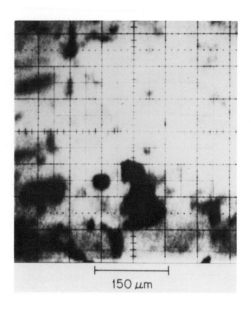

150 μm

Fig. 7. Photoluminescence scan of the epi-layer of InP grown with substrate cover and without In melt-back.

likely since it is consistent with growth experiments in which only a portion of the substrate was etched with In and substructures observed in association with thermal etch pits [16].

OPTICALLY-INDUCED DEGRADATION OF InGaAsP EPI-LAYERS

In the past, photoluminescence studies have proved quite useful in the characterization of GaAs/GaAlAs double-heterostructure laser materials [30,31]. It has also been shown that the microstructural features introduced in these materials be optical pumping at low levels are similar to those observed in degraded lasers [32]. These studies have now been extended to InGaAsP layers grown on (001) InP by LPE, and demonstrate that spatially resolved photoluminescence (SRPL) is also useful in the evaluation of these materials [33]. In addition, the layers have been found to undergo optically-induced degradation. We have addressed the detailed nature of this degradation [34].

An optical micrograph depicting the surface morphology of a degraded layer is reproduced in Fig. 8(a). Occasionally, certain portions of the as-grown layer did not luminesce during SRPL examination. For comparison, an optical micrograph from such a layer has been included as Fig. 8(c), whereas Fig. 8(b) shows the boundary between the degraded and non-luminescent regions. In addition to growth terraces, Fig. 8(a) shows short horizontal lines are normal to the [110] direction. The observed crystallography of these features is consistent with {111} slip. Since the lines are observed only in the degraded regions, they could be optically-induced slip bands.

To assess the origin of non-luminescence in Fig. 8(c), TEM examination of the epi-layer was carried out and these results are shown in Fig. 9. The

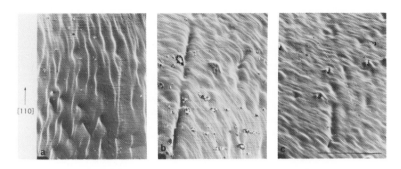

[110]

Fig. 8. Optical micrographs obtained from optically degraded and non-luminescent regions of an InGaAsP epi-layer: (a) degraded region, (b) degraded and non-luminescent regions and (c) non-luminescent region. The marker represents 0.5 mm.

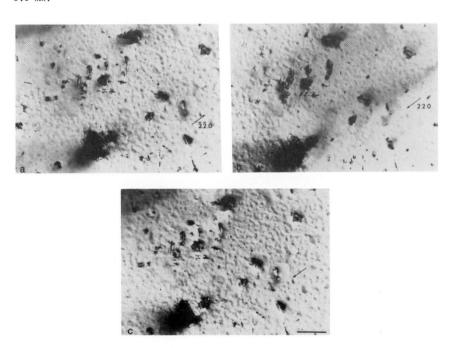

Fig. 9. TEM micrographs showing inclusiosn and associated dislocation clusters observed in an InGaAsP epi-layer. (a) and (b) are bright-field micrographs and illustrate the effect of changing the sign of the operating reflection on black and white lobes associated with different inclusions, whereas (c) is a dark-field micrograph. The plane of the micrographs is ~ (001), and the marker represents 1 um.

474

micrograph in Fig. 9(a) is characterized by a high density of inclusions surrounded by dislocation clusters. It is clear from Fig. 9(a) that black and
white lobes are associated with some inclusions and two interchange their
positions on changing the sign of the operating reflection [compare Figs. 9(a)
and (b)]. This implies that these inclusions impose a stress on the surrounding matrix. To ascertain the nature of this stress, a dark-field image of the
same area was obtained and this result is reproduced as Fig. 9(c). Since the
vector connecting the black and white lobes points in the direction of the
operating reflection, it is inferred that the inclusions impose a compressive
stress on the adjoining matrix. However, the chemical nature of these inclusions as well as their origin need to be elucidated.

Several of the degraded areas were examined in detail by TEM. Two situations have been chosen to illustrate the evolution of substructural features
associated with the degraded regions. Figure 10 depicts one such area imaged

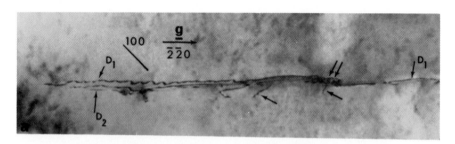

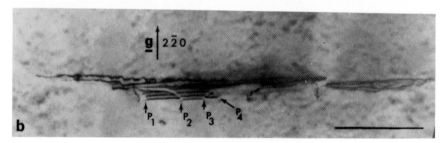

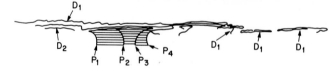

$$\underline{b}_{D_1} = \pm \frac{1}{2}[011], \quad \underline{b}_{D_2} = \pm \frac{1}{2}[110], \quad \underline{b}_{P_1} = \underline{b}_{P_2} = \underline{b}_{P_3} = \pm \frac{1}{6}[\bar{1}12] \text{ AND } \underline{b}_{P_4} = \pm \frac{1}{6}[21\bar{1}]$$

c   HABIT PLANE OF FAULTS = $(1\bar{1}1)$

Fig. 10. Electron micrographs showing an optically degraded region in an
InGaAsP layer under two different operating reflections. The plane of the
micrographs (a) and (b) is ~ (001). (c) Schematic of the situation shown in
(a) and (b). Burgers vectors and crystallography of different structural
features are identified.

under two different operating reflections. The principal characteristics of these micrographs are: (i) the presence of dislocations, see the arrowed regions in Fig. 10(a), which extend towards the substrate, a feature exhibiting sausage-link structure [see the doubly arrowed region in Fig. 10(a)] and faulted regions defined by partials P_1, P_2, P_3 and P_4; (ii) the alignment of dipole and imperfections along the [110] direction; (iii) the jaggedness of the top dislocation segment in Fig. 10(a). In order to establish the crystallography of the observed features, the region was examined using different reflections. It was thus ascertained that the habit plane of the faults shown in Fig. 10(b) is (1$\bar{1}$1) and the Burgers vectors of P_1, P_2, P_3 and P_4 are ±1/6 [$\bar{1}$12] and ±1/6 [21$\bar{1}$], respectively. Also, on the basis of the diffraction contrast experiments, ±1/2 [011] and ±1/2 [110] Burgers vectors can be assigned to D_1 and D_2 dislocations, respectively. For clarity, these results are schematically illustrated in Fig. 10(c).

Figure 11 illustrates an advanced stage in the development of substructure that is oriented along the [110] and [1$\bar{1}$0] directions. Dislocations are closely packed in bands S_1 and S_2, whereas individual shear loops are resolved in band S_3. The line joining the end points of these shear loops delineate a trace along the [110] direction. Furthermore, dislocations constituting S_3 are cusped. The same region was examined using $\tilde{g}$ = $\bar{2}$20, and only the disloca-

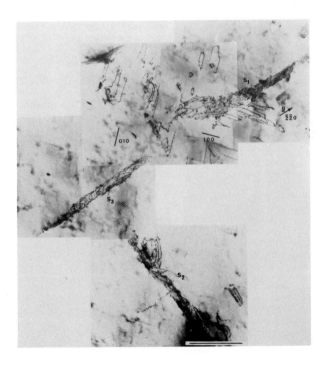

Fig. 11. Electron micrograph showing an optically degraded region in an InGaAsP layer. The plane of the micrograph is ~ (001), and the marker represents 1 μm.

tions defining S_1 and S_3 exhibited residual contrast. This implies that S_1 and S_3 could evolve as a result of the ±1/2 [110] glide on the ($1\bar{1}1$) and ($\bar{1}11$) planes. The trace generated by the end points of shear loops present in S_3 is consistent with this assessment. A few dislocation segments which project along the [100] and [010] directions are also observed. It is also emphasized that in no case were the preceding microstructural features seen in the nondegraded regions.

The central question that needs resolution concerns the process that governs dislocation multiplication during optical pumping of InGaAsP layers. Consider the simple situation depicted in Fig. 10 and see if a possible explanation based on glide can be developed for its evolution. The observed dipoles could form by the glide of jogged D_1 dislocations on the ($1\bar{1}1$) and ($11\bar{1}$) planes. Since the faults lie on the ($1\bar{1}1$) planes and are bounded by Shockley partials, it is reasonable to assume that slip occurs only on the ($1\bar{1}1$) planes. Assuming the occurrence of the 1/2 [011] ($1\bar{1}1$) slip, there are four possible low energy orientations which the 1/2 [011] dipole may assume: (i) 60° dipole oriented along the [110] direction, (ii) 60° dipole oriented along the [$10\bar{1}$] direction, (iii) the perfect edge dipole oriented along the [$21\bar{1}$] direction, and (iv) the screw dipole. However, the last situation would be unstable becuase the two dislocation segments constituting the dipole would cross glide and annihilate each other. Since the observed dipole is aligned along the [110] direction, it is inferred that it is 60° dipole. Furthermore, the observed sausage-link contrast in Fig. 10 may arise from the tendency of a dipole to break up into small loops. This process requires climb and could be effected by the attractive elastic interactions between the opposite segments of the dipole.

Referring to Fig. 10(a) it is apparent that some portions of the top jagged segment project close to the [100] direction. It can be shown that a dislocation oriented along the [$10\bar{1}$] direction would project along the [100] direction lying in the (001) plane. The observed jaggedness may develop due to oscillation of the 1/2 [011] dislocation between the two energetically equivalent orientations. Also, the development of the faulted configuration shown in Fig. 10 can be rationalized by invoking interactions between D_1 and D_2 dislocations. The short traces observed in Fig. 8 and detailed structure of S_1, S_2 and S_3 are also consistent with the occurrence of glide.

It is envisaged that, when the layers are optically pumped, the threading and inclusion-generated dislocations (Fig. 9) undergo extremely rapid glide on the {111} planes. As a result, dislocation clusters or bundles develop along the <110> and <100> directions. It is implicit in the suggestion that, when these covalently bonded crystals deform either at a high strain rate or at temperatures which are considerably low than their melting points, a major portion of the debris is in the form of 60° dislocations and dipoles. This is consistent with the recent work on deformed silicon [35].

Recently, it has been observed that InP crystals, oriented for single slip, can be deformed plastically when the deformation temperature is around 400°C [36], and we may assume that a similar result will hold for the quaternary layer. However, it is highly unlikely that the local temperature would have arisen to such a level during optical pumping [37]. Therefore, to rationalize our observations we have to invoke that the observed glide is facilitated by nonradiative recombination [38]. In addition, during cool-down following LPE the interfacial stress may develop due to the difference between the thermal expansion coefficients of InP and InGaAsP materials [39], and this would also favor glide.

As shown above, the development of substructural features observed in optically degraded InGaAsP epi-layers can be rationalized in terms of glide alone. It is relatively easy to visualize that once the dislocation network has been initiated and evolved by glide, substructures resembling those observed in degraded GaAs/GaAlAs double heterostructure lasers [40,41] can be obtained by invoking the occurrence of climb. This glide-climb approach for GaAs/GaAlAs materials has several advantages over explanations based exclusively on climb [40,41]. First, under the influx of point defects a grown-in dislocation of arbitrary orientation should climb more or less uniformly without the formation of dipoles. This assessment is consistent with the shape of precipitate colonies commonly generated by climbing dislocations in Si [42,43]. However, the formation of dipoles during glide is quite common. Second, the observed alignment of the 1/2 [011] dipoles along the [100] and [2$\overline{1}$0] directions lying in the (001) plane can easily be rationalized in terms of the glide-climb approach. Third, it appears that the formation of dark line defects is sensitive to resolved shear stress [44]. This is again consistent with the glide hypothesis.

ACKNOWLEDGMENTS

The author is grateful to numerous colleagues for fruitful collaborations. In particular, he would like to thank W. A. Bonner, A. K. Chin, V. G. Keramidas, D. C. Miller and H. Temkin for sustained interactions.

REFERENCES

1. M. Horiguchi and H. Osanai, Elec. Lett. 12, 310 (1976).

2. D. N. Payne and W. A. Gambling, Elec. Lett. 11, 176 (1976).

3. L. W. James, G. A. Antypas, R. L. Moon, J. Edgecombe and R. L. Bell, Appl. Phys. Lett. 22, 270 (1973).

4. C. C. Shen, J. J. Hsieh and T. A. Lind, Appl. Phys. Lett. 30, 353 (1977).

5. A. G. Dentai, T. P. Lee, C. A. Burrus and E. Buehler, Elec. Lett. 13 484 (1977).

6. M. A. Pollack, R. E. Nahory, J. C. DeWinter and A. A. Ballman, Appl. Phys. Lett. 33, 314 (1978).

7. Y. Seki, J. Matsui and H. Watanabe, J. Appl. Phys. 47, 3374 (1976).

8. Y. Seki, H. Watanabe and J. Matsui, J. Appl. Phys. 49, 822 (1978).

9. A. K. Chin, H. Temkin and R. J. Roedel, Appl. Phys. Lett. 34, 476 (1979).

10. W. A. Bonner, Mat. Res. Bull. 15, 63 (1980).

11. T. Iizuka, J. Elect. Chem. Soc. 118, 1190 (1971).

12. G. B. Mullin, A. Royle, B. W. Straughan, P. G. Tufton and E. W. Williams, J. Crystal Growth 13/14, 640 (1972).

13. R. C. Clarke, D. S. Robertson and A. W. Vere, J. Mat. Science 8, 1349 (1973).

14. A. Huber and N. T. Linh, J. Crystal Growth 29, 80 (1975).

15. K. Akita, T. Kusunaki, S. Komiya and T. Kotani, J. Crystal Growth 46, 783 (1979).

16. S. Mahajan and A. K. Chin, to appear in J. Crystal Growth (1981).

17. A. K. Chin, S. Mahajan and A. A. Ballman, Appl. Phys. Lett. 35, 784 (1979).

18. S. Mahajan, W. A. Bonner, A. K. Chin, D. C. Miller and H. Temkin, unpublished work (1981).

19. V. V. Galavanov, S. G. Metreveli, N. V. Siukaev and S. P. Starosel´tseva Sov. Phys. Semicond. 3, 94 (1969).

20. R. M. Kundukov, S. G. Metreveli and N. V. Siukaev, Sov. Phys. Semicond. 1, 765 (1967).

21. A. Hooper and B. Tuck, Solid State Electron. 19, 513 (1976).

22. B. Tuck and A. Hooper, J. Phys. D. 8, 1806 (1975).

23. J. Chikawa, Proc. of the Conf. on Defects in Semiconductors, Boston (1980).

24. A. J. R. deKock, W. T. Stacy and W. M. van de Wijgert Appl. Phys. Lett. 34, 611 (1979).

25. P. D. Wright, Y. C. Chai and G. A. Antypas, IEEE Trans. Electr. Dev. ED-26, 1220 (1979).

26. P. D. Wright, E. A. Rezek, M. J. Ludoise and N. Holonyak, Jr., IEEE J. Quantum Elect. QE-13, 637 (1977).

27. R. C. Goodfellow, A. C. Carter, I. Griffith and R. R. Bradley, IEEE Trans. Electr. Dev. ED-26, 1215 (1979).

28. H. Temkin, V. G. Keramidas and S. Mahajan, to appear in J. Elect. Chem. Soc. (1981).

29. R. F. C. Farrow, J. Phys. D: Appl. Phys 7, 2436 (1974).

30. W. D. Johnston, Jr. and B. I. Miller, Appl. Phys. Lett. 23, 192 (1973).

31. F. R. Nash, R. W. Dixon, P. A. Barnes and N. E. Schumaker, Appl. Phys. Lett. 27, 234 (1975).

32. P. M. Petroff, W. D. Johnston, Jr. and R. L. Hartman, Appl. Phys. Lett. 25, 226 (1974).

33. W. D. Johnston, Jr. G. Y. Epps, R. E. Nahory and M. A. Pollack, Appl. Phys. Lett. 33, 992 (1978).

34. S. Mahajan, W. D. Johnston, Jr., M. A. Pollack and R. E. Nahory, Appl. Phys. Lett. 34, 717 (1979).

35. A. T. Winter, S. Mahajan and D. Brasen, Phil. Mag. A. 37, 315 (1978).

36. D. Brasen and F. A. Thiel, Bell Laboratories, unpublished work (1977).

37. M. Lax, J. Appl. Phys. 48, 3919 (1977

38. D. V. Lang and L. C. Kimerling, Phys. Rev. Lett. 33, 489 (1974).

39. R. Bisaro, P. Merenda and T. P. Pearsall, Appl. Phys. Lett. 34, 100 (1979).

40. P. M. Petroff and R. L. Hartman, Appl. Phys. Lett. 23, 469 (1973).

41. P. W. Hutchinson and P. S. Dobson, Phil. Mag. 32, 745 (1975).

42. D. Bialas and J. Hesse, J. Mat. Sci. 4, 779 (1969).

43. K. V. Ravi, Met. Trans. 4, 681 (1973).

44. T. Kamejima, K. Ishida and J. Matsui, Jap. J. Appl. Phys. 16, 233 (1977).

ELECTRONIC DEFECTS IN METALORGANIC GaAlAs

N. M. JOHNSON, R. D. BURNHAM, D. FEKETE, and R. D. YINGLING
Xerox Palo Alto Research Centers, Palo Alto, California 94304

ABSTRACT

Electronic defect levels have been measured in n-type epitaxial films of $Ga_{1-x}Al_xAs$ ($0 \leq X \leq 0.33$) which were grown by metalorganic chemical vapor deposition. Electron traps were characterized by transient capacitance spectroscopy on Schottky-barrier diodes. The thermal activation energy for electron emission from the near-midgap defect level is found to be larger in $Ga_{1-x}Al_xAs$ than the value of 0.83 eV generally observed in GaAs.

INTRODUCTION

Metalorganic chemical vapor deposition (MO-CVD) is of current technological interest as a growth technique for fabricating $Ga_{1-x}Al_xAs$ heterostructure lasers [1]. With MO-CVD, $Ga_{1-x}Al_xAs$ can be epitaxially grown with layer thickness uniformity and doping control which are superior to those of the conventional liquid-phase epitaxial (LPE) growth technique. In addition, large-scale cost-effective production of $Ga_{1-x}Al_xAs$ heterostructure lasers stongly favors the MO-CVD technique. These considerations have motivated a detailed study of electronic defects in state-of-the-art material in order to provide information for selecting future directions in materials development and identifying possible future issues in device performance. This paper presents results from a characterization of electronic defects in n-type $Ga_{1-x}Al_xAs$ ($0 \leq X \leq 0.33$) with emphasis on the near midgap defect level.

The near-midgap level has been observed in previous studies. In GaAs this level is commonly designated EL2 [2] and has a thermal activation energy for electron emission of 0.83 eV, as determined in bulk material [3] and in material grown by both vapor-phase epitaxy [4,5] and MO-CVD [6]. In $Ga_{1-x}Al_xAs$, grown by MO-CVD, it has recently been reported [7] that this activation energy is independent of aluminum content over the investigated range of $0 \leq X \leq 0.35$. In the study reported here it was found that the activation energy for the near-midgap level does vary in $Ga_{1-x}Al_xAs$ ($0 \leq X \leq 0.33$) from its value in GaAs.

SAMPLE PREPARATION

The MO-CVD process used in this study is similar to that reported by Dupuis and Dapkus [1]. Physical vapor-phase mixtures of trimethylgallium, trimethylaluminum, and arsine were pyrolyzed in H_2 at $750°C$ or $825°C$ to form epitaxial layers of $Ga_{1-x}Al_xAs$ with thicknesses of 5-10 μm. The electrical conductivity of the material was controlled by doping during deposition to obtain net dopant concentrations of $\leq 4 \times 10^{16}$ cm^{-3}. The dopant source was hydrogen selenide for n-type Se doping. The films were (hetero-) epitaxially grown on (100)-oriented GaAs substrates, which were degenerately doped with Si (2×10^{18} cm^{-3}) for n$^+$-type conductivity.

The electroptical quality of the MO-CVD material was characterized by photoluminescence (PL). The wavelength of the luminescence peak was used to determine alloy composition [8], and the peak intensity was compared to $Ga_{1-x}Al_xAs$ standards grown by LPE to establish relative PL efficiency. The comparisons were made for the same net dopant concentrations [9]. The PL efficiency of the MO-CVD grown material generally exceeds that of LPE material. Also, the MO-CVD material has been used to fabricate laser diodes with threshold currents that are lower than those of comparable LPE diodes [10].

Schottky-barrier diodes were fabricated on the $Ga_{1-x}Al_xAs$ epilayers for electrical measurements. The rectifying electrode consisted of a 35-nm thick layer of chromium vacuum deposited onto the $Ga_{1-x}Al_xAs$ surface, with a 350-nm overlayer of gold to support pressure contact for testing. Ohmic back contacts were formed by depositing 300-nm of Au/Ge alloy (12% Ge by weight) over the back surface of the specimens and sintering prior to front-surface metallization.

MEASUREMENT TECHNIQUES

Transient capacitance spectroscopy was used to characterize electronic defect levels in the heteroepitaxial layers of $Ga_{1-x}Al_xAs$. The basic measurement technique used in this study is termed deep-level transient spectroscopy (DLTS) [11]. The measurement was performed in the constant-capacitance mode [12] and used to obtain thermal emission spectra, activation energies for thermal emission of trapped charge, and defect concentrations. Since the Schottky-barrier diode is a majority-carrier device, for n-type conductivity DLTS detects electron traps that are essentially in the upper half of the semiconductor bandgap. Thermal emission spectra were obtained by scanning the diode temperature between 80 to 400K and recording the DLTS signal for specific emission rate windows e_0, which are set by the spectrometer. Emission peaks appear at temperatures for which the emission rate e_n of a given trap equals the rate window. From the principle of detailed balance, the emission rate is

$$e_n = \sigma_n v_n N_C \exp(-G_t/kT), \tag{1}$$

where σ_n is the electron capture cross-section, v_n is the mean thermal velocity for electrons, N_C is the effective density of states in the semiconductor conduction band, G_t is the (Gibbs) free energy required to emit an electron, k is Boltzmann's constant, and T is the absolute temperature. Activation energies E_A were obtained from an Arrhenius analysis of the temperature dependence of the emission rate. This procedure yields the change in enthalpy, E_A, rather than the change in free energy, G_t, of the emission process, for a constant capture cross-section [13]. The two thermodynamic quantities are related by $G_t = E_A - TS_t$, where S_t is the entropy of the defect center. The quoted energies are corrected for the assumed T^2 dependence of the product $v_n N_C$ in the prefactor of e_n. Defect concentrations N_t were computed from the constant-capacitance DLTS signal as described in Ref. 12, for uniform trap-filling voltage pulses. The defect densities were obtained by using the activation energies as representative of trap depths relative to the semiconductor conduction band minimum. The quoted values are for essentially uniform trap densities within $1\mu m$ of the semiconductor surface.

RESULTS AND DISCUSSION

Electronic defects in MO-CVD GaAs are evaluated in Figs. 1 and 2. In Fig. 1 are shown two spectra, which were recorded with different emission rate windows in order to expand the range of detectable energy levels for the given temperature range. The rate windows are 347 and 6.9 sec^{-1} for the low-temperature and high-temperature spectra, respectively. The low-temperature spectrum reveals defect concentrations in the range of 10^{12} cm^{-3} for electron traps in the upper half of the GaAs bandgap. The high-temperature spectrum is dominated by a single electron emission center with a trap density of 3×10^{14} cm^{-3}. The emission peak identifies

Fig 1. Electron Emission Spectra for MO-CVD GaAs. The thermal activation energy E_A and trap concentration N_t are listed for the high-temperature emission peak.

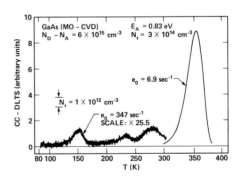

a defect level near midgap as the dominant electron trap in GaAs. In Fig. 2 (a) the high-temperature spectrum is displayed on an enlarged temperature scale, and the Arrhenius analysis for the emission center is presented in Fig. 2 (b). The temperature dependence of the emission rate was determined by recording the diode transient response, after a trap-filling voltage pulse, with a fast digital voltmeter and performing a least-squares fit (LSF) to obtain the emission time constant over a range of fixed temperatures. The Arrhenius analysis yields a

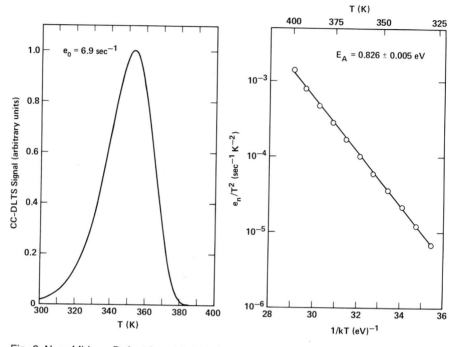

Fig. 2. Near-Midgap Defect Level in MO-CVD GaAs: (a) electron emission spectrum and (b) Arrhenius plot of the emission rate.

thermal activation energy of 0.826 eV, with the deviations of the data from the fitted curve in Fig. 2 (b) giving an uncertainty of 0.005 eV. Thus, within experimental error this is the 0.83-eV defect level generally observed in GaAs [2-6].

Emission spectra for $Ga_{.86}Al_{.14}As$ are shown in Fig. 3. The low-temperature spectrum displays two well resolved emission peaks with activation energies of 0.14

Fig. 3. Electron Emission Spectra for MO-CVD $Ga_{1-x}Al_xAs$ with 14% Al.

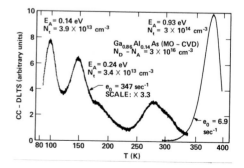

and 0.24 eV and defect densities in the range of 10^{13} cm^{-3}. These activation energies were obtained by recording the temperature of the emission peak for different emission rate windows. The high-temperature spectrum reveals an emission center with an activation energy of 0.93 eV and a trap concentration of 3 X 10^{14} cm^{-3}. While the trap concentration for the near-midgap defect level essentially equals that for the GaAs layer (Fig. 1), the activation energy differs from the value for GaAs (Fig. 2 (b)) by an amount which exceeds experimental error; the overall uncertainty in the energies is estimated to be ± 0.010 eV.

The near-midgap emission center in MO-CVD Ga$_{1-x}$Al$_x$As is further examined in Figs. 4 and 5. High-temperature emission spectra for select aluminum concentrations are shown in Fig. 4. The epilayers with 0, 5, and 14% aluminum were deposited at 750° C, and the layers with 25 and 33 % aluminum were grown at 825° C. Each spectrum is normalized to its peak intensity in order to clearly demonstrate the shift to higher temperatures of the emission peak, for a constant rate window, as the aluminum concentration increases. The temperature of the emission peak shifts by ~40 K as the aluminum concentration varies from 0 to 33%. The Arrhenius plot for each diode is shown in Fig. 5. The activation energies where determined with the LSF analysis described above. The range of detectable emission rates was limited by experimental conditions: the upper limit was set by the maximum calibrated temperature (400 K) of the diode temperature sensor and the lower limit was set by a practical minimum emission rate window of approximately 1 sec^{-1}. The more restricted ranges for the higher activation energies (i.e., for 25 and 33% Al) contribute to an increased uncertainty of ± 0.015 eV. However, these results clearly demonstrate that the activation energy for the

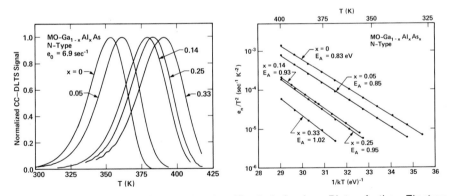

Fig. 4. Electron Emission Spectra for the Near-Midgap Defect Level in MO-CVD Ga$_{1-x}$Al$_x$As. Each spectrum is normalized to its peak intensity.

Fig. 5. Arrhenius Plot of the Electron Emission Rates for the Near-Midgap Defect Level in MO-CVD Ga$_{1-x}$Al$_x$As.

near midgap defect level in $Ga_{1-X}Al_XAs$ $(X > 0)$ does not in general equal the value for GaAs.

CONCLUDING REMARKS

In the present study it was found that in $Ga_{1-X}Al_XAs$ grown by MO-CVD the near-midgap defect level has an activation energy which increases with aluminum concentration from its value of 0.83 eV in GaAs. This finding is in disagreement with Ref. 7 in which it is reported that this energy is independent of aluminum content. A mechanism that could contribute a large uncertainty in the determination of the activation energy is the effect of high electric fields, which are known to considerably enhance the electron emission rate of the EL2 level in GaAs above its low-field "thermal" value [14]. In the present study to minimize this effect diodes were fabricated with low net dopant concentrations ($\leq 4 \times 10^{16}$ cm^{-3}), and low average electric fields ($< 1 \times 10^5$ V cm^{-1}) were maintained in the diode depletion layer during measurement. Thus, although a systematic study of the electric-field effect was not conducted, these precautions should have insured minimal perturbation of the results presented here.

ACKNOWLEDGEMENT

The authors express their appreciation to W. Mosby and J. Tramontana for technical support.

REFERENCES

1. R. D. Dupuis and P. D. Dapkus, IEEE J. Quantum Electron. QE-15, 128 (1979).
2. G. M. Martin, A. Mitonneau, and A. Mircea, Electron. Letters 13, 191 (1977).
3. F. Hasegawa and A. Majerfeld, Electron Letters 11, 286 (1975).
4. D. V. Lang and R. A. Logan, J. Electron. Mat. 4, 1053 (1975).
5. A. Mircea and A. Mitonnau, Appl. Phys. 8, 15 (1975).
6. P.K. Bhattacharya, J. W. Ku, S. J. T. Owen, V. Aebi, C. B. Cooper, III, and R. L. Moon, Appl. Phys. Letters 36, 304 (1980).
7. E. E. Wagner, D. A. Mars, G. Hom, and G. B. Stringfellow, J. Appl. Phys. 51, 5434 (1980).
8. H. C. Casey, Jr., and M. B. Panish, Heterostructure Lasers, Part A-Fundamental Principles, (Academic, New York, 1978).
9. G. B. Stringfellow and H. T. Hall, Jr., J. Electron. Mat. 8, 201 (1979).
10. R. D. Burnham (unpublished).
11. D. V. Lang, J. Appl. Phys. 45, 3023 (1974).
12. N. M. Johnson, D. J. Bartelink, R. B. Gold, and J. F. Gibbons, J. Appl. Phys. 50, 4828 (1979).
13. G. L. Miller, D. V. Lang, and L. C. Kimerling, Annu. Rev. Mater. Sci. 7, 377 (1977).
14. S. Makram-Ebeid (these proceedings).

INVESTIGATION OF DEFECT CONCENTRATION DISTRIBUTIONS IN ION-IMPLANTED AND
ANNEALED GaAs

K.L. Wang and G.P. Li
University of California, Los Angeles, CA 90024
and
P.M. Asbeck and C.G. Kirkpatrick
Rockwell International, Thousand Oaks, CA 91360

ABSTRACT

Uncapped and Si_3N_4-capped annealing of GaAs grown with the horizontal
Bridgman technique was investigated with deep-level transient spectroscopy.
Electron trap concentration distributions were measured with a reduced noise
DLTS system to ensure reliable data. Ion implantation using Se ions both prior
to capping and through a Si_3N_4 cap was carried out. The evolution of defect
energy levels and the changes in concentration distributions with anneal temperature were studied. It is concluded that the defects residing in the probed
space-charge region can be annealed out with a Si_3N_4 cap at a temperature higher
than 750 C.

INTRODUCTION

Direct ion implantation into GaAs stubstrates provides several advantages,
such as simplicity, greater controllability and reproducibility in fabrication
of GaAs integrated circuits.[1-3] Currently problems exist in annealing GaAs
materials, however, particularly in the qualification of semi-insulating substrate materials.[4] Qualification procedures are conventionally implemented to
eliminate substrates whose surface becomes conducting after annealing. The
presence of impurities and other defects appears to contribute to this conversion problem. Since the production and migration of defects play an important role in controlling GaAs material properties during annealing, it is
essential to characterize and identify the origin and evolution of the deep-
level defects. Although deep-level defects in GaAs have been investigated
with various techniques for some time,[5-7] the origins of the defects are not
well known. This is partly due to the greater complexity of defect formation
in GaAs compared with that in silicon. With a detailed analysis of defect
spatial concentration distributions using deep-level transient spectroscopy
(DLTS),[8] the evolution of defect redistributions may be studied. Previously,
because of the presence of noise involved in profiling deep-level concentration
distributions, the results obtained were often unreliable.

In this paper, a study of the deep-level defects and their evolution in
horizontal-Bridgman grown GaAs is described. The results of annealing control,
uncapped and Si_3N_4-capped samples are reported. The defect spatial distributions obtained by using a DLTS profiling system are discussed. The profiling
system incorporates both hardware and software enhancements for averaging in
order to obtain reliable distributions.

EXPERIMENTS

Undoped (n-type) and Cr-doped (semi-insulating) horizontal Bridgman GaAs
used in this study were obtained from Crystal Specialties, Inc. Test Schottky
diodes were used for DLTS experimentation. In fabrication Schottky diodes, a
Au/Ge alloyed ohmic contact was formed at 450 C, followed by electron-beam Al
or Ti/Au deposition to form Schottky gates. In capped annealing, a Si_3N_4 film

was sputtered onto the substrate and the resulting capped sample was heated in H_2 ambient at a specified temperature for 30 minutes. The Si_3N_4 was subsequently removed with plasma etching prior to the application of the alloyed ohmic contact and Schottky gates. The annealing temperature for the capped samples ranged from 450 to 950 C while the annealing temperatures of the uncapped samples were limited to 550 C since beyond 650 C the GaAs surface tends to decompose. Two ion implantation procedures using Se ions were carried out for comparison: one at 125 keV prior to capped annealing at 850 C and the other at 400 keV after the Si_3N_4 cap was deposited. The energies were so chosen that the projected ranges in GaAs for both implants were almost identical. The deep-level defect distributions were obtained using a technique described elsewhere.[9-10] The sample temperature was held constant to within 0.1 C while the voltage pulse changed from the bias V_b to a desired V_p in a step-by-step fashion. The voltage pulse cycle was repeated many times (200 in this case) while the transient capacitance difference at two time apertures, $\Delta C(t_2) - \Delta C(t_1)$, was taken. The defect spatial distribution $n_T(X)$ can be calculated using the following expression,

$$n_T(X) = \frac{2\epsilon_s^2 N_D(X_s) \, \Delta C}{(X_1^2 - X_2^2) \, c^3}$$

where C and X_s are the capacitance and shallow dopant depletion width at bias V_b respectively, X_1 and X_2 are the deep-level depletion lengths during the capture when the pulse voltages V_{p1} and V_{p2}, are applied, respectively, and $X = (X_1 + X_2)/2$. The deep-level depletion widths X_1 and X_2 may be evaluated as described elsewhere.[9]

RESULTS

The electron-trap DLTS spectrum obtained for undoped as-grown material (after the alloying process used to make ohmic contacts) is shown in Figure 1. Six major electron-trap levels were observed. These major levels along with capture cross sections, which were calculated from the prefactors of the Arrhenius plots, are tabulated in the first column of Table I. In most cases the values compare well with previously reported traps in bulk-grown n-type GaAs.[5] These results may be compared directly with those of the capped, unannealed sample. This capped control sample was capped with a Si_3N_4-film and then stripped without subjecting to any additional high temperature treatment. As indicated in Table I, some new peaks appear from the capping process. For capped samples subjected to additional annealing, it was found that new peaks appear and old ones disappear as the annealing temperature increases. The new peaks are shaded in Table I and a dashed line connects the energy levels which are believed to be due to the same defect. In Table I

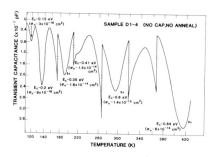

Fig. 1.   DLTS spectrum of an unannealed sample showing major electron traps.

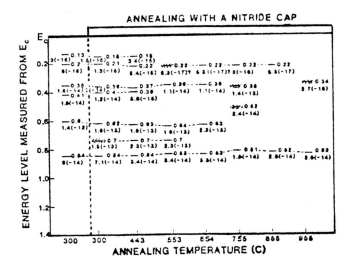

Table I Electron trap energy levels for capped annealing at different temperature. The capture cross section is shown immediately below the level, e.g. 1 x 10^{-14} abbreviated as 1(-14) in cm^2.

and in the subsequent discussion, the energy levels of defects have not been corrected for any temperature dependence of the capture cross-section.

(1) Capped Annealing. Comparing the first two columns for unannealed samples of the capped and the as-grown materials, one observes that the peak of E_c-0.7 eV was introduced as a result of the capping process. This additional peak is similar to that observed in electron-irradiated material and may be attributed to impurity-vacancy complexes.[10] The evolution of this and other defect energy levels and the approximate concentrations are shown as a function of annealing temperature in Table I and Figure 2, respectively. For example, the levels at E_c-0.15 eV and E_c-0.4 eV become insignificant after a 443 C annealing, while the capping-induced peak, E_c-0.7 eV disappears after a 553 C annealing. On the other hand, the E_c-0.83 eV level appears to persist through a wide range of annealing temperatures. Most defects appear to anneal out beyond 750 C. After 850 C annealing, a very few defects remain in the space charge region probed with DLTS. It should be pointed out that the defect concentrations are in general not uniformly distributed in the space-charge region. The depletion width in the Schottky diode is limited to about a few μm by breakdown, which in turn depends on the doping density. The defect concentration distributions were obtained by using the deep-level profiling technique described earlier. The defect distributions of the control sample are illustrated in Figure 3. The detailed description of the spatial distributions for the annealed samples will be published elsewhere owing to the limited space in these Proceedings.

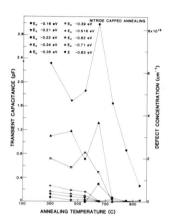

Fig. 2. Approximate concentration change for each energy level
for capped annealing at different temperatures.

(2) <u>Uncapped H_2 Annealing</u>. The defect levels after annealing in hydrogen
ambient are similar to those annealed with Si_3N_4 capping. The energy level
data along with the approximate concentrations as a function of the annealing
temperature are shown in Figure 4. The annealing temperature in this case was
limited for the reason that at high temperature the surface morphology becomes
too rough to have meaningful Schottky characterization. In this limited range
of temperature however, the defects of the uncapped samples were shown to have
roughly similar annealing behavior to that of capped-annealed samples. The
level of $E_C-0.6$ eV appears to anneal out at a lower temperature in the uncapped
annealing.

(3) <u>Ion-Implanted Samples</u>. Figure 5 shows two DLTS spectra for 125 keV and
400 keV Se-implanted samples after 850 C annealing, respectively. The residual
defect spectrum shown in the dashed line for the sample implanted at 125 keV
prior to capping is similar to that of unimplanted samples. On the other hand,
the solid-line spectrum indicates the residual defects for the case of implanting through the cap at 400 keV. In the latter case, a new peak at $E_C-0.32$ eV
is shown to appear near the surface, possibly due to the recoil induced defects,
although the level at $E_C-0.83$ eV remains the dominant defect. The spatial
distribution of these levels is shown in Figure 6. The distribution of the
defects at $E_C-0.83$ eV is roughly similar to that of the implanted dopants.

DISCUSSION

From the results of defect annealing and the changes in concentration
distributions, surface and bulk defects may be distinguished. In addition,
the diffusion of the defects may be studied to provide clues for identifying
some of the defect levels. For example, the levels at $E_C-0.35$ eV, $E_C-0.62$ eV,
and $E_C-0.83$ eV for the capped samples are predominantly located in the bulk
although slight redistribution near the surface was observed. The capping-
induced defects at $E_C-0.71$ eV are at the surface and are possibly due to
radiation damage during the sputtering. A close inspection of the data for
the capped and uncapped control samples indicates that at the energy level of
$E_C-0.21$ eV there are apparently two different kinds of defects.

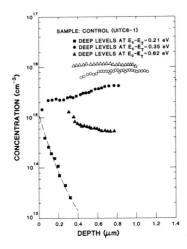
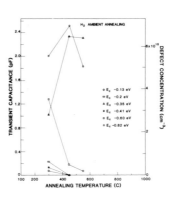

Figure 3. Defect spatial distributions for control sample, i.e. uncapped and unannealed.

Figure 4. Major electron traps and approximate concentration changes as a function of the annealing temperature in uncapped annealing.

The level observed in the temperature range below 550 C has a capture cross section about 2 orders of magnitude larger than the apparently same level occurring after 550 C annealing. Since the concentrations shown in Figure 2 do not show a significant change at 650 C, the disappearance of the first kind of defect may be coupled to the growth of the second kind of defect. Indeed, the concentration distributions obtained support the assertion that the first kind of defect is of the surface type while the second one is in the bulk. Examination of the DLTS spectra of the capped and uncapped samples suggests that the surface defect level was introduced during the capping. The defects of the control samples at $E_c-0.83$ eV are shown to out-diffuse and to pile up at the surface as the annealing temperature increases. The surface pile up differs from the diffusion characteristics previously observed for this center in uncapped anneals.[5] This level cannot be due to Cr-related defects for two reasons. First, the chemical analysis of similar material indicated a Cr concentration of less than 1×10^{15} cm^{-3}, not high enough to account for the observed high defect concentration. Secondly, the Cr out-diffusion observed with secondary ion mass spectroscopy (SIMS) indicated a depletion of Cr below the surface,[4] different from the DLTS profiles observed here. The results of this work are consistent with the previously advanced hypothesis that this level (EL2 from reference 5) is due to a complex involving a Ga vacancy. It should be noted that the surface pile-up of these and other defects occurs in a region that is depleted in typical MESFET applications, and thus the defects will not influence the channel mobility of these devices.

492

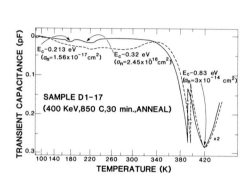

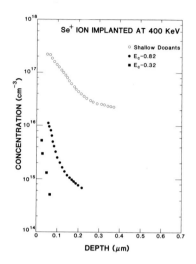

Figure 5.   DLTS Spectra for 125 keV (dashed line) and 400 keV
(solid line) Se-implanted samples, which correspond
to implanting prior to capping and through the cap,
respectively.

Figure 6.   Deep-level spatial distribution profiles for implanted
and annealed samples. Solid data points refer to
these implanted through a cap.

SUMMARY

    In summary, defect energy levels and their evolution after both capped and
uncapped annealing have been investigated using DLTS. The results indicate
that the defect concentrations in the space charge region become negligible
after 750 C annealing with a Si_3N_4 cap. The deep-level spatial distributions
were obtained and used to assist characterization of the defects. With further
experimentation, the nature of the defects may be clarified when the deep-
level spatial concentration distributions are compared with those obtained
from chemical analysis. For device applications, it is apparent that by a
suitable annealing procedure, the defects in the space charge region of both
implanted and unimplanted samples may be reduced.

ACKNOWLEDGMENT

    The group at the University of California, Los Angeles would like to
acknowledge that the work was partly supported by National Sciences Foundation
under a grant ESC 8011037. The group at Rockwell International would like to
acknowledge that the work was supported, in part, by the Defense Advanced
Research Projects Agency of the Department of Defense and monitored by the Air
Force Office of Scientific Research under Contract No. F39620-77-C-0087.

REFERENCES

1. For example, see R.C. Eden, B.M. Welch, R.Zucca, and S.I. Long. IEEE ED-26, 299 (1979).

2. J.A. Higgins, R.L. Kuvas, F.H. Eisen and D.R. Ch'en. IEEE ED-25, 587 (1978).

3. R.Zuleeg, J.K. Notthoff and K. Lehovec, IEEE ED-25, 628 (1978).

4. P.M. Asbeck, J. Tandon, B.M. Welch, C.A. Evans, Jr., and V.R. Deline, IEEE Electron Device Letters, vol EDL-1, 35 (1980).

5. A.Mircea and D. Bois, Inst. Phys. Conf., Serv. 46, 82 (1979) and references therein; D. Pons, A. Mircea, A. Mitonneau and G.M. Martin, ibid., 352 (1979); A. Mircea, A. Mitonneau, L. Hallan and A. Briere, Appl. Phys. 11, 153 (1976); A Mircea and A. Mitonneau, Appl. Phys. 8, 15 (1975).

6. D.L. Patin, J.W. Chen, A.G. Milnes and L.F. Vassamillet, J. Appl. Phys. 50, 6845 (1979).

7. C.H. Henry, J. of Electronic Materials 4, 1037 (1975); D.V. Lang and R.A. Logan, ibid, 1053 (1975); D.V. Lang, Inst. Conf. Ser. No. 31, p. 70 (1977).

8. D.V. Lang, J. Appl. Phys. 45, 3014 (1974); G.L. Miller, D.V. Lang and L.C. Kimerling in Ann. Rev. Mater. Sci. (Annual Rev. Inc., Palo Alto, CA 1977) vol. 7.

9. K.L. Wang, Appl. Phys. Letters, 29, 700 (1976), and H. LeFeure and M. Schultz, Appl. Phys. 12, 45 (1977).

10. D.V. Lang and L.C. Kimerling, "A new technique for defect spectroscopy in semiconductors." Application to one MeV electron irradiated n-GaAs", in Lattice defect in semiconductors (Institute of Physics, London, 1974) pp. 581-588.

EFFECT OF ELECTRIC FIELD AND CURRENT INJECTION
ON THE MAIN ELECTRON TRAP IN BULK GaAs

S. MAKRAM-EBEID
Laboratoires d'Electronique et de Physique Appliquée
3, avenue Descartes, 94450 Limeil-Brévannes (France)

ABSTRACT

The dark capacitive transients related to the main electron trap in bulk and VPE GaAs have been studied in detail.
The low temperature low field transients are induced by
electrons injected from the metal counter electrode. Both hot
electron capture and impact ionization effects are observed.
At high fields, a phonon assisted tunneling model accounts
for the observations and yields the values of the Franck
Condon Shift $S\hbar\omega$ and the energy $\hbar\omega$ of the phonon coupled to
the level. These values are found to be in good agreement
with other independent estimates.

INTRODUCTION

Deep levels in a semiconductor are often characterized by the electrical
transients they produce. When these transients are due to levels in the depleted
region of a reverse biased Schottky or junction diode, the influence of the carriers injected by the reverse current is commonly neglected. We show in this
paper that even modest reverse bias current densities may considerably modify
the deep level capture and emission kinetics.
Another effect which is frequently underestimated is the electric field ionization of deep levels. This type of ionization has been shown to be particularly
important for electron trapping levels in GaAs [1-3] and GaP [1,4] where the
electron effective mass is small. This small mass together with strong coupling
to the lattice vibration is liable to considerably enhance the electron emission
rate above its low field "thermal"value. This may cause orders of magnitude
changes in the electron emission rates. As a consequence, DLTS peaks are seve-
rely altered in position, shape and amplitude for fields as small as 10^5 V cm^{-1}
[2,3].
To illustrate the above mentioned effects, we give in this paper a detailed
account of the transients associated to EL2, the main electron trap in bulk and
VPE GaAs under different conditions of carrier injection and electric field.

DIFFERENTIAL ANALYSIS OF DEEP LEVEL TRANSIENTS

The experimental study has been performed on reverse biased Au-Schottky barriers made on n-type GaAs Czochralski crystals ($n = 1.4$ to 1.8×10^{17} cm^{-3})
cut along different crystallographic directions. Capacitive isothermal transients have been recorded using a Boonton model 72B capacitance bridge monitored
by a calculator. Electrical and optical DLTS studies have shown that EL2 is the
only observable level in the material used.
To study in details the kinetics of EL2 as a function of electric field and
of reverse bias current, we made use of an experimental technique briefly described in a previous paper [1]. In this technique, the reverse bias wave shape
shown in Fig. 1a is applied to the Schottky diode. The pulses of height V_p and
$V_p + \Delta V_p$ serve to set the initial conditions of the deep levels in the depleted

region. The capacitive transient following the pulse of height V_p is measured and stored in the memory of the calculator. This is repeated after the pulse of height $V_p + \Delta V_p$. The calculator then computes the difference between the two transients (Fig. 1b) and this differential transient is subsequently automatically plotted with the time differences on a log-scale (Fig. 1c). This semilog-plot allows us to decide whether the transients are associated to a wide or narrow time constant distribution [5]. Two key quantities are deduced from the semilog plot of Fig. 1c, these are the transient amplitude ΔC_t and the time $\Delta t_{1/2}$ necessary to reach one half of this amplitude.

<u>Differential transient calculations</u>. In the following, it is assumed that the semiconductor contains a constant volume density N_D of shallow donors and a much smaller density N_T of a deep level having a free ionization energy E_T which lies below the Fermi level in the semiconductor volume.

Fig. 2 shows the energy band diagrams in the Schottky diode when the pulse of height V_p is applied and when it is removed. Before application of the pulse V_p, we assume that the occupancy ratio of the level E_T is at its steady state value $f_\infty(X)$ at any point X of the semiconductor. Detailed balancing gives :

$$f_\infty (X) = e^*_p/(e_n + e^*_p) \quad \dots \quad (1)$$

with $e^*_p = c_n n + e_p$ and where e_n and e_p are the electron and hole emission rates and $c_n n$ the electron capture rate. The rate $c_n n$ is due to the capture of electrons injected from the metal electrode or to the capture of the free electrons in the Debye tail of the bulk electrons.

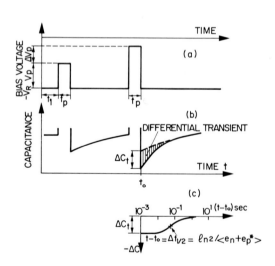

Fig. 1. Sketches of the waveshapes of the diode bias voltage and the corresponding capacitive transients.

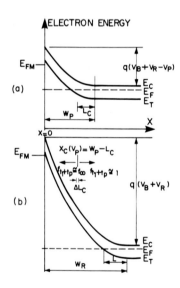

Fig. 2. Energy band diagrams during pulse V_p (a) and after pulse (b).

When a bias pulse of height V_p is applied , the depleted region width shrinks to W_p (Fig. 2a). We assume that this bias pulse is applied for a sufficiently long time t_p such that

$$t_p \ (\sigma_n \ N_D \ v_{th}) \gg 1,$$

where σ_n is the capture cross-section for electrons having the thermal velocity v_{th}. Due to the above condition, all levels lying in the region $X > W_p$ will have practically unity occupancy ratio at the end of the pulse. However, some levels will also be filled in the region $X < W_p$ due to the Debye tail of the bulk electrons during the pulse time. A length L_c can be defined such that for $X = X_c(V_p) = W_p - L_c$, the occupancy ratio at the end of the pulse is middway between unity and the steady state value $f_\infty(X)$. The length L_c is easily put in the form

$$L_c = W_R \sqrt{\frac{kT}{q(V_R + V_B)}} \quad \ln \ (1.44 \ x \ \sigma_n v_{th} \ N_D \ t_p) \ldots \quad (2)$$

k being the Boltzman constant, q the electron charge, T the absolute temperature. The other symbols are as defined in Figs. 1 and 2. For $X < X_c(V_p)$ the electron occupancy of the level will remain as it was before the application of the pulse V_p and will not contribute to the subsequent transients. For $X > X_c(V_p)$ the electron occupancy ratio will be practically unity. The transition between steady state and unity occupancy ratio takes place over a length ΔL_c which can be safely neglected.

With the above simplifications, it is easily seen that the differential transient defined in Fig. 1 is due exclusively to levels in the inverval

$$X_c \ (V_p + \Delta V_p) < X < X_c \ (V_p), \quad (3)$$

with a unity initial electron occupancy ratio.

Equation (1) and a straightforward solution of Poisson's equation yield the differential capacitance transient amplitude ΔC_t in the form

$$\Delta C_t / C_R = \langle e_n/(e_n + e^*_p) \rangle \cdot (N_T/2N_D) \cdot [(X^2_c \ (V_p) - X^2_c(V_p + V_p)] / W^2_R \quad (4)$$

where C_R and W_R are the average capacitance and depleted region width during transient recording. The brackets indicate averaging over the interval defined by inequality (3). The value of $(e_n + e^*_p)$ in the middle of this interval can be related to the time $\Delta t_{1/2}$ (Fig. 1) by the equation

$$\langle e_n + e^*_p \rangle = \ln 2/\Delta t_{1/2}. \quad (5)$$

Strictly speaking, this relation is true when $(e_n + e^*_p)$ varies slightly in interval (3). When $(e_n + e^*_p)$ varies considerably, the numerical factor $\ln(2)$ has to be modified but the error involved can be shown to be of the order of 10 %. It is thus seen that the differential transient measurements yield $\langle e_n + e^*_p \rangle$ and $N_T \ \langle e_n /(e_n + e^*_p) \rangle$.

EXPERIMENTAL RESULTS

Fig. 3 shows typical experimental results obtained for a reverse bias of (4V) at two different temperatures and for a Schottky diode made on the (111) Ga surface. In Fig. 3, we note the rapid increase of $\langle e_n + e^*_p \rangle$ at high electric field is accompanied by a saturation of the product $N_T \ \langle e_n/(e_n + e^*_p) \rangle$ to a temperature independent value. A little thought shows that this value is nothing else than N_T ($N_T \cong 10^{16} cm^{-3}$) ; $\langle e_n \rangle$ and $\langle e_n + e^*_p \rangle$ can then be evaluated separately.

The value of $\langle e_n + e^*_p \rangle$ in Fig. 3b can be observed first to decrease as F increases, to go through a plateau and finally to increase very rapidly.

In Fig. 3, the region defined by $(W_R - X) < L$ corresponds to levels below the semiconductor Fermi level. The quantities $\langle e_n + e^*_p \rangle$ and $\langle e_n/(e_n + e^*_p) \rangle$ are seen to vary much more slowly than would be expected if we assumed that the free electron gas in this region is in thermal equilibrium with the lattice. Injection of reverse bias current seems to considerably increase the effective temperature of the electron gas in the Debye tail of the bulk electrons.

The role of the reverse bias current. We have studied the reverse bias current I_R as a function of reverse voltage V_R and of the temperature T. The slight temperature changes of I_R and its fast exponential dependence with V_R suggest that it is due to electron tunneling from the metal counter electrode to the semiconductor conduction band. This current was observed to vary by more than one order of magnitude between Schottky diodes made on different crystal-lographic planes. The largest currents were associated with the (111) Ga surfaces. The differences may be attributed to the formation of oxides with different thicknesses; a change of the order of one Ångström in thickness accounts for a difference in one order of magnitude in the tunneling current [6].

Fig. 4 shows the relation between the plateau value of $\langle e_n + e^*_p \rangle$ versus reverse bias current I_R. The linear relation demonstrates that only the injected carriers play a significant role. The only mechanisms involved seem to be hot electron capture and impact ionization. We also note that the anisotropy previously observed for the EL2 transients [5] can be attributed to anisotropy in the reverse current I_R (Fig. 4).

We now attempt to evaluate the effective capture and impact ionization cross-sections σ_c and σ_e for the hot carrier injected. If v_x is the electron velocity normal to the Schottky barrier and n the electron density, then

$$I_R = qA \overline{(n\, v_x)}$$

where A is the diode area and the bar indicates statistical averaging. If v is the magnitude of the electron velocity, we certainly have $\overline{nv} > \overline{nv_x}$. When e_n and e^*_p are exclusively due to injected carriers, we should define the cross-sections σ_e and σ_c as the ratios $e_n/\overline{nv}$ and $e^*_p/\overline{nv}$ respectively. Unfortunately, we cannot have a direct measure of $\overline{nv}$. However, rough estimates or rather upper bounds for σ_e and σ_c can be evaluated from

$$\sigma_e \sim e_n/\overline{nv_x} = e_n \cdot qA/I_R \quad \text{and} \quad \sigma_c \sim e^*_p \cdot qA/I_R.$$

The above estimates σ_e and σ_c are plotted vs. 1/T in Fig. 5 together with the conventional cross-section σ_n of EL2 after Mitonneau et al. [7]. We note that σ_c and σ_e have a temperature dependence similar to that of σ_n, that σ_e and σ_c have comparable magnitudes and are consistently smaller than σ_n.

Phonon assisted tunnel emission. At high electric fields, the electron emission rate is observed to increase very rapidly above the injection plateau of Fig. 3b. For these high fields, no anisotropy in the electron emission rate could be detected. The mechanism responsible for the high field electron emission has been demonstrated already for EL2 [1], E3 [1,3] and the Cr level [2] in GaAs and for the ZnO center in GaP [1]. The phenomenon involved is electron tunneling from the trapped state to the conduction band. The tunneling transition rate is greatly enhanced by coupling to lattice phonons of energy $\hbar\omega$ with a linear Huang-Rhys coupling constant S. The values of S and $\hbar\omega$ can be estimated by fitting the experimental electric field and temperature dependence of e_n with theoretical predictions. A semiclassical model has already been published [3] but we use here a more refined quantum model [8] which yields slightly different estimation for the S and $\hbar\omega$ values obtained with the previous model.

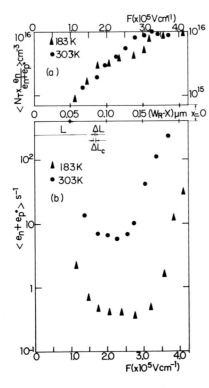

Fig. 3. Differential transient results for a (111) Ga face.

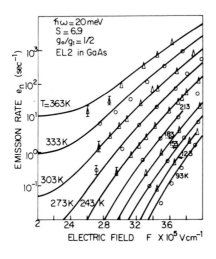

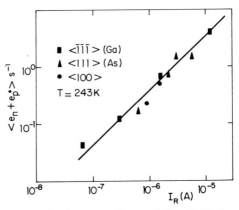

Fig. 4. Dependence of the plateau value of $\langle e_n + e^*_p \rangle$ on reverse current. $A = 6 \times 10^{-3} cm^2$.

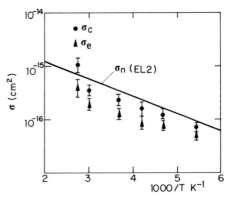

Fig. 5. Temperature dependence of σ'_c, σ'_e compared to that of σ'_n after [7].

Fig. 6. High-field electron emission rate for EL2 after substraction of injection component. The points are experimental and the curves theoretical.

In the new version, the electron emission rate e_n can be written in the form

$$e_n = e_{no} + \sum_{p=-\infty}^{+\infty} W_p . \Gamma (\Delta_p)$$

where e_{no} is the zero field electron emission rate,

$$\Delta_p = E_c - E_T + p\hbar\omega$$

$$W_p = \exp\left[(p\hbar\omega /2kT) - S.coth (\hbar\omega /2kT)\right] . I_p\left[S.csch(\hbar\omega /2kT)\right]$$

and $\Gamma(\Delta_p)$ is the elastic tunneling ionization rate for a localized state lying at depth Δ_p below the conduction band at the trapping site. In the calculation of $\Gamma(\Delta_p)$, we take into account the non parabolic E-k relation in the forbidden band.

The result of the fit and the best fitting parameters are shown in Fig. 6. A sensitivity analysis yield the following range of probable values

$$S\hbar\omega = 140 \pm 10 \text{ meV} \qquad \text{and} \qquad \hbar\omega = 20 \pm 5 \text{ meV}$$

The above parameter values should be compared with those obtained by fitting σ_n vs 1/T for EL2 with the low temperature asymptotic expression of Ridley [9]. The corresponding values are

$$S\hbar\omega = 110 \pm 60 \text{ meV} \qquad \text{and} \qquad \hbar\omega = 20 \pm 3 \text{ meV}$$

Furthermore, photoionization cross-section measurements by Bois et al. [10] yield $S\hbar\omega$ = 120 meV. It is seen that the above estimates are coherent together.

SUMMARY AND CONCLUDING REMARKS

The carrier emission and capture kinetics of EL2, the main native electron trap in bulk and VPE GaAs has been studied in detail as a function of electric field and hot electron injection.

At low temperatures and electric fields, the EL2 kinetics in a reverse biased Schottky diode is dominated by a combination of hot carrier capture and impact ionization. Even very small reverse currents (10^{-5} A.cm^{-2} say) may produce observable effects. The previously reported anisotropic behaviour of the EL2 related transients seems to be due to these effects. The corresponding capture and ionization cross-sections have comparable magnitudes yielding average electron filling ratios near to ½ and hence random fluctuations in the levels occupancy are highly probable. This type of process may thus produce excess noise in GaAs devices.

At high electric field the EL2 kinetics is dominated by a phonon assisted tunnel emission of electrons. The value of the parameters S and $\hbar\omega$ needed to fit the experimental results are found to be in good agreement with two other independent estimates.

ACKNOWLEDGMENTS

The author is grateful to A. Mitonneau for providing the samples and for interesting remarks and to A. Mircea for helpful discussions.

This work is partly supported by DGRST contract n° 79 0725.

REFERENCES

1. S. Makram-Ebeid, Appl. Phys. Lett. 37, 464 (1980).
2. S. Makram-Ebeid, G.M. Martin and D.W. Woodard, 15th Int. Conf. on the Phys.

of Semicond., Kyoto (Sept. 1980) - To be published Inst. Phys. (London).

3. D. Pons and S. Makram-Ebeid, J. de Phys. Lett. (F) $\underline{40}$, 1161 (1979).
4. D.V. Lang, J. Appl. Phys. $\underline{45}$, 3014 (1974).
5. A. Mircea and A. Mitonneau, J. de Phys. Lett. (F) $\underline{40}$, L31 (1979).
6. I. Giaever, in Tunneling in Solids, E. Bernstein and S. Lundquist Editors, Plenum (1969) p. 19.
7. A. Mitonneau, A. Mircea, G.M. Martin and D. Pons, Rev. de Phys. Appl. (F) $\underline{14}$, 853 (1979).
8. S. Makram-Ebeid and M. Lannoo, to be published.
9. B.K. Ridley, J. Phys. C, $\underline{11}$, 2323 (1978).
10. D. Bois, A. Chantre, G. Vincent and A. Nouaillat, Inst. Phys. Conf. Ser. n° 43, 295 (1979).

IMAGING OF DEFECTS IN CADMIUM TELLURIDE USING HIGH RESOLUTION TRANSMISSION
ELECTRON MICROSCOPY

F. A. Ponce, T. Yamashita, R. H. Bube, R. Sinclair
Department of Materials Science and Engineering, Stanford University
Stanford, California USA

ABSTRACT

The defect structure of cadmium telluride has been investigated using high resolution transmission electron microscopy. The variation of the TEM images with the defocus
value is discussed, and defect symmetry considerations are
used to correlate the image contrast characteristics with
the lattice structure. Experimental micrographs of stacking faults and dislocations in the structure are analyzed.

INTRODUCTION

Most semiconductor materials have tetrahedrally coordinated structures. The
structure of elemental semiconductors (silicon and germanium) is diamond cubic,
whereas most binary compounds have either the sphalerite or the wurtzite structure. The sphalerite structure (also known as zinc blende) is a derivative
of the diamond cubic where the two elements occupy alternate {111} layers each.
Faults in the structure are likely to adversely affect the electronic characteristics of the material [1] and so it is useful to obtain fundamental information concerning the structure. The defect structure in these materials can
be observed in sharp detail using multibeam, high resolution transmission
electron microscopy (HRTEM). Point resolutions of 3.0 A or less allow the
interpretation of some features of the image in terms of direct structural
considerations. This technique has been applied to the study of defects in
silicon [2-5] and germanium [6,7].
We have carried out a study of the defect structure in cadmium telluride, a
compound semiconductor with the sphalerite structure which is a candidate for
solar cell applications. In this paper we discuss the observation of several
types of faults in this material, the behavior of the image with defocus, and
use defect symmetry to correlate image contrast characteristics with the lattice
structure [5].

EXPERIMENTAL

The specimens were prepared from a single crystal by cutting a thin disc
with a flat oriented along a 110 direction. The disc was mechanically thinned
and polished and then sputtered with an Ar+ ion beam to obtain electron transparency. The observations reported here were made with a Philips EM400 electron
microscope equipped with a high resolution stage with a spherical aberration
coefficient of 1.1 mm. The images were recorded under axial illumination
with a beam divergence of less than 2.1 mrad. Under these conditions the
resolution limit of the instrument is about 2.9 A. The images shown in this
article are from very thin areas of the specimen, probably less than a few
hundred Angstrom units in thickness.

IMAGING OF THE SPHALERITE STRUCTURE

The <110> projection is best suited for the study of defects in the sphalerite structure. Since two of the four {111} slip planes are seen edge-on
along this projection, there is a 50% probability for observing edge-on defects
present in the material. In cadmium telluride, the <110> projection exhibits
columns of Cd and Te atoms stacked in an array such as show in Figure 1(a).
For each atom, two bonds lie in the plane of the projection with a length of
2.81 A, whereas the other two make an oblique angle with the plane and appear
shorter with a length of 1.62 A. This distance is below the resolution limit
of most instruments and is seldom resolved. Most often the image consists of
bright spots located at the intersections of the two {111} planes present in
this projection.

The interpretation of the resulting images requires the association of the
contrast characteristics with positions in the lattice. Generally this task
is not straightforward; whether bright spots in the image represent atomic
sites (large electronic scattering centers) or channels in the structure must
be determined. An accepted procedure is to do image calculations which include

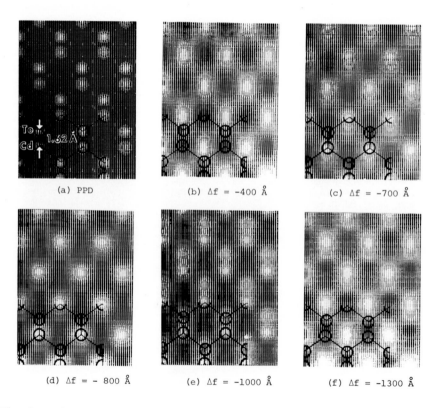

(a) PPD (b) Δf = -400 Å (c) Δf = -700 Å

(d) Δf = - 800 Å (e) Δf = -1000 Å (f) Δf = -1300 Å

Fig. 1. Calculated images for a perfect CdTe crystal. (a) Projected electrostatic potential distribution. (b)-(f) Images for a foil thickness of 80 A, with
a beam divergence of 2.1 mrad and Δ = 40 A, for different values of the defocus.

the main experimental parameters. These image calculations are then matched to the experimental micrographs.

Experimentally, the value of the defocus, Δf, can cause a large change in the characteristics of the image. Typically a change in Δf causes shifts in the position of the bright spots as shown in Figure 1. In Figure 1(a) the projected electrostatic potential (PPD) is shown. This plot includes 51 Fourier coefficients [8], closely resembles the electron density projection on the lower surface of the specimen, and acts as the object in the image calculations. Figures 1(b)-(f) are calculated images for a beam divergence of 2.1 mrad, and an envelope function given by the current and voltage stability parameter $\Delta = 40$ A; corresponding to several values of the defocus [9]. At $\Delta f = -400$A, the bright spots appear at the channel positions, whereas for $\Delta f = -700$ and -800 A the spots appear at the Cd position. At $\Delta f = -1000$ A, the bright spots are elongated and appear in the midposition between the Cd and the Te positions. At $\Delta f = -1300$ A the position of the bright spots is similar to that for $\Delta f = -400$ A, this is due to a periodic behavior of the image with defocus, in this case the period being ~ 900 A.

IMAGING OF DEFECTS

In perfect materials the shift in the positions of the image due to changes in defocus have very little effect on the image characteristics due to the periodic nature of the lattice. In the presence of defects, where departures from perfect periodicity exist, shifts due to defocus often have an important influence on the image.

Intrinsic stacking faults. These faults consist of the absence of a plane in the stacking sequence. In Figure 2(a) the bright spots are placed in between atomic positions (middle of the short segments in the diagram). At the fault, three spots appear stacked up vertically. In Figure 2(b) the spots are at the Cd positions (lower end of the short segments), and the rows of spots exhibit a "switchback" or "kink" at the fault position. A similar array would result if the spots were at Te positions (upper end of short segments) or at the

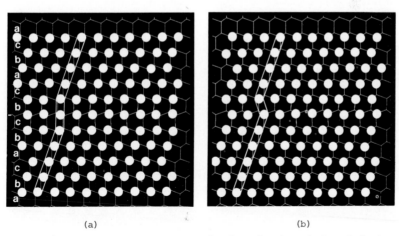

(a)   (b)

Fig. 2. Intrinsic stacking fault in the [110] projection of the sphalerite structure. (a) Bright spots at center between atomic positions, (b) at a single atomic position.

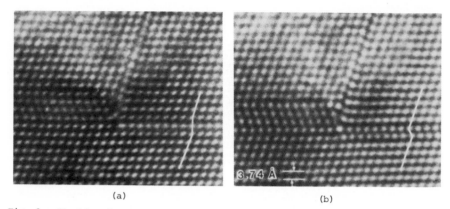

(a)   (b)

Fig. 3. Stacking faults converging to a point. (a) $\Delta f \simeq -1000$ A, (b) $\Delta f \simeq$ -700 A.

channel position, thus not being distinguishable by simple defect symmetry considerations. In Figure 3 a region is seen where several stacking faults converge. There are two intrinsic stacking faults and a three-layer microtwin. A Burger's circuit about the intersection gives a Burger's vector of $a_o[110]$. The intrinsic stacking faults exhibit a variation as described above. Figure 3(a) corresponds to $\Delta f \simeq -1000$ A, thus bright spots are at the mid position between atoms as seen in Figure 1(e) and the defect geometry is as in Figure 2(a). Figure 3(b), at $\Delta f \simeq -700$ A, corresponds to Figure 1(c) and matches the geometry of Figure 2(b).

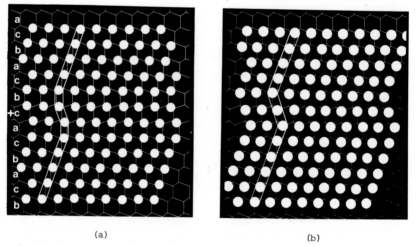

(a)   (b)

Fig. 4. Extrinsic stacking fault. (a) Bright spot at center between atomic positions, (b) at a single atomic site.

Twins. The behavior of the image of a twin also follows a similar analysis. The twin boundaries of the microtwin in figure 3(a) exhibit mirror symmetry about the mid positions between rows of bright spots, whereas in Fig. 3(b) the plane of symmetry coincides with a row of bright spots. This behavior is consistent with the position of the image bright spots relative to the lattice as can be observed by constructing a diagram similar to Figure 2 and using the information in Figure 1.

Extrinsic stacking faults. These faults consist of an extra plane in the stacking sequence. Following the same analysis as above, in Figure 4(a) the spots appear at mid position between atoms, whereas in Figure 4(b) the spots are at an atomic site. In Figure 5 this behavior is observed experimentally. An extrinsic stacking fault exists due to the presence of a Frank loop. The strain field from the Frank partial dislocation affects the overall image and the interpretation is not as simple as for strain free material. The trend is however followed at the center of the fault. In Figure 5(a), $\Delta f \simeq -1000$ A and the fault exhibits soft corners whereas in Figure 5(b) for $\Delta f \simeq -800$ A the corners are sharp, as observed in Figure 4. Another noticable effect is in the position of the fault boundaries. In Figure 5(a) the termination of the fault occurs at different planes, the left side being on a plane immediately lower than the one corresponding to the right side termination. In Figure 5(b) the fault terminates at the same level on both sides. Thus the position of the plane termination is ambiguous in the images and arises from the large distortion associated with a dislocation affecting the position of the terminating fringe.

The interpretation of faults in highly strained regions such as dislocation cores is far from trivial and extensive calculations are needed [4].

Schockley partial dislocations. Figure 6 shows a microtwin bound on both sides and a stacking fault that terminates at the right. The crystal at the right is perfect, and the faults start with the presence of a Shockley partial for the case of the intrinsic stacking fault (below) and two partials in the case of the microtwin (above). This micrograph suggests that deformation in the material occurs by slip along {111} planes generated by the motion of partial dislocations.

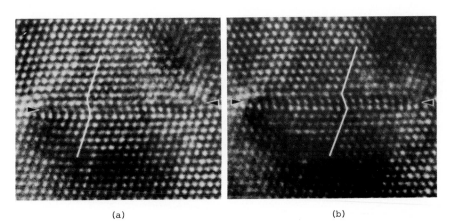

(a)   (b)

Fig. 5. Dislocation loop bound by two Frank partials. The mid section consists of an extrinsic stacking fault. (a) $\Delta f \simeq -1000$ A, (b) $\Delta f \simeq -800$ A.

508

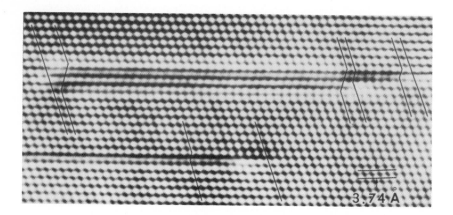

Fig. 6. Microtwin and intrinsic stacking fault terminating in Shockley partial dislocations.

SUMMARY

A variety of defects in cadmium telluride have been imaged using high resolution TEM. The images have been interpreted by using calculated images to determine the relative position of the contrast features with respect to the lattice. This approach agrees well with analysis based on the symmetry of the defects.

ACKNOWLEDGMENTS

We gratefully acknowledge support from the Basic Energy Sciences Division of DOE (FAP, RHB, and RS) and from the Xerox Corporation (TY).

REFERENCES

1. V. F. Petrenko, R. W. Whitworth, Phil. Mag. A 41, 681 (1980).

2. J. C. H. Spence, H. Kolar, Phil. Mag. A 39, 59 (1980).

3. M. Sato, K. Hiraga and K. Sumino, Japan. J. Appl. Phys. 19, L155 (1980).

4. A. Olsen and J. C. H. Spence, Phil. Mag. (1980) in press.

5. J. C. H. Spence, Proc. 38th Annual Meeting EMSA (San Francisco, 1980), 282.

6. S. W. Chiang, C. B. Carter and D. L. Kohlstedt, Phil. Mag. A 42, 103 (1980).

7. A. Bourret and J. Desseaux, J. de Phys. C6, 7 (1979).

8. P. Pirouz, Phil. Mag. A 41, 33 (1980).

9. T. Yamashita, F. A. Ponce, P. Pirouz and R. Sinclair (to be published).

Published 1981 by North-Holland, Inc.
Narayan, and Tan, eds.
Defects in Semiconductors

SCANNING CATHODOLUMINESCENCE MICROSCOPY OF POLYCRYSTALLINE GaAs

JACK P. SALERNO, RONALD P. GALE AND JOHN C. C. FAN
Lincoln Laboratory, Massachusetts Institute of Technology
Lexington, Massachusetts 02173

JOHN VAUGHAN
Department of Materials Science and Engineering
Massachusetts Institute of Technology
Cambridge, Massachusetts 02139

ABSTRACT

Scanning cathodoluminescence microscopy (SCM) has been
used for nondestructive characterization of the optoelectronic properties of heavily Zn-doped, Bridgman-grown polycrystalline GaAs. Grain boundaries can either show no
cathodoluminescence contrast, appear as dark lines, or
appear slightly brighter than the surrounding matrix.
Boundaries with similar surface morphologies can show different contrast. Spectral analysis data indicate that many
of the observed features are due to local variations in
impurity concentration.

INTRODUCTION

The utilization of polycrystalline GaAs for low-cost solar cells is not
yet practical because the grain boundaries (GBs) in this material have adverse
effects on device performance. With the objective of developing procedures
for reducing these effects, we are investigating the relationship between the
structure and optoelectronic properties of GBs and the performance of devices
fabricated in polycrystalline GaAs. This paper reports the use of scanning
cathodoluminescence microscopy (SCM) for an analysis of the optoelectronic
properties of heavily Zn-doped ($p \sim 10^{19}$ cm^{-3}), Bridgman-grown polycrystalline
GaAs. Interest in this material was prompted by its use as a substrate for a
13% (AM1) efficient shallow-homojunction solar cell prepared by chemical vapor
deposition [1].

Most previous investigations of GB effects in GaAs employed techniques
such as Hall and resistivity measurements [2,3], diode characteristics [4,5],
scanning photocurrent and photovoltage measurements [6], and deep level
transient spectroscopy [7]. These studies were limited in scope, however.
They dealt either with average measurements made over many grains or, when the
properties of single boundaries were investigated, with nominally undoped
material.

In an earlier study [8] we demonstrated the effectiveness of
electroluminescence imaging for characterization of individual GBs in heavily
doped GaAs. The present investigation extends that work by employing
cathodoluminescence (CL) analysis. By imaging the CL produced by the electron
beam of a scanning electron microscope (SEM), the SCM technique permits
observation of spatial variations in radiative recombination, an effect
reciprocal to the photovoltaic process, without the necessity for a p-n
junction or sample contacts. In addition, CL spectral analysis permits

measurement of majority carrier concentration and minority carrier diffusion length with a spatial resolution of a few diffusion lengths. Thus SCM permits nondestructive evaluation of potential materials for photovoltaic device applications.

EXPERIMENTAL PROCEDURE

Polycrystalline GaAs slices were lapped, mechanochemically polished in clorox to a thickness of 125 µm, and sawed into squares 2.5 mm on a side. Samples containing GBs with various surface morphologies, as observed by optical Nomarski microscopy, were selected for SCM analysis. All experiments were performed with the sample at ambient temperature in an SEM capable of supplying accelerating potentials ranging from 0–50 kV. The CL collection is accomplished with an optical microscope whose objective lens is coaxial with the incident electron beam. SCM imaging is accomplished by replacing the microscope eyepiece with a photomultiplier assembly, while spectral analysis is performed by directing the light onto a grating monochromator with resolution of 1 nm. A photomultiplier tube with a type S-1 photocathode is used for both imaging and spectral analysis. Image resolution in the SCM mode varies greatly with both carrier diffusion length and incident electron excitation volume. In this system SCM can typically be performed over a magnification range from 75 to 1000X.

All SCM and SEM images presented here were taken with a 35 keV electron beam at normal incidence, yielding a penetration depth of approximately 3.5 µm, and a typical beam current of 60 nA. Carrier concentration determinations were made by spot-mode spectral full-width-at-half-maximum measurements at 30 kV using the data of Cusano [9]. The CL peak for the material studied is located at approximately 890 nm. A method similar to that of Wittry and Kyser [10] was employed to determine diffusion lengths by measuring the CL intensity at 930 nm as a function of accelerating potential from 15 to 50 kV. This sub-bandgap wavelength was selected for diffusion length determinations in order to prevent intensity measurement errors due to self-absorption of CL by the sample.

RESULTS AND DISCUSSION

Figure 1, which compares SEM and SCM composite micrographs of the same sample area, demonstrates the capability of SCM to reveal defects not visible by surface microscopy. The only prominent features in the SEM micrograph are two parallel straight GBs running diagonally across the field of view and three closely spaced linear features at the upper left. In contrast, the SCM micrograph reveals a complex grain structure. Note that the two straight boundaries are not clearly defined in the SCM micrograph but contain discrete features (for example, those indicated by A) along their length. Also note the presence of striated features.

We have found that GBs can show three different types of SCM contrast. These are illustrated in the SCM micrograph of Fig. 2, where boundary A exhibits dark-line contrast, boundary B is slightly brighter than the surrounding matrix, and the two boundaries designated C show essentially no contrast, except for irregularly spaced dark spots that may be precipitates. Note the bright regions running along both sides of boundary A. Most, but not all, of the boundaries appearing as dark lines showed this feature. The strong dark boundary to the upper left of C does not appear in the SEM image. The network structure at the upper left of the SCM micrograph is probably a dislocation network, perhaps associated with a sub-grain boundary.

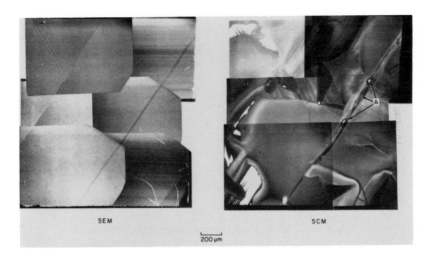

Figure 1. Composite SCM and SEM micrographs of the same area of a polished polycrystalline GaAs sample.

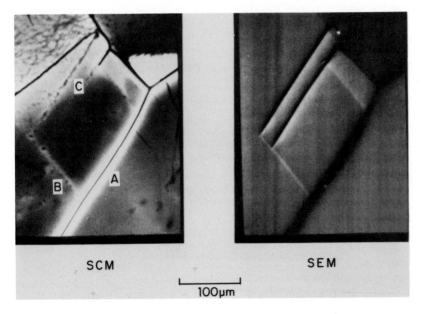

Figure 2. Composite SCM and SEM micrographs of the same sample area showing three types of grain boundary cathodoluminescence contrast.

The origin of the bright regions adjacent to the dark boundaries was investigated by making local carrier concentration and diffusion length measurements. In Fig. 3 the prominent curved boundary displays dark-line contrast, although the CL from the adjacent bright regions is of such intensity that the dark line is barely visible. The carrier concentration and diffusion length were measured at point A in the adjacent bright region and at point B located 50 µm further into the grain. At A the carrier concentration is 9×10^{18} cm^{-3} and the diffusion length is 1.3 µm, while at B the respective values are 2×10^{19} cm^{-3} and 0.6 µm. These data show that dopant impurities are depleted from the bright regions bordering the GB and suggest that these impurities are concentrated at the GB itself. The straight boundaries in Fig. 3 show little or no CL contrast, but many bright features (at C, for example) occur along their length and extend into adjacent grains. The contrast of these features is also attributed to impurity segregation effects.

Two additional significant observations have been made in the SCM study. First, different regions of a GB with constant misorientation angle but changing boundary plane can have different CL properties. This is illustrated by Fig. 4, where there is a marked difference in CL contrast between sections A and B of the same curved grain boundary, probably due to variation in impurity concentration. Thus, misorientation angle alone is not a sufficient parameter for correlating grain boundary structure and properties. Second, GBs with similar surface morphologies can have quite different CL properties. This is illustrated by Fig. 5, where boundary A shows dark-line CL contrast while boundary B shows little or no contrast, although their SEM images are similar.

The SCM results strongly indicate that both defect structure and impurity effects influence the optoelectronic properties of GBs in GaAs. Grain boundaries that show no CL contrast are believed to have little misfit between adjoining grains. Consistent with this idea, transmission electron microscopy has shown one boundary without contrast to be a {111} twin boundary with no visible dislocation structure. Bright boundaries are probably low-misfit boundaries where the impurity concentration in the adjacent regions has been reduced by segregation to the boundaries. The GBs exhibiting dark-line contrast are probably high-misfit boundaries. Although nonradiative recombination can be expected to occur at the misfit defects, this effect is insufficient to account for the observed darkening, since the width of the dark lines (typically 2-3 µm) is much greater than the width of the GBs themselves. Both an increase in impurity concentration and the formation of impurity precipitates probably contribute significantly to the reduction in CL intensity along these boundaries.

We believe that a close correlation will be found between the CL properties of GBs in GaAs and the effects of the GBs on the properties of photovoltaic devices. Thus, it seems likely that boundaries showing no CL contrast are electrically inactive and will not have a significant effect on device characteristics. The existence of such inactive GBs has been demonstrated by our electroluminescence observations [8]. Boundaries that appear dark are associated with non-radiative recombination centers that can be expected to reduce short-circuit current and increase series resistance. Hopping transport might occur along such boundaries that cross a device junction, causing a reduction in open-circuit voltage and fill factor. In addition, impurities concentrated along GBs could diffuse into epitaxial layers grown on heavily doped substrates, affecting device performance by altering bandgap states at the grain boundary or by changing the doping of the

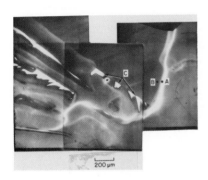

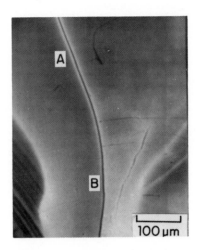

Figure 3. SCM micrograph illustrating the effects of local variations in impurity concentration. Carrier concentration and diffusion length measurements were made at A and B.

Figure 4. SCM micrograph showing variation in cathodoluminescence contrast with change in grain boundary plane.

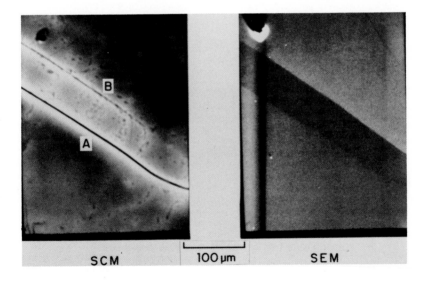

Figure 5. SCM and SEM micrographs of the same sample area showing different CL contrast from grain boundaries with similar surface morphologies.

device structure. These effects are consistent with the low open-circuit voltages commonly observed for polycrystalline GaAs solar cells.

CONCLUSION

The use of SCM as an effective tool for nondestructive characterization of polycrystalline GaAs has been demonstrated. We believe that SCM will also be useful for evaluation of other photovoltaic materials. Grain boundaries in heavily doped GaAs can either show no contrast, appear as dark lines, or appear slightly brighter relative to the surrounding matrix. The presence of GBs and other imperfections not affecting surface morphology is clearly revealed by SCM. In addition, there is no obvious correlation between boundary surface morphology and optoelectronic properties. Many GB features are probably due to local variations in impurity concentration. In achieving an understanding of the properties of GBs and their effects on device performance it will be essential to take account of the intimate relationship between defect structure and impurity effects.

ACKNOWLEDGEMENTS

We are grateful to A. J. Strauss and A. F. Witt for many valuable discussions and to B. DiGiorgio and S. Duda for technical assistance. The Lincoln Laboratory portion of this work was supported by the Department of the Air Force.

REFERENCES

1. G. W. Turner, J. C. C. Fan, R. P. Gale, and O. Hurtado, Conf. Record of the 14th IEEE Photovoltaic Specialists Conference, 1980 (IEEE, New York, 1980), p. 1530.

2. J. J. J. Yang, P. D. Dapkus, R. D. Dupuis, and R. D. Yingling, J. Appl. Phys. 51, 3794 (1980).

3. M. J. Cohen, J. S. Harris, Jr., and J. R. Waldrop, Inst. Phys. Conf. Ser. No. 45, 263 (1979).

4. W. Hwang, H. C. Card, and E. S. Yang, Appl. Phys. Lett. 36, 315 (1980).

5. M. J. Cohen, M. D. Paul, D. L. Miller, J. R. Waldrop, and J. S. Harris, Jr., J. Vac. Sci. Technol. 17, 899 (1980).

6. R. M. Fletcher, D. K. Wagner, and J. M. Ballantyne, Solar Cells 1, 263 (1980).

7. M. Spencer, R. Stall, L. F. Eastman, and C. E. C. Wood, J. Appl. Phys. 50, 8006 (1979).

8. G. W. Turner, J. C. C. Fan, and J. P. Salerno, Solar Cells 1, 261 (1980).

9. D. A. Cusano, Solid State Commun. 2, 352 (1964).

10. D. B. Wittry and D. F. Kyser, J. Appl. Phys. 38, 375 (1967).

EFFECTS OF DAMAGE-IMPURITY INTERACTION ON ELECTRICAL PROPERTIES OF Se+-IMPLANTED GaAs

D. K. SADANA, J. WASHBURN, M. D. STRATHMAN
Lawrence Berkeley Laboratory, University of California, Berkeley, CA 94720

G. R. BOOKER
Department of Metallurgy, University of Oxford, Oxford, England

M. H. BADAWI
Department of Electronics and Electrical Engineering, University of Surrey,
   Guildford, Surrey, England

ABSTRACT

Interaction of impurities with the "visible defects" in hot implanted Cr doped semi-insulating (100) GaAs has been investigated. The defects studies were performed using transmission electron microscopy (TEM) and MeV He+ channeled Rutherford backscattering. The defects distribution was obtained by 90° cross-sectional TEM (XTEM). The atomic concentration profiles of Se, and carrier-concentration and mobility profiles were obtained by secondary ion mass spectrometry (SIMS) and Hall measurements in conjunction with chemical stopping, respectively. Comparison of defects, atomic and electrical profiles, showed the formation of secondary defects at and beyond the projected range (R_p), a significant amount of Se+ diffusion beyond R_p, and compensation of electrical carriers caused mainly by the point defects present in hot implanted GaAs.

INTRODUCTION

The effect on electrical properties of the ion implantation induced damage which is left after subsequent furnace annealing in Si and GaAs has been investigated earlier [1-3]. The depth profiles of visible defects as obtained by cross-sectional transmission electron microscopy (XTEM) and electrical profiles obtained by Hall measurements in conjunction with layer stripping has demonstrated that both in Si and GaAs, there are strong correlations between the structural and electrical properties of these materials. In Si, a significant number of impurity atoms were found to segregate near the visible defects (defects visible by TEM) causing a pronounced distortion in the distribution of the implanted impurity. This distortion was also reflected in the electrical profiles. Similar correlations for ion-implanted GaAs are not available at present [3]. The aim of this work was therefore to understand the structural-electrical properties correlation in Se+-implanted GaAs. For this purpose, cross-sectional electron microscopy, MeV He+ channeling, secondary ion mass spectrometry (SIMS), and electrical measurements were performed on the same sample. It has been observed that the effect of the visible defects on the distribution of the implanted impurity (Se) is not straightforward. It has been deduced from the findings that the point defects and their complexes play the dominant role in determining the electrical properties of Se+-implanted GaAs.

EXPERIMENTAL

Semi-insulating Cr-doped (100) GaAs were implanted in a non-channeling direction with Se$^+$ ions at 450 KeV to doses of $1.0 \times 10^{14}/cm^2$ and $5.0 \times 10^{14}/cm^2$. The substrate temperature was kept at 200°C during the implantation. The resulting specimens were encapsulated with a 5000 Å thick Al layer and annealed at temperatures in the range 400-800°C. The results from only 600°C anneals have been included in the paper. The damage was examined both by the TEM and channelled Rutherford backscattering methods. The data for visible defects-depth profiles was obtained from 90° cross-sectional specimens using the strong-beam bright-field TEM method. For the channeled RBS measurements, a 1.6 MeV He$^+$ beam was used. The atomic distribution of the implanted ion in the annealed material was obtained by the SIMS measurements using oxygen ions. The electrical profiles were obtained by measuring the Hall coefficient and resistivity by the van der Pauw method in conjunction with chemical stripping. For the mobility curves, the ratio $\mu r = \mu/\mu_0$ was plotted, where μ is the measured mobility and μ_0 is the value taken from standard data for the mobility of the appropriate electrical carrier concentration in unimplanted GaAs layers.

RESULTS

A comprehensive study of the damage for various ion doses and annealing temperatures has been reported elsewhere by Sadana and Booker [5]. Figures 1 and 2 show results from the 600°C annealed samples with doses of $10^{14}/cm^2$ and $5 \times 10^{14}/cm^2$, respectively. Figure 1a showed two discrete bands of damage containing mainly dislocation loops of mean diameter--300 Å. The bands were 1400 Å and 1200 Å wide, respectively, and were located at mean depths of 1700 Å and 3400 Å, respectively. The surface region contained entangled dislocations [5]. The 2.0 MeV He$^+$ channeling results showed that the surface was damaged. The relative heights of Ga and As peaks in the channeling spectrum showed that As loss occurred at the surface. The higher dechanneling with a hump at the lower energy side of the channeling spectrum indicated that a significant amount of disorder was present in the deeper regions of the material. The atomic distributions of Se^{78} obtained from the same specimen is shown in Fig. 1c. The theoretically calculated atomic distribution for Se under the experimental conditions used is also shown by the dotted line in Fig. 1c. The carrier concentration profile exhibited a pronounced asymmetry over the depth range 1100 to 2900 Å (Fig. 1c) that corresponded very closely to the position of the first damage band in the XTEM micrograph of Fig. 1a. The mobility progressively increased over the depth range of 1000 to 2800Å (Fig. 1c), the overall increase corresponding to a factor of 3. The lowest mobility value corresponded to the position of the first-damage-band edge which was nearer to the specimen surface.

Figure 2 shows the results for the $5 \times 10^{14}/cm^2$ implant. The XTEM micrograph showed a 2400 Å wide band of damage clusters (Fig. 2a) that was located at a mean depth of 3000 Å. The 1.6 MeV He$^+$ channeling method gave complementary results (Fig. 2b). The atomic distribution of Se^{78} obtained from the same specimen is shown in Fig. 2c. Significant diffusion in the vicinity of visible defects was observed (note the position of the damage band shown as the shaded region in Fig. 1c). These results were in disagreement with the earlier findings of Evans et al. [6], where no appreciable movement of Se atoms occurred up to 800°C. The carrier concentration profile again exhibited a pronounced asymmetry, including a minimum over-the-depth range of 1000 to 3000Å. The portion of the profile showing asymmetry corresponded approximately to the position of the damage band (Fig. 2a). The mobility profile showed a dip corresponding to the upper edge of the damage band in Fig. 2a, followed by a progressive increase over the depth range 1000 to 3000Å. The overall increase in the deeper parts was by a factor of 5. The hypothetical Gaussian shaped profile (dotted lines over the carrier concentration profiles in Figs. 1c and 2c, respectively) that might have arisen if there were no asymmetry in the carrier concentration profiles indicated an overall background compensation being ~X10 for the $10^{14}/cm^2$ sample and ~50 for the $5 \times 10^{14}/cm^2$ sample.

TEM CROSS-SECTION MICROGRAPH

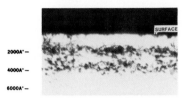

Figure 1. $Se^+ \rightarrow (100)$ GaAs, $\overline{Ti=200}°C$ 450 keV, $10^{14}/cm^2$, annealed at 600°C for 15 minutes with Al encapsulated.

(a) TEM cross-section micrograph showing two discrete layers of dislocation loops.

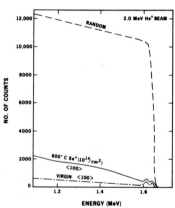

(b) 2.0 MeV He^+ channelling spectrum from the same specimen.

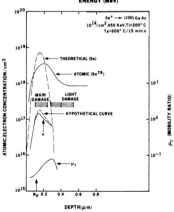

(c) Atomic Se (SIMS), carrier concentration and Hall mobility profiles from the same specimen.

518

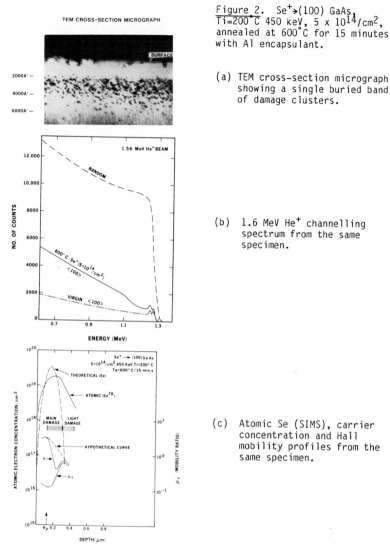

TEM CROSS-SECTION MICROGRAPH

Figure 2. Se$^+$→(100) GaAs, Ti=200°C 450 keV, 5 x 10^{14}/cm^2, annealed at 600°C for 15 minutes with Al encapsulant.

(a) TEM cross-section micrograph showing a single buried band of damage clusters.

(b) 1.6 MeV He$^+$ channelling spectrum from the same specimen.

(c) Atomic Se (SIMS), carrier concentration and Hall mobility profiles from the same specimen.

DISCUSSION

The asymmetry in the electrical profiles at depths corresponding to the damage band observed by XTEM showed that the visible damage adversely affects the electrical properties. However, an overall background compensation of the electrical carries by a factor of 50 and a significant amount of deep diffusion of Se as observed by the SIMS measurements indicates that hot implantation of Se probably causes the formation of mobile point defects in the implanted region and beyond it. The prsence of such point defects is further confirmed by the formation of secondary defects observed as either damage clusters or small dislocation loops at low annealing temperatures (>400°C) at depths up to $X3R_p$ or $X4R_p$ [5]. It is also likely that the visible defects are surrounded by a cloud of point defects. It is therefore suggested that the decrease in the carrier concentration at the band of visible defects could be due to a high density of point defect—impurity complexes present in the vicinity of these defects in GaAs.

CONCLUSIONS

The following conclusions can be drawn from the present study. (1) Deep diffusion of Se occurs in the as—implanted GaAs sample at the implantation temperature of 200°C. (2) Formation of secondary defects at and beyond R_p, significant amounts of Se diffusion beyond R_p and compensation of electrical carriers are associated with the mobile point defects present in the hot implanted GaAs.

ACKNOWLEDGMENTS

The authors would like to thank Dr. Brian Sealy of Surrey University, England, for Se implantation, and Dave Steidl of Argonne National Laboratory for the SIMS measurements of Se. We would also like to acknowledge the financial support of the U.S. Department of Energy through the Materials and Molecular Research Division, and Engineering Department, of the Lawrence Berkeley Laboratory under Contract No. W—7405—ENG—48.

REFERENCES

1. D. K. Sadana, M. Strathman, J. Washburn, C. W. Magee, M. Maenpaa, and G. R. Booker, App. Phys. Lett. 37, 615 (1980).

2. D. K Sadana, J. Fletcher, and G. R. Booker, Electronics Lett. 13, 632 (1977).

3. D. K. Sadana, G. R. Booker, S. S. Kular, B. J. Sealy, K. G. Stephens, and M. H. Badawi, Rad. Eff. 49, 183 (1980).

4. D. K. Sadana, M. Strathman, J. Washburn, and G. R. Booker, J. App. Phys. (in press).

5. D. K. Sadana and G. R. Booker, Rad. Eff. 42, 35 (1979).

6. A. Lidow, J. F. Gibbons, V. R. Deline, and C. A. Evans, App. Phys. Lett. 32, 15 (1978).

TEM AND SEM STUDIES OF DEFECTS IN GaInAs AND GaInAsP EPITAXIAL-LAYER DEVICE-TYPE STRUCTURES

M.M. AL-JASSIM, M. HOCKLY AND G.R. BOOKER
Department of Metallurgy and Science of Materials, Parks Road, Oxford OX1 3PH

ABSTRACT

TEM and SEM studies have been carried out on a series of GaAs/GaInAs, InP/GaInAs and InP/GaInAsP device-type structures grown either by vapour phase epitaxy (VPE) or liquid phase epitaxy (LPE). By using the TEM high voltage electron microscope operating at 1000kV, a determination was made of the nature, origin and three-dimensional distribution of structural defects in these specimens (mainly dislocations and stacking faults). SEM EBIC and CL examinations were also performed to study the electrical and luminescent properties of these layers. The ultimate aim is to correlate structural information with electrical and luminescent properties, and with device performance.

INTRODUCTION

GaInAs ternary and GaInAsP quaternary alloys are becoming increasingly important because of their role in the fabrication of infra-red emitters and detectors operating in the 1-2μm wavelength range. However, since they are hetero-epitaxially grown, lattice mismatch is often incurred and dislocations are frequently formed as a by-product of the mismatch. These layers are either grown with a compositionally graded layer between a substrate and the required constant-composition layer, or grown with a fixed composition to match a commercially available binary substrate.

In a previous investigation [1] we reported the results of TEM and SEM studies of GaAs/Ga$_{0.83}$In$_{0.17}$As VPE graded-layer structures suitable for infra-red emitters ($\lambda = 1.06\mu m$). The present paper describes recent structural and electrical studies of GaAs/GaInAs, InP/GaInAs and InP/GaInAsP device-type structures grown either by VPE or LPE. For each device-type structure, specimens that gave good and poor devices were generally examined. Particular attention was given to the poorer specimens in order to understand better how defects were generated and subsequently affected the material and device properties.

EXPERIMENTAL

TEM examinations were performed on plan-view and 90° cross-sectional thinned specimens using an AEI EM7 high voltage electron microscope operating at 1000kV. The plan-view specimens were prepared by the jet-chemical thinning technique using chlorine-methanol solution, while the cross-sectional specimens were prepared by either chemical or ion-beam thinning [2]. The examination of cross-sectional specimens was found to be essential for the investigation of dislocation generation mechanisms at substrate/epilayer interfaces. SEM EBIC and CL examinations were carried out on bulk specimens, and these gave micro-graph-type images that revealed the defects in the surface regions of the layers as dark spots or lines, the contrast arising because the defects acted as either electrical or non-radiative recombination centres. Data for devices subsequently fabricated on some of the structures were available.

RESULTS

GaAs/GaInAs VPE structure for 1.06μm emitters

This structure consists of a 2μm thick GaAs epilayer grown on a (100) GaAs substrate, followed by an n-type GaInAs layer with composition graded from GaAs to $Ga_{.83}In_{.17}As$ over ∿12μm thickness, an n-type layer of this composition ∿8μm thick and finally a p-type layer ∿4μm thick of the same composition. TEM cross-sectional examination (Fig.1a) showed: a) an irregular dislocation network (Fig.1b) at the initial growth interface (which seems to be dependent on surface preparation); b) a dense three-dimensional dislocation network (Fig.1c) in the graded region (associated with misfit strain); c) two types of linear defect running along directions close to [100] from the graded-region network to the surface (and therefore through the p-n junction), these consisting of threading dislocations and another type of defect which gives anomalous weak TEM contrast (Fig.1a); and d) a planar cross-grid network of dislocations (Fig.1d) with spacing 0.2-2μm, ∿0.5μm below the surface. (This network arises because of an abrupt compositional change which occurs during the termination of layer growth, and can be removed chemically after growth).

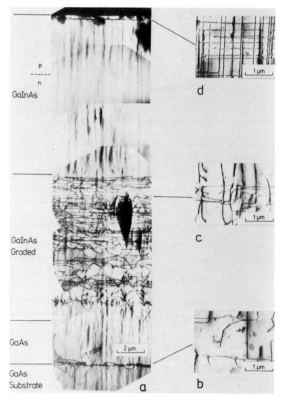

Fig.1 VPE graded GaInAs emitter. TEM micrographs. a) Cross-section. b) Plan-view of initial interface. c) Plan-view of graded region. d) Plan-view of shallow dislocation network.

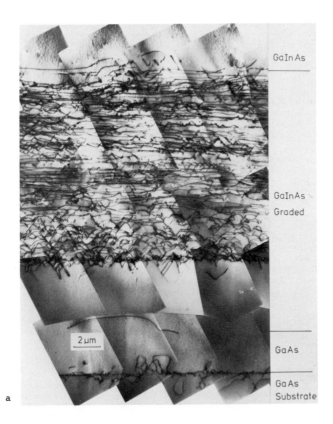

a

Fig.2 VPE graded GaInAs detector. a) TEM cross-section showing dislocation networks at initial interface (irregular) and in graded region (three-dimensional).

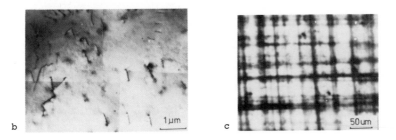

b

c

Fig.2 VPE graded GaInAs detector. b) Tilted TEM plan-view near p-n junction
showing threading dislocations. c) SEM EBIC plan-view of surface showing dark
spots (threading dislocations) and dark lines (shallow network).

GaAs/GaInAs VPE structure for 1.3μm detectors

This structure is similar to the above emitter structure, except that compositional grading occurs over a greater thickness (∿25μm) and the final layer composition is $Ga_{.63}In_{.37}As$. TEM cross-sectional examination (Fig.2a) revealed a distribution of dislocations similar to that of the emitters, the main difference being that the detectors did not contain the anomalous contrast defects seen in the constant composition region of the emitters. A TEM plan-view montage (Fig.2b) of this region in a detector, obtained after tilting the specimen through a large angle in the TEM, shows the presence of threading dislocations only (up to $6x10^7$ cm^{-2}). The effect of these dislocations as recombination centres can be seen in the plan-view SEM EBIC micrograph of the surface region (Fig.2c), where small groups of these dislocations show up as dark spots. The cross-grid of dark lines is caused by the planar network of dislocations just below the surface, which arises as in the emitters.

InP/Ga$_{.47}$In$_{.53}$As VPE structure for 1.65μm detectors

Lattice-matched $Ga_{.47}In_{.53}As$ layers 3-4μm thick were grown on (100) InP substrates. TEM examination of a cross-sectional specimen using a 400 reflection (Fig.3a) showed that numerous pairs of dislocations (∿10^8 cm^{-2}) were generated at the interface and threaded through the layer to the specimen surface, along directions close to [100] or lying in {111} planes. Similar examination using an 022 reflection (Fig.3b, adjacent but not identical area) showed strong "wavy" contrast, not visible with the 400 reflection, which tended to mask the dislocation contrast. This indicated the presence of severe locally varying strain which is minimal along the [100] growth direction and maximal perpendicular to it. A few specimens also contained stacking faults, originating at the interface and lying along the inclined {111} planes (Fig.3c). TEM examination of cross-sectional specimens after tilting through a large angle (Fig. 3d) revealed a dense planar cross-grid dislocation network (mean spacing 800Å) at the InP/GaInAs interface, indicating significant lattice mismatch at the growth temperature.

InP/Ga$_{.47}$In$_{.53}$As LPE structure for 1.65μm detectors

Lattice-matched $Ga_{.47}In_{.53}As$ layers 10-15μm thick were grown on (100) InP substrates. In general, no threading dislocations, anomalous TEM contrast defects or interfacial dislocation networks were observed by TEM. In one specimen, however, TEM cross-sectional examination showed several cross-grid mismatch dislocation networks lying in (100) planes at different depths in the

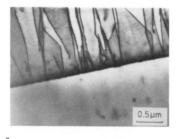
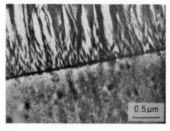

a   b

Fig.3 VPE lattice-matched GaInAs detector. TEM cross-sections showing interface. a) $\underline{g}$ = 400. Threading dislocations clearly visible. b) $\underline{g}$ = 022. Strong locally varying strain contrast masks the threading dislocations.

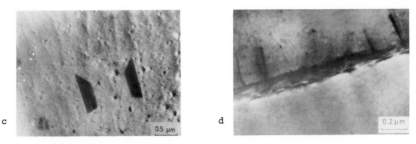

Fig.3 VPE lattice-matched GaInAs detector. c) TEM plan-view of layer showing stacking faults. d) Tilted TEM cross-section showing interface network.

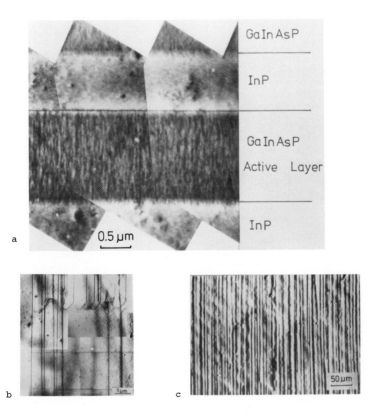

Fig.4 LPE lattice-matched GaInAsP laser. a) TEM cross-section showing disloca-tion network at active layer interface. b) TEM plan-view of active layer inter-face showing asymmetric dislocation network. c) SEM EBIC plan-view of the same interface as b), showing only one set of dark lines due to asymmetry.

layer, indicating that several abrupt changes in composition had occurred during the growth of this particular layer.

InP/GaInAsP LPE multilayer structure for 1.3μm double-heterostructure lasers

This structure is lattice-matched and comprises an n-type InP layer 2μm thick grown on a (100) InP substrate, followed by an undoped $Ga_{.29}In_{.71}As_{.65}P_{.35}$ active layer 1.4μm thick, a p-type InP confining layer 1.2μm thick, and a p-type $Ga_{.19}In_{.81}As_{.43}P_{.57}$ contacting layer 0.5μm thick. TEM examination of cross-sectional specimens (Fig.4a) sometimes showed a planar cross-grid network in the critical active layer/confining layer interface. TEM plan-view examination (Fig.4b) showed that this network was asymmetric, with a mean dislocation spacing of 1.7μm. SEM EBIC micrographs (Fig.4c) showed the deleterious effect of this network. The dark lines indicate that the dislocations act as electrical recombination centres, their unidirectionality confirming the asymmetric nature of the network.

DISCUSSION

Investigation of a number of good and poor specimens for each type of structure led to the following observations. For the two VPE graded layer structures, the important defects were the threading dislocations present in the constant composition region, these being generated by mismatches in the graded region. These dislocations were electrically active. Their density for both the emitters and detectors was in the range 10^8 to $< 5x10^6$ cm^{-2}, the lower end of this range corresponding to good devices. Other threading defects exhibiting a weaker anomalous TEM contrast were sometimes also present in the constant composition region. The threading dislocation density could be further decreased by several procedures, e.g. by decreasing the grading rate.

For the VPE lattice-matched structure, the important defects were again the threading dislocations present in the layer, these being generated at the substrate/layer interface, probably by substrate surface irregularities immediately prior to growth. The threading dislocation density was in the range 10^8 to $< 5x10^5$ cm^{-2}. TEM contrast indicating severe localised strain was often observed in these layers, this being thought to arise from local changes in layer composition in directions parallel to the layer surface. For the analogous LPE lattice-matched structure, the threading dislocation density was generally $< 5x10^5$ cm^{-2}, and the anomalous contrast defect was not observed.

For the LPE lattice-matched laser structure, the important defect was the dislocation network that sometimes occurred at the active layer/confining layer interface. This was considered to arise mainly by a small lattice mismatch being present at the growth temperature, this generating dislocations at the layer surface, and these dislocations then moving down to the interface. This network was electrically active. Structures without this network gave good devices.

ACKNOWLEDGEMENTS

The authors thank their colleagues at Plessey Research Centre, Towcester (R. Nicklin and S. Hersee); Post Office Research Centre, Martlesham (M. Faktor); and Royal Signals Radar Establishment, Baldock (M. Astles); for specimens, information and discussions; and acknowledge financial support from the Iraqi Government (MMA); CSIR (MH); DCVD, Ministry of Defence; and SRC.

REFERENCES

1. M. Hockly, M.M. Al-Jassim and G.R. Booker, J. Microscopy 118, 117 (1980).

2. J. Fletcher et al., Inst.Phys.Conf.Ser. No. 52 (1980) pp.153-6.

MICROSTRUCTURE OF POLYCRYSTALLINE SIC CONTAINING EXCESS SI AFTER NEUTRON
IRRADIATION

S.D. HARRISON* AND J.C. CORELLI
Dept. Nuclear Engineering, Rensselaer Polytechnic Institute, Troy,NY 12181 USA

ABSTRACT

   The microstructure of commercially available reaction-
bonded polycrystalline SiC containing 8-10 wt% excess silicon was studied after irradiation by reactor neutrons uitliz-
ing transmission electron microscopy (TEM) and scanning electron microscopy (SEM). For TEM studies on samples irradiated
below 473°K we observe dislocation tangles near grain boundaries or impurities plus isolated dislocations throughout the
remainder of the grain, whereas for irradiation temperatures
of $\sim$1373°K the material exhibits copious quantities of "black
spot" defects (2-5nm size) and a lower concentration of tangles
than 473°K irradiation. The SEM studies of surfaces of samples fractured at 1473°K indicate that the mode of fracture
is predominantly transgranular.

INTRODUCTION

   The ceramic material SiC has some unique properties which make it useful in
high temperature nuclear fusion reactor applications (eg: first wall, limiter,
etc.). The advantages as well as the disadvantages in the use of SiC for nuclear fusion applications have been described in quite some detail by Rovner
and Hopkins [1] and will not be stated here. Since the structures used in fusion reactors will be subjected to intense neutron fields (14MeV neutrons), as
well as ions,X-rays, and gamma radiation it is important to study the response
of the material to simulated radiation fields. In recent years there have been
a number of studies on the fracture, swelling,and thermal properties of various
forms of SiC material [2,3,4,5,6]. However, detailed studies of the microstructure response of irradiated commercially available silicon carbide have not been
reported.
   In this paper we shall report on the microstructural changes in reactor neutron irradiated SiC for which we also have made studies on the response of the
mechanical properties (fracture strength, swelling) following neutron irradiation.

EXPERIMENTAL METHODS AND PROCEDURES

Materials, sample preparation and microscopes
   The material examined in this study is a reaction bonded polycrystalline silicon carbide produced by the Norton Co., Worcester, Mass., and designated as
NC-430. This complex material is formed by reacting molten Si with a compacted
mixture of alpha phase silicon carbide (α-SiC) grains and graphite. The resulting structure is 88.5% SiC, 8-10% Si metal and the remaining weight percent consisting of 0.4% Fe and 0.1%A with traces of other impurities such as Co, Cr,Zn
[7,8]. This type of manufacture results in a bimodal distribution of grain
sizes, polytypes and crystal structures. The alpha phase grains range in di-

---

*
Now at Physics Dept., U.S.Military Academy, West Point, NY 10996

ameter from 20μm to 150μm and have a hexagonal close packed lattice structure and have varying stacking sequences designated 2H, 4H, 6H, 8H, 15R, etc. (109 have been identified), with 6H polytype being most prevalent [9]. The grains formed in the reaction between the silicon and graphite are of the beta phase (β-SiC) 3C polytype, have a FCC structure, are less than 10μm in diameter and transform to an alpha phase in many instances[10]. The grains observed in TEM are predominantly alpha phase.

Samples were prepared for TEM by two different techniques. The first method involved ion milling to perforation 250μm thick chips prepared by standard cutting and grinding techniques. The ion mill utilized 6kV argon ions with a 100 μAmp beam current incident on the sample at 25°. In the second method TEM samples were prepared by extracting chips of SiC (∿5μm diameter) from fracture surfaces of material using acetate tape and transferring the chips onto a carbon substrate. The ion milled samples were examined with a JEOL 100S 100kV electron microscope, while the samples prepared from extracted chips of SiC being ∿1 to 2μm thick necessitated the use of 1200kV electrons from an AEI 1200kV HVEM type EM7 electron microscope.

The fracture surfaces of samples used in mechanical property studies were examined using a 20kV SEM. The samples were prepared by evaporating a thin gold film (∿200Å) onto the fracture surface in order to prevent charge buildup during viewing. All samples were examined after they had been exposed to 1473°K temperatures in air for ∿10 minutes. The SEM used in our studies had a maximum practical spatial resolution of 200Å.

Irradiations

The irradiations were conducted in the V-15 core position of the High Flux Beam Reactor (HFBR) at the Brookhaven National Laboratory. Samples were irradiated to neutron fluences (with E>1MeV) of $2\times10^{24}n/m^2$ (0.44 dpa), $4\times10^{24}n/m^2$ (0.87 dpa) and $12\times10^{24}n/m^2$ (2.6 dpa) where dpa represents the computed displacements per atom in the HFBR neutron spectrum. Sample temperature during irradiation was $420 \pm 30°K$, and for one experiment the temperature was $1373 \pm 50°K$.

RESULTS AND DISCUSSION

In Figure 1 is shown a TEM photomicrograph with {220} reflection of a sample irradiated to a fluence of $2\times10^{24}n/m^2$ (E>1MeV). The sample temperature during irradiation was 420°K and had not yet been annealed to higher temperature. The diffraction pattern for the {220} reflection 3C polytype SiC is shown in Figure 1. The dislocation tangles evident in Figure 1 have been labeled with the letter D, are radiation-induced and are not observed in unirradiated material. Heat treatment to 1373°K of the sample shown in Figure 1 does not significantly alter the dislocation tangle structure. This can be seen in the 1200kV HVEM photomicrograph ({1$\bar{2}$10} reflection) of 6H polytype SiC irradiated at 420°K and then given a 30 minute anneal at 1373°K, where one can still observe dislocation tangles near grain boundaries. The dislocation tangles are marked by the letter D and also shown is the diffraction pattern. We also have additional results on this sample (not shown here) in which individual dislocation loops ranging in size from 30nm to 110nm are observed.

A dramatic change in the microstructure of this NC-430 SiC material is observed if the sample temperature during irradiation is kept at 1373°K. In Figure 3 is shown the resulting microstructure after irradiation of SiC to a fluence of $4\times10^{24}n/m^2$ (E>1MeV) while its temperature is kept at 1373°K during irradiation. The photomicrograph in Figure 3 shows the presence of copious quantities of "black spot" defects ranging from 2 to 5nm in size. Although not shown in Figure 3 we also observe dislocations and dislocation tangles but at much lower concentration than for samples irradiated at lower temperature(420°K). The black spot defects apparently result from the collapse of dislocation tan-

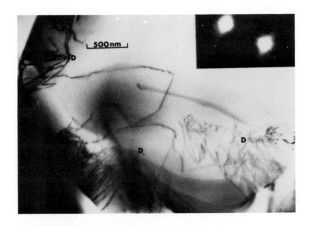

Fig. 1 TEM photomicrograph with {220} reflection of 3C polytype SiC irradiated at 420°K to 2X10²⁴n/m² E>1MeV (0.44 dpa) and not annealed showing dislocation tangles near grain boundaries.

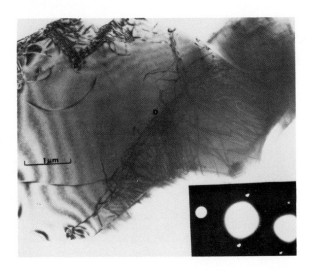

Fig. 2 HVEM photomicrograph with {1̄210} reflection of 6H polytype SiC irradiated at 420°K to 4X10²⁴n/m² E>1MeV (0.87 dpa) and annealed at T = 1373°K showing dislocation tangles near grain boundaries.

gles during irradiation at high temperature (1373°K).

The results shown in Figures 1-3 do not indicate the presence of radiation-induced voids. This finding is consistent with the results given by Price[6,11] who reported that voids are induced in β-SiC at irradiation temperatures of ∿1600°K, but Price did not observe voids for samples irradiated at temperatures of 1373°K or below. At high temperatures (∿1600°K) it is reasonable to assume that vacancies can coalesce to form voids, but at lower temperature ≲1370°K the coalescence of multiple vacancies to form voids does not occur.

The fracture surfaces on samples of SiC used in fracture tests were examined by SEM. The data [5,12] which we have on irradiated NC-430 indicate that within experimental error the fracture strength exhibits no significant decrease when measured at 1474°K after irradiation to fluence levels of $1.2 \times 10^{25} n/m^2$ (E>1MeV) in which the sample temperature is kept at 420°K. Scanning electron microscope photomicrographs were taken of surfaces from both irradiated and unirradiated samples to allow comparisons of the fracture mode to be made. In Figures 4 and 5 we show SEM photomicrographs of unirradiated and irradiated NC-430 SiC samples respectively. The irradiated sample (Fig. 5) was exposed to a fluence of $1.2 \times 10^{25} n/m^2$ (E>1MeV) at an irradiation temperature of 420°K. A comparison of the results shown in Figures 4 and 5, which are typical of this material indicates that there is little change observed in the predominantly transgranular mode of fracture for both irradiated and unirradiated material. The SEM results on material irradiated at 1373°K also indicate transgranular mode of fracture but in addition we observe a small increase in the size of the α-phase grains which is caused by a combination of the radiation and temperature (1373°K). Although the temperature is lower than that known (10) for the α to β transformation, it is reasonable to conclude that the combination of neutron radiation and temperature may be causing the α to β transformation in the present case. Detailed SEM examinations on all samples did not reveal the presence of microcracks at the surface.

SUMMARY AND CONCLUSIONS

The microstructure of NC-430 SiC after reactor neutron irradiation reveals similar SEM patterns on fracture surfaces for both irradiated and unirradiated material from which we conclude that the mode of fracture is predominantly transgranular. Additionally, the temperature for observing the α to β transformation in SiC is reduced from ∿2300°K (10) to ∿1370°K by the concomitant action of the radiation with neutrons and temperature (1370°K). From TEM studies on samples irradiated below 473°K we observe dislocations and dislocation tangles which pile up near grain boundaries. If the material is irradiated at 1473°K we observe "black spot" defects 20-50Å size and a lower concentration of dislocations. We conclude that these "black spot" defects are static Frank loops formed at high temperature by inhibiting vacancy-interstitial recombination to form large dislocation loops.

Based on our results on microstructure and fracture strength we conclude that reaction bonded SiC may be a promising material for nuclear fusion applications from the point of view of its ability to withstand deleterious neutron radiation effects. However, this conclusion still remains to be tested on material irradiated to fluences ∿100 higher than used in the present study.

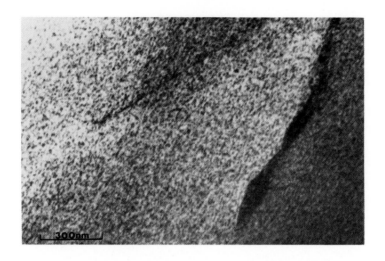

Fig. 3   HVEM photomicrograph of 3C polytype SiC irradiated at T = 1373°K to 4×10^{24} n/m^2 E>1MeV (0.87 dpa) and annealed at 1373°K showing "black spot" defects.

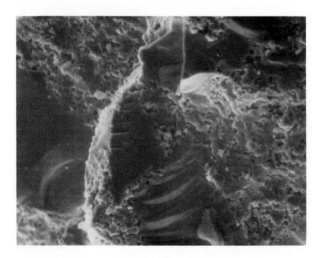

Fig. 4   SEM of unirradiated SiC (M = 500X), note the large α grains with cleavage type fracture surfaces. Granular regions are β-SiC free silicon metal.

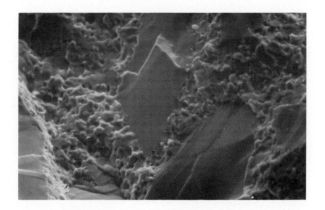

Fig. 5 SEM of irradiated SiC (M = 500X) fluence $1.2 \times 10^{25} n/m^2$ E>1MeV(2.6dpa).

ACKNOWLEDGEMENTS

The assistance of Dr. A. Rathkowski of the New York State Health Department and Mr. J.W. Malloy of RPI is acknowledged with thanks. The excellent work in typing the manuscript by Jean Mulson is affectionately acknowledged by the authors. This research was supported by EPRI under contract number RP-992-3.

REFERENCES

1. L.H. Rovner and G.H. Hopkins, Nucl. Tech. 29, 274 (1976).

2. R.J. Price, J. Nucl. Mat. 33, 17 (1969).

3. R.B. Matthews, J. Nucl. Meth. 51, 203 (1974).

4. R. Blackstone and E.H. Voice, J. Nucl. Tech. 39, 319 (1971).

5. R.M. Matheny, J.C. Corelli and G.G. Trantina, J. Nucl. Mat. 83, 313 (1979).

6. R.J. Price, J. Nucl. Mat. 48, 47 (1973).

7. G.R. Sawyer and T.F. Page, J. Mat. Sci., 13, 889 (1978).

8. M.L. Torti, J.W. Lucek and G.Q. Weaver, Article 780071, Soc.of Auto.Engs. Inc.,SAW Cong. & Expo., Feb. 27-March 3, 1978, Detroit, Michigan.

9. D.J. Smith, N.W. Jepps and T.F. Page, J. of Microscopy 114, 1 (1978).

10. A.H. Herrer, G.A. Fryburg, L.N. Ogbiyi, T.E. Mitchell and S. Shinozaki, J. Am. Cer. Soc., 61 406 (1978).

11. R.J. Price, Nucl. Tech. 35, 320 (1977).

12. J.W.Malloy, L.Fulop and J.C.Corelli,Bull.Am.Phys.Soc. 25,#3,178 (1980).

Author Index

534

Subject Index

A-center,241
Annealing,
  capped,487
  electron beam,393
  flame,191
  laser (see Laser annealing)
  mechanisms,191
  thermal,191
Amorphous clusters,135
Arsenic cluster,451

Bimolecular annihilation,333

Capacitance junction spectros-
  copy,85,495
Cathodoluminescence,457,509,521
CdTe defect structure,503
Conduction electron spin reson-
  ance (CESR),1
Constitutional super-cooled
  cell,431
Convection induced cell,431
Convergent beam pattern,279
Cristobalite, modified,381
Crystallite morphology,439
Cryatal growth,309
Crystal growth sequence,209

Damage,
  displacement,409
  ion-implantation,173,179,185,
  191,209,515
  -impurity interaction,515
  neutron,191
Deep level transient spectroscopy (DLTS),79,85,241,273,481,
  487
Defect,
  black-spot,527
  bonding,445
  -boron complex,71
  electronic level,481
  formation,355
  intermediate,1,163,173
  negative-U,21
  point,31,55,151,163,185
  residual,451
  rod or rod-like,163,173,209
  secondary, effect of orientation on,209
  structure transition,393
  sulfur-related,79
  symmetry,1,279,503
  swirl,309

Diffuse scattering,
  asymptotic,151
  Huang,151,
  X-ray,151
Diffusion,
  concentration enhanced,31
  gold in Si,31,55
  mobility enhanced,31
  oxidation enhanced (OED),55
  333,355
  oxygen out-,367
  Si self-,31,55
  thermally activated,31
Di-interstitial,185
Dislocation,131,151,303,503
Dislocation,
  defect states,273
  doping effect on velocity of,
  251
  electrical behavior of,273
  electronic structure of,251
  Frank partial,163
  glide set partial,251,279
  mobility of,317
  ninety degree,163
  nucleation from point defect
  condensation,163,173
  partial,297
  Schockley partial,291
  suffle set partial,257,279
  sixty degree, 163,173,257,279
  tangles,527
  undissociated,163,173,179
Dopant profile broading,409
Displacement,
  knock-on,225
  γ-recoil,225

E-center,241
EFG Si ribbon,279,303
Electroluminescence,85
Electron beam induced current
  (EBIC),297,303,521
Electron microscopy,
  cross-section,173,179,393,421,
  515
  high resolution(HREM),173,179,
  185,279,285,503
  high voltage (HVEM),297
  scanning-transmission (STEM),
  457